AVIAN COGNITION

Exploring the Intelligence, Behavior, and Individuality of Birds

AVIAN COGNITION

Exploring the Intelligence, Behavior, and Individuality of Birds

Debra Herrmann
Biological Computing Initiative
North Bethesda, Maryland, USA

CRC Press
Taylor & Francis Group
Boca Raton London New York

CRC Press is an imprint of the
Taylor & Francis Group, an **informa** business
A SCIENCE PUBLISHERS BOOK

CRC Press
Taylor & Francis Group
6000 Broken Sound Parkway NW, Suite 300
Boca Raton, FL 33487-2742

Printed on acid-free paper
Version Date: 20150928

International Standard Book Number-13: 978-1-4987-4842-1 (Hardback)

Visit the Taylor & Francis Web site at
http://www.taylorandfrancis.com

and the CRC Press Web site at
http://www.crcpress.com

Dedication

This book is dedicated to all the birds of the world. May they fly free and be happy, healthy, and safe forever. My hope is that *Avian Cognition* will serve as a catalyst for increased public understanding of avian intelligence and behavior, better treatment of wild and pet birds worldwide, and more research in this area.

And to a new American robin (*Turdus migratorius*) family I had the pleasure of observing from 4/15/2013–5/28/2013. Between April 15th and May 1st the mother robin built four successive nests. For each she carried a half quart (liter) of mud—one beakful at a time—up to the site and pressed it into a cup, before adding the outer twigs. The first three nests were blown down by spring thunderstorms. But she did not give up. The fourth nest stayed put and she began nesting May 1st. Two weeks later three tiny nestlings hatched. The mother ferried food back and forth to them every 15–20 minutes from sunrise to sunset. On May 28th over a 6 hour period all three baby robins fledged. The robin returned to the same nest in 2014 and reinforced it. On 5/20/2014, again over a 6 hour period, all three baby robins fledged.

And to a common grackle (*Quiscala quiscala*) nestling that was part of our lives from 6/17/2013 –7/7/2013, until it was ready to be returned to the wild.

The determination and resourcefulness of the mother American robin and common grackle nestling were daily reminders of why this book needed to be written.

Photograph D.1 New Beginnings. © D. Herrmann

Preface

What is *Avian Cognition*?

It is appropriate that this book was begun during the International Year of Biodiversity. No species exhibits more biological diversity than the 9,672 species of birds that currently populate the Earth.[152]

Birds are masters of senescence; for their tiny body mass birds have achieved one of the longest life spans of all the animals on Earth.[130] Birds also have one of the largest brain size to body mass ratios.[127, 176] Parrots in particular, out of all the bird families, have both of these traits in spades. Birds have been living on this planet about 10,000 years longer than humans. Suffice it to say that birds have learned a few things that humans have not.

Avian Cognition explores the fascinating world of avian intelligence, behavior, and individuality. The preferences, choices, motivation, and habits of species, flocks, and individual birds are discussed and compared. *Avian Cognition* investigates who birds are and why they do what they do. Daily, seasonal, and play activities, creativity, reasoning ability, problem solving skills, adjustment to changes in their environment, social interaction, life stages, vocalizations, and communication patterns are described. An important and often overlooked distinction is made between vocalizations that are learned versus those that are inherited.

Avian Cognition looks at wild birds living in their normal outdoor habitat and pet birds living indoors. A variety of birds are examined, such as Green herons (*Butorides virescens*), Great blue herons (*Ardea herodias*), black-crowned night herons (*Nycticorax nycticorax*), Yellow-crowned night herons (*Nyctanassa violacea*), Canada geese (*Branta canadensis*), Mallard ducks (*Anas platyrhynchos*), Wood ducks (*Aix sponsa*), Ruby-throated hummingbirds (*Archilochus colubris*), Cardinals (*Cardinalis cardinalis*), Zebra finches (*Peophila guttata castanotis*), and several members of the parrot family. Many of the birds were observed for five, ten, or more years. Some birds were observed for their entire lives.

This allowed:

- Specific behaviors to be put in context
- Behaviors that were unique to an individual or flock to be identified
- Behaviors that were repeated or enhanced as part of the learning process to be recognized
- Many new behaviors and vocalizations to be identified

Avian Cognition distinguishes between biological intelligence, genetic intelligence, and applied intelligence. Contemporary comparisons between human intelligence and other species are challenged. For the first time the relationship between adaptation, instinct, intuition, and intelligence are probed.

Avian Cognition will appeal to a broad worldwide audience that includes birders, ornithologists, pet bird hobbyists, animal trainers, avian veterinarians, professors, researchers, and students of animal cognition and behavior, naturalists, and others who are curious about the world around them.

Why Was *Avian Cognition* Written?

There isn't another book that examines avian intelligence, behavior, and individuality to the depth and breadth as *Avian Cognition.* Unlike other books entire chapters are devoted to a single species. More than 300 individual birds and 15 species of birds native to 6 different continents are discussed. Original photographs, original audio recordings, and statistics support the new observations and findings. With its expanded scope and depth *Avian Cognition* takes ornithology into the 21st century and opens many new topics that have not been dealt with before.

Avian Cognition was undertaken to fill an important gap in the literature by focusing on *who* birds are, not what they are, and *why* they engage in certain behaviors, not just what they do. As a result many new observations are made that have not been reported elsewhere. As Cruickshank so aptly stated in 1938:[45]

> "Since birds are forever shifting ranges and even modifying ancestral habits, no matter how much is learned about them there is always something new to be discovered."

Most bird books contain a one or two paragraph description by species of a bird's appearance and its behavior. While accurate at the level written, these descriptions hardly scratch the surface or tell the whole story; nor do they convey much information at all about the broad range of avian intelligence, behavior, or individuality. To really understand avian intelligence, behavior, and individuality it is essential to delve much deeper.

Other books about bird behavior have a different weakness. They may have a chapter devoted to migration, courtship, or nest building. But these books lump all 9,672 species of birds together. Within a given paragraph there may be one sentence about Mallard ducks and one sentence about Ruby-throated hummingbirds. With this arrangement it is difficult, if not impossible, to develop: (1) an understanding of who a Mallard duck or Ruby-throated hummingbird is; or (2) an appreciation of which behaviors are common to a species or unique to a flock or individual. To illustrate, whether you are talking about biology or behavior a Ruby-throated hummingbird has about as much in common with an Emperor penguin (*Aptenodytes forsteri*) as a Zebra finch has in common with an Ostrich (*Struthio camelus*).

Approach

To remedy these limitations *Avian Cognition* has taken a new approach. The information presented in *Avian Cognition* is based on firsthand observations. Fifteen species of birds and over 300 individuals native to 6 different continents were observed over a period of 30 years. In order to discern the widest spectrum of behavior the mix of birds included:

- Well known birds and those that are not well known
- Birds that are predators and those that are not
- Birds that migrate and those that do not
- Birds that are kept as pets and those that are not

Daily observations were recorded in a standardized manner. The observations were made at varied times throughout the day, in all seasons and all types of weather, in the same locations year after year. This way the behavior could be compared and analyzed, and patterns, trends, and unique occurrences could be recognized. Likewise, similarities and differences in behavior could be identified between and within species. The observations were captured in text, photographs, and sound recordings.

Numerous new observations are made, which have not been reported before. For example in Chapters 2 through 11 readers will learn about:

- A flock of Canada geese that practices nesting dispersal
- Zebra finches that weave. (Zebra finches were moved from *Taeniopygia guttata* to *Peophila guttata castanotis* some time ago because supposedly they do not weave.)
- Cardinals that engage in play activities
- Budgerigars (*Melopsittacus undulatus*) who change roommates
- Mallard ducks that take day trips
- A 14-year old Indian ringneck parakeet (*Psittacula krameri manillensis*) who learns new native vocalizations
- Green herons that use bait and leverage the movement of Mallard ducks to catch their food. (Previously the use of bait was only attributed to green-backed herons in Asia.)
- A Meyer's parrot (*Poicephalus meyeri*) who figures out how to determine whether or not a sunflower seed is inside without removing the hull
- Canada geese and Mallard ducks who recognize individual humans, even with a change of clothes
- A Budgerigar who turns a physical disability into an advantage
- A Great blue heron that camouflages itself to avoid snapping turtles
- A Mitred conure (*Aratinga mitrata mitrata*) that uses tools
- Elderly parrots that sneak in a nap when no one is looking
- Male Mallard ducks that quack. (Supposedly only female Mallard ducks do that.)
- Juvenile Canada geese that jump off the top of 16-story buildings for fun
- Blue jays (*Cyanocitta cristata*) that weigh peanuts before deciding which one to take

Controlled experiments serve an important function, particularly in the hard sciences like chemistry and physics. In contrast, there are questions about the validity of using controlled experiments to study animal behavior and concerns about reading too much into the results. Specifically, how can one expect to observe the normal behavior of a wild bird who is used to: (a) total freedom of movement (it may migrate 2,000 miles or more each year); and (b) constant interactions with other living entities, plant and animal (it may travel around in flocks of 100 or 1,000 birds), when it is confined indoors in a lab, in a tiny cage by itself, under artificial light instead of the natural sunlight it needs to see clearly, eating strange food, and surrounded by unfamiliar sights and sounds? In short, if the objective is to identify normal behavior, the species must be observed in its normal environment. Hence, this book relies primarily on observational studies.

Several of the birds discussed in this book were given names. This practice acknowledges the fact that, although birds are members of a species and a flock, they are separate living entities. This practice also made it easier to record observations. For example, it was easier to refer to Squeaky or Mr. Rumpletail than try to remember if a given bird was Mallard duck #19 or Mallard duck #27.

Organization of the Book

Avian Cognition is scientifically sound but written in plain English, rather than the traditional dry, stilted, boring style of academe. Instead of saying, "Other atypical behaviors were infrequently observed, contrary to Smth97, Hwth01, and McFud11", *Avian Cognition* simply says, "Once in a while it was noted that ..." The purpose of *Avian Cognition* is to share information, not engage in verbal buffoonery. Over the years I have noticed that students and other people actually read and remember books that are written in an open engaging style. That is not true for dry academic tomes.

Chapter 1 begins by defining intelligence and other closely related concepts. Humans have struggled for centuries to define and measure human intelligence. Even today there is no universal definition or measure of human intelligence, or the closely related concepts of learning, thought, emotion, and consciousness. Hence, the even greater struggle to define and measure animal intelligence.

Avian Cognition creates order out of chaos by developing a standardized definition of intelligence that is applicable to not just humans, but all life forms. There are a variety of types of life forms; hence, there are a variety of types of intelligence. *Avian Cognition* acknowledges the distinction between applied intelligence, biological intelligence, and genetic intelligence. *Avian Cognition* defines the closely related concepts of learning, thought, emotion, and consciousness. In doing so *Avian Cognition* presents the first model which integrates intelligence, consciousness, and behavior.

This foundation establishes a framework for discussing and evaluating avian intelligence, behavior, and individuality in Chapters 2 through 11. This framework is essential; it is what has been missing from previous studies of animal and avian intelligence. Simply put, before anything can be explored, observed, analyzed or compared it must first be defined. This is particularly true for properties and attributes that cannot be measured in the traditional sense with rulers, scales, or laboratory equipment. Without a clear definition of what is being assessed, there is no basis for making an assertion, such as claiming that birds are or are not intelligent, do or do not feel emotions, and do or do not experience consciousness.

Chapters 2 through 11 are each devoted to a particular bird species or closely-related group of species. To understand intelligence, behavior, and individuality, it is necessary to examine all aspects of a bird's life in total. Hence, unlike other books, everything about an individual bird or bird species is presented in the same chapter.

Chapters 2 through 11 adhere to a consistent format. Each chapter begins with general background information about the species. This is followed by interesting historical notes and the population status. A discussion of characteristics that are common to all members of the species follows, like physical appearance, diet, range, habitat, and vocalizations. Then family life, daily activities, and common behaviors are discussed.

Next individual and unique behaviors that were observed are presented. This is the longest section of each chapter. These descriptions include behaviors that were unique to one individual in a flock and behaviors that were unique to the flock under observation that are not normally associated with this species. Many new examples of individual and unique behaviors are reported in each chapter. In addition, Chapters 7 through 11 compare the behavior of birds that are kept as pets to their wild cousins. That has not been done before.

Individuality has not been acknowledged in regard to animals until very recently. All Mallard ducks or all tree squirrels (*Sciuridae sciurini*) were assumed to behave in an identical manner. The first book to break ranks was published in 1952: *Birds as Individuals* by L. Howard. There was silence until 1977 when a book published by Dr. T. Lafeber again mentioned the individuality of birds. In 1996 Dr. Skutch, one of the leading field ornithologists of the last century, raised the issue to a higher level. In 2005 the documentary, *The Wild Parrots of Telegraph Hill*, based on M. Bittner's book of the same title, caught the public's attention as he demonstrated the individuality of wild (escaped or released) parrots living in San Francisco. *Avian Cognition* presents the first serious discussion of avian individuality. Most studies and books view birds as a group from a distance. In contrast, *Avian Cognition* views birds up close as individuals.

In the last section of each chapter (Discussion) common, individual, and unique behaviors are analyzed, compared and interpreted. Rather than just say what a bird did, *Avian Cognition* explains why a bird engaged in a particular behavior. Each behavior is put in context. The significance of each behavior is examined for indications of intelligence and individuality.

Many chapters conclude by highlighting ideas and inventions humans appropriated from a particular species. This important step is often overlooked when assessing avian intelligence. After all, if an idea was good enough for humans to appropriate from birds, it must have been rather intelligent.

There are many cross references between chapters. However, Chapters 2 through 11 are self-contained and can be read in any order.

The book concludes with a discussion of avian communication patterns in Chapter 12. This chapter provides background information for the two audio CDs that accompany the book. Through these CD's readers are given an opportunity to hear the birds and experience the variety and complexity of vocalizations within and among species as they go about their daily lives. Three tracks are linked to specific photographs in the text.

Current bird CDs consist of a 15 to 20 second recording of the signature call from a species. While this is useful for identifying a particular species in the field, it does not convey much information about the variety or complexity of avian communication patterns. In contrast, *Avian Cognition* captures original recordings of the full range of vocalizations and complete conversations among multiple birds in different settings. In doing so many new vocalizations were identified.

The text is supplemented by numerous original photographs which illustrate biological intelligence, genetic intelligence, applied intelligence, and specific behaviors. No special equipment or spotting scopes were used. Rather, the photographs were taken at short range for the simple reason that it is difficult to notice or understand the subtle nuances of behavior from a distance. These first-of-a-kind photographs capture many attributes that have not been featured in bird books before, such as images illustrating:

- The serrated rim of a Canada goose's bill
- The lamellae of a mallard duck
- The inner and outer eyelids of a Canada goose
- A father mallard duck escorting his 1-day old ducklings. (According to the literature only female mallard ducks do that.)
- A Canada goose with a white forehead
- The development of goslings and ducklings from the time they are 1-day old until they are full size 8 weeks later

The photographs reflect events as they occurred. The photographs are not layered; nor were any special effects used. Only minor editing has been performed, such as cropping, adjusting the lighting of an image that was shot at dawn or dusk, or deleting a bird dropping here and there. Often several images are grouped on one page to illustrate a sequence, progression, or variation in behavior. All photographs are accompanied by an explanatory caption and referred to in the text, for example "See Photograph 9.5".

Annex A presents a glossary of terms, acronyms, and abbreviations as they are used in *Avian Cognition.*

Annex B contains a complete bibliography of the sources consulted. Annex B is organized in four parts: publications, videos, websites, and correspondence.

Scientific and lay publications were consulted. Often more is written about lesser known birds in the lay press than in the scientific press. This is particularly true of the parrot family. Pet bird hobbyists have spent more time observing certain types of parrots than scientists. Hence, lay sources of information about bird behavior should not be ignored.

It was noted that older publications, especially those published in the late 1800's and early 1900's, contain more accurate descriptions of birds and their behavior than contemporary publications. That is probably true because those authors relied on personal field observations more than paper studies.

Most books written about birds today fail to include any citations. That is a serious oversight. Consequently, *Avian Cognition* includes citations. Not the cumbersome McFudd11—which is difficult to read over—but the less intrusive."27" Citations accomplish two important things. First, they give credit where credit is due. Second, they allow readers to locate the original source of information in Annex B, should they so desire.

Annex C describes the locations in which the birds discussed in this book were observed. Information is also provided about the specific populations observed and the timeframes in which they were observed.

Annex D is a tribute to Perky, my first budgerigar. This annex presents a positive way for children to be introduced to and learn about pet birds. Hopefully this will help stem the tide of impulse buying which results in so many innocent pet birds—particularly budgerigars—being dumped en masse on parrot adoption and rescue organizations.

Annex E identifies external factors avian veterinarians should consider when giving a bird a medical exam, consistent with the mind/body connection.

Acknowledgments

Anyone who has sat through the credits at the end of a movie is aware that the creative collaboration of a large cast of people with a wide variety of skills is required to produce a successful film. The same is true for a book. Accordingly, with gratitude I would like to acknowledge the leading characters who contributed to this book.

The 300-plus individual birds who are discussed in this book are center stage. While I had the pleasure of observing them over a 30-year period and the opportunity to transcribe and critique their performances, it is the birds' story that is told.

Many thanks to Jennifer Tinnell and Bill Pranty of the Florida Fish and Wildlife Conservation Commission who provided information about the wild flock of budgerigars who live in Florida.

Many thanks to Dr. Yossi LeShem, of Tel Aviv University and the International Center for the Study of Bird Migration, for sharing his research, observations, and insights. His work is directly responsible for reducing the number of bird/aircraft collisions by 76% in Israel—the major bird migration route between Europe, Asia, and Africa.

Many thanks to John Shurky, who shared his observations and the history of the flocks at Pond B with me.

Many thanks to Paul Schmidt, Acting Assistant Director of the Migratory Bird Program at the U.S. Fish and Wildlife Service, for sharing information about the North American Waterfowl Management Plan, duck population trends, and the risk to migratory birds from the 2010 BP Gulf oil spill.

Many thanks to the U.S. Center for Veterinarian Medicine for sharing insights about avian nutritional requirements and information about how the pet food industry is regulated.

Many thanks to the World Parrot Trust for sharing insights about wild and captive parrot behavior. Their 25 years of accomplishments in the conservation, rehabilitation, and protection of wild parrots are impressive. May they have even more success in the next 25 years!

Several of the birds in this study were adopted, primarily from Phoenix Landing Foundation. The parrot adoption, rescue, and education organizations around the world deserve special mention for their tireless work—mostly by volunteers —to enrich the lives of captive parrots by increasing the public's understanding of birds' intellectual, social, physical, and nutritional needs.

Many thanks to Shannon Follin, General Manager, Adam Stamper, Edit/Mix Engineer, and Brian Carroll, Assistant Engineer of Omega Recording Studios of Rockville, MD. This studio provides professional sound engineering services to many big names in the music industry. Yet, they took time to painstakingly distill several hours of outdoor audio recordings into the most unique and interesting bird CD tracks. As a result readers can not only read about the birds' behavior, but also hear and experience the variety and complexity of avian communication patterns.

Many thanks to the many veterinarians and veterinary technicians who cared for the birds discussed in Chapters 7–12 over the years, in particular: Robin Byrd, Senior Veterinary Technician at the Veterinarian Holistic Center, Wendy Miller, DVM of the Town and Country Veterinary Clinic, Sneha Parna, DVM of Banfield Pet Hospital, and Gary Schwartz, DVM of Wheaton Animal Hospital.

Many thanks to Liberty Air Charter Service of Fredrick, Maryland for flying the birds across country and back in style. The birds were much less stressed by these 4.5 hour nonstop flights in the cabin than they are by a half hour car ride.

Contents

List of Tables

List of Figures

List of Photographs

About the Author

Debra Herrmann is a Principle Research Scientist with the Biological Computing Initiative. She leads interdisciplinary research that examines the parallels between avian intelligence, behavior, learning, reasoning, problem solving, and communication skills and computer intelligence, artificial intelligence, neural nets, computer heuristics, as well as other biological computing models found in nature. Earlier she was employed as a scientist by the U.S. federal government for many years.

Debra has been active in the federal and academic research communities in the Washington, D.C. area for the past 30 years, where she competed for and received from $50K to $4M in research funding. Because of her expertise Debra has been asked to sit on panels evaluating research proposals submitted to the U.S. government for funding.

This book includes input from the U.S. Fish and Wildlife Service, the U.S. Center for Veterinarian Medicine, the International Center for the Study of Bird Migration, the U.S. Federal Aviation Administration, and the Florida Fish and Wildlife Conservation Commission.

Debra led or participated in all the observational field studies reported in the book. She has participated in birding activities locally and on three continents (so far) over a span of 30 years. Her lifelong fascination with birds began as a child.

Debra teaches graduate, undergraduate, and professional short courses. She is a frequent guest speaker at annual conferences, monthly educational seminars, and information committee meetings hosted by scholarly organizations and general membership societies. *Avian Cognition* is her fifth book.

Her hope is that *Avian Cognition* will serve as a catalyst for increased public understanding of avian intelligence and behavior, better treatment of wild and pet birds worldwide, and more research in this area.

1

What is Intelligence Anyway?

This chapter introduces the topics of intelligence, learning, thought, emotion, consciousness, and behavior. These topics have been studied and debated for centuries. A complete dissertation could be written on any one of these topics and in fact many such excellent dissertations have been written. In contrast the purpose of this short chapter is to:

- Introduce these topics
- Present some novel insights
- Stimulate new thinking in these areas
- Establish a framework for discussing and evaluating avian intelligence, behavior, and individuality in Chapters 2–12 that follow.

1.1 Definitions, Models, and Theories

Before anything can be explored, observed, analyzed or compared it must first be defined. This is particularly true for properties and attributes that cannot be measured in the traditional sense with rulers, scales, or laboratory equipment. Without a clear definition of what is being assessed, there is no basis for making an assertion, such as claiming that birds are or are not intelligent, do or do not feel emotions, or do or do not experience consciousness. Consequently, we will begin by defining intelligence and other closely related concepts.

People have tried to define intelligence—or at least human intelligence—for centuries. Yet there is no consensus on what intelligence is or how to measure it. Many different theories of intelligence have been put forward. However, none of them have been proven in the absolute sense; nor is any single theory universally accepted.

In the absence of a standard definition this book defines intelligence as:

> The ability to acquire, discern, assimilate, generate, retain, and apply facts and knowledge, some of which are constants and some of which are variables, that are needed for problem solving, especially when confronted with a novel situation, by an individual and species to survive, prosper, and progress in its ever changing environment. This includes the ability to distinguish irrelevant data and trivia from useful information.

There are four common threads in the multitude of definitions of intelligence that exist today:[46, 72, 179, 203]

- Intelligence is a property of the mind
- A key aspect of intelligence is the ability to respond adaptively to novel situations

- A distinction is made between crystalized intelligence (knowledge or the constants mentioned above) and fluid intelligence (mental flexibility or the variables mentioned above)
- Intelligence is composed of both verbal and non-verbal (quantitative, analytical, logical) criteria

When we switch from the noun 'intelligence' to the adjective 'intelligent' additional parameters pop up like good judgment and sound thought.[179] Synonyms for intelligent are: clever, alert, and quick-witted.[179]

Intelligence is not static. Rather, intelligence is linked to culture, geography, and time. There are cultural variations in cognitive competencies.[72] This does not mean one culture is smarter or dumber than another. It simply reflects the fact that different cultures approach the acquisition of knowledge and problem solving tasks from different perspectives. Some cognitive domains are universal, while others like music or mathematics are valued and encouraged or discouraged and restricted to certain cultures or small groups of people within a culture.[72]

This situation points to the phrase in the definition of intelligence about an individual and species surviving, prospering, and progressing in *its* environment. To illustrate, suppose that an investment banker from Manhattan and a Xavante elder from Brazil suddenly changed places. Neither would be able to function or likely survive on their own in each other's circumstances. Not because they were "stupid", but because the intelligence they possess is linked to their environment. In comparison a Berber individual and a Bedouin individual could suddenly trade places and they both would likely be alright after a short adjustment period.

Even contemporary universal competencies are somewhat specialized. For example consider the skills necessary to drive an automobile. A person who grows up and learns how to drive in Texas acquires similar—but different—skills than someone who grows up and learns how to drive in Minnesota. The Texas driver would not have any knowledge or experience driving in snow and ice. The Minnesota driver would not have any experience driving in sand storms or understand that the roads in Texas are very slippery when it rains, because the oil dripped from automobiles accumulates between the infrequent rainfalls.

Intelligence is also linked to time. If we took our Manhattan investment banker and left him in Manhattan but turned the clock back 500 or 1,000 years he probably would not be able to function or survive. All of his knowledge about commodities, derivatives, futures trading, smart phones, subways, ordering carry out, etc. would be of no use. Someone from that time period would not be able to function on his own in today's world either.

Beginning in the 1950's western societies became fascinated with the notion of measuring human intelligence. A plethora of tests and measurements were produced in an attempt to ascertain the ever elusive intelligence quota (IQ). These standardized tests were applied to school children from first grade through high school, purportedly to identify the potential for future success, and the fame of the Number 2 pencil spread overnight. By the 1970's and 1980's questions were raised about what these tests actually measured and whether or not the design of the tests contained cultural and/or gender biases. Regardless, standardized tests have become a fact of life as evinced by the omnipresent SAT, GRE, LSAT, MCAT, and cottage industry of professional certification exams.

Some unusual theories were developed in the early days of human intelligence testing in an attempt to explain the spectrum of results seen in the general population. It is interesting to note that although it is acknowledged that there is a wide range of intelligence in the human population, the tendency is to assign a single intelligence rating to an entire non-human animal species...

One argument focused on whether genetics or environmental enrichment played a larger role in determining one's intelligence. Another argument favored brain size as an indicator of intelligence. Naturally this was meant to "prove" that humans are superior to animals. An extreme version of this argument went so far as to claim that the ratio of brain weight to body weight was an indicator of intelligence. That of course would mean that skinny people are smarter than heavy people! If we look beyond our own species we find that ravens and dolphins have the same brain weight to body weight ratio as humans.[127] More telling is the fact that a sparrow's brain represents 1/20th of its body weight, while a human's brain only represents 1/50th of its body weight![176]

Even so new theories about intelligence continued to emerge. Before long there was a debate about whether or not intelligence was a singular property, as in 'general intelligence', or a collection of multiple diverse properties. Some experts argued that the categories listed below are autonomous human intellectual competencies; others claimed the categories represented different skill sets or subsets within general intelligence:[72]

- Linguistic: listening, speaking, writing, and reading
- Musical
- Logical, analytical, mathematical
- Spatial
- Movement, athletics, dance
- Inter- and intra-personal

As society tuned into touchy feely topics the term emotional intelligence was coined to amplify the concept of inter- and intra-personal intelligence. This intelligence was said to reflect the skills needed for successful leadership and social awareness.[203] In essence one's emotional intelligence was a barometer of one's emotional self-efficacy.[203]

In some circles the realm of intelligence was not limited to animal life forms. Theories were advanced that various forms of plant life also possessed a degree of intelligence, due to their ability to adjust their daily and seasonal life activities based on perceived environmental and climatic conditions and changes in these conditions or patterns.[203]

The earliest work on intelligence was done primarily by philosophers. Before long psychologists, particularly educational psychologists, became firmly entrenched in the field. Today a whole spectrum of specialties are studying various aspects of intelligence and several new disciplines have emerged like cognitive science, cognitive psychology, cognitive ethology, cognitive systems, neurobiology, and neuroscience.

1.2 Learning

The ability to learn is a prerequisite for intelligence. The act or process of learning takes many forms for humans, from formal classroom study to observation, practice, memorization, reflection, and introspection. The outcome of learning is understanding and comprehension. Learning involves reasoning, pattern recognition, relating symbols and objects in space and time, and discerning causality.

Memory is an active part of learning as is attention to detail. Once something is learned it is stored in memory. Each learning token is then used as a building block to acquire additional knowledge that is: (1) a more refined or extended version of the current learning token, (2) a learning token about a similar area of knowledge, or (3) a totally new type of learning token. As knowledge is acquired through the learning process learning tokens are compared and organized. Hence, learning tokens from the past and present are assimilated to create learning tokens in the future.

This process follows the association theory of learning. To illustrate, one does not start out in life by learning differential calculus. Instead a child begins by learning cardinal and ordinal numbers. That knowledge forms the foundation for learning addition and subtraction. Multiplication, division, algebra, geometry, and trigonometry follow, each learning token building upon the former ones. After this information is assimilated, one is ready to learn differential calculus. Along the way parallel mathematical skills may be acquired as well, like how to make change or balance a check book, understand the concepts of distance and weight, and appreciate timing or sequencing in a musical score.

Today the term 'lifelong learning' is frequently heard. The notion of lifelong learning is not limited to employability. Rather, it applies to life in general. As just mentioned, learning is an ongoing cumulative process built around acquiring, comparing, and organizing new learning tokens with existing learning tokens. If a person ceases to acquire new learning tokens, by definition they also cease accessing, comparing, and organizing existing learning tokens. Eventually learning and the capacity for learning cease and the functional knowledge retained decreases. Knowledge that isn't accessed and used frequently is lost. In short, the brain is subject to atrophy just like a muscle.

People learn in different ways. Some people are visual learners and are more apt to grasp a new concept by seeing a diagram, map, schematic, or photograph. Some people are verbal learners and will grasp a new concept faster if it is explained to them verbally or in writing. Other people are tactile learners. Their preferred method of acquiring a new concept is to touch a real life object. This latter group of people is often referred to as hands-on learners. Instead of learning how to fix a flat tire by watching someone else do it or reading a manual, they learn by changing the flat themselves.

There are four methods of learning:

- Learning from experience
- Learning by imitation
- Trial and error or experimentation learning
- Learning by insight

People can learn from experience. For example, the first time a child rides his bike through a mud puddle he learns that he will get sprayed. Once acquired this learning token is stored for future reference and mud puddles are avoided, unless the child wants to get wet.

People can learn by imitation. Most athletic and artistic endeavors are learned by imitating a coach or instructor. The teacher illustrates how to hit a tennis ball or play a violin and the student attempts to imitate the movements and style of the teacher. Imitation learning usually involves a fair amount of practice before the student masters the technique and makes it her own.

Human kind as a whole has historically demonstrated skill at trial and error or experimentation learning. Most knowledge about mathematics, science, engineering, and technology was acquired through experiments or trial and error learning. Consider the humble wheel. The idea for a wheel probably came about by watching a rock or tree roll down a hill. The first wheels appeared in 3500 B.C. and were solid wood; by 2000 B.C. spokes were added.[46] Eventually the wheel was separated from the axle. Ox carts, wheel barrows, Roman chariots, covered wagons, horse drawn carriages, bicycles, and trains followed. Fast forward to the early 1900's. Automobiles replaced the metal wheels of carriages with rubber tires to provide a smoother ride. Flat tires became a common problem and cars slid off the road whenever there was rain, snow, or ice. These problems had not been experienced with wood or metal wheels. Through experimentation and trial and error learning tires were eventually reinforced with tread and separated from inner tubes. The whole sequence from the wooden wheels of 3500 B.C. to the car tires of today represents the gradual accumulation of knowledge through observation, experimentation, and trial and error learning.

Insight is the fourth mode of learning. This mode of learning is more cerebral than the other three modes. It consists of thinking about a problem, then devising a correct solution purely through mental insight. In the vernacular, insight learning is described as the proverbial 'light bulb' turning on in one's brain. Arriving at the solution may involve analyzing previous experiences and/or knowledge for similarities and differences to the current situation. But the new knowledge is created without any experimentation or trial and error learning. Or, to put it another way, the first trial and error experiment is correct. From a mathematical sense one could argue that some degree of luck, chance, or randomness caused the first trial and error learning experiment to be correct.

In the final analysis experience, imitation, trial and error learning or experimentation, and insight represent a continuum of learning modes—not four discrete learning modes. Likewise, all four learning modes employ an understanding of historical and contemporary cause and effect relationships. Learning is an example of applied intelligence.

1.3 Thought

Thought is a precursor to learning. A being has to be able to think before it can learn. Thought or the process of thinking is referred to as cognition. Thought is fluid and not bounded by time. Thought may involve reflection, recollection, or remembrance of a person, place, object, or event from the past. This could be something a person experienced firsthand, heard about from someone else, or read about. Thought may involve foresight or anticipating an encounter with a person, place, object, or event in the future. Or, thought may involve a serious consideration of past events or experiences and how they may or may not be perceived to apply to the present. Thought includes the ability to imagine various alternative scenarios, conceive ideas, draw inferences, and make judgments about the relative merits of each. Thought often highlights or informs intentions, plans, strategy, opinions, or beliefs. However, humans being humans, not all thought is logical or serious. On the contrary, human thought can be wildly illogical or simply directed toward entertainment purposes, such imagining a humorous incident.

The thought process consists of several steps, such as evaluating and interpreting a collection of competing and conflicting current and historical information, developing a hypothesis, and arriving at a conclusion or solution. The conclusion or solution may be: (1) consistent with previous conclusions or solutions and applied to a similar situation, (2) consistent with previous conclusions or solutions and applied to a new situation, (3) a novel conclusion or solution applied to a similar previous situation, or (4) a novel conclusion or solution applied to a new situation.

Culture influences thought to some degree. Moving beyond an individual to society at large, thought can be said to reflect the intellectual products, organized views, principles, and values of a particular time period, place, and community.[179] Although, there is less social cohesion today than in the past.

Thought invokes higher mental processes. Depending on the cognitive complexity and type of thought or task being solved (numerical, navigation, speech, movement, simple recall) different areas of the brain are activated. This is true in humans as well as several species of animals that have been studied.[82] This area of neurobiology has received a lot of attention recently given the current ability to create images of brain activity. This aspect of thought gave rise to the expression of a person being left- or right-brained, to indicate that their thought patterns tend to be either more artistic or more analytical.

It is not unusual for a person to claim that he or she is "deep in thought" and doesn't want to be disturbed. The person may appear to be inattentive or preoccupied. He or she may close their eyes to emphasize the fact that they are concentrating on something and have blocked all sensory input and interruptions. A person is aware that they are deep in thought. At the same time a person is not cognizant of all the mental activities that are taking place during a period of deep concentration. This situation is analogous to computers that are designed with the ability to enable, disable, and prioritize the processing of interrupts.

Three main theories attempt to explain how knowledge is acquired:

- Empiricism
- Rationalism
- Skepticism

According to empiricism all knowledge is acquired through observation, experiment, experiences, and sensory input. Empiricism embodies the experience, imitation, and trial and error modes of learning. Empiricism is sometimes referred to a logical positivism.[179]

In contrast rationalism posits that knowledge can be established by reasoning alone, without the need for observation, experiments, experiences, or sensory input. Historically rationalism was used to establish the validity of various religious truths that could not be proven in the scientific sense.[179] Rationalism relies solely on the insight mode of learning.

A third school of thought, skepticism, raises doubts that humans are capable of acquiring or generating true knowledge. Facts, figures, and theories are viewed through the lens of uncertainty.

Today the scientific community relies primarily, but not exclusively, on empirical thought. Empiricism invokes logic in order to arrive at the correct inference or reasoning when analyzing facts, observations, and events. Human thought is capable of engaging in multiple types of logic, as listed below, in response to the specific circumstances and goal being pursued. These models attempt to reason about something through various logical constructs. Several types of logical reasoning have parallels in modern computer science.

Logic models that apply to humans and computers

- Many value or tri-state logic: yes/no/maybe, positive/negative/zero, positive/negative/don't care
- Predicate logic (or quantification theory): some/all/none, less than/equal to/greater than
- Propositional logic: and/or/not, if/then/else, case statement/otherwise clause
- Temporal logic: past/present/future

Logic models that only apply to humans

- Deontic logic: imperatives, obligations (taxes, confession, fasting)
- Epistemic logic: beliefs
- Circular logic: argument, proof, and conclusion are all the same
- Syllogistic logic: major/minor premises point to a conclusion

All of these logic models are examples of the deductive reasoning of which Sherlock Holmes was so fond. Some people appear to have an innate ability to think logically. Other people can be taught to think logically, such as learning how to construct a legal argument or reason mathematically. Even Dr. Watson was able to improve his deductive powers.

It is interesting to note that the four logic models that cannot be represented in computer programs do not stand up to the test of Boolean or modern symbolic logic. Rather, these four models allow humans to convince ourselves of the logic of certain views or opinions about the world around us and our place in it. These four models attempt to explain the unexplainable in a philosophical rather than scientific sense.

It is not surprising that these four logic models were developed. Humans have a tendency of: (a) not handling negative and contrary information very well, and (b) avoiding information that contradicts currently held beliefs—like the earth is flat.[46] Sometimes this reluctance reflects an individual trait. An individual may simply discard information for financial, family, professional, personal, or emotional reasons. This situation was captured by Upton Sinclair over a century ago when he said "It is difficult to get a man to understand something when his salary depends on not understanding it".[77] Other times societal norms enforce conformity of thought, for example during the Inquisition in the past and withholding promotions, tenure, and research funding in the present.

In addition to the "art of deduction" there is the world of inductive reasoning. Inductive reasoning is used to solve problems that cannot be solved by deduction alone. This is the realm of probability, statistics, and Bayesian belief networks.

Few would deny that philosophy, theology, and psychology are intertwined. Even so, that connectedness has created some problems over the years when studying—or attempting to explain—intelligence, learning, and thought. Most people today agree that intelligence, learning, and thought belong to the domain of the mind. However, that raises questions about the relationship between the mind and the body. Is the mind just the brain, something other than the brain, or something operating in conjunction with the brain? Is the mind physical or not? These questions have been particularly perplexing for theologians over the centuries because of the supposed linkage between the mind and the soul. This linkage was used by Descartes to deny that animals could possibly think. As a classic example of 1600's rationalism and circular logic his supposition went as follows:

1) Animals do not have souls
2) Therefore animals do not have minds and cannot think

The mind, body, soul trichotomy was not the only problem philosophers, theologians, and psychologists found perplexing in regard to intelligence, learning, and thought. The concept of free will was (is?) also a sticky issue. From a theological perspective free will means humans possess the freedom to choose to do or not do whatever they want, whenever they want. The notion of free will is the cornerstone of moral responsibility. The notion of free will is also consistent with the properties of intelligence, thought, and consciousness, all of which reflect the aspect of volition. However, the notion of free will or volition conflicts with the theological concept of predestination. On the other hand, if there is no free will are humans really sentient beings that think, feel, and experience consciousness?

These questions have been debated for over five centuries. As a result they will not be answered in this short chapter! These questions are raised here because they influence how humans understand animal intelligence and thought. As we see in Table 1.1 human perception of animal intelligence and thought has changed throughout history. This is not surprising since our perception of human intelligence and thought has been evolving as well.

Table 1.1 Historical Perspectives on Animal Thought (compiled from multiple sources).

Supposition	**Theorist**	**Time frame**
Animals have no thought	Descartes	1595–1650
Animals have perception, memory, and reason, but no abstraction	Locke	1632–1704
Animal inferiority is a matter of degree	Hume	1711–1776
Animals are not machines	Kant	1724–1804
Animals have understanding and will without language	Schopenhauer	1788–1860
The difference in mind between man and the higher animals is one of degree	Darwin	1809–1902
Animal thought consists of conditioned reflexes	Pavlov	1849–1936
Animal thought consists of stimulus/response scenarios	Thorndike von Frisch Lorenz Tinbergen	1874–1949 1886–1982 1903–1989 1907–1988
Behaviorism	Watson Hull Skinner	1878–1958 1884–1952 1904–1990
Animals have beliefs, intentions, conscious perception, and introspection	Armstrong	1968–
Beginning of cognitive ethology	Griffin	1976–

The debates about the mind/body relationship continue into our day. It is given that the central nervous system arbitrates the complete suite of mental states, such as desires, beliefs, memories, anticipation, observation, analysis, decisions, questions, concept formulation, pattern recognition, and so forth.[82] The question is how do the central nervous system and/or mind do this? Is thought or reasoning based on internal mental logic, or are other currently unknown mental images, scenarios, and models at work?[46] Are the central nervous system and/or mind hardwired for specific functions (like an application specific integrated circuit or ASIC) or a flexible general purpose entity (like a laptop computer)?[72] Why does each individual have some common and some unique reasoning and intellectual capabilities?[72]

If so many questions remain unanswered about human thought and how it works, how can anyone be adamant that animals do not think? Or, to start where Descartes left off one could assert:

1) Animals do not think
2) Humans are animals, therefore humans do not think!

1.4 Emotion

Emotion has the ability to attenuate thought for better or worse. Emotion reflects the affective part of consciousness.[179] Unlike thought emotion straddles both the psychic and physical domains of a being. Emotion is considered to be a complex psycho-physiological experience that simultaneously mirrors and reacts to changing environmental inputs and biochemical states.[46, 203] Emotion gives rise to subjective responses and unreasoned opinions and beliefs that are often driven by instinct or unidentifiable inner ("gut") impressions.[46, 179, 203] While emotion doesn't entirely determine, it certainly influences one's mood, temperament, disposition, and motivation. Like intelligence and thought emotion has been studied for centuries without much consensus, except for the following two points:

- It is important to distinguish between an emotion (what is felt) and an emotional state, which reflects the expression or embodiment of an emotion.[189, 190, 203]
- It is important to separate the sequence and duration of emotions in order to be able to differentiate a single episode (an outburst) from a general tendency (one's personality).[203]

Most people identify with the term emotion even if at times they are unable to articulate exactly what they are feeling. It is even more difficult to understand what emotions others are feeling, apart from the physical expression of that emotion. Consider the all too common scenario when one person is experiencing an intense emotion and another somewhat obtuse person keeps telling them to just calm down. Emotions are felt not thought. Emotions are a reaction to certain stimuli, not a cognitive outcome (unless one is a professional actor). The verb 'to feel' also refers to tactile sensation. So we could say that emotions touch our psyche.

Emotion can be an individual or group experience. People who go through the same experience may or may not have the same emotional reaction. The depth or range of emotions experienced varies from person to person. So does the expression of those emotions. Some people are quite reserved; other people are very demonstrative. Each culture has established norms for what is and isn't considered acceptable modes of emotional expression at home and in public. Like individuals, some cultures are quite reserved when it comes to expressing or even talking about emotion. Other cultures expect people to be very demonstrative.

Then there is the crowd or mob phenomenon. Individuals in a crowd can be spun up into a frenzy to do or say something or act in a manner they would never even consider doing on their own. People in the crowd feed off each other's emotional energy or the group psyche.

Emotion has been described as reaction specific energy.[5] In this model energy, both emotional and physiological, accumulates in response to a condition; then it is released as a specific behavior.[5] However, while there is a definite link between mind and body, there is not an automatic or consistent one-to-one relationship between a person's physical state and their emotional state. A person can be tired and happy: he or she just completed a long journey or difficult project. Or, a person can be tired and grumpy. A person can cry when he or she is sad, in response to a personal loss or setback. Or, a person can cry when they are happy, like when seeing someone they haven't seen for a long time. Furthermore, the same condition does not always produce the identical emotion, nor does the same emotion always produce the identical emotional expression. In essence human emotion and the expression of it are inconsistent and somewhat unpredictable.

Like thought emotion transcends time. A significant emotion from the past can be recalled and experienced just as intensely in the present. Conversely, a person can have a strong emotional reaction to some anticipated future event. Hence, one's current emotional state may have nothing to do with the present.

Emotion and emotional responses change with age. What scared us as children usually doesn't scare us as adults. On the other hand several things children don't fear are of concern to adults: financial collapse, disease, war. Children experience and express more basic emotions than adults. As people become adults more energy is spent trying to control and conceal emotions. Actually humans cannot control their emotions—a person cannot stop an emotion (love, hate, jealousy, fear) from popping into their head. A person can only control their expression of and reaction to that emotion. As a result an adult's emotions are not so honest or easy to read. In contrast, birds and other animals are rather good at reading each other's emotions.

Emotion originates from a primitive part of the brain and is induced through the endocrine system.[5] Emotions are often triggered by the situations identified below; this explains why hesitation or the lack of choice often intensifies emotion:[5]

- An adequate or appropriate response to a given situation is not available
- Instinct does not have an immediate outlet
- The internal state of a being is out of harmony with its behavior or that of others
- The being is conflicted
- There is some impediment or obstacle to correct action

Extreme emotions can cause excitability, agitation, and disturbances in sensory and bodily functions, like an accelerated heart rate, nervous stomach, or trembling. In the past the body's physical response to stressful situations was a survival tactic. For example, sudden acute fear triggered the formation of blood clots. Sudden acute fear usually meant a person was about to be attacked by a wild animal or have a spear thrust through his side. The clots were a protective measure to prevent bleeding to death. People today still experience fear and stress, though not too many of us are chased by wild animals. Yet our physiological response has not changed; hence the high incidence of heart disease.

The interconnectedness between feeling and physiology that is the essence of emotion is highlighted in the amygdala. The amygdala supervises the processing of stimuli that trigger emotions, the emotional response to these triggers, and the storage of memories related to emotional events. The amygdala consists of groups of nuclei located in the medial temporal lobes of the brain. The human amygdala shares an evolutionary history with that of most complex vertebrates. The amygdala sends signals to the:[203]

- Hypothalamus for activation of the central nervous system
- Thalamic reticular nucleus for sharper reflexes
- Nuclei of the trigeminal (facial) nerves

The amygdala is responsible for stimulating the release of dopamine, norepinephrine, and epinephrine (adrenalin), all of which are neurotransmitter hormones, and the processing of olfactory sensations (scent). Interestingly, the amygdala also plays a role in consolidation of memory.[203] Perhaps this explains why emotional events are remembered vividly, although not always exactly as they occurred.

The amygdala and its functions illustrate a bridge between human emotion and that of other animals. Animals experiencing various emotional events send the same physiological signals to the brain via neurotransmitter hormones as do humans.[189, 190] For example, the fear response in most complex vertebrates, including humans, is an ancient survival instinct that is primarily controlled by the amygdala, bypassing active thought.[189, 190] This is not surprising since the brain physiology of humans and other animals is similar.[5, 189, 190] Another interesting point to consider is that on average the amygdala of birds is larger proportionately to that of humans.[128]

A few people have put forward a theory that animals cannot express or experience emotions because they lack language. A variation of this theory states that humans cannot comprehend animals' emotions because animals lack language. On the contrary, as we have just seen, emotion is a psycho-physiological experience. Humans may or may not use language to *express* an emotion (for example slamming a door); but language is not required to *experience* an emotion—listening to instrumental music can evoke emotions. One can observe young children from any culture who have not yet learned to speak and easily discern their emotions, such as an infant smiling. At the same time it is important to remember that different species have different or expanded sensory perceptions, social orders, and life styles. As a result, different species may experience emotions and express emotions in ways that are unfamiliar to humans.

The emotional states of the birds discussed in Chapters 7 through 11 were monitored. The term emotional state is used instead of emotion to indicate the expression of an emotion (which is observable), rather than an emotion that is felt (which is difficult to observe). In particular the birds in Chapters 7 through 11 were monitored for indications of 22 different emotional states. These states were grouped into positive and negative emotions. Positive emotions are a standard set of emotions that are experienced within the normal ups and downs or challenges of everyday life. Negative emotions are those which if left unchecked can lead to self-destructive behavior and/or disease. (See Table 1.2.)

Table 1.2 Emotional States for which the Birds Were Monitored.

Positive Emotions	**Negative Emotions**
• Excited	• Anger
• Happy	• Hatred
• Sad	• Jealousy
• Confident	• Grumpy
• Timid, unsure	• Overwhelmed
• Lonely	• Stressed out
• Curious	• Possessiveness
• Puzzled	• Self-pity
• Playful, eager	• Depression
• Anxious, fearful	• Boredom
• Aggressive, impatient	
• Gentle, patient	

**Note*: These are the 22 emotional states for which the birds were monitored. This list does not imply that every bird or any given bird exhibited all of these emotional states.

Because of the long observation period changes in an individual bird's emotional state were discernable. Individual birds of the same species are more or less demonstrative than the others of the same species. Individual birds exhibit various emotional states at different times during their lives, even within the same day. Much like reading body language in people, the activity level, social interaction, posture, and eyes change to reflect various emotional states. Vocalizations may increase or decrease or change in volume. A transition from a state of high alertness to total relaxation also reflects a change in emotional state, as does a change in the level of curiosity or playfulness. The breathing rate may be faster or slower. Feather position is also a good indicator of emotional state. As Dr. T.J. Lafeber stated in 1977:[99]

> "... people working with birds see an assortment of personalities ranging from a nice average to both extremes of friendship and hostility. This reconfirms our knowledge that birds are emotional. Without feelings, there could be no fear, hostility, or the opposite—friendly, outgoing, attention seeking attitudes."

1.5 Consciousness

Consciousness dwells in the domain of the mind. Consciousness enables a being to participate in and experience the reality of their existence. Consciousness is characterized by perceiving thought, emotion, and sensory input, and exercising volition.[179] Consciousness reflects total awareness (albeit subjective) of oneself, others, and the external world.[46, 179, 203] Sometimes consciousness is referred to as the executive control system of the mind.[203] Consciousness permits the perception of time and reflection about the past, present, and future; this in turn informs plans, goals, and intentions.[82] Within the legal and theological communities the awareness attributed to consciousness is used to infer responsibility and the ability to know right from wrong.

Consciousness serves several supervisory functions that are necessary for a being to live:[203]

- Defining and establishing contexts (situational awareness)
- Adapting and learning
- Making decisions
- Prioritizing action
- Organizing and coordinating motor and mental responses before taking action
- Identifying and correcting errors

Consciousness makes individuality possible.[82] Expressed in other terms individuality demonstrates consciousness. That is because consciousness involves will, intention, and choice—things that are decided or determined on an individual basis.

There are four levels of consciousness, each with its own characteristics, mental tasks, and activities:

- Full consciousness
- Semi-consciousness
- Sub-consciousness
- Unconsciousness

The highest level is full consciousness. At this level one is awake and alert and aware of one's thoughts, emotions, and self, as well as others and the external world. A conscious being thinks, feels, perceives sensory input, and is able to make decisions and take action. Some consider the ability to recognize the reflection of one's own image an indication of consciousness. If this is so, then by definition the ability to recognize an image that is not one's own is also an indication of consciousness.

The next level down is referred to as semi-consciousness. At this level a being has a partial or incomplete awareness. There are two variations of semi-consciousness. A being can be totally in a semi-conscious state; for example when they are starting to wake up, but are not fully awake. Or, a being can be engaged in a combination of conscious and semi-conscious tasks at the same time. Typing and driving are two common examples of semi-conscious activities. To illustrate, if I want to type this paragraph I read the first sentence, "The next level down ...". My eyes focus on the letter images. My brain interprets the letters, words, and punctuation, and perhaps even catches a spelling or grammar error. But, my conscious mind does not tell which finger to push what key or when. That mental activity is handled at the semi-conscious level. I am aware that I am typing, but not of the instructions sent to each finger.

The third level down is referred to as sub-consciousness. At this level mental activities are occurring, but the being is not aware of them like they are when they are in a conscious or semi-conscious state. Autonomic processes and biofeedback loops take place at the subconscious level. For the most part beings are not aware of these processes. Higher mental activities also take place at the subconscious level. For example, a person may have been unable to find an answer to a particular problem for a period of time. Then he or she wakes up one morning and suddenly has the answer. Or, a person may have suppressed certain thoughts, emotions, or memories associated with a traumatic event. These thoughts, emotions, and memories are not readily available to the conscious mind. But they are very much available in the subconscious and can be involuntarily launched into the conscious mind should an unusual circumstance propel their release. This scenario can happen regardless of the time gap between the present and when the traumatic event occurred. Other examples of subconscious mental activity include receiving subliminal messages and perhaps telepathy.

The fourth and lowest level is unconsciousness. In this state a being is asleep and not responsive to sensory input or external stimuli. There is no awareness of self, thought, or feelings. Even so, certain brain activities are taking place. During the sleep cycle the brain alternates between deep sleep (delta or low wave sleep) which lasts about 90 minutes and rapid eye movement (REM) which lasts about 30 minutes.[46] It is as if the brain is organizing and storing new information received during the day. Sleep of course is when dreams occur. Dreams may or may not have something to do with the day's events.

The subconscious mind is active at all times, regardless of whether one is conscious, semi-conscious, or unconscious. And, as noted earlier one can be conscious and performing a semi-conscious task at the same time. In summary, one's consciousness consists of four compartmentalized levels that can be dynamically reconfigured into four distinct operating modes as needed:

Consciousness Operating Modes and Levels

- Subconscious + unconscious
- Subconscious + semi-conscious
- Subconscious + conscious
- Subconscious + conscious + semi-conscious

1.6 Behavior

Behavior is a broad topic that is not well understood. Contemporary references to behavior frequently focus on the negative aspects: "conduct unbecoming ..., your behavior lately ..., quit behaving like a ..." References to irresponsible, compulsive, and addictive behavior are common. In contrast references to exemplary and altruistic behavior are uncommon. Perhaps, like a self-fulfilling prophecy, that is why exemplary and altruistic behavior are so rare.

Behavior can be positive and constructive, negative and destructive, or neutral and neither constructive nor destructive. Which category a specific set of behavior falls into depends on many things, such as: the particular circumstances and time frame, what the behavior accomplished compared to what it was intended to accomplish, and whether the behavior was that of an individual or a group. Cultural norms and values also help determine whether or not behavior is considered to be positive, negative, or neutral.

Behavior falls into the domain of the body. Behavior is something an individual or group does versus what an individual or group thinks. Ideally thought and intelligence inform behavior. But humans frequently act without thinking, or they do not think a situation all the way through before acting and the law of unintended consequences kicks in.

Any activity a living being engages in constitutes behavior. Daily activities, seasonal activities, play activities, family activities, and so forth are all different types of behavior. Certain behaviors may follow a regular routine and exhibit a high degree of consistency and predictability. Other behaviors may seem novel or inconsistent and unpredictable. That is because an individual does not always behave the same way in the same or a similar situation. Furthermore, an individual may simply decide to do something different. Behavior may be associated with an individual, a group, or a species. Some behaviors are linked to geography and the environment. Behavior may be initiated at the conscious level or the semi-conscious level, like a habit.

Behavior must be put in context in order to be understood correctly. There is the old story about how student A constantly bullies student B and the teacher doesn't notice. Then one day student A bullies student B and student B fights back, verbally or physically, just as the teacher walks in. The teacher disciplines student B while student A laughs.

This story illustrates the point that behavior is a continuum and not just isolated events. As a result, to understand behavior correctly it is necessary to examine the totality of an individual's behavior in context, not just isolated events. This is particularly important when trying to understand animal behavior; hence, long periods of observation are needed.

Attempts have been made to reduce behavior, human and animal, to a series of stimulus response scenarios. The majority of advertising is based on this premise. However, in depth analysis indicates that a lot more is going on. A stimulus may be a memory or some combination of sensory input. Between the stimulus and a given response a whole series of processing takes place—cognition. Each stimulus has to be decoded, interpreted, analyzed, and accepted or discarded, along with a multitude of others. Then a decision is made about how to respond. Viktor Frankl was one of the first to point out that between stimulus and response lays this important mechanism called choice.

To use an analogy, computer operations are governed by inputs, processing, and outputs. Inputs are received and processed by a combination of algorithms (computer programs or software), which produce an output. Computers are not simply input (stimulus)/output (response) machines. It is the processing power that makes computers useful. That is the difference between a manual typewriter and a laptop computer.

If we put all the pieces together we see that behavior is an outcome of several steps, not just an isolated stimulus, as illustrated in Figure 1.1.

An individual's consciousness acts like a computer operating system (OS) and is continually monitoring and arbitrating between the ongoing cycle of inputs, processing, and outputs.

An individual has two sources of inputs: memory and sensory input. Humans, like many animals, have five types of sensory input: touch/tactile, taste, smell, hearing, and vision. For example an individual may hear a random sound or someone speaking. An individual may view the flowers blooming in a garden or read a book. Different species rely more (or less) on different types of sensory input than other species.

Memory and sensory inputs are processed simultaneously. Sensory inputs go through a whole series of processing steps, like interpreting, analyzing, comparing, aggregating, validating or discarding, and prioritizing, to convert the raw data into something that can be processed. Memory and sensory inputs may be processed rationally, through thought, reasoning, pattern recognition, and contemplation. Memory and sensory inputs may also be processed emotionally. There is always some degree of tension between the processing of thoughts and emotions. Thought processing enables learning, both the ability to: (1) compare prior learning experiences with the present situation, and (2) acquire new learning. The learning process enables flexible decision making.

The output or outcome of processing emotions, thought, and learning is intelligence and behavior. An individual's intelligence, including their emotional intelligence, guides and moderates their behavior. And an individual's behavior reflects their intelligence (at least most of the time ...). Hence, the sequence of events depicted in Figure 1.1 is not that different for animals and humans.

1.7 Language, Culture, and Technology

Language, culture, and technology are the three artifacts most often cited as proof of human intellectual superiority. The hypothesis is that because humans have these three artifacts, they are intelligent and other animals are not. We would do well to remember that the same language, culture, and technology arguments were used in the past for other reasons. During the era of the explorers, colonialism, slavery, and the westward expansion into Native American lands the argument was made that certain groups

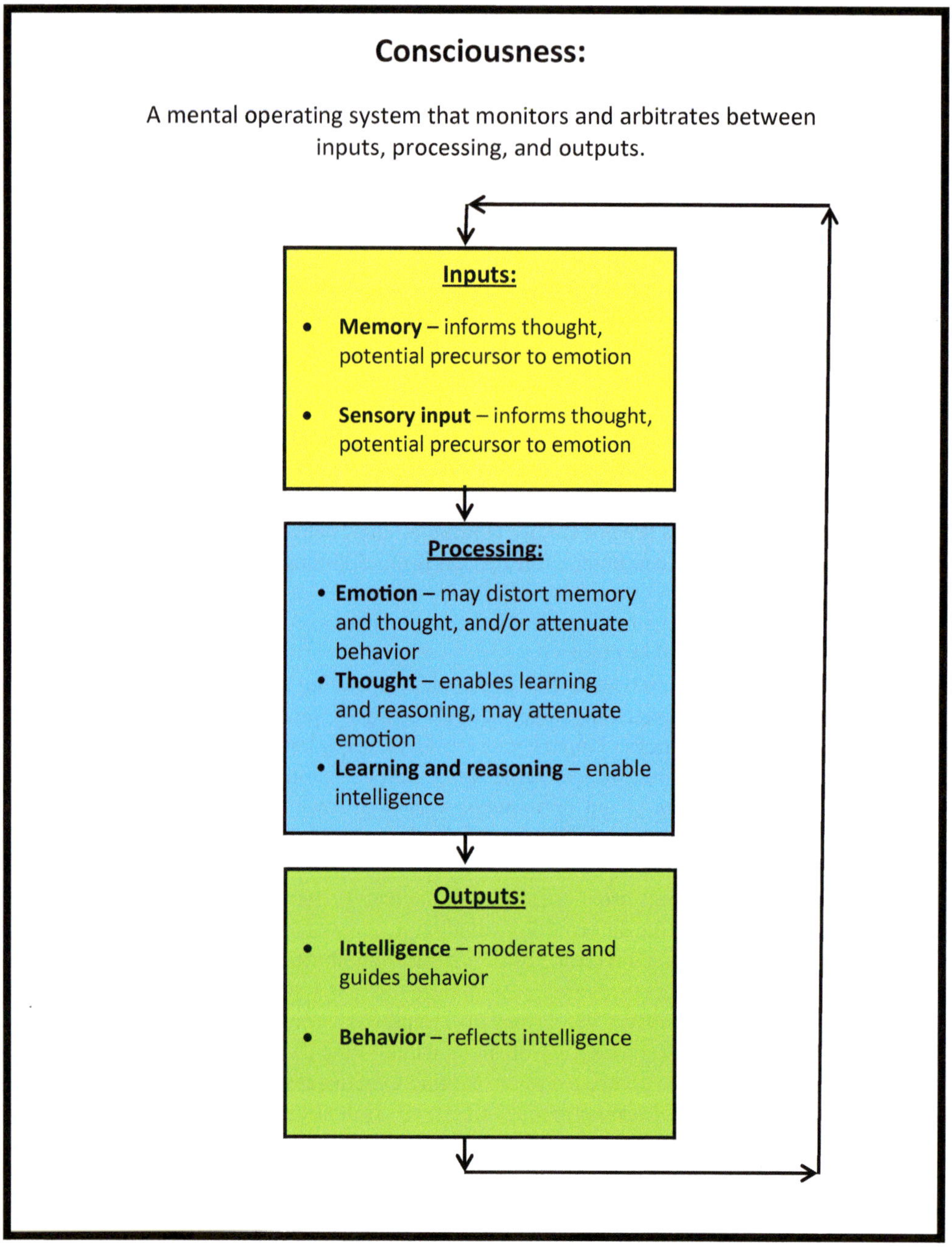

Figure 1.1 Intelligence, Consciousness, and Behavior in Perspective. © D. Herrmann.

of people were "subhuman" because they lacked a written language, belonged to primitive cultures, and possessed inferior technology. Hence, it was alright to subjugate them. This example of circular logic was in reality nothing more than a convenient socioeconomic model.

Language

Language is the means by which ideas, thoughts, and feelings are expressed and exchanged. Language takes many forms from sounds to notations to gestures, each of which can be used to express a particular meaning. There are many categories of languages from human languages, to languages that are used to program computers, to the various systems by which animals communicate.[179] In addition there is body language, which can communicate additional thoughts and/or conflict with what is spoken.

Within the scope of human languages, there are a multitude of natural languages that are spoken and/or written. A current estimate states that there are between 5,000 and 7,000 natural languages in use today.[46] There are also a number of natural languages that have become extinct over the years. Languages are not static; rather they are continually evolving. New words are added. Old words that are no longer used are dropped. Pronunciation changes, perhaps even the script used to represent the language changes. Eventually a language is replaced by one or more new languages, like how Latin formed the foundation for many modern European languages.

Some languages are only spoken; other languages are only written. For example, the ancient Egyptians used hieroglyphics to record events—they did not speak in hieroglyphics. Sign language is used to communicate within the hearing impaired community, as well as in some sporting events, like between a pitcher and catcher during a baseball game. Braille is used as a system of communicating by touch within the visually impaired community. Humans also communicate ideas through mathematical notation systems and musical notation systems which have evolved over the centuries. Unlike natural languages, mathematical notation systems and musical notation systems are recognized across multiple cultures. In the past humans used various visual and acoustical systems to communicate long distance, such as smoke signals and drum beats, prior to the advent of the telegraph, telephone, and Internet.

While language allows ideas, thoughts, and feelings to be expressed, it also limits or constrains the scope of ideas and thoughts.[72, 82] Some ideas, thoughts, and feelings simply cannot be expressed in words. Some ideas, thoughts, and feelings can only be expressed in one natural language. Different languages allow the expression of different ideas because language is tied to history, culture, and geography.[72] This is why it is nearly impossible to produce an exact translation of a speech or document when converting it from one language to another. There are limits to transliteration too, since certain sounds only exist in one language. As one of my professors used to say, "You don't really know a foreign language until you are able to think in that language without going through the translation process." This inexact translation problem has led to frequent miscommunication, misunderstandings, and tension between people and countries throughout history and continues today.

Human language is structured around letters, alphabets, symbols, words, sentences, syntax, punctuation, and grammar which can take years to learn. Birds, on the other hand, have a more compact and efficient style of communication.[14] This is due in part to their unique syrinx organ, through which two sounds can be produced at the same time.[18, 74, 137] In addition birds have more complex hearing than humans; they can hear a range of frequencies that is both higher and lower than humans can hear.[14, 18, 74, 154] Birds hear sounds from a greater distance than humans. Birds have more intricate abilities for the time resolution of sound.[14] To illustrate, birds have the ability to isolate or distinguish individual sounds arriving in rapid succession ten times faster than humans.[14] What we hear as a single sound, a bird may hear as ten discrete tonal variations.[14] Feathers cover the ear canal to protect a bird's ears and hearing from turbulence on windy days and during flight.[74] This is similar to putting a shield over a microphone to prevent it from picking up wind or a person breathing. But humans didn't develop this insight until the late 1980's.

Birds vary their pitch, volume, and frequency (high to low, wide to narrow band) depending on whom they want to communicate with and the environment they are in.[74, 137] By varying these parameters a bird determines whether or not the source and location of the message can be determined and how far the message will travel.[74] For example, long distance vocalizations can travel 50 to 200 meters.[74] By adjusting these parameters a bird also compensates for water, vegetation, and other environmental factors that might cause interference, absorption, distortion, scattering, and attenuation of the message being vocalized.[74, 137] This process is not unlike how modern telecommunication systems adjust signal strength and the signal to noise ratio (SNR) to ensure reliable communications.

By varying pitch, duration, volume, timbre, and intensity birds communicate complete thoughts in a few notes. What appears to be a single note or two to our ears embeds a complete idea, like today's computer function keys and mouse clicks or the hieroglyphics and pictograms of ancient cultures. Instead of reflecting a vowel or consonant, a note reflects a complete sentence or paragraph. Bird song is a more rapid and accurate form of communication; one that is rarely misunderstood or mistranslated due to cultural or stylistic nuances, or over use of idiomatic expressions, acronyms, jargon, and slang.

A common example illustrates this point. Consider a solitary loud, medium to high pitched shriek—an alarm call. This sound could be made by almost any bird, yet all the birds within hearing distance, along with some small mammals, will understand its meaning. In contrast, humans would have to use 51 words and 30 seconds to convey the same message. And, the message would have to be translated into multiple languages to make sure everyone present understood it.

Bird alarm call:
"Shriek!"

Human warning message:
"Excuse me. Excuse me. Ladies and gentlemen may I have your attention. There is a dangerous and hungry predator in the neighborhood. We all need to get out of here quickly and safely. Please do not panic. Depart through the nearest exit in an orderly fashion. I repeat, do not panic."

Car horns, police sirens, and fire alarms, which approximate a bird's alarm call, are one example of the many ideas humans appropriated from birds.

A key goal of *Avian Cognition* is to explore the variety and complexity of avian communication patterns within and among species. Unlike humans avian communication is rarely 1-to-1. Rather avian communication is mostly 1-to-many or many-to-many group conversations. Furthermore, other birds (not the message's intended recipients) are always listening in. Since avian communication takes place outdoors, there are multiple simultaneous sources of audio information that a bird must isolate, distinguish, and act upon in addition to direct vocalizations. As a result, it is important to understand the context of avian communication patterns—what else is going on—and how these communication patterns fit within the interactions of multiple birds and multiple species of birds.

To this end *Avian Cognition* captures original audio recordings of the full range of vocalizations and complete conversations among multiple birds in a variety of settings on two CDs. The recordings feature dialogs among multiple birds of the same species and their interaction with other avian species in their midst as they go about their daily lives. In doing so many new vocalizations have been identified.

A second goal of *Avian Cognition* is to explore which vocalizations are inherited versus learned among the five species of parrots examined. There is a concern that not only are parrots being lost from the wild, but those species that have learned-only or combined inherited and learned vocabularies are losing their natural vocabularies in captivity. As will be seen in Chapters 7 through 11:

- Young pet parrots often end up speaking a form of parrot gibberish since they have no one from which to learn their native language
- The native vocabulary of pet parrots increased dramatically in proportion to the number of parrots present, especially in mixed age groups.

- The tendency of parrots to mimic human speech is not a cute trait, but rather an attempt by a solitary parrot to learn its language and communicate

When it comes to body language, the 9,672 species of birds that currently populate the Earth are more adept at and have more sophisticated means of communicating via visual displays than any other creature living on the planet.[5, 12, 14, 18, 74, 152, 153, 154, 160, 161] This raises an interesting question. If birds do not think or have self-awareness, why do they expend so much effort to communicate through complex and varied vocalizations and elaborate visual displays?

Culture

Culture is the term used to describe a particular group's lifestyle and their common beliefs, values, and behavior.[46] In the past, cultures were ensconced in a particular political, economic, and social order. Today for the most part culture is primarily linked to a particular social order, since political and economic structures tend to change frequently. Culture goes beyond a day to day existence or earning a living. For example there are several groups of subsistence farmers around the globe; but they each belong to a different culture.

There are ancient cultures, some of which predate recorded human history. There are contemporary cultures. Some cultures are nomadic; others cultures are stationary. There are indigenous cultures living apart or within a contemporary culture. Cultures often mirror religious traditions and geographical conditions. Some cultures are more aggressive and warlike; other cultures are more passive and peace loving. Some cultures place a high value on politeness or scholarship and knowledge; other cultures value athleticism or business acumen. Some cultures value tradition; other cultures are always seeking new or better ways of doing things. Within any culture there may be subcultures or regional variations. Even corporations are said to have different corporate cultures. One culture is not "good" and another culture "bad". They are just different.

Culture is reflected in a people's art, music, literature, cuisine, dress, and architecture. Cultural artifacts are used to establish, maintain, and justify a collective identity and social order. In the not so distant past when a confrontation occurred between one culture and another, the victors promptly destroyed the losers' religious temples, libraries, and other cultural symbols in order to eradicate any sign of the former culture. This act was an early and not so subtle form of psychological warfare.

Some cultures simply collapsed of their own accord, in the absence of violence, and were subsumed or supplanted by another culture. Sometimes this was due to a numbers game. It is difficult for a culture of 10,000 people to maintain its identity and way of life when it is surrounded by 2 million people. Other times the culture simply ran out of ideas or could not adapt to changing circumstances. Historically, from ancient times through the 19th century, there was an alternating cycle of Eastern (Asian) cultures rising as Western (European) cultures were stagnating and declining, and vice versa. For example, while Europe was in the Dark Ages, many Eastern cultures were at their zenith. This cycle was a phenomenon of the European and Asian continents whose people were in touch with each other. Likewise, there was a succession of cultures in Central America: Olmec (1200–600 B.C.), Maya (250–900 A.D.), Toltec (900–1150), and Aztec (1400–1521). Some of these cultures were conquered by warfare; others declined of their own accord.

These facts raise some questions. If culture is a sign of intellectual superiority:

- Why do cultures, even advanced ones, collapse?
- Why are so-called advanced cultures often overthrown by so-called barbarians?
- Why is it necessary to "change the corporate culture" to make a company more progressive and competitive?

Isn't culture really just a means of ensuring conformity through a group identity and thereby controlling or moderating behavior? If so, is human culture really that different from the sophisticated social orders of many animal species and their concomitant accepted behavioral norms? Birds have the most complex social orders after humans; this is particularly evident in cooperative mixed species flocks.[5, 14, 74]

Technology

We humans are very fond and proud of our technology, from the spear to the sextant to the steam engine to today's smart phones. Owning the latest technology has become the ultimate fad. Technology refers to a combination of tools, equipment, materials, techniques, and energy sources used by humans.[46] Humans have developed and used technology from ancient times—like a mortar and pestle—to the present. The goal of developing and using technology throughout human history has been to make certain tasks easier and life more comfortable.[46, 179] Much of this effort has been focused on trying to control, dominate, or harness the forces of nature.

It is good to step back for a moment and ask: what has technology really brought us? The answer is that technology has brought humankind as many negative consequences as positive results. Part of the problem is that humans are not very good at thinking long term. They fail to see that although the short term results of a new discovery are positive, the long term consequences are often negative. Another problem is that humans fail to comprehend that the long term negative aspects of technology don't just affect humans, but other life forms as well. Nor do humans appreciate the fact that the decline of other life forms ultimately affects humans, like the sudden rapid decline of insects that pollinate our food crops.

A quick review of recent technological developments indicates that many of the benefits of technology were rather short-lived:

- In the 1800's mercury was used as a miracle cure for a variety of ailments. Mercury was later found to be toxic, especially cumulative doses.
- Antibiotics were the miracle cure of the 1900's. Today many germs have mutated and are resistant to antibiotics.
- Many of the products recently created to increase our food supply and its shelf life, like pesticides, herbicides, and preservatives, have been found to be carcinogenic.
- We have surrounded ourselves with a cornucopia of devices that are dependent on energy, most of which comes from a finite set of fossil fuels.
- The ever present air and water pollution has caused many people to drink filtered water and buy special air purifiers for their homes.
- The problems with asbestos and lead paint are well known.
- People have been driving automobiles for over 100 years. During that time period a number of safety devices have been added to automobiles. Yet a high number of traffic deaths continue to occur each year.
- The sedentary lifestyle our technology has brought us has created the need to go to a gym to exercise and fostered the two leading causes of death in advanced societies: cancer and heart disease.
- Massive landfills are bloated with items that cannot be recycled and are not biodegradable, like disposable diapers, car seats, old televisions, and yesterday's toys and CD's. (It would be interesting to know how archeologists 1,000 years from now will interpret these landfills—probably not as a sign of intelligence!)

So, we have to ask, is human technology really a sign of an advanced or superior intellect?

Or, is human technology evidence of a failure of human adaptation? All other living organisms on earth have learned how to live with nature as part of the natural world. At some point in time humans took a wrong turn and mistakenly started thinking of themselves as apart from nature and the natural world.

1.8 Adaptation, Instinct, and Intuition

The accomplishments of non-human animals are frequently dismissed as being nothing more than simple adaptation or instinct.

Adaptation

The word adaptation is fairly new; it first appeared in 1610.[179] Adaptation refers to a living entity adjusting to the ever changing environmental conditions in which it lives. Adaptation may involve physiological changes and/or behavioral changes.[46] Adaptation may involve the adjustment of certain sensory organs to specific conditions, such as the ability to see ultraviolet light or in extreme darkness, or the ability to hear certain sound frequencies.[179] Adaptation may involve modification of physical and/or behavioral traits that enhance a living entity's chance of survival and make it more fit for its environmental niche, such as the ability to withstand extreme heat or cold, tolerate an abundance of water or drought, hibernate, or migrate to a different region at certain times of the year.[46, 179] The process of adaptation fine tunes where a species lives, how a species lives, and even what it eats. Adaptation is an example of biological intelligence. (See Photograph 1.1.)

Adaptation is evident in the plant and (non-human) animal kingdoms. Over the centuries both plant and animal life forms have diversified, specialized, and at the same time adjusted to environmental changes. For example today there are some 9,762 species of birds and 1,600 species of cacti, each of which is ideally suited for its environmental niche.[126, 152] The gradual process which resulted in such biological diversity is referred to as adaptive radiation. This term is new as well; it was coined in 1902.[179] Adaptation occurs as the result of an organism adjusting or modifying its physical and/or behavioral traits. Sometimes adaptation occurs as the result of cooperation between two species and a symbiotic relationship develops, like bees or other insects pollinating plants and birds distributing seeds.

Adaptation is a work in progress. We can trace back to the beginnings of various adaptive features or behaviors. But, the process of adaptation is never finished. To survive, prosper, and progress each species must continue to adapt to its ever changing environment, just like all the plants and animals around it are doing, each in their own way.

Adaptation is the opposite of human reliance on and even obsession with technology. The central thrust of adaptation is for a species to improve *itself* by modifying, enhancing, and specializing its physical and/or behavioral traits. The focus of adaptation is internal—how a species improves itself to better fit into the ever changing world around it. Technology's focus is external—how to change the external world to (supposedly) make life easier for humans. But, as we have just seen the benefits of many technological "advances" are very short-lived, while the negative consequences are permanent.

Adaptation is reflected in a species' biological tools. A biological tool is a biological system or structure that has become specialized over time to facilitate a species ability to survive, prosper, and progress within its particular ecosystem niche. Examples of biological tools abound within the world of birds. To name a few the bill of a hummingbird is perfect for extracting nectar, as is its ability to hover. The extensible tongue of a woodpecker is great for capturing tree ants. The webbed feet of ducks and geese make swimming a breeze. Feathers are intricately designed to facilitate flight and thermo-regulation in a variety of climates. Biological tools obviate the need for developing and

Photograph 1.1 Adaptation. (Top) If a bird's primary food source is under water and it does not swim, it's a good idea to have a long beak like this Lesser Yellow Legs (*Tringa flavipes*), who was migrating through the Mid-Atlantic area on 11/19/2011. (Bottom) Duck and geese feathers are designed to be waterproof; they do not change color or appear wet. Parrots are not water birds. As a result their feathers change color when they are wet, like the Mitred Conure (*Aratinga mitrata mitrata*) shown here after his bath. © D. Herrmann

carrying inanimate tools external to oneself. In contrast, humans, who do not have any biological tools, are dependent on external inanimate tools which can be lost, stolen, damaged, become obsolete, and/or require external energy sources which may be depleted or become prohibitively expensive.

Through adaptation plants and animals have diversified and specialized so that they are not all competing for the same finite resources. For example, the 9,762 species of birds live in different regions, build their nests with different materials and in different locations, fly at different altitudes and speeds, and eat different food. Some birds migrate; some do not. Some birds swim and dive. Some birds only swim. Some birds do not swim or dive. In essence, the 9,762 species of birds have spread out, diversified, and specialized through adaptation so they are not in constant competition with each other. To illustrate, a ruby-throated hummingbird (*Archilochus colubris*) does not compete for food, housing, or flying space with a cardinal (*Cardinalis cardinalis*) even though they live in the same habitat 6 months of the year.

In contrast, humans have been unsuccessful at adaptation, diversification, and specialization. Instead, the trend is toward homogenization or de-adaptation. As a result humans are in constant competition for the same finite territory, housing, food, and energy resources. This competition—or failure of adaptation—is a root cause of the non-stop conflict throughout human history.

The definition of intelligence included the ability to adapt to novel situations and an ever changing environment. Plants and non-human animals have shown a marked ability for adaptation. In contrast, humans have not. We let our appendix go to waste, when we now know eating more fruits and vegetables is healthier. Except for a few elite dancers and gymnasts, our toes aren't all that useful since they are encased in shoes all day. And, like the early or primitive vertebrates, all our toes still point forward.[137] Our teeth develop cavities and have to be pulled because of the diet we eat. Our physiology is not adapted or specialized to live in any natural environment found on the planet.

Therefore, if adaptation is so simple, why haven't humans been able to improve our physiology or behavior through adaptation?

Instinct

Instinct is another topic that is talked about frequently but rarely defined. It is not uncommon to hear the phrase, "My instinct tells me ...". We all identify with the concept of instinct, but few of us have ever taken the time to determine what instinct really is or is not.

In theory instinct refers to what is believed to be an inherited, unlearned, and supposedly unalterable tendency to behave in a certain way.[46, 179] The implication is that instinct determines a living being's reaction to a specific situation. And, that a living being's reaction to the exact situation will always be the same—it cannot be unlearned.

Some theories claim that instinctive behavior is controlled by inputs (stimulus) and outputs (responses) alone. But as we have seen, the brain doesn't work that way. A lot of processing is performed between inputs and outputs. Between stimulus and response there is a mechanism called choice.

The decision-making process that guides instinct and instinctive behavior originates at the subconscious level. If we remember from our discussion of consciousness, the subconscious is always active regardless of whether a being is conscious, semi-conscious, or unconscious. As a result mental activities are involved in instinct and instinctive behavior. We are just not as aware of them as we are with conscious and semi-conscious mental activities.

Instinct reflects knowledge that has been acquired and refined through the ongoing process of adaptation. So-called instinctive behavior patterns have an evolutionary history.[158] Even so-called instinctive behavior becomes more efficient with practice and experience.[158] Hence, instinct is not

Photograph 1.2 Flexible Decision-Making. (Top) Great blue heron (*Ardea herodius*) are known for wading in ponds and eating fish. Many great blue heron migrate, but not this one that was near Pond B on 1/19/2012. It stays put during winter. When the ponds are frozen over this great blue heron can be found up in the trees eating tree bark and dried vegetation. (Bottom) Canada geese (*Branta Canadensis*) are water birds. Canada geese seem to understand that ice is frozen water. When a Canada goose lands on water its tail and webbed feet are tilted up, like water skis. To land on ice a Canada goose adjusts its landing procedure. The tail and the nails on the webbed feet are pointed down to provide traction, like this goose at Pond C on 1/16/12. Interestingly, the distance it takes to stop is the same on ice as it is on water. © D. Herrmann

Photograph 1.3 Individuality and Choice. (Top) This Canada goose (*Branta Canadensis*) nested at the new Pond D for the first time in 2011. That nest was on the ground 7 feet (2.13 m) from the water's edge. (See Photograph 3.11.) The same goose made her nest 10 feet (3.05 m) up in the air on the Pond D retaining wall in 2012. Her feathers blend in with the retaining wall, which provides protection from predators and the prevailing NW winds. (Bottom) A bird chiseled these three marks 0.5 x 1.25 inches (1.27 x 3.17 cm) above the entrance to a tree hole long before the branch fell to the ground. What is the purpose of these marks? Are they decorative, like hanging a wreath on a door? Are the marks a location identifier, like a house number? Or, do they mark territory like a name on a mail box? Downy woodpeckers (*Picoides pubescens*), red-bellied woodpeckers (*Melanerpes carolinus*), and an occasional pileated woodpecker (*Dryocopus pileatus*) live in the area. It is unclear which bird made the marks or why. But this was an act of individual expression. No other woodpecker made similar marks during the 8 years this area was under observation. © D. Herrmann

static. For example birds have the ability to delete old or obsolete information through neuronal replacement.[92] Instinct may include information like basic survival skills or historical memories. Originally this information was acquired through the learning process. Over time this information became more precise, accurate, and predictable. Then it was distilled and stored as instinct for more efficient processing, storage, and transmission.

Consider an analogy. Instinct is equivalent to a pre-stored computer program or application specific integrated circuit (ASIC). In other words, the logic for how a particular input should be processed has already been determined. Because the "program" is pre-stored the thinking cycle is very fast. Hence, the processing time is nanoseconds versus seconds or minutes and appears to be instantaneous. The speed at which the output is generated creates the misconception that no thought is involved.

In reality instinct and instinctive behavior can be overruled. Instinct and instinctive behavior pop into a being's brain very quickly. But, before any action is taken a choice is made at the conscious or semi-conscious level. A decision may be made to go with the instinctive option. Or, a decision may be made to pursue another option. Different decisions may be made in response to the identical situation at different points in time. That is because the subconscious is never active by itself. The subconscious is always active in conjunction with another level of consciousness. As Tudge notes, what is usually referred to as instinct is really instinct plus flexible decision-making.[169] Instinct plus flexible decision-making reflects intelligence and individuality. (See Photographs 1.2 and 1.3.)

If instinct could not be overruled, it would be impossible to tame or train any animal from a budgerigar to a macaw, dog, dolphin, horse, lion, tiger, or bear. Nor would people be able to overcome their fears. Furthermore, it has been observed that hybrid birds often retain the so-called instinctive behavior of both species; over time they learn which behavior is more appropriate in a given situation.[158]

In short, instinct is an efficient means of processing certain inputs, but it is not an end unto itself.

Intuition

With the advent of graphical user interfaces for computers intuition has become the ultimate buzz word and compliment. Software and people are referred to as being "so intuitive". Although in the case of software sometimes you have to ask—intuitive for whom?—since people think in different ways. Intuition refers to instant insight or perception, cognition that is arrived at without any rational thought.[179] Supposedly an intuitive person knows the answer without even thinking! In the old days this phenomenon was called a hunch.

Let's see ... Intuition is good or superior because people are intuitive. Instinct is bad or inferior because animals are instinctive. Actually, in the final analysis instinct and intuition are two sides of the same coin because in each case the initial/potential answer, response, or output originates at the subconscious level. Instinct and intuition are examples of genetic intelligence.

2 Heron

Butorides virescens, Ardea herodias, Nycticorax nycticorax, Nyctanassa violacea

2.1 Background

Humans in the western world have been aware of herons since at least the 14th century. At that time names for heron began appearing in English (heiroun), French (hairon), German, and Greek.[179] Much earlier herons were associated with the Egyptian god Osiris, and the Iliad mentions that Odysseus and Diomedea received a heron from the goddess Athena.[127]

Herons live on every continent except Antarctica.[153] A wading bird, herons are mainly found in the tropics. Herons belong to the family *Ardeidae*, which consists of 21 genera and 64 species. Eight genera and 13 species live in the southern two-thirds of North America, mostly in coastal and interior wetlands.[153]

Different species of herons share several common characteristics. Herons are noted for their long thin bills, long necks, long legs, wide wingspans, upright stance, and soft plumage. Unlike most birds, herons are able to extend and retract their necks because of special neck vertebrae.[44, 127] In fact herons fly with their necks retracted to reduce the drag coefficient. Male and female herons have the same coloring.[83] Herons may engage in a brisk walk or jump short distances, but they do not run or hop.[83] Unlike ducks and geese, herons have a full size back toe or hallux.[83] Herons strong toes allow them to securely grasp various types of vegetation or other surfaces as they travel around the edges of ponds.[83] This is an important feature since herons would not be able to lunge at their prey if they had a weak grip. Herons do not have a crop since they primarily consume animal matter.[83] The indigestible parts of items consumed are regurgitated as pellets, similar to owls.[83, 128]

Herons have a unique patch of powder down feathers on their breast or rump.[44, 83, 127, 153] The powder down feathers dissolve periodically, producing a fine powder. Herons use this powder to clean, waterproof, and remove oil from their feathers.[44, 83, 127, 153] This feature is very handy given the environment in which herons live. Powder down feathers replace the preening oil (or uropygial) gland many other birds have.[83] In order to distribute this fine powder effectively, herons have what is referred to as a pectinate middle toe.[18, 44, 83, 137, 153] The nail of the middle toe is serrated like a fine comb to facilitate preening.[18, 44, 83, 137, 153] (See Photograph 2.1.)

Herons are seasonally monogamous and first breed when they are 2 or 3 years old.[153, 203] Herons have altricial young; only about 30% survive their first year.[153] The annual survival rate for adults is estimated to be 69%.[18] The oldest heron on record is a great blue heron that lived to be 23 years

Photograph 2.1 Green Heron Preening. Herons, like this green heron, use a combination of their bill, special powder from their powder down feathers, and their pectinate middle toe to preen. The nail on a heron's pectinate middle toe resembles a small comb and is specially designed to distribute the powder. Herons use this special powder to clean, waterproof, and remove oil from their feathers.[44, 83, 127, 153] © D. Herrmann

old.[18, 137, 153] Herons are considered by some to be a biological indicator of the health of wetland ecosystems.[153] They are also one of the oldest living bird families on the planet.[127] Fossils of herons, storks, and ibises date back approximately 40 million years.[83]

The main threats to herons today are wetland drainage, urban expansion, agricultural encroachment, and pesticide residues.[83, 95, 153] Extensive use of DDT after World War II caused reproductive problems like thinning egg shells, especially for black-crowned night herons.[95, 153] DDT is still a problem for herons that live or winter in Central and South America where the use of DDT has not yet been banned.[95, 153]

A myth has developed and been perpetuated that green herons, great blue herons, and black-crowned night herons stab their prey with their long thin bills. In fact they do not. Instead, herons use their bill like a pair of chopsticks. Herons snap their bill shut on their prey with lightning speed. Many attempts to photograph green heron, great blue heron, and black-crowned night heron as they lunged toward their prey produced blurred images, even though the shutter speed was set at 1/500th of a second. Once caught the prey is perpendicular to the heron's bill. Next a heron maneuvers the prey—without dropping it—so that the prey is aligned with its bill. Then a heron swallows the prey whole head first.[45, 95] This entire process requires an extreme amount of dexterity, since all a heron has to work with is its bill and tongue. Conversely, consider the improbability of accomplishing this feat if a heron were to stab its prey first. (See Photographs 2.2 and later 2.12.)

Another legend that has entered heron lore concerns herons choking to death when trying to swallow something that is too large. Two websites even mention this story. One website linked the

Photograph 2.2 Green Heron Lunch. This green heron caught a small frog that jumped at the wrong time. The speed and force with which the green heron snapped its bill shut must have caused the frog's immediate demise, because the frog did not struggle to escape afterward. It took only 0.44 seconds for the heron to catch the frog, reposition, and swallow. © D. Herrmann

story to great blue herons; the other linked the story to black-crowned night herons. Both websites referred to choking on a large bull frog. Let's examine the logic of these assertions.

1. As just explained, herons catch their prey so that it is perpendicular to their bill. Herons then rotate the prey so that it is aligned with their bill. Afterward they swallow the prey whole, head first. If the prey is too large to swallow, a heron would not be able to align the prey parallel to its bill.
2. A heron's bill is narrower than their throat. If the prey is too large to swallow, it wouldn't fit in their bill.
3. Herons use a technique called gular fluttering to stay cool during extreme heat. This involves expanding and contracting the flexible skin on their neck. If the skin on a heron's neck can be stretched to cool off, it can be stretched to prevent choking.

4. Herons feed their young by regurgitation. If they can expel partially digested food to feed their young, they can expel an item rather than choke on it.
5. Great blue herons partially retract their neck to facilitate swallowing. This shortens the distance and widens the passage the prey travels on the way to the digestive system. (See later Photograph 2.18.)

In summary, a logical analysis of heron behavior and anatomy disproves the choking legend. If a heron or any other water bird catches prey that is too large to swallow, the prey is simply released. Or, as in the case of yellow-crowned night herons, the prey will be broken into smaller bite-sized pieces.

Next we are going to examine four species of heron in detail:

- Green heron
- Great blue heron
- Black-crowned night heron
- Yellow-crowned night heron

The herons behaviors are presented as they occurred. The analysis, comparison, and interpretation of these behaviors are reserved for Section 2.6 Discussion. These observations took place at the locations and during the timeframes described in Annex C.

2.2 Green Heron *Butorides virescens*

It is appropriate that we begin with a discussion of green heron, since they are one of over 30 species of birds that are known to use tools.[137]

Very little is written about herons. That is surprising given: (1) the wide distribution of herons around the world, and (2) how long humans have been aware of herons. Some of the sources consulted confused green herons with the green-backed or mangrove herons (*Butorides striatus*) that are found in Asia. A few sources failed to acknowledge that green herons exist!

People occasionally get hung up on taxonomies. They forget that taxonomies are a human invention to simplify the classification and naming of species as *we* become aware of them. The birds (or other animals) may or may not see themselves as belonging to such families, genera, or species. Historical classifications of birds are based on behavior and appearance. But, appearances can be deceiving. Recent DNA studies of 169 species of birds found new and different genetic links between various birds.[96] For example, it was found that falcons have more in common genetically with parrots than eagles or hawks.[96]

As mentioned earlier herons are one of the oldest living bird families on the planet.[127] It is likely that green herons are related to the Early Pleistocene *Butorides validipes* fossils found in Florida.[203] This is plausible since green herons live in North and Central America, primarily the eastern half of the United States from Maine to Florida, along the Gulf coast, and in a narrow area along the west coast.[29, 95, 203] As a result, green herons are occasionally referred to as American green herons.[83]

Green herons reside near small freshwater wetlands, such as streams, riverbanks, lakes, ponds, marshes, and swamps that are lined with vegetation.[29, 95, 203] Wetlands provide food for the green herons. As a result green herons living in the northern latitudes migrate south before the ponds and such freeze over and their food supply becomes inaccessible. Migration from these areas may begin as early as mid-September.[203] The green herons discussed in this chapter were gone by October 1st each year. Green herons migrate as far south as Panama.[95] Green herons return by mid-April or early May. Green herons living in the lower latitudes, particularly Central America and the West Indies, are permanent residents year round.[95, 203]

Green herons do not engage in elaborate preparations for migration; at least not visible ones like flying in a V-formation. Instead the pre-migration activities of green herons are as private as their daily activities. Green herons are simply here one day and gone the next.

Small fish, like minnow and sunfish, are the main components of a green heron's diet.[95, 171, 203] Green herons also consume frogs, tadpoles, crabs, crayfish, and aquatic insects; once in a while they eat snakes, grasshoppers, crickets, snails, leeches, or earth worms.[95, 171, 203] Undoubtedly this variable diet has contributed to the green herons success.

Green herons seek out their prey along the shallow edges of waterways, walking very slowly until an item of interest is spotted. Next they stand absolutely still, watch intently, and wait for the unsuspecting prey to come closer. As the prey comes within range, a green heron slowly lowers its body from an upright position to being parallel to the water.[44] At the ideal moment a green heron lunges forward with lightning speed, rarely ever missing, and it is lights out for the prey. An alternate approach is to stand on a low branch that overhangs a waterway and lunge from there. Herons also have the ability to hover above water briefly in order to catch their prey.[203] (See Photographs 2.3 and 2.4.)

Photograph 2.3 Green Heron Wading and Fishing. Green herons wade along the edges of ponds as they seek out their prey. Green herons walk along the vegetation that borders a pond and in the water up to the top of their legs. Green herons move in a quick deliberate unobtrusive manner which does not attract the attention of their prey or would be predators. © D. Herrmann

Photograph 2.4 Green Heron Preparing to Lunge. Green herons are totally focused on their prey prior to lunging. Green herons lower their body in a variety of different positions in preparation for striking, depending on whether they are in the water, on the shore, or above the prey on a rock or branch. Throughout this process they never lose sight of the prey.[83] Green herons wait for the optimum moment, then lunge their head, neck, and bill forward with lightning speed, while relying on their strong toes to maintain their balance. Green herons rarely miss. This high success rate indicates that green herons have a keen understanding of the speed with which they can lunge and the speed with which different types of prey can flee. © D. Herrmann

Green herons were never observed lunging toward prey and missing. They seem to have an innate understanding of when they will or will not catch prey and don't waste energy on iffy lunges. This behavior indicates that green herons understand the speed with which they are able to lunge *and* the speed with which different types of prey are able to flee. Green herons are able to correlate this information in real time to assess the probability of success for each unique situation and act upon it in a timely manner.

For the most part green herons search for food alone. Because of that green herons are referred to as being "solitary, secretive, patient, and cunning".[44, 95] There are two times during the year when green herons do not fish alone. Bonded pairs may fish together as part of the courtship ritual prior to nesting. Also, one or both parents stay with a fledgling until it learns how to find food on its own.

Green herons don't always just sit and wait for the prey to come to them. Often they are more creative and proactive, by dropping small items into the water like bait to lure fish.[30, 44, 95, 137, 171, 203] Green herons use small objects that are found near the water's edge, and hence, would be familiar to the fish.[44] Green herons have been observed dropping a variety of objects as bait, such as sticks, feathers, bread, insects, flowers, corn, berries, and leaves.[30, 44, 95, 137, 171, 203] This behavior has also been attributed to green-backed heron in Japan.[12] On occasion green herons will snap twigs from vegetation bordering a waterway, if they can't find anything else suitable.[44] Because of this behavior green herons are considered by some to be "among the world's most intelligent birds".[203] Devising a tool to use oneself reflects one level of accomplishment. Devising a tool to use to lure another living entity is a higher level of accomplishment. The latter involves planning ahead and understanding cause and effect.

The practice of using bait to lure fish begins early in life. As shown in Photograph 2.5, a juvenile is preparing to drop a leaf in Pond C. Note that it still has some fuzzy feathers on its head. This photograph was taken on July 28, 2011. It is unlikely that the juvenile figured out how to do this on its own by then. Green heron parents probably teach their young how to use bait in the few weeks after the juveniles fledge, but before they become independent.

Green herons are the smallest North American herons.[141] When it comes to appearance, green herons are quick change artists. Green herons range from 15 to 22 inches (38 to 56 cm) in length, depending on whether their neck is fully retracted, partially extended, or fully extended.[29, 203] Green herons weigh about 8.6 ounces (240.8 gr) and have a wingspan of 28 to 30 inches (71 to 76 cm). The legs are 7 to 8 inches (17.7 to 20.3 cm) in length and are yellowish. They may be brighter during breeding season.[29, 137] Three toes face forward and one backward. The middle or pectinate toe is longer than the other two. The eyes are yellow and outlined by a thin yellow membrane.[44] A small yellow patch runs from the eyes to the base of the bill. The long thin bill is mostly black with some yellow streaking on the bottom.[44] (See Photograph 2.6.)

The feather coloring of an adult green heron depends on the angle of the sun and the objects the heron is near, as shown in Photographs 2.7 and 2.8. Green herons have a crest which is about 2 inches (5 cm) high when raised. The crest may appear to be black or slate blue. The sides of the head and neck may appear to be chestnut or a dark crimson. The back and wings may appear to be slate blue, a bluish black, or a bluish green, with subtle white accents. Only the eyes, legs, and bill remain the same color. Hence, it is a bit curious that green herons were named green herons.

The head and neck of a juvenile green heron is rust and white streaked. This pattern is particularly noticeable when the neck is extended. For a time the top of the head is still fuzzy.[172] The back and wings are mostly brown. Juvenile green herons can easily be confused with least bittern (*Ixobrychus exilis*). (See Photograph 2.9.)

Green heron hatchlings are covered with light gray down on the head, sides and back, while the underside is white.[203]

The extensible neck of a green heron comes in handy for more than fishing. It also helps a green heron to cool off on hot humid days. To do so a green heron opens its bill wide and vibrates its thin stretchy throat membranes.[137] This process, referred to as gular fluttering, exposes a large area of skin to air, thereby facilitating heat loss.[137] It is the green heron equivalent of panting.

The vocalization green herons are most known for is a somewhat loud sudden medium-pitched "kyow" or "skyow" sound.[29, 203] This sound is heard frequently when a green heron suddenly flees an area. Green herons also make a series of "kuk" sounds when perching. When stalking prey near the shoreline green herons are silent.

Photograph 2.5 Juvenile Green Heron Using a Leaf as Bait. Green herons are one of the few birds known to use tools, in this case dropping bait on the water's surface to lure fish. This practice begins early in life. Here a juvenile green heron retrieves a leaf. Next the juvenile holds the leaf flat. Then the juvenile folds the leaf so that it will release more kinetic energy when it hits the water. Lastly, the juvenile heron drops the leaf and lunges toward an unsuspecting fish. The total time elapsed from the first to the fourth photo was 0.30 seconds. © D. Herrmann

Green herons have a specialized vocabulary to go with their elaborate courtship rituals.[95] Three different sounds can be heard while a male is in flight: "raah-raah", "whoom-whoom-whoom", and "roo-roo".[203] A fourth sound, "aaroo-aaroo", is heard after the male has landed.[203]

Green herons tend to return to the same breeding grounds each year.[203] The male chooses the nest site, which is always near water.[95, 203] The nest is started by the male, but finished by the female using materials the male supplies.[95, 137] The nest, which is hidden from view, consists of a platform of sticks.[95, 141, 171, 203] The nest is usually in a tree or bush, 5 to 30 feet (1.5 to 9.1 m) above ground.[83, 95, 203] Once in a while green herons nest on the ground, if it is determined to be safe.[95, 203]

Three to six pale green or greenish blue eggs are laid, 36 to 48 hours apart.[18, 29, 95, 141, 203] Green heron eggs are round and about 1.45 by 1.10 inches (3.6 by 2.7 cm) in size.[141] Both parents take turns incubating the eggs, which hatch in 19 to 21 days.[83, 95, 171, 172, 203] Green heron parents take care to

Photograph 2.6 Green Heron Postures. Green herons are very flexible because of the structure of their legs, neck, and back. As a result green herons easily assume a variety of different postures as they go about their daily activities. © D. Herrmann

shade their eggs if there is a concern that they are receiving too much sunshine/heat.[171] This behavior indicates that green herons understand that their eggs must be kept warm—but not too hot—to hatch successfully. Both parents feed the hatchlings, mostly regurgitated fish.[83, 95, 172, 203]

Green herons are described as being colony breeders. Yet no more than two green herons were ever observed at the same time. Most often only one green heron was spotted. However, a few great blue herons, black-crowned night herons, and yellow-crowned night herons also live at Pond B. All the herons share the same nesting site at Pond B, which could be considered a small mixed heron colony.

Green heron hatchlings have an unusual characteristic. When they are 16 to 17 days old green heron hatchlings begin to climb around on the branches near their nest; they may even jump a short distance from branch to branch.[95] Through this activity green heron hatchlings practice their foot-eye

Photograph 2.7 Green Heron Perching on Branches. Herons are water birds, part of the group known as long-legged waders. That doesn't stop green herons from perching on branches of various sizes to get a better view of what is going on above, behind, or below them. Green herons were observed fishing from branches 2–3 feet (.61–.91 cm) above the water and roosting on branches 6–10 feet (1.83–3.05 m) above ground. © D. Herrmann

coordination and strengthen their toes, both of which are essential for fishing. Green heron hatchlings don't fledge until later, when they are 21 to 23 days old.[95, 172, 203] Green heron hatchlings continue to be fed by their parents until they are 30 to 35 days old; after that they are independent.[95, 203] Green herons in northern climates have 1 brood per year; those in the warmer climates may have 2 broods per year.[172, 203]

Green herons were observed in one's and two's. There were never more than that number present at any given time. In the spring, just prior to nesting time, pairs flew and fished together around Pond B. The green herons were more vocal at this time than the rest of the year. During July two green

Photograph 2.8 Green Heron Climbing on Rocks. Green herons are very adept at scampering along rocks while searching for prey at the edge of a pond. Their strong front and back toes are well suited for gripping the surface of a rock. Note the full-size back toe and the extra-long middle toe. © D. Herrmann

herons were frequently seen at Pond C. Generally this was an adult with a juvenile. The two herons would start fishing in the southeast corner of the pond. Then the adult would fly to the northern side of the pond, while the juvenile continued fishing along the southern side. Periodically the adult would fly back to check on the progress of the juvenile. Then the adult returned to the northern side of Pond C. When the juvenile was finished, it would fly to the western side of the pond to roost.

The literature asserts that green herons are most active at dawn and dusk. That was not the case. Instead green herons were seen most frequently between mid-morning and mid-afternoon. Table 2.1 summarizes the distribution of green heron observations. Green herons were seen at Pond C more often (59% of the time) than at Pond B (38% of the time), even though the green herons live and nest at Pond B. Only 12% of the total sightings of green herons were made in the morning hours.

Photograph 2.9 Juvenile Green Heron. Juvenile green herons are mostly brown, with vertical brown and white streaks on the throat and upper breast. These juvenile green herons at Pond C still had some fuzzy head feathers on 7/31/2011 (top) and 7/28/2011 (bottom). Note the alertness as one juvenile checks out an unfamiliar noise coming from above, while the other monitors activity around the pond. © D. Herrmann

Table 2.1 Summary of Green Heron Sightings (Spring 2009–Spring 2012).

1. Location Observed[1]			
• Pond A	0%		
• Pond B	38%		
• Pond C	59%		
• Pond D (old)	3%		
			100%
2. Time of Day Observed[2]			
AM	% of AM	% of Total	
10:00	75%	9%	
11:00	25%	3%	
	100%		12%
PM	% of PM	% of Total	
12:00	10%	9%	
1:00	20%	18%	
2:00	23%	20%	
3:00	27%	23%	
4:00	10%	9%	
5:00	3%	3%	
6:00	7%	6%	
	100%		88%
			100%
3. Time of Year Observed			
• April through September			
• Earliest sighting: 4/21			
• Latest sighting: 9/26			
4. Observations by Month			
• April	3%		
• May	9%		
• June	21%		
• July	32%		
• August	26%		
• September	9%		
			100%

Notes:

1—Green herons were only observed at the original Pond D. No herons visited the new Pond D after it was redesigned in 2010.
2—The time slot reflects the entire hour. For example 10 a.m. refers to sightings that occurred between 10:00 and 10:59 a.m.

Green herons were most likely to be seen between 12 p.m. and 3 p.m., the time of day when 80% of the total observations were made. Green herons were seen most often in July (32% of observations) and August (26% of observations).

The green herons waded in the shallow water along the edge of Ponds B and C. Sometimes they climbed the large rocks that line part of Pond B. Other times the green herons hid or perched in the vegetation that borders Ponds B and C. The green herons flew directly over the pond about 10 feet (3 m) above the water's surface. Flights between Ponds B and C were a little higher. The flight of green herons is rapid and direct. Their standard cruising speed is 2.5 wing beats per second.[18] At no time did the green herons wander farther than 2 feet (61 cm) from the edge of the ponds. And, the green herons stayed clear of turtles and falcons.

Some of the sources consulted inferred that green herons flicker their tail and crest nervously when fishing. Green herons were not seen flicking their tails at all. Green herons were observed raising and lowering their crests. But it was not a nervous flickering. Rather, a green heron raised its crest briefly when it spotted some particularly interesting prey in the water. The raised crest reflected the excitement of the chase, not nervousness.

The resemblance between the appearance of a green heron's crest and a Mohawk haircut is hard to miss. But the similarity doesn't stop there. The traditional Mohawk haircut was only worn by warriors going off to war,[203] like a green heron only raising its crest in the middle of the hunt. (See Photograph 2.10.)

Photograph 2.10 Green Heron Crest. Green herons have a distinctive black crest which runs along the top of their head from front to back, like a Mohawk haircut. They can raise and lower their crest at will. The crest, which is about 2 inches (5.08 cm) high when raised, comprises the darkest feathers a green heron has and contrasts with the yellow eyes and bill. Green herons raise their crest quickly for a short period of time when they are excited about spotting some particularly desirable prey. The top photograph shows a green heron from the back as it starts to raise its crest. The bottom photograph shows a side view of the same green heron with a fully raised crest .04 seconds later. © D. Herrmann

It is possible to observe Canada geese for an hour or more at a time. Canada geese are large birds. They swim, graze, sit, and go about their daily routine out in the open. That is not the case with green herons. Observations of green herons are normally only 15 to 20 minutes in duration. Green herons are much more private as they go about their day to day lives. When fishing green herons attempt to blend in with the vegetation by using the reflective properties of their feathers and standing very still. This behavior increases their fishing success and, at the same time, does not attract predators. The green herons at Ponds B and C allowed a person to walk within 3 feet (.9 m) of them, as long as there was sufficient vegetation for cover and the person walked slowly, without making any sudden movements. If not, the green heron was airborne instantly in one quick silent motion and made the famous "skyow" sound as it darted across the pond.

The green herons that visited Pond C went a step beyond using bait to lure fish. They learned that when ducks and geese entered or exited the pond that activity often caused the small frogs and small fish in that area to scatter. As a result the green herons at Pond C were frequently seen hanging out near the mallard ducks and Canada geese. This arrangement was acceptable to all the birds involved, as shown in Photograph 2.11. This practice demonstrates knowledge which was acquired by keen

Photograph 2.11 Good Neighbors Have Benefits. The small fish and tadpoles near the shore scatter whenever the ducks and geese enter or exit a pond. The green herons observed this behavior and learned how to benefit from it. As a result green herons often position themselves near the Canada geese (*Branta canadensis*) and mallard ducks (*Anas platyrhynchos*) to make it easier to fish. Note the differences in size among these three species. © D. Herrmann

observation of the actions of other birds and discerning the impact on potential green heron prey. The green heron did not cause the geese or ducks to enter or exit the pond; yet they learned how to benefit from this activity. The green herons did not engage in this behavior at Pond B. Nor was this listed in the literature as a common behavior of green heron.

2.3 Great Blue Heron *Ardea herodias*

Great blue herons are closely related to the common or gray heron *Ardea cinerea* found in Europe and Asia.[41, 83] An all-white slightly larger great blue heron subspecies *Ardea herodias occidentalis* is only found in southern Florida and the Keys; this subspecies has yellow legs.[45, 157, 203] At one time it was thought that the white feathers were just a color phase that this particular group of herons went through. On 5/25/2015, at 5:30 p.m., a Great blue heron was spotted at a small pond 2 miles (2.66 km) north of Pond D. This bird had black legs and the bill coloring matched that of a Great blue heron. Its long wing and tail feathers were slate blue, but the short feathers on its body were white. It is unknown whether this bird was: (a) a hybrid, or (b) just going through a molting phase.

Great blue heron populations decreased dramatically in the late 1800's and early 1900's due to hunting, habitat loss, and habitat fragmentation.[198] To illustrate, half of the wetlands in New York state have been drained since the arrival of European settlers and the breeding areas along the coastal lowlands of New York have been totally eliminated.[198] Great blue heron populations began to recover by the mid 1900's.[198] Various studies conducted since then indicate that great blue heron populations have been increasing by 1.4% per year since 1966.[198]

Great blue herons are commonly found in fresh water and salt water locations, such as lakes, ponds, rivers, marshes, streams, wetlands, swamps, and coastal bays.[29, 45, 95, 157, 203] Great blue herons are found throughout most of the U.S., except high mountains, deserts, and areas that experience extreme cold.[29, 95, 157, 203] The northern populations migrate to Mexico, central America, or the Caribbean for the winter.[29, 95, 157, 203] The southern populations and those residing along the western coast of the U.S. are permanent residents year round.[29, 95, 203] On rare occasions great blue herons have been reported visiting Spain, England, Greenland, the Azores, and Hawaii.[203] The reasons for this are unclear.

Small fish are the main component of a great blue heron's diet.[29, 45, 203] Great blue herons also consume lesser quantities of frogs, salamanders, small turtles, snakes, shrimp, crabs, aquatic insects, and small reptiles.[29, 95, 157, 203] If these items are in short supply, great blue herons will eat vegetable matter.[157] This varied diet allows great blue herons to reside in a larger range and survive in a broader spectrum of climates.[95]

Great blue herons seek out their prey during the day as well as at night.[45, 95] In this process they are aided by exceptionally keen eye sight.[3] Great blue herons wade along the edges of ponds and lakes to fish, with a slow steady stride about 9 inches (22.8 cm) in length.[203] If need be, they can walk rather quickly without appearing to do so. Great blue herons will wade in water that is as deep as the top of their legs. Once something of interest is spotted, great blue herons stand motionless to lure their prey into a false sense of security.[45, 95] When the prey comes within range, a great blue heron lunges rapidly. (See Photograph 2.12.)

Great blue herons are quiet, cautious, and calm. The size of a great blue heron affords it a certain degree of confidence. At the same time, great blue herons are alert and aware of their surroundings. As a predator themselves, great blue herons are quite familiar with the tactics other predators use. Great blue herons are normally solitary birds outside of nesting time. They do, however, mingle at will among other water birds. (See Photograph 2.13.)

Great blue herons are the largest North American herons.[45, 95, 136, 203] They stand about 4 feet (1.21 m) tall and have a wingspan of 6 feet (1.8 m) or more.[29, 45, 141, 157, 198, 200, 203] Great blue herons achieve this height through their long thin legs, which range from 18 to 20 inches (45.7 to 50.8 cm),

Photograph 2.12 Great Blue Heron Preparing to Lunge. A great blue heron prepares to lunge vertically from above toward some prey. The bill is wide open so that it can be snapped shut at just the right moment. Great blue herons tend to do vertical lunges when wading and horizontal lunges when standing on land. Because of their height great blue herons rarely lunge from branches. Note the front and back plumes and rust thigh patches. © D. Herrmann

Photograph 2.13 Great Blue Heron Wading. (L) A great blue heron wades in Pond B near some mallard ducks. The upper leg is straight, the lower leg is bent forward, and the foot is relaxed. The great blue heron's neck and body are in a low curved position, which is more conducive to fishing. (R) A great blue heron wades in an upright position among the algae in Pond C. Note the head, front, and back plumes. Head plumes are only displayed on rare occasions. © D. Herrmann

and their long thin neck. The actual body of a great blue heron is quite small. As a result, great blue herons only weigh between 4.0 and 4.8 pounds (1.79 and 2.15 kg).[18, 198, 203] The maximum recorded life span of a great blue heron is 23 years.[18, 137] (See Photograph 2.14.)

Great blue herons have an interesting leg design which makes their legs more efficient, especially given the length. Great blue herons have an upper joint just below their rust colored thigh patches. This joint allows great blue herons to extend their legs forward and backward. A second joint is lower on the leg. The lower joint is fixed and only allows great blue herons to move their lower legs and feet forward. The distance between the upper and lower joints is the same as the distance between the lower joint and the foot.

Photograph 2.14 Great Blue Heron Standing Tall. An alert great blue heron stands up tall and stretches its neck straight while checking out unusual activity at Pond B (L) and Pond C (R). © D. Herrmann

The design and range of motion of a great blue heron's leg and foot is closer to a human arm and hand, than a human leg and foot. This range of motion has several advantages. Great blue herons lift their feet out of the water when wading. This minimizes resistance, making it easier to walk, and increases their stride. When seated on the ground, great blue herons are in a ready-to-go position. The feet and toes are already facing forward. As a result great blue herons can stand-up and take flight in one quick movement. (See Photograph 2.15.)

Great blue herons have a delicate appearance. They are mostly covered with slate blue feathers.[157, 203] The head is white, with narrow black stripes running from the base of the bill along the top of the eyes to the back of the head.[157, 203] The long neck varies from white to light gray and has a few long plumes which hang over the breast.[136, 157] Fortunately, great blue herons only have a

Photograph 2.15 Great Blue Heron Walking. A great blue heron walks on the small island in the middle of Pond B at dusk. The upper leg is straight, the lower leg is bent forward, and the foot is relaxed. © D. Herrmann

few dull colored plumes. Consequently they were not hunted to near extinction like many of their relatives, whose exotic plumes were plundered by the fashion industry in the early 1900's.[95] The legs, eyes, and long thin bill are yellow.[29, 157, 203] The legs of a great blue heron are duller outside breeding season and may almost appear gray. The bill is about twice as long as the head.[18] A noticeable feature of great blue herons is the bold black circular shoulder patches, about 2 inches (5 cm) in diameter, on top of each wing. This patch is bordered by a smaller rust brown patch which is about 0.5 inch (1.27 cm) in diameter. (See Photographs 2.16 and 2.17.)

For the most part great blue herons are quiet birds as they go about their daily business. Once in a while they make what is referred to as a hoarse guttural squawk, croak, or "grak" sound.[29, 157, 203] This sound functions similar to an alarm call, since it is heard when a great blue heron is disturbed and flees an area.[157] Great blue herons were not heard making vocalizations at any of the ponds.

Great blue herons return to the same breeding grounds each year, given the opportunity.[45] In the northern climates great blue herons nest as single pairs or a few pairs together.[141] In the southern climates, where great blue heron populations are higher, they nest in colonies.[141, 157, 203] These colonies may have anywhere from 5 to 500 pairs and include other types of herons.[203]

Great blue herons nest in tall trees anywhere from 20 to 60 feet (6 to 18 m) above ground.[83, 95, 136, 141, 157] Once in a while great blue herons nest on the ground, if it is considered a predator-free area.[95, 157, 203] The male selects the nest site. The female builds the nest from materials supplied by the male.[95, 137, 157] The nest consists of a bulky platform of sticks and twigs.[29, 95, 141, 157, 203] Great blue heron nests range in size from 20 to 30 inches (50.8 to 76.2 cm) in diameter.[198] Nest building activities may begin anytime from February through April, depending on the latitude.[157]

Three to five pale greenish blue eggs are laid, which are 2.5 by 1.5 inches (6.3 by 3.8 cm) in size.[29, 95, 141, 157, 203] The eggs are laid 36 to 48 hours apart and have a chalky surface.[18, 137] Both parents

Photograph 2.16 Great Blue Heron Profile. A great blue heron assumes a common posture, with the neck slightly curved and the back, tail, and wings visible. Note that the great blue heron's slate blue feathers blend in with the discarded tree bark. © D. Herrmann

Photograph 2.17 Great Blue Heron: Funny from the Front. Great blue heron appear beautiful, even exotic, when viewed from the side, back, or while in flight. But, from the front great blue heron appear shaggy, unkempt, and funny looking. Note the white underside and short round tail. © D. Herrmann

take part in incubating the eggs, which hatch asynchronously in 25 to 30 days.[83, 95, 157, 203] Both parents feed the hatchlings by regurgitation.[83, 95, 171, 203] Great blue herons have 1 brood per year.[95]

Great blue heron hatchlings may be taken by nest predators like turkey vultures (*Cathartes aura*), corvids, hawks, and raccoons.[203] Great blue herons may be attacked by bald eagles (*Haliaeetus leucocephalus*), great horned owls (*Bubo virginianus*), and red-tailed hawks (*Buteo jamaicensis*).[203] Human disturbances cause great blue herons to abandon their eggs and hatchlings.[203]

Great blue heron hatchlings weigh about 2 ounces (56 gr) upon hatching.[171] Their body is covered with uneven pale gray down; initially the wings do not have any feathers.[171] Great blue heron hatchlings are able to stand when they are 14 days old.[171] They begin to walk around the nest area when they are 21 days old, much like green heron hatchlings.[171] Great blue heron hatchlings start to flap their wings for practice when they are only 4-weeks old; yet they don't fledge until they are 7 to 8 weeks old.[95, 171] After fledging great blue heron hatchlings continue to be fed by their parents for another 2 to 3 weeks until they master the intricacies of fishing.[83, 171]

Great blue heron juveniles have the same coloring as adults, only duller.[41, 203] Great blue heron juveniles lack the black stripes on the sides of their head and the bold black circular shoulder patches.[41, 203]

The flight of great blue herons has been described as being "powerful, dignified, and majestic".[45, 157] During flight great blue herons fold their head back onto their shoulders and hold their feet out straight together behind them.[29, 45, 95, 157] Great blue herons flap their long wings in a slow graceful motion at 2.5 wing beats per second.[18, 95] Yet they are able to achieve cruising speeds up to 30 mph (45.5 kph).[45]

Flight is one of three instances when great blue herons retract their neck. During cold weather great blue herons retract their neck against their body to stay warm.[172] Unlike green herons, great blue herons do not retract their neck when they are walking or wading. Instead, great blue herons hold their neck in an S-like position.[41] Great blue herons lower and collapse their necks when swallowing a meal. This shortens the distance the food has to travel and widens the passage a bit. (See Photograph 2.18.)

Photograph 2.18 Great Blue Heron Swallowing. A great blue heron swallows its catch after a successful lunge. To do so it lowers and partially retracts its neck. This minimizes the distance the food has to travel and widens the passage a bit. At the same time the heron moves its body into a horizontal position. This posture is only maintained for a few seconds. © D. Herrmann

Great blue herons can and do assume a variety of different postures, thanks to their flexible neck and efficient leg design. Great blue herons can stand-up tall with their neck straight, or they can bend over and lunge. Great blue herons can hold their neck in an S-like position, twist it to the left or right, and retract it. Great blue herons can walk and wade with their body in an upright vertical position, a horizontal position, or somewhere in between. The posture changes depending on what the great blue heron is trying to do and how quickly it wants to accomplish that task. For example, a great blue heron's daily activities include: standing, walking, wading, scanning for prey, lunging, sitting, swallowing, flying, roosting, and preening, all of which entail different postures. (See Photograph 2.19.)

Great blue herons are normally spotted in the northern part of Pond B, the area that the Canada geese prefer. The great blue herons frequently wade near the shore of the northern part of Pond B, on both the eastern and western sides, and to and from the small island in the middle. Unlike the green herons, the great blue herons do not climb about in the rocky sections of the pond.

Great blue herons usually navigate with a slow steady stride. If people get too close the herons simply increase their speed and continue on their way. The great blue herons waded in water of various depths, up to the rust colored feathers on their thighs. The great blue herons spent the majority of the time wading in the water. They spent a lesser amount of time walking along the edge of a pond. In fact, the great blue herons spent most of the time they fished from land doing so from the small island in the middle of Pond B.

Presumably herons stand on one leg to rest the other like most birds, including flamingos. However, neither the great blue herons nor any of the other herons were seen resting on one leg on the ground during the spring, summer, or fall. Whenever the great blue herons were spotted they were always busy: walking, wading, scanning for fish, lunging, preening, or flying. There are several possible explanations. Great blue herons have fewer rest breaks. Great blue herons rest breaks are in private where they are concealed from view. Or, great blue herons do not take rest breaks on the ground.

Photograph 2.19 Great Blue Heron Sitting (and Swimming?). (L) A great blue heron appears to be swimming across Pond B. Its body is horizontal to the water, like a Canada goose swimming. The heron's legs are not visible. Because a heron's legs and toes are long, thin, and not webbed, they are not very useful for swimming. This heron might just be wading a little deeper than normal; however, the ripple effect on the water is wider than usual for a heron. (R) The same great blue heron sits down for a short rest on the island in the middle of Pond B after its "swim" on 6/29/2011 at 6:33 p.m. This is the only time a great blue heron was observed sitting on the ground. © D. Herrmann

It is not unusual to see Canada geese, mallard ducks, and wood ducks (*Aix sponsa*) drink water. In contrast, great blue herons were not observed drinking water; nor were any of the other herons. Perhaps they take in a sufficient quantity of water when they capture their prey. Or, perhaps they only consume water at specific times during the day.

Great blue herons are predators. However, they have to be on the lookout for others who might view them as prey. The list of potential predators includes the usual birds of prey and land-based predators. But, that is not everyone; to ignore water-based predators could be fatal.

On the afternoon of 7/4/2011 an attempt was made to locate the great blue heron that frequents Pond B. The bird was not in any of the usual locations where it is normally found. Nor was it at Pond C. A second sweep of Pond B revealed the great blue heron's whereabouts. It was in the middle section of Pond B, near the eastern shore where the mallard ducks and wood ducks reside. But, it was not wading or fishing. Instead, the great blue heron had taken extreme measures to conceal itself in the midst the woody branches of a tall bush. The heron's legs, feet, and slate blue feathers blended in perfectly with the vegetation. (See Photograph 2.20.)

The great blue heron had a good reason for carefully concealing itself. A huge snapping turtle (*Chelydra serpentina*) was trolling only 4 feet (1.21 m) away! The snapping turtle's neck and tail were each at least 8 inches (20.3 cm) long. The snapping turtle's neck was wider than a person's fist. Its shell was at least 18 inches (45.7 cm) long. The snapping turtle's powerful jaws could have easily snapped the heron's legs in two, rendering it helpless and subject to being eaten. The great blue heron's alertness, keen eye sight, and resourcefulness prevented that from happening.

Photograph 2.20 It Pays to Be Alert #1. A great blue heron uncharacteristically conceals itself amidst some bushes that line Pond B to avoid the powerful jaws of a huge snapping turtle. Its feathers and legs blend in perfectly. Only the yellow eyes and bill are noticeable. The great blue heron's alertness, keen eye sight, and resourcefulness allowed it to detect and avoid the snapping turtle even though it was trolling several inches under water. The great blue heron also demonstrated an ability to distinguish between nuisance threats (a photographer who got too close) and a life-threatening threat (the snapping turtle). The great blue heron chose to focus exclusively on the life-threatening threat, and ignore the nuisance threat, until this incident was over. Hence, it survived what could have been a fatal encounter. © D. Herrmann

The great blue heron was not the only one who took notice of the snapping turtle that afternoon. A small brown turtle, with a shell about 6 inches (16.2 cm) long, also assumed a defensive posture. The small brown turtle was just the right size for the snapping turtle to munch. The small brown turtle understood that. Instead of trying to out swim the snapping turtle—which had little probability of success—the small brown turtle quickly climbed up on the highest log it could find, while keeping its eyes focused on the snapping turtle. This was a smart maneuver and it worked, since snapping turtles generally troll below water. Snapping turtles rarely raise more than their eyes and nose above the surface of the water. (See Photograph 2.21.)

Photograph 2.21 It Pays to Be Alert #2. A small turtle (top) also takes evasive action to avoid becoming the snapping turtle's lunch (bottom). Both the great blue heron and the small turtle wisely focused all their attention on the snapping turtle until this life threatening danger had passed. © D. Herrmann

Both the great blue heron and the small brown turtle were totally focused on the immediate life threatening danger posed by the snapping turtle. Throughout this incident they both ignored a pesky human who got too close to take their photographs. This behavior also demonstrated intelligence, specifically the ability to distinguish between threats that posed different degrees of danger. One threat was life threatening. The other was a nuisance. The great blue heron and the small brown turtle smartly chose to ignore the nuisance threat and focus on the life-threatening situation. As a result, they both survived.

The coloring of great blue herons, combined with their quiet nature, does not attract attention. Their slate blue feathers blend in easily with branches, tree bark, and dry vegetation. Other than the encounter with the snapping turtle, the great blue herons made no attempt to hide. The great blue herons were always alert and aware of their surroundings, including human spectators. The presence of humans did not make the great blue herons nervous, as long as there was a minimum separation distance of 6 feet (1.8 m). On one occasion a great blue heron flew from one side of Pond B to the other, placing it much closer to the human observers.

Green herons are not the only ones that have learned the advantages of paying attention to their neighbors. On the afternoon of 10/15/2011 a great blue heron was wading around and fishing in the northern part of Pond B. Several times it stopped, bent over in preparation for lunging, then decided not to. After not having any luck on its own, the great blue heron waded over to where a migratory flock of mallard ducks was swimming back and forth. The great blue heron positioned itself about 3 feet (.9 m) from the ducks. It stood still, waited, and finally lunged. This time it was successful. The pattern the mallard ducks were swimming in flushed some unsuspecting prey right into the great blue heron's bill. In spite of the sudden aggressive movement of a much larger bird, the mallard ducks continued going about their business as if nothing unusual had happened. Apparently mallard ducks also recognize the advantages of neighbors. (See Photograph 2.22.)

Great blue herons were only seen one at a time. They were never seen in pairs, not even during breeding season. Observations were 10 to 20 minutes in duration. By then a great blue heron had finished what it was doing and flew somewhere else. Great blue herons flew 10 to 15 feet (3 to 4.5 m) above ground if they were only flying from one part of a pond to another.

Great blue herons were frequently observed flying from Pond B to Pond C and vice versa at an altitude of 50 feet (15 m) or more. Often a great blue heron was observed at Pond C, then it would leave flying toward Pond B. Not long afterward the great blue heron that was at Pond B flew toward Pond C. It was almost as if the herons were taking turns fishing at the different ponds. When the great blue herons visited Pond C, they waded along the northern and western sides of the pond. Only on rare occasions were great blue herons spotted on the southern or eastern sides of the pond—the sides which are easiest for humans to access. (See Photograph 2.23.)

Great blue herons were seen at Pond C more often (70% of the time) than at Pond B (30% of the time), even though Pond B is home and that is where the great blue herons nest. All of the sightings of great blue herons were made in the afternoon. The great blue herons were most likely to be seen between 12:00 p.m. and 2:30 p.m. The great blue herons were seen most often during the colder months (December and January). That seems a bit odd, but the explanation is simple. The great blue herons stay in the area much longer than the other herons. Food is scarce then and they need to eat more because of the cold. Hence the great blue herons have to spend more time looking for food. Since there is less vegetation in winter the great blue herons are more visible. Even on the far side of Pond C the tall slender silhouette of the great blue heron was unmistakable, whether or not any other birds were present. (See Table 2.2.)

Great blue herons were always the last to leave Pond C once winter set in. For example, in the fall of 2010 the great blue herons were still around 3 weeks after the local belted kingfishers (*Megaceryle alcyon*) and hooded mergansers (*Lophodytes cucullatus*) had left and the last migratory flocks of Canada geese and mallard ducks had passed through. This occurs in part because of the difference in the way

Photograph 2.22 Great Blue Heron Lunging. A great blue heron lunges toward a potential meal in Pond B on 10/15/2011. The knees are bent forward, the wings are partially spread, and the head and neck are underwater. The footing did not change during the lunge. Only 0.08 seconds earlier the great blue heron was standing upright, scanning for prey. In spite of this sudden aggressive movement by a much larger bird, the mallard ducks did not change course. © D. Herrmann

each of these birds search for food. Belted kingfishers and hooded mergansers dive, particularly in the deep part of a pond. Canada geese and mallard ducks upend. These maneuvers are difficult if not impossible to execute once a pond is mostly frozen. In contrast, a great blue heron only needs a small opening in which to insert its long narrow bill. The great blue herons seem to understand that fish will come to the small open area, especially on sunny days, because of the increased air circulation. (See Photograph 2.24.)

Because of its strong toes and nails the great blue heron did not have any trouble slipping on the ice; it could fly and land wherever it pleased. On warm days the great blue heron usually stayed on the north side of the pond, in particular the northwest corner. On 12/9/2010, when Pond C first became covered in sheets of ice, the great blue heron stayed put. The great blue heron continued to wade and fish in small narrow areas near the edge of the pond; sometimes the openings were only inches (centimeters) wide and 2 to 3 feet (60 to 91 cm) long. When it was really cold the great blue heron moved to the northeast corner of the pond to face the sun. One day the great blue heron flew to the southwest corner of the pond. From there it walked along the western edge of the pond, which is covered in heavy vegetation. The great blue heron stood up tall and appeared to be biting the branches. Eventually it settled down to roost in a small thawed area. The great blue heron was last spotted at Pond C on 1/5/2011 that winter. The great blue heron reappeared at Pond B on 4/3/2011, five days before the great (white) egret (*Casmerodius albus*) made its first appearance that spring.

On 4/3/2012 a solitary great blue heron was spotted for the first time at a manmade pond in downtown Washington, D.C. This small L-shaped pond is no more than 10 feet (3 m) wide by

Photograph 2.23 Great Blue Heron Flight. Great blue herons flap their long wings in a slow graceful motion at 2.5 wing beats per second; yet they are able to achieve speeds up to 30 mph (40 kph).[18] Here a great blue heron prepares to takeoff, makes a short flight across Pond C, and lands. © D. Herrmann

20 feet (6 m) long and 3 feet (.9 m) deep. It was created in 2006 when the Museum of the American Indian opened. This solitary great blue heron popped in between 12:30 and 1:00 p.m. on a Tuesday afternoon and immediately started fishing, unconcerned by the tourists who were watching it. The great blue heron caught 3 small (2.5 inch or 6.35 cm) fish in quick succession. Then it stalked a larger fish (6 inch or 15.24 cm) for a while before catching it. In typical fashion the great blue heron caught the fish perpendicular to its bill. Then the great blue heron stood up and slowly rotated the fish around with its tongue until the fish was aligned with its bill. Then the great blue heron swallowed the fish

Table 2.2 Summary of Great Blue Heron Sightings (Spring 2009–Spring 2012).

I. Location Observed		
• Pond A	0%	
• Pond B	30%	
• Pond C	70%	
• Pond D (old and new)	0%	
		100%
2. Time of Day Observed[1]		
PM	% of Total	
12:00	23%	
1:00	30%	
2:00	23%	
3:00	6%	
4:00	6%	
5:00	6%	
6:00	6%	
		100%
3. Time of Year Observed		
• April through January		
• Earliest sighting (spring): 4/3		
• Latest sighting (fall/winter): 1/19		
4. Observations by Month		
• April	6%	
• May	-	
• June	6%	
• July	6%	
• August	-	
• September	-	
• October	6%	
• November	6%	
• December	58%	
• January	12%	
		100%

Note:

1—The time slot reflects the entire hour. For example 10 a.m. refers to sightings that occurred between 10:00 and 10:59 a.m.

head first. The great blue heron continued to scan for fish in a quick deliberate manner. It caught one more small fish, then a large one (8 inches or 20.32 cm). The great blue heron held the large fish in its bill for a moment, then let it go because it was too large to swallow. Next the great blue heron took off flying from the pond. It headed west then circled counter-clockwise to the north as it gained altitude and continued its migratory journey north and home.

This great blue heron was hungry. The great blue herons at Ponds B and C only eat one large fish or two small fish at a time. This great blue heron ate 5 fish! How did the great blue heron know that this small manmade urban pond in the midst of mid-rise office buildings, historical monuments, and museums contained fish? Did the great blue heron find the pond that day by accident? Or, had the great blue heron discovered the pond during the fall or an earlier migration and remembered that it contained fish? Either way this implies that a great blue heron's vision so acute that it can spot fish while flying overhead at its usual migrating altitude and cruising at speeds up to 30 mph (45.5 kph).

Four months later in August 2012 a solitary great blue heron was spotted 40 miles away fishing in a small ornamental manmade pond near an office complex. The pond was surrounded on three sides by 10 to 15 story glass office buildings. A 6-lane major road was on the fourth side. Four mallard ducks were also visiting the pond which contained aquatic plants. The great blue heron was quite

Photograph 2.24 Great Blue Heron in Winter. Some great blue heron do not migrate. (R) To stay warm they grow extra thick plumes on the front and back and maintain a compact posture with their neck retracted, like this great blue heron at Pond C on 11/19/2011 (top) and at Pond C on 11/28/2012 (bottom). (L) On extremely cold days great blue heron tuck their long bill, which is sensitive to temperature, under their plumes. This great blue heron at Pond B on 12/3/2011 tucked its head and bill under its plumes; then looked up briefly about every 40 seconds. Note that the heron's slate blue feathers blend in with the tree bark. Winter is the only time great blue heron were seen standing on the ground on one leg. © D. Herrmann

successful catching the brightly colored 4 to 6-inch (10.16 to 15.24 cm) goldfish (*Carrasius auratus*) in the pond. Since it was August, the great blue heron would not have been migrating. Yet, the same questions are raised. Did the great blue heron find the pond that day by accident? Or, had the great blue heron discovered the pond earlier remembered that it contained fish? Again, this implies that great blue herons have acute vision and good memories. Neither downtown museums nor office complexes near major roadways are their natural habitat.

2.4 Black-crowned Night Heron *Nycticorax nycticorax*

Black-crowned night herons are simply referred to as night herons in Europe and Asia.[203] The scientific name for black-crowned night herons, *Nycticorax nycticorax*, translates into night raven and refers to their habit of fishing at night.[203]

Black-crowned night herons experienced serious population declines during the 20th century. Habitat loss from droughts and the draining of wetlands, the use DDT and other pesticides which interfered with reproduction, and changes in water quality were the main culprits.[41, 95, 201] Black-crowned night herons that live and/or winter along the Gulf of Mexico also suffer from the recurring oil spills into the Intra-coastal Waterway.[129]

Black-crowned night herons are killed in substantial numbers at fish hatcheries.[195] One would think that the people managing a hatchery would understand that fish are the mainstay of a heron's diet. Tropical fish hobbyists routinely cover their tanks to prevent the fish from jumping out and predators (like house cats) from getting in. It would be cheaper, more effective, and more responsible for fish hatcheries to simply put a screen over their tanks.

Because of all of this black-crowned night heron populations are listed as declining in many parts of North America, such as:[201]

- Ontario
- Hudson river area
- Delaware
- Florida
- Midwestern prairie states
- Mountainous part of western North America
- Pacific coast
- Southern breeding areas

Black-crowned night herons inhabit fresh and salt water marshes and coastal lagoons, rivers, ponds, lakes, swamps, and wooded streams.[29, 95, 136, 157, 203] Black-crowned night herons prefer water locations that are surrounded by dense vegetation.[41, 157]

Black-crowned night herons are spread out in a larger range around the world than other types of herons.[41] Black-crowned night herons live on every continent except Australia and Antarctica.[45, 95] Black-crowned night herons live throughout most of North America, except high mountains, deserts, and areas that experience extreme cold.[29, 95] They also live in most of South America, Central America and the West Indies.[157] Likewise black-crowned night herons are found throughout the Indian subcontinent and central and southern Asia.[157] Black-crowned night herons have been observed at altitudes up to 6,200 feet (1.8 km) in the Nepal Valley.[157]

Black-crowned night heron populations that live in the higher latitudes migrate; those that live in the lower latitudes are permanent year round residents.[95, 157, 203] North American black-crowned night herons winter in Mexico, Central America, and the West Indies.[41, 95, 203] European and Asian black-crowned night herons winter in tropical Africa and Southern Asia.[203]

Black-crowned night herons primarily eat small fish.[41, 83, 95, 136, 157, 203] They supplement their diet with other items when they are available, such as: squid, crustaceans, aquatic insects and their larvae, frogs, mollusks, worms, snakes, clams, and mussels.[41, 83, 95, 136, 157, 203] Black-crowned night herons are most active fishing at night, at dawn, at dusk, and on cloudy days.[29, 45, 83, 95, 157, 203] They are usually only seen fishing during the day when there are hatchlings to feed.[95, 136, 157]

Like other herons, black-crowned night herons fish by wading stealthily near the shore, then standing very still, and lunging at the opportune moment.[41, 95, 203] As mentioned earlier, black-crowned night herons do not stab their prey. Instead, they pursue a "stalk and snatch" policy.[137] There has been some speculation that black-crowned night herons fish at night to avoid competition with other larger herons.[41] That theory is inconsistent with the fact that green herons, who are smaller than black-crowned night herons, primarily fish during the day. (See Photograph 2.25.)

Photograph 2.25 Black-crowned Night Heron Wading. Black-crowned night herons are focused and alert when wading—their next meal depends on it! (Top) A black-crowned night heron assumes a horizontal position while wading around a pond with long strides. (Bottom) A black-crowned night heron assumes an upright position to scan a pond before deciding in which direction to wade. A solitary white head plume is visible on both birds. © D. Herrmann

Black-crowned night herons are alert, patient, solitary birds. Like other herons they go about their daily activities on their own. Black-crowned night herons are somewhat cautious about people.[157] That is, until they recognize certain people performing similar tasks on a regular basis, like an individual photographing them repeatedly. For example, Vornberger reported that black-crowned night herons became used to the humans (and their dogs) that frequent the Harlem Meer area in Central Park.[172] In fact, these black-crowned night herons have become more than just used to the humans present. They have noticed that Harlem Meer has a catch and release fishing policy, for humans anyway.[172] The black-crowned night herons "let" the humans catch the fish; then, when the fish are released, the black-crowned night herons re-catch and eat them![172] This indicates that black-crowned night herons are capable of recognizing and remembering individuals and their previous patterns of behavior.

Like a great blue heron, a black-crowned night heron retracts its neck to stay warm.[137, 172] Black-crowned night herons also tuck their bills, which are sensitive to cold, into their breast feathers for warmth.[137] When roosting and napping, black-crowned night herons rest on one leg at a time. This posture only occurs when black-crowned night herons are in trees. Black-crowned night herons do not rest on one leg while on the ground.

Black-crowned night herons are considered to be a medium-sized heron.[29] As their name implies, black-crowned night herons have a black crown and back.[29, 157, 203] The wings are light grey and the underside is bright white.[29, 157, 203] During breeding season the black crown is adorned with a couple long white plumes.[29, 157] Black-crowned night herons have yellowish legs and a long thin black bill.[29, 157, 203] The eyes are bright red.[157, 203] There is a small white area at the base of the bill and around the eyes.[157] Black-crowned night herons have keen eyesight; this allows them to spot prey that is below the water's surface, at dawn, dusk, and throughout the night.[41]

Black-crowned night herons range from 22 to 28 inches (55 to 71 cm) in length, depending on whether the neck is retracted or extended.[29, 41, 136, 157, 195, 200, 203] Black-crowned night herons have a 41 to 44 inch (1 to 1.1 m) wingspan.[29, 41, 200] They weigh between 24 to 28 ounces (672 to 784 gr).[41, 203] Male and female black-crowned night herons look the same.[83, 157]

Immature black-crowned night herons have dull brown feathers that are speckled with white and light grey.[29, 83, 171, 203] Juveniles do not acquire adult plumage until their third year.[83, 195]

Black-crowned night heron hatchlings are covered with grey down. They have 1-inch (2.54 cm) long white-tipped down feathers on their head.[171]

For the most part black-crowned night herons are quiet birds. They primarily vocalize when they are flying or in the evening.[83, 95, 141, 203] The most frequently heard black-crowned night heron vocalization has been described as a harsh "quawk", "wok", or "kwaack".[29, 95, 141, 157] This may be some sort of contact call. As a result, black-crowned night herons are often called "quarks" in the Falkland Islands.[203] Black-crowned night heron hatchlings are known to make a vigorous persistent begging call similar to a "click-click-click" sound.[95, 141] Black-crowned night heron vocalizations were not heard at any of the ponds.

Black-crowned night herons do not breed until their third year.[83, 195] Given the opportunity, they will return to the same nesting grounds year after year.[95, 141] Black-crowned night herons nest in dense vegetation and tall trees near water.[29, 157] The male chooses the nest site, which is usually 10 to 40 feet (3 to 12 m) above ground.[95, 157, 171] The platform nest is built by the female of sticks, twigs, and reeds that the male provides.[29, 41, 95, 136, 141, 157, 203] Black-crowned night herons nest as single pairs or in colonies with other herons.[29, 41, 83, 95, 136, 141, 157, 171, 203] Colonies of up to ten nests in a single tree have been reported.[83] Individual colonies have been reported to be intact for multiple generations in the same location over a span of 50 years.[171]

Three to five pale blue green eggs are laid.[29, 41, 83, 95, 141, 157, 203] The oblong eggs measure about 2.0 by 1.4 inches (5 by 3.5 cm).[141, 157] Both parents incubate the eggs, which hatch in 21 to 26 days.[41, 95, 157, 171] It is not uncommon to see the parents reinforcing the nest while they are incubating the eggs.[171] The young hatch asynchronously in up to 2 day intervals.[41] Black-crowned night heron parents provide shade for their eggs and hatchlings if there is a concern that they are being exposed to too much heat.[83] Black-crowned night heron parents will defend their eggs and hatchlings aggressively against intruders.[195] Black-crowned night heron pairs that live in the northern latitudes only have 1 brood per year. Those that live in the warmer latitudes may have 2 broods a year.

Both parents feed the hatchlings by regurgitation.[41, 83, 95, 157] This exchange is a bit more elaborate than other birds feeding their hatchlings. When a black-crowned night heron parent arrives at the nest it bows to its hatchlings.[41] The parent makes sure the hatchlings see its black crown and white plumes.[41] This process serves as more than a greeting ceremony; the intent is to confirm the parent's identity to the hatchlings.[41] If a parent fails to engage in this behavior, the black-crowned night heron hatchlings will attack it.[41]

This behavior is unique and intelligent. It implies that at an early age a black-crowned night heron hatchling: (1) can distinguish between its parents and other living entities, and (2) understands the need to protect and defend itself from would be predators. In contrast most hatchlings have no concept of fear or self-preservation. Many hatchlings do not mind being held. Survival behaviors usually don't develop until after a young bird has fledged. This behavior also infers that black-crowned night heron hatchlings, unlike most hatchlings, have keen eye sight at an early age. The white plume of an adult black-crowned night heron is long, but very narrow.

Black-crowned night heron hatchlings develop quickly in stages. Their eyes are open almost immediately after they hatch.[171] At about 10 days of age black-crowned night heron hatchlings begin walking around in the nest.[171] By the time they are 4 weeks old black-crowned night heron hatchlings begin to walk around and make short hops to and from branches near their nest.[83, 95, 171] This practice behavior strengthens their legs and feet and improves their coordination. These capabilities will be essential once the young black-crowned night herons begin to fish on their own. Black-crowned night heron hatchlings fledge at about 6 weeks of age.[41, 95, 171] They follow their parents around for a couple of weeks, as they learn how to fish and where the best fishing areas are located.[95, 171]

The flight of black-crowned night herons has been described as being strong and direct.[157] Black-crowned night herons are able to become airborne instantly, in one quick motion, regardless of whether they are standing on the ground, wading, or perching in a tree. Their quiet rapid wing beats allow them to travel from one side of a pond to another or depart an area quickly. When traveling locally, black-crowned night herons fly 12 to 15 feet (3.65 to 4.57 m) above the water's surface. Their landings are as quick and quiet as their takeoffs.

Black-crowned night herons are normally spotted at Pond B. They can be found anywhere in and around Pond B: the northern part, the middle, the southern part, or napping in some tall trees that border the pond. The black-crowned night herons do not seem to have a preferred area to fish. They are spotted on one side of Pond B as often as the other. Some days the black-crowned night herons visit the rocky areas; other days they do not. The black-crowned night herons frequently visit a small inlet on the north side of Pond B in late April and early May, just as numerous tadpoles are hatching there. (See Photograph 2.26.)

Black-crowned night herons were seen at Pond B more often (73% of the time) than at Pond C (27% of the time). That makes sense since Pond B is home and that is where the black-crowned night herons nest. When black-crowned night herons were spotted at Pond C, they were always in the northeast corner. Black-crowned night herons were never seen at Ponds A or D.

The black-crowned night herons were seen more often during the afternoon (82% of the time), than in the morning (18% of the time). The black-crowned night herons were most likely to be seen between 12:00 p.m. and 3:30 p.m. Seventy-eight percent of the sightings occurred then.

These observations seem inconsistent with the profile of black-crowned night herons as being nocturnal. Again the explanation is simple. The majority (73%) of the sightings were made during June, the time of year when black-crowned night heron parents are busy in the daytime obtaining food for their hatchlings. (See Table 2.3.)

Photograph 2.26 Black-crowned Night Heron Fishing. An enterprising black-crowned night heron fishes by monitoring a small waterfall. It is leaning over, ready to lunge. This astute black-crowned night heron learned that small fish and tadpoles are often washed over the edge after a heavy rain. Black-crowned night herons fish at the edge of a pond in direct sun and in the shade. Notice the long back toe (hallux) and the long nails for gripping hard surfaces. © D. Herrmann

The black-crowned night herons did not expend much, if any, effort to conceal themselves in or behind vegetation. Perhaps that is because their coloring doesn't really allow them to blend into the scenery. Sometimes the black-crowned night herons waded in the water while fishing; other times they scanned for prey from the shore. The black-crowned night herons maintained a certain rhythm while they were fishing. A black-crowned night heron would take two or three quick steps, pause to scan the area for prey, then take two or three more quick steps and pause, repeating this process over and over. Black-crowned night herons waded in water up to a depth that covered the top of their legs.

Table 2.3 Summary of Black-crowned Night Heron Sightings (Spring 2009–Spring 2012).

1. Location Observed				
• Pond A	0%			
• Pond B	73%			
• Pond C	27%			
• Pond D (old and new)	0%			
				100%
2. Time of Day Observed[1]				
AM	% of AM		% of Total	
10:00	100%		18%	
11:00	0%		0%	
		100%		18%
PM	% of PM		% of Total	
12:00	11%		9%	
1:00	22%		18%	
2:00	11%		9%	
3:00	34%		28%	
4:00	-		-	
5:00	11%		9%	
6:00	11%		9%	
		100%		82%
				100%
3. Time of Year Observed • May through September • Earliest sighting: 5/23 • Latest sighting: 9/17				
4. Observations by Month				
• April	0%			
• May	9%			
• June	73%			
• July	9%			
• August	0%			
• September	9%			
				100%

Notes:

1—The time slot reflects the entire hour. For example 10 a.m. refers to sightings that occurred between 10:00 and 10:59 a.m.

While pausing to scan for prey, the black-crowned night herons assumed a variety of different postures, from standing up tall with their neck extended, to leaning over the water horizontally with their neck retracted, and bending from side to side. The black-crowned night herons often made a smooth flowing transition from one posture to another. Black-crowned night herons did not stand still very long unless they were on the verge of lunging.

Black-crowned night herons are alert and aware of their surroundings. The black-crowned night herons at Pond B acknowledged the presence of human observers, but were not unduly alarmed by them. These black-crowned night herons were comfortable with a minimum separation distance of 4 feet (1.2 m) when they were fishing and 8 feet (2.4 m) when they were roosting or napping in a tree. A much larger minimum separation distance was required when the black-crowned night herons were spotted fishing at Pond C, usually at least 12 feet (3.6 m). Perhaps this was because Pond C is not home and the black-crowned night herons are more wary of predators there. (See Photograph 2.27.)

Photograph 2.27 Peeking Through the Fence. An alert black-crowned night heron that was wading in Pond B stops for a moment to check out a noise behind it on the other side of the fence. Instead of automatically flying away, the black-crowned night heron stretches its neck and turns its head to assess the situation. This behavior indicates an ability to distinguish between nuisance threats and serious threats. © D. Herrmann

Black-crowned night herons were only seen one at a time. Even during breeding season only solitary birds were spotted. Observations were 20 to 30 minutes in duration. During that time a black-crowned night heron could travel a fair distance around the pond. The black-crowned night herons go about their daily activities on their own. However, they were frequently seen wading in and among mallard ducks, wood ducks, and Canada geese. The ducks and geese did not seem to mind at all.

The black-crowned night herons rarely flew away upon seeing a human, as long as the human was quiet and did not make any sudden menacing movements. The exception to the rule occurred on the morning of 5/23/11. While leaving Pond B a white object was spotted about 12 feet (3.6 m) up in a tree. It turned out to be a black-crowned night heron who was napping at 11:13 a.m. The black-crowned night heron awoke and figured out that it was being watched. For some reason it did not want to be watched that morning. In one quiet smooth methodical fluid motion the black-crowned night heron unfolded from its sleeping position, stretched and preened a little (or at least pretended to), and took off like a flash. The eyes were constantly scanning the human intruder. The only hint that the heron was about to depart was the feet tightening their grip on the branch it was standing on before takeoff. The black-crowned night heron headed toward another part of Pond B to continue its nap in an area that has thick vegetation and is inaccessible to humans. (See Photograph 2.28.)

Another day (7/2/11) a black-crowned night heron was spotted napping on a branch that overhangs a narrow stream which feeds into Pond B. It was 6:13 p.m. This time the black-crowned night heron did not leave. It merely monitored the human's presence, then went back to sleep. (See Photograph 2.29.)

2.5 Yellow-crowned Night Heron *Nyctanassa violacea*

Yellow-crowned night herons are often referred to as American night herons in Europe and Asia.[203] In one of his early reports Audubon noted that the Creole people in southern Louisiana called yellow-crowned night herons "Cap-cap".[191] Yellow-crowned night herons are biologically related to the Bermuda night heron, which became extinct after the arrival of European settlers on the islands.[203]

Photograph 2.28 Morning Nap. A black-crowned night heron awakes from its nap on a tree above Pond B at 11:13 a.m. on 5/23/11. The body is compact and the middle toe is loosely gripping the branch. Slowly in one smooth fluid motion, the black-crowned night heron extends its neck, leans forward, scratches, stands up tall, and tightens its grip in preparation for departure. The black-crowned night heron was airborne 0.06 seconds after the fourth photograph was taken. © D. Herrmann

During the 1800's through the early 1900's yellow-crowned night herons were hunted extensively by local populations, like the Creole people, for food and their plumes which were prized by the hat industry.[191, 197] Yellow-crowned night heron populations had barely recovered by mid-century when they began to face a new crisis: habitat destruction and loss. The development craze of the last half century caused wetlands to be drained and coastal areas to be developed, which wiped out many traditional breeding areas.[197] As a result, today yellow-crowned night herons are listed as being:[197]

- threatened in New Jersey, Pennsylvania, Illinois, and Kentucky
- endangered in Indiana
- a species of concern in Connecticut and Virginia

Photograph 2.29 Afternoon Nap. A black-crowned night heron settles in for a late afternoon nap at 6:13 p.m. on a branch overhanging a narrow stream that feeds into Pond B. The neck is retracted, the body is compact, and it is resting on one leg. The black-crowned night heron still has one short plume even though it is July 2nd. Individual black-crowned night herons sported plumes of various lengths. Some black-crowned night herons had one plume, others had two. The black-crowned night herons lost their plumes at different times. © D. Herrmann

Yellow-crowned night herons live in fresh and salt water marshes and tidal pools, mangroves, barrier islands, wooded swamps, bayous, streams, rivers, ponds, and coastal thickets.[29, 95, 197] Parts of the Mississippi river, cypress swamps, and water sources surrounded by dense vegetation are favorite locations.[95, 197]

Yellow-crowned night herons live in the mid-Atlantic and southern states of the eastern half of the United States, as well as Central America and the West Indies.[29, 95] Yellow-crowned night herons occasionally wander up to the lower Great Lakes region.[203] Yellow-crowned night herons were introduced in Bermuda to fill the ecological niche previously filled by the now extinct Bermuda night heron.[95] Populations living in the lower latitudes are permanent year round residents.[95, 203] Populations living in the higher latitudes migrate and spend their winters in Mexico, Central America, and the West Indies.[95, 203]

Prior to the 1900's yellow-crowned night herons were uncommon in the northeastern U.S.[197] The first official breeding record for New Jersey was made in 1927.[197] The first definitive record of yellow-crowned night herons breeding in New York state was made in 1938.[45] According to this report, there were two nests with two eggs each.[45] In recent years yellow-crowned night herons have been reported nesting (out of necessity) near humans in residential parks and camp grounds, particularly in New Jersey and the Tidewater area of Virginia.[197]

Yellow-crowned night herons eat a variety of items, such as crabs, crayfish, mollusks, frogs, small fish, and aquatic insects.[95, 203] However, their preferred specialty is crabs, including land crabs.[83, 171]

Yellow-crowned night herons swallow smaller prey whole. Medium and larger sized crustaceans are shaken until the legs and pincers fall off, then they are crushed into bite-sized pieces.[83, 171] This behavior eliminates the problem of not being able to eat something because it is too big to swallow.

Contrary to their name yellow-crowned night herons do not fish exclusively at night and may be seen on occasion during the day.[83] Yellow-crowned night herons fish by walking slowly, standing still once they've spotted something interesting, then lunging quickly at the opportune moment.[83, 95] (See Photographs 2.30 and 2.31.)

Photograph 2.30 Yellow-crowned Night Heron Walking. A yellow-crowned night heron walks along the shore of Pond B while scanning for prey. The yellow-crowned night heron appears to be staring at its own image in the water. The brown plumage and yellow eyes indicate that it is a juvenile. A juvenile quickly learns to distinguish between itself and its reflection when scanning for prey. The yellow-crowned night heron walked along the grassy parts, but did not set foot on the rocks. © D. Herrmann

Yellow-crowned night herons are by nature quiet, cautious, furtive birds; in other words the exact opposite of a publicity seeker. Yellow-crowned night herons go about their daily activities as individuals. Even during breeding season only one yellow-crowned night heron is seen at a time. On the other hand, yellow-crowned night herons will vigorously attack predators and defend their nest with their bill and strong toes and nails.[191]

Yellow-crowned night herons are medium-sized herons. They range from 24 to 30 inches (61 to 76 cm) in length and have a wingspan of about 44 inches (1.1 m).[29, 157] Yellow-crowned night herons do not extend and retract their neck as much or as often as other herons. Yellow-crowned night herons weigh about 1.4 pounds (560 gr).[203]

The appearance of male and female yellow-crowned night herons is the same. The narrow pale yellow crown is bordered by horizontal black stripes that run above and below the eyes. There is a small white cheek patch between the black stripes. The second or lower stripe, along with the height, can be used to distinguish a yellow-crowned night heron from a great blue heron at a distance. Yellow crowned night herons have orange-red eyes and a long thin black bill. During breeding season the crown sports a couple long thin pale yellow plumes.[197]

Photograph 2.31 Yellow-crowned Night Heron Fishing. An adult yellow-crowned night heron positions itself in the middle of a small stream that flows out of Pond B. This stream often carries small fish, tadpoles, and baby turtles with it. Clams and snails are commonly found here too. The heron bent over and leaned in all directions, following the activity around it. But, the heron did not move its legs or feet. At this moment the heron is getting ready to lunge. © D. Herrmann

Yellow-crowned night herons have a long slender body. The neck, breast, underside, and back are slate grey.[29, 157, 203] Yellow-crowned night herons have a short round tail. Fine detailing is visible when the wings are retracted. The long thin legs and feet are a medium to dull yellow, depending on whether or not it is breeding season. The legs of a yellow-crowned night heron are about 20% longer than those of a black-crowned night heron. As a result, yellow-crowned night herons are taller.

Juvenile yellow-crowned night herons have brown and white mottled plumage and yellow eyes.[171, 197] Juveniles acquire adult plumage by the time they are 2 years old.[197]

Yellow-crowned night herons rarely vocalize unless they are disturbed.[197] When that happens, yellow-crowned night herons make a "quawk" or "whoc" sound that resembles that of a black-crowned night heron, but at a slightly higher pitch.[29, 197] Yellow-crowned night heron vocalizations were not heard at Pond B.

Yellow-crowned night herons nest in trees or thickets 30 to 40 feet (9 to 12 m) above ground.[29, 95] On rare occasions yellow-crowned night herons nest on the ground, if they have determined that it is a predator-free area.[29] Yellow-crowned night herons nest as single pairs or in colonies with other herons.[29, 141, 203] The higher the number of herons living in the area, the more likely they are to nest in colonies. The nest of a yellow-crowned night heron consists of a platform of sticks and twigs lined with leaves.[29, 95, 203] Again, the male selects the nest site and the female builds the nest from materials the male provides.

On average, three to five pale blue green eggs are laid.[29, 95, 141, 203] The eggs are very similar in size, shape, and color to black-crowned night heron eggs.[141] Both parents take turns incubating the eggs, which hatch in 21 to 25 days.[95] Both parents take turns feeding the hatchlings by regurgitation.[83, 95]

Yellow-crowned night heron hatchlings fledge at about 6 weeks of age.[95] Like other heron hatchlings they follow their parents around for a few weeks until they have mastered the art of fishing.

The flight of yellow-crowned night herons has been described as being strong, direct, and graceful, as is evident from their slow steady wing beats.[197]

Yellow-crowned night herons were only seen at Pond B. They were never spotted at Ponds A, C, or D. Yellow-crowned night herons were observed resting and fishing in a variety of different locations in and around Pond B. Yellow-crowned night herons were always near dense vegetation—never out in the open—in case they needed to make a quick retreat. Yellow-crowned night herons perched on branches that overhang the rocky parts of Pond B. But, they were not seen walking or wading among the rocks. (See Photograph 2.32.)

Photograph 2.32 Yellow-crowned Night Heron Resting. An adult yellow-crowned night heron rests on a sheltered branch that overhangs a narrow stream that feeds into Pond B. The neck is retracted. The body is more compact and relaxed than usual. This yellow-crowned night heron is resting in almost the same spot as the black-crowned night heron shown in Photograph 2.29. © D. Herrmann

The yellow-crowned night herons were only seen between April and September. Most of the observations were during June. Their departure coincided with the first cold weather each fall.

The yellow-crowned night herons were primarily seen during the late afternoon. Adults were most likely to be spotted between 5:00 p.m. and 6:30 p.m. Juveniles occasionally made an appearance earlier in the afternoon.

The slate gray plumage of yellow-crowned night herons does not attract attention and in fact blends in nicely with dried vegetation, tree bark, and branches, much like that of great blue herons. Yellow-crowned night herons take advantage of this feature. During the day they often conceal themselves by standing still near vegetation or on branches. Usually the only clue that gives them away is the two black stripes on their head. (See Photograph 2.33.)

Yellow-crowned night herons walk and wade with a slow steady yet somewhat tentative pace. As they walk yellow-crowned night herons are constantly scanning in every direction, both for prey and potential predators. When yellow-crowned night herons stand still their legs and feet do not move.

Photograph 2.33 Yellow-crowned Night Heron Hiding. A yellow-crowned night heron conceals itself at dusk among some tree branches about 12 feet (3.65 m) above ground near Pond B. Its slate gray feathers blend in perfectly with the tree bark, yet it maintains an alert status. © D. Herrmann

However, their body is constantly fidgeting in all sorts of directions. This movement contrasts with the smooth fluid movement of a black-crowned night heron or the lightning lunge of a green heron and great blue heron. In comparison, the movement of a yellow-crowned night heron seems to be more jittery or restless.

The yellow-crowned night herons at Pond B do not like to be observed. The time of day did not make any difference, nor did the distance. A person could be 6, 10, 20 feet (1.8, 3.0, 6 m) or more away. The yellow-crowned night herons still fidgeted nervously without taking their eyes off the human. The distance that was considered too close varied from day to day. When that threshold was crossed, the yellow-crowned night heron quickly flew to the other side of Pond B without making a sound. Juvenile yellow-crowned night herons were more tolerant of human observers until mid-August; after that they responded in the same manner as an adult yellow-crowned night heron. (See Photograph 2.34.)

Yellow crowned night herons were only seen one at a time. Yellow-crowned night herons go about their daily activities on their own. Even during breeding season only solitary birds were spotted. Observations were rarely more than 15 minutes in duration. By that time a yellow-crowned night heron had finished what it was doing and moved on. Or, the yellow-crowned night heron had become wary of the human observer by then and departed.

The yellow-crowned night herons at Pond B do not mingle with the mallard ducks, wood ducks, Canada geese, or other herons. However, they nest in the same area of Pond B as the other herons.

Photograph 2.34 Yellow-crowned Night Heron on Alert. A juvenile yellow-crowned night heron fluffs its feathers while standing in a tree. This posture and behavior indicate a nervous restlessness, as the heron contemplates whether it should stay put or flee. Note the yellow liner on the inside of the eye near the bill. © D. Herrmann

2.6 Discussion

Now let's see what these behaviors tell us about green heron, great blue heron, black-crowned night heron, and yellow-crowned night heron intelligence, behavior, and individuality. In particular, let's see which of these behaviors were:

- Common to all four species of herons
- Unique to one species of heron or an individual heron
- Indicators of individuality and intelligence

Common Behaviors

All four species of herons experience life as individuals. Except for a couple months each year when they are nesting and tending to their hatchlings, herons go about their daily activities (walking, wading, fishing, preening, resting, flying, etc.) as individuals. Young herons are self-sufficient and on their own a few weeks after they fledge.

There is no flock mentality, yet, each individual heron identifies with other herons. When it comes time to nest, herons nest in colonies if multiple pairs live in the same area. The colonies are often composed of multiple species of herons. For example, there are a small number of green herons, great blue herons, black-crowned night herons, and yellow-crowned night herons living at Pond B. They could easily spread out and have private nests. But, year after year, the four species come together to nest in the same small area. And no territoriality or aggression is displayed.

Herons living in the northern climates migrate for the winter. Herons living in the southern climates do not migrate. Herons migrate as individuals or as a small group of individuals. This behavior indicates that on an individual basis each heron decides whether or not it needs to migrate, when to migrate, where to migrate, and what route to follow. Likewise, since herons do not migrate as a flock, each heron acquires the knowledge needed about when, where, why, and how to migrate on its own, as well as information about stopover locations and alternate routes.

All of the herons observed exhibited a quiet, alert demeanor. The yellow-crowned night herons and green herons spent time and effort to conceal themselves among the vegetation at the water's edge. The black-crowned night herons and great blue herons did not; they walked and waded freely out in the open. The green herons and black-crowned night herons walked and waded among the rocks. The great blue herons and yellow-crowned night herons did not; size may have been a contributing factor. None of the four species sought to draw attention to themselves.

All four species of herons assumed a wide variety of postures while walking, wading, and fishing. The black-crowned night herons, great blue herons, and green herons made these transitions in one smooth fluid movement. This was particularly true for the black-crowned night herons. In contrast, the movements of the yellow-crowned night herons appeared tentative and jittery. Green herons and black-crowned night herons were observed fishing and lunging from branches that overhung a pond. Great blue herons and yellow-crowned night were not observed doing this.

None of the herons were ever observed lunging and missing. A few times an individual heron was ready to lunge, but at the last minute decided not to. This behavior indicates that herons have:

- An understanding of the speed with which they are able to lunge and the speed with which different types of prey are able to flee
- An ability to correlate this information in real time to assess the probability of success for each unique situation
- An ability to act upon this correlated information in a timely manner

All four species of heron fish according to a 3-step process: walking and wading, standing still, then lunging at the opportune moment. The lunges may be horizontal or vertical. Fish are the mainstay of the diet of green herons, great blue herons, and black-crowned night herons. Yellow-crowned night herons prefer crabs. All the herons consume a variety of other supplemental items; the lists of supplemental items consumed are similar, but not exactly the same.

Green herons, great blue herons, black-crowned night herons, and yellow-crowned night herons are seasonally monogamous. This behavior is consistent with their individual lifestyle. In all cases the male selects the nest site, which is usually in the same breeding grounds each year. The female builds the nest from materials provided by the male. The nest itself is a platform of sticks. The size and height of the nest varies by species and individual pair. In all four species of herons both parents incubate the eggs and feed the hatchlings. Unlike most hatchlings, heron hatchlings walk and jump on the branches around their nest several days before they fledge. This behavior allows the hatchlings to develop the leg and foot strength and coordination that are essential for successful fishing.

Unique Behaviors

The most notable difference in behavior among the four species of herons was the distinct range of temperaments. Great blue herons, with their calm, cautious, steady temperaments, were at one end of the spectrum. Yellow-crowned night herons, with their furtive, fidgety, restless temperaments, were at the other end of the spectrum. Green herons and black-crowned night herons were somewhere in the middle. They were neither completely calm, nor nervous. Rather, green herons and black-crowned night herons seemed to be constantly in a state of high alert; perhaps that was due to their smaller size.

The difference in temperament was reflected in other aspects of behavior. For example, the four species of herons exhibited different gaits when they were walking and wading. Great blue herons moved with a slow steady gait and long strides. Green herons walked and waded quickly, and varied their neck height along the way to scan for prey. But they did not stop until something of interest was spotted. For example, a green heron might walk the entire eastern side of Pond C without stopping. The black-crowned night herons moved as quickly as the green herons, but they stopped to scan the water every 2 or 3 steps. They could also cover quite a distance in a short amount of time. In contrast, the yellow-crowned night herons steps were always tentative, last minute decisions. The yellow-crowned night herons never walked a long distance. They only walked 10 feet (3 m) or less then flew to the next fishing location.

The difference in temperament was also reflected in what each species of heron considered a safe minimum separation distance. The herons became used to being observed and photographed in a short amount of time. After that they would allow a human observer to walk up to them, as long as a minimum safe separation distance was maintained and the person didn't make any sudden threatening movements. This minimum safe separation distance varied by species, location, and activity as shown below:

- Green heron: 3 feet (.9 m)
- Great blue heron: 6 feet (1.8 m)
- Black-crowned night heron:
  Pond B: 4 feet (1.2 m) when fishing, 8 feet (2.4 m) when napping or roosting
  Pond C: 12 feet (3.6 m) when fishing
- Yellow-crowned night heron: at least 15 feet (4.5 m), preferably more!

Four ponds, a creek, and a golf course water trap are in the observation area. All are a short flight from one another. Yet, all of the herons consider Pond B home and that is where they all nest. None of the herons were ever seen at Pond A or the golf course water trap. Green herons were observed at the old Pond D on a few occasions, but never at the new Pond D. Green herons, great blue herons, and black-crowned night herons were observed visiting Pond C. Green herons were spotted at Pond C frequently during July and August. A great blue heron was spotted at Pond C often during November and December. A black-crowned night heron visited Pond C a few times each spring. Yellow-crowned night herons did not visit any of the other ponds; perhaps because of their slightly different diet. It is likely that all or most of the herons also fished along the creek. The creek is closest to Pond B and very secluded.

Green herons were the only herons to be seen wading or flying in pairs. This behavior was limited to the courtship and nesting season.

Black-crowned night heron parents are the only species of herons that are known to go through a pseudo greeting ceremony upon arrival at the nest. A black-crowned night heron parent bows to its hatchlings to make sure that the hatchlings see its black crown and white plumes.[41] This behavior is intended to confirm the parent's identity before it attempts to feed the hatchlings.[41] If a black-crowned night heron does not identify itself first, the hatchlings will attack it.[41] The hatchlings behavior indicates that from a very early age they recognize that they are vulnerable to predators. Very few hatchlings or young fledglings have that awareness or sense of danger.

Before swallowing a crab yellow-crowned night herons shake it until the legs and pincers fall off.[83, 171] If the prey is too large, a yellow-crowned night heron breaks it into bite-sized pieces first.[83, 171] This behavior eliminates the problem of prey being too big to swallow. Yellow-crowned night herons were the only one of the four heron species to do this.

Individuality and Intelligence

Several behaviors provided indicators of individuality and intelligence. In particular individual herons demonstrated knowledge that they had acquired through their own observation, analysis, and experience. Here are some examples.

A black-crowned night heron learned to fish by monitoring a small waterfall within Pond B. The black-crowned night heron noticed that small fish and tadpoles are frequently washed over the edge after a heavy rain and applied that information to simplify its fishing activities. (See Photograph 2.26.)

A great blue heron uncharacteristically concealed itself among vegetation near the edge of Pond B and remained very still to avoid becoming a snapping turtle's lunch. Obviously the great blue heron had seen the large snapping turtle trolling before—several ducklings are lost to the turtle each year—and learned to stay out of the snapping turtle's way. The great blue heron's behavior demonstrated the ability to distinguish between a life threatening situation and a nuisance threat. The great blue heron wisely focused all its attention on the snapping turtle until the life threatening situation had passed. (See Photograph 2.20.)

A black-crowned night heron learned that Central Park has a "catch and release" fishing policy by observing and remembering human behavior.[172] This heron lets humans catch a fish. Then, the heron re-catches the fish when it is released. This observation and knowledge saves the heron a lot of time and work.

Individual green heron at Pond C and great blue heron and black-crowned night heron at Pond B learned to pay attention to and benefit from the movements of their mallard duck and Canada geese neighbors. Whenever the ducks and geese enter or exit a pond, or swim in a certain pattern, the fish and tadpoles nearby head in the opposite direction. Individual heron have noticed this and tend to position themselves so that the fish and tadpoles swim right into lunging range. (See Photograph 2.11.)

If the minimum safe separation distance was violated or some other movement or activity disturbed a heron, the majority of the time it launched into flight immediately in one quick movement. Individual black-crowned night herons at Pond B were the exception. In some situations they too took to flight. In other situations a black-crowned night heron paused to analyze the situation first before interrupting its business and expending energy on a sudden departure. This indicates that the black-crowned night heron understands the difference between a nuisance threat and a serious threat, and that a different response to each situation is acceptable. (See Photograph 2.27.)

A juvenile green heron at Pond C learned to use bait to lure fish. This particular green heron didn't just drop a leaf to attract the fish's attention. It folded the leaf up first, so that the leaf would have more kinetic energy when it hit the water's surface. In other words, this heron improved on a technique attributed to green heron. It is unclear whether the juvenile green heron learned this on its own or from its parents. This juvenile was the only green heron observed using bait. This juvenile green heron only used bait when it was fishing at Pond C. (See Photograph 2.5.)

Each year a solitary great blue heron stayed several months after the other herons had left on migration. During this time the great blue heron spent more time at Pond C than at Pond B. Pond C receives more direct sunshine during the winter. This creates small areas along the shore line that don't freeze completely. The great blue heron noticed that and the fact that the fish tend to congregate in these small open areas because of the increased air circulation. Also, the great blue heron expanded its winter diet to include vegetation—something that it didn't eat during the warner months. (See Photograph 2.24 and earlier Photograph 1.2.)

Ideas and Inventions Appropriated by Humans

Lastly, in order to do a complete assessment of intelligence we need to identify the ideas and inventions that humans appropriated from herons. After all, if an idea was good enough to appropriate, it must have been rather intelligent.

Herons snatch a fish or other prey between their bill. Once the prey is secure, it is rotated to be aligned with the heron's bill and swallowed. It is likely that the sight of herons dining in this manner prompted the invention of chopsticks. It is probably no accident that chopsticks, and the speed and precision with which they can be used to move food, resemble the long slender bill of a heron.

Herons and other long-legged waders have legs whose length is out of proportion to their body. This feature allows herons to wade in deep water, up to the top of their legs, without getting their feathers wet. The long thin legs of herons glide through water easily, allowing them to walk at a brisk pace with long strides. It is likely that the sight of herons and other long-legged waders cruising along at the edge of ponds gave rise to the invention of stilts. This observation certainly gave rise to the idea of putting houses near water on stilts.

Walking stilts are a sort of mobile ladder that only has one step. The word ladder appeared in human language in the 12th century.[179] The noun stilt, as in the name of a species of birds, appeared in the 15th century.[179] But, the idea of raising something or someone up on stilts did not appear in human language until the mid-17th century.[179]

Herons use powder from their powder down to cleanse their feathers. This powder is very effective at absorbing dirt and oil. It is possible that the sight of herons using powder as a cleansing agent, instead of the more common preening oil, inspired the development of similar products for human use.

Humans have used spears since prehistoric times. However the use of spears was limited to hunting and deterring the aggression of mammals and other humans. In contrast, the concept of fishing with spears or harpoons is fairly recent. The word harpoon did not appear until the early 1600's.[179] It is highly likely that the idea of spear fishing and harpoons came about from watching herons fish and the rapid lunging movement involved. (Herons do not actually spear fish, but that distinction wasn't made until the late 20th century.)

Green herons have a black crest that covers the top of their head. This crest can be raised and lowered at will. A green heron raises its crest when it is excited about spotting some particularly interesting prey. In this situation a green heron raises and lowers its crest once or twice quickly before lunging. When a green heron's crest is raised it looks like a Mohawk haircut. Or, to be more precise a Mohawk haircut looks like a green heron's crest. (See Photograph 2.10.)

Also, it is hard to miss the similarity of machinery, camera lenses, Slinky™ toys, and other devices that extend and retract, and herons that extend and retract their necks; especially green herons and black-crowned night herons that extend and retract their necks a considerable distance. Surely this too is a borrowed idea. (See Table 2.4.)

And what ideas and inventions have herons deemed worthy to appropriate from humans:? None.

Table 2.4 Ideas and Inventions Humans Appropriated from Herons.

Ideas and Inventions	**Observations that Led to the Ideas and Inventions**
Chopsticks	• Long thin bill of herons • Speed with which herons snap their bill shut to catch prey
Stilts	• Long thin legs of herons • Ability of herons to wade in water without getting their body wet
Using powder as a cleansing agent to absorb oil and remove dirt	• Herons cleaning their feathers with powder from their powder down feathers
Spear fishing, harpoons	• Long thin bill of herons • Speed and accuracy with which herons lunge toward their prey
Mohawk haircut	• Green herons raising and lowering their crest when they are excited about pursuing potential prey
Extensible machinery and equipment	• Herons extending and retracting their necks when fishing

3
Canada Geese
Branta canadensis

3.1 Background

The term Canada geese originated in 1772,[179] about the time the colonists started to notice the geese's regular migration patterns. Undoubtedly, Native Americans were familiar with Canada geese migration patterns long before then.

Canada geese are native to North America as the name implies. Canada geese have spread to Japan, eastern China, and the Kamchatka peninsula in eastern Siberia; they have been introduced in Europe and New Zealand.[41, 83, 95, 157, 203] There are 11 recognized subspecies of Canada geese.[208] The largest subspecies, *Branta canadensis maxima*, is rated endangered by the IUCN.[83] It was thought to be extinct until a small population was discovered near Rochester, MN in 1962–63.[83] The mean North American Canada goose population between 2001 and 2003 was estimated to be 5.2 million.[209] Canada geese were first reported to be breeding north of the U.S. border in 1915.[141]

Government efforts to conserve Canada geese began in the late 1800's. At the time Canada geese faced serious population declines from over-hunting, agricultural expansion, and habitat destruction, as more and more immigrants arrived in North America.[83, 95, 203, 208] Then, as now, the North American governments emphasis has been to ensure that there are enough birds for hunters to shoot—not ensuring the species' survival for its own intrinsic value.[177]

The decline of waterfowl populations was particularly noticeable during the Dust Bowl days. This led to passage of the Migratory Bird Conservation Act of 1929.[177] The Bureau of Biological Survey, the predecessor to the U.S. Fish and Wildlife Service, who was responsible for enforcing the act, adopted the Canada goose as its mascot for many years.[177, 206] Displaying a rugged resilience equal to that of the pioneers, Canada geese populations recovered as they adapted to the metropolitan lifestyle of the 21st century. Some flocks have eschewed migration for permanent residency in locales that sport mild winters and year around availability of food. (See Table 3.1.)

Now let's examine Canada geese behavior. Behaviors are presented as they occurred in Sections 3.2 through 3.5. The analysis, comparison, and interpretation of these behaviors are reserved for Section 3.6 Discussion. These observations took place at the locations and during the timeframes described in Annex C.

Table 3.1 Chronology of North American Government Actions to Protect Migratory Birds (compiled from multiple sources).

Year	Activity
1885	Congress established Branch of Ornithology within the Division of Entomology, as part of the U.S. Department of Agriculture
1886	Branch of Ornithology becomes the Division of Economic Ornithology and Mammalogy
1905	Division of Economic Ornithology and Mammalogy becomes the Bureau of Biological Survey
1916	Cooperative migratory bird treaty signed with Canada
1918	Migratory Bird Treaty Act of 1918, 16 USC 703-712; Ch. 128, implements treaty with Canada
1929	Migratory Bird Conservation Act, 16 USC 715–715d, 715cf, 715f–715r, passed; amended: 1976 —P.L. 94–215, 1978—P.L. 95–552 and P.L. 95–616, 1983—P.L. 98–200, 1984—P.L. 98–548, 1986 —P.L. 99–645
1946	Migratory Bird Treaty Act amended to add Mexico
1950's	Waterfowl flyway management plans developed
1960's	Species level management plans developed
1970's	Regional habitat concept plans developed
1974	Migratory Bird Treaty Act amended to add Japan
1978	Migratory Bird Treaty Act amended to add U.S.S.R.
1982	U.S. National Waterfowl Management Plan
1986	North American Waterfowl Management Plan (Canada, Mexico, U.S.)
1989	North American Wetlands Conservation Act, P.L. 101–233, passed
1998	Fines for violating the Migratory Bird Treaty Act raised to $100,000 for individuals and $200,000 for organizations; fines for misdemeanor violations raised to $15,000
2000	Neo Tropical Migratory Bird Conservation Act, P.L. 106–247 passed
2004	1st update of North American Waterfowl Management Plan
2012	2nd update of North American Waterfowl Management Plan
2013	List of species protected under Migratory Bird Treaty Act expanded, see www.fws.gov/migratorybirdprogram
2014	Revised objectives (first of seven): An Addendum to the 2012 North American Waterfowl Management Plan; see www.nawmprevision.org for latest information

3.2 Characteristics

Several characteristics are common to all Canada geese, such as their:

- Diet
- Physical features
- Vocalization
- Aggression, territorial and defensive postures

Diet

Canada geese are herbivores, preferring green vegetation, grains, silt, and aquatic plants.[83, 95, 157, 203] On occasion Canada geese eat berries.[157] The bill of a Canada goose is specially designed for eating vegetation.[41] The outer rim of a Canada goose's bill is serrated on the top and bottom; up close it looks like tiny off white pinking shears 0.125 inch (0.317 cm) high. When grazing a Canada goose snaps its bill shut on a stem or leaf. Then it twists its head quickly, while the serrated rim or lamellae snips off a bite. The tip of a Canada goose's bill has a sharp point which aids in cutting or slicing a leaf or stem. Canada geese have a light pink tongue. The tongue does not extend beyond the bill, rather it helps to grasp or hold onto vegetation. (See Photographs 3.1 and 3.2.)

Photograph 3.1 A Snip Here and a Snip There. The outer rim of a Canada goose's bill is serrated on the top and bottom. (Top) To eat a Canada goose snaps its bill shut on a leaf or blade of grass. Then it twists its head quickly, while the serrated rim or lamellae snips off a bite. (Bottom) See the small thin lines just inside the lower bill. The serrated rim of a Canada goose's bill functions like a comb when preening, as shown later in Photograph 3.26 bottom right. © D. Herrmann

Canada geese do not normally pull plants up by the roots. Instead, they just trim off what they need. This allows a plant to grow back and produce more food for a later meal. Regardless of flock size Canada geese do not over-graze an area, as mammals are prone to do. Canada geese eat a little bit here and a little bit there and move on. While Canada geese do not store food, their grazing habits ensure the availability of more food in the future.

Canada geese adjust their diet to the seasonal availability of food. For example, on 1/21/2011 a few Canada geese at Pond C were observed pulling up dormant grass and eating the roots. On 2/17/2011 at Pond C a few Canada geese were observed breaking off small branches and stems of plants and bushes and eating them, even though they were dry. Canada geese do not engage in either of these behaviors in spring and summer when the vegetation is lush and green. In addition, during the winter of 2010/2011 Canada geese at Pond B were observed eating seeds (millet and milo) and peanuts that had been put out for songbirds and squirrels (*Sciuridae sciurini*).

Two other dietary behaviors were also observed that have not been reported elsewhere. On 2/20/2011 at Pond C one goose carefully picked out 3 small pebbles from the edge of the pond. Each pebble was less than 0.5 inch (0.127 cm) in diameter. The goose dunked each pebble in the surface water a couple times to wash it, then swallowed it. On 5/11/2011 at Pond B an adult goose stopped

Photograph 3.2 Drinking. A Canada goose lowers its long neck and head into the water to drink. Next a Canada goose fills its lower bill with water. Then a Canada goose slowly raises its neck and tilts its head upward, while using gravity to assist the swallowing action. Geese frequently take two or three drinks at a time. Geese drink water while standing at the edge of a pond and while swimming. Canada geese only drink clean surface water, and ignore deeper water where sediment and debris have settled. © D. Herrmann

to eat some white sand that had been spilled on the lawn. The 5 young goslings with it did likewise. They each took 2 or 3 bites of sand before continuing on their way. Pebbles and sand help Canada geese macerate or pre-digest their vegetarian diet.

Physical Features

A Canada goose's feathers are an exquisite mixture of earth tones. The head, bill, neck, legs, and feet of a Canada goose are all jet black. A bright white patch runs from ear to ear under the bill, like an upside down pair of earmuffs.[203] The crop area is solid light beige. The remainder of the front, underside, back, and wings are a variegated beige pattern similar to a sandstone mosaic. The light beige pattern continues on the underside to the legs. From the legs to the tail the underside is white. A Canada goose has a short stocky black tail and a wide wingspan, varying from 4 to 6 feet (122 to 183 cm).[41, 203] Male and female Canada geese look identical. Adult female Canada geese weigh 5.5 to

12 pounds (2.4 to 5.3 kg), while adult male Canada geese weigh 7 to 14 pounds (3.13 to 6.27 kg).[203] The largest Canada geese are about 36 inches (0.91 m) tall.

Canada geese are supposed to have black heads and more than 99.99% of them do. However, on 6/3/2011 a Canada goose with a white patch on its head was spotted. It was part of a group of 17 local Canada geese that were visiting Pond D that afternoon. This goose had a white patch about 1 inch (2.54 cm) high by 2 inches (5.08 cm) wide on its forehead just above the bill. It would be interesting to know how this mutation developed and whether it will be passed on to the next generation. Luckily this unique goose was captured in a photograph. (See Photograph 3.3.)

Photograph 3.3 White Forehead. Here is a rare sight: a Canada goose with a white patch on its forehead! This goose was spotted among a flock of 17 geese that were visiting Pond D on 6/3/2011. The white patch, just above the bill, was about 1 inch (2.54 cm) high by 2 inches (5.08 cm) wide. © D. Herrmann

The eyes of a Canada goose are jet black, alert, and about 0.75 inch (1.9 cm) in diameter. When Canada geese blink their eyes a dark inner membrane slides outward horizontally from the inside of their eye near the bill, rather than up and down. Canada geese blink so rapidly this inner membrane is rarely visible. To capture the movement of the inner eyelid the photograph below had to be shot at 1/500th of a second. The inner eyelids are smooth and moist; they keep the eyes lubricated. The inner eyelids may also help ducks and geese see underwater.[137]

Canada geese also have an outer eye membrane that covers their eyes when they are sleeping. The outer eyelid is white and fluffy, like the white feathers that wrap around the ears and under the chin. This bright white membrane serves two purposes. First, this membrane protects and shields the eyes of a Canada goose when it is asleep, much like human eyelids. Second, this membrane serves a security function. The head of a Canada goose is jet black; this membrane is bright white. From a distance a bold white dot on a jet black head creates the impression that a goose is awake when it is asleep. This feature helps to deter predators, which is handy since Canada geese sleep on the ground. The outer eye membrane of a gosling is more of an ivory or beige color to provide contrast with their hatchling and later juvenile feathers. (See Photograph 3.4.)

The rich plumage of Canada geese is an example of biological specialization that is consistent with their lifestyle. A Canada goose's feathers are airtight and watertight, which facilitates buoyancy and thermo-regulation.[83] Through an interlocking system of feather barbs and barbules this plumage keeps Canada geese warm when flying at high altitudes, dry when swimming in cold water, and cool

Photograph 3.4 Here's Looking at You. The jet black alert eyes of a Canada goose are specially designed for life outdoors. Canada geese have both inner and outer eyelids. (Top) The inner eyelid is a thin moist dark membrane that keeps the eyes clean and lubricated. This membrane blinks so rapidly, it is hardly visible. Here it appears as a dull film covering half of the eye. (Bottom left) The outer eyelid protects and shields the eyes when a goose is asleep. The exterior of the outer eyelid is white and fluffy, like the feathers that wrap around the ears and under the chin. This white outer eyelid is also a security feature—it creates the impression that a goose is awake when it is asleep. This is a classic example of biological intelligence. © D. Herrmann

in summer. When a Canada goose finishes swimming and returns to land its feathers do not look wet. On the contrary, any excess water beads up on its feathers. In contrast, a parrot's feathers look drenched after it bathes. Some parrots' feathers don't even appear to be the same color when wet because they reflect light differently. A Canada goose lives part of its life in water; a parrot does not. Hence the different feather structures. (See earlier Photograph 1.1.)

The feathers of young goslings are not as waterproof as their parents. On 5/13/2011 a family with six 1-week old goslings was observed at Pond D. Unlike other days the goslings' front feathers looked soaked and were very muddy. The goose family was returning to Pond D from a hike elsewhere. It is unclear why the goslings' feathers appeared soaked and muddy this day and not at other times.

Down provides insulation to keep a bird warm and regulate its body temperature year round. Hatchlings grow down before they grow feathers. Birds that live in climates that have three or four distinct seasons molt and grow new down that is appropriate for each season. Through a combination of genetic and biological intelligence birds ascertain impending seasonal changes, and adjust their down accordingly, often rather quickly. Summer down is light and airy so birds do not over heat. Winter down is dense, with extra barbules and barbicels, to keep birds warm and dry during freezing temperatures. Birds have the ability to vary the angle of their down in relation to their body to increase or decrease the insulation value. For example, a bird that is all fluffed up is maximizing heat retention. Even pet birds kept indoors molt and grow new down in response to sudden changes in temperature. In essence, down is the biological equivalent of dressing in layers.

A Canada goose's neck is about 10 inches (25.4 cm) in length. It is an example of a specialized multi-purpose biological tool. The neck is extremely flexible and can be bent and coiled in all sorts of ways: backward, forward, and to the side. This capability comes in handy when preening, searching for food on land and water, or standing guard on lookout for predators. It also gives Canada geese a 360° view of their world. On the other hand, this flexibility must be countered with concentration to keep their necks perfectly straight during long flights. A long thin neck is hard to keep warm during winter. To compensate Canada geese grow extra thick neck feathers. At times it looks like they are wearing a chenille turtleneck sweater. These feathers begin thinning in March. Canada geese lay half of their neck flat along their back, with the other half rolled forward, when they are most content. They assume this posture when resting on land or water. It seems to reflect a half awake, half asleep state during the day. (See Photograph 3.5.)

Canada geese have 3 front toes and one smaller back toe for stability that is a bit higher up on the back of the leg. The front toes range from 3 to 4 inches (7.62 to 10.16 cm) in length and the back toe or hallux is 1 to 1.5 inches (2.54 to 3.81 cm) long. Each toe has a sharp nail at the end. When the front toes are spread the width ranges from 3 to 4 inches (7.62 to 10.16 cm) at the widest point. The toes are flexible and can be bent or curled as needed. The back toe can be moved to face straight back or toward the center of the body. The legs of a Canada goose are about 6 inches (15.24 cm) long. The legs and feet are covered with scales. The scales on the legs and toes are about 0.25 inch (0.635 cm) in size. The scales on the webbing between the toes are much smaller. The feet of Canada geese touch sometimes when they are standing.

The thin webbing between the three front toes resembles a shiny flexible black rubber cloth. In direct sunlight the webbing looks like a fine black mesh. The webbing between the three front toes of a Canada goose represents biological specialization and serves multiple purposes. Canada geese spread their toes on the backstroke for maximum thrust when swimming. On the forward stroke Canada geese clench their toes together.[83] As a result they do not have to take their "oar" out of the water. Canada geese employ the same open and closed toe movements when walking on land as they do when swimming. (See Photograph 3.6.)

Webbed feet make it easier for Canada geese to navigate, pivot, and change directions in water. Canada geese look very peaceful when they are swimming, almost as if they are floating effortlessly. In reality their webbed feet are continually peddling and rotating at various angles like a precision rudder. Because of this, Canada geese can swim forward, turn to the left or turn to the right 45°, 90°, 180°, or full circle if they so desire. When swimming forward, Canada geese swing both feet backward at the same time. If they wish to drift, Canada geese keep their feet extended backward for a while. To turn left or right, they pivot their feet toward the desired direction.

Photograph 3.5 The Ultimate Flexible Neck. The neck of a Canada goose is very flexible and can be bent and coiled in all sorts of ways. This flexibility gives Canada geese a 360° view of their world. It also makes scanning for dangers, preening, and retrieving food on land or underwater a snap. Canada geese lay half their neck flat along their back, with the other half rolled forward, when they are most content. © D. Herrmann

On 6/7/2011 an adult Canada goose was observed limping at Pond C. This goose walked slowly and cautiously. The goose could not bend its right leg up toward its body, like a goose normally does when walking. The goose attempted to keep its weight on its left leg and mainly used the right leg for balance or as a crutch. This injury undoubtedly makes swimming more difficult too. During winter when Pond C was frozen one goose slipped when climbing up the icy bank. The goose fell on its right leg. Once the goose was upright again, it limped. It is possible that it was the same goose in both instances.

Webbed feet make water takeoffs and water landings possible and assist Canada geese during long migratory flights. Canada geese can land and takeoff from both land and water, and they can do so rather quickly if the situation demands it. A water takeoff consists of stretching the neck and body up tall while leaning forward, flapping fully out stretched wings, and what appears to be walking on top of the water. Actually, the legs have been extended and the webbed feet are pressing down hard and fast on the water until a Canada goose can flap its wings for lift. By the third "step" the geese are airborne. A similar process of running along the ground is used to takeoff from land. Landing, on water or land, is also an act of precision as the geese gradually slow the wing flapping, tilt to the left, tilt to the right, tilt upright, and gracefully descend to the exact intended location. In this instance Canada geese use their webbed feet like water skis to slow down when landing on water.

Photograph 3.6 These Feet Are Made for Walking, Swimming, and Flying. Scales cover the legs and feet of Canada geese. The scales on the legs and toes are larger than those on the webbing between the toes. The toes and webbing are flexible and can be bent, curled, contracted, or spread as needed when walking, swimming, or flying. The webbing makes swimming and flying more efficient and prevents geese from sinking in the mud at the edge of a pond. © D. Herrmann

Canada geese are strong quiet flyers. Unlike mourning doves (*Zenaidura macroura*) and other birds their flight makes little or no sound. Even when a goose or group of geese fly within 10 feet (3 m) of a person, they are unlikely to be heard. Instead, the airborne chatter of Canada geese makes their presence known. During migration Canada geese generally fly at 40 mph (53 kmph) and cruise at an altitude of 3,300 feet (1,006 m), occasionally going as high as 12,000 feet (3,658 m).[41] Canada geese stretch their neck, back, and feet out straight in a horizontal position when flying. Webbed feet assist a Canada goose while airborne by acting as an air foil and contributing to horizontal stability during flight.

Webbed feet provide another benefit which is not often acknowledged. Canada geese spend a lot of time walking along muddy shorelines. Webbed feet distribute the weight of a Canada goose and prevent it from sinking, using the same principle as snowshoes. If Canada geese did not have webbed feet they would sink in the mud, at least a few inches (centimeters) because of their weight.

This feature also comes in handy when Canada geese are walking on sandy beaches or in snow. In fact, the foot prints of Canada geese in snow and mud only show the three front toes, not the webbing. (See Photograph 3.7.)

Photograph 3.7 Footprints. Webbed feet distribute the weight of a Canada goose and prevent it from sinking in mud, sand, and snow, as shown in these footprints. This is the same principle that inspired the development of snowshoes. © D. Herrmann

Vocalizations

A Canada goose's vocalizations consist of a range of melodic honks of different pitch, volume, and duration that convey various messages. Canada geese are known for their melodic honks. These honks are resonant, which allows the sound to travel a distance without being overly loud. This is an important feature for intra-flock communication when the flock is airborne during migration or spread out in and around a pond, since water dampens sound. Members of a flock recognize each other's honks. Likewise immediate and multi-generational family members recognize each other's unique honks. Even young goslings can identify their parents' vocalizations. It is unclear when young geese first learn to honk. Only adults were heard honking, not any goslings or juveniles.

The sound traditionally associated with Canada geese is the medium-pitched drawn out "e-hink" that is repeated over and over when individuals and pairs are flying overhead. Pairs often make this sound as a duet.[160] This sound functions as a navigation call and a roll call, especially when the flock is flying long distances.

Six other Canada geese vocalizations were heard, as shown in Table 3.2. These Canada geese vocalizations can be heard on Disk 1 Tracks 2 and 3.

On several occasions a group of geese were swimming peacefully in a pond or the golf course water trap. All of a sudden there was some quiet bantering among some but not all of the adults present. This continued for about 90 seconds. Then all of a sudden the entire flock became airborne. Sometimes the flock only flew a short distance, like to the lawn. Other times the flock continued their southern migratory journey. When multiple flocks were swimming together everyone was able to recognize their flock's unique round-up call. This call was primarily heard during the fall and winter when migratory flocks were passing through the area.

Table 3.2 Canada Geese Vocalizations.

Sound	Who Makes It	Attributes and Usage
I. Traditional Vocalizations		
"e-hink"	Adults	• Medium pitched drawn out sound that is repeated over and over • Pairs and individuals make this sound when flying overhead • Functions like a combined navigation and roll call • Pairs often make this sound as a duet
II. New Canada Geese Vocalizations Heard During This Study		
bantering	Adults	• A quiet quick low sound that is repeated for about 90 seconds • Exchanged among multiple adults immediately before the flock becomes airborne • Occurs whether flock is flying short or long distance • Heard most often when flock was in the water together, before they took off flying
"e-yonk"	Adults	• Medium pitched drawn out sound that is repeated over and over • Pairs and individuals make this sound just before landing, like a navigation call • Frequently heard when several small flocks are coordinating turns landing at the same location, like Pond C
"honk"-lite	Adults	• Short high pitched honk that is repeated several times • Made by several adults when greeting a nearby flock that is flying overhead or has just landed
"ruh"	Adults	• A medium pitched nervous sound that is repeated several times • Used in situations when one adult is uneasy about something or the pair has become separated accidently • Call is communicated a short distance to the other member of the bonded pair who is nearby on the ground
"honk"	Mother	• Soft, quiet medium pitched honk • Mother goose makes this sound once when a gosling has strayed too far away from the family on land or in water, or when a person gets too close • Call summons the goslings to move closer to the parents
"peep"	Goslings	• A soft high pitched sound that is repeated twice • Individual goslings make this sound when they become separated from their parents and family group • Mallard ducklings make an almost identical sound under similar circumstances

Note: (1) Canada geese vocalizations can be heard on Disk 1 Track 2.
(2) Canada goslings vocalizations can be heard on Disk 1 Track 3.

Just as bantering announces an imminent take-off, another vocalization announces an imminent landing. That is the medium-pitched drawn out "e-yonk" sound. Individuals and pairs of Canada geese repeat this sound just before they land, like a navigation call. In particular this sound is heard when several small flocks or sub-flocks are coordinating turns landing at the same location.

This behavior was observed frequently at Pond C during the fall/winter migration. A large group of Canada geese would arrive at Pond C and took turns landing. They all flew in a circular pattern around the pond in the same direction and one small group landed each time the group circled. Within a few minutes all the geese had landed. The "e-yonk" vocalization was used to announce which small group was landing. This process worked with the precision of a busy major airport on 1/22/2011 when over 100 geese landed at Pond C.

Canada geese are social birds and interact with members of their own flock as well as members of other flocks. A special "honk"-lite vocalization is used to greet members of another flock. This vocalization is a short high-pitched honk that is repeated quickly many times. Multiple geese make this sound at the same time asynchronously to greet a new flock that is flying overhead or one that has just landed nearby. This message must mean something like: "this is a nice place, you are welcome to join us." On several occasions a flock was observed flying overhead. Once the flock heard the loud "honk"-lite greeting below, they circled back, landed, and joined the flock that was already on the ground. On one occasion the flock on the ground had been there for 2 days before the second flock appeared.

Canada geese are normally calm, cool, and collected. Only on rare occasions do they become nervous or frazzled. One such case was observed on 5/13/2011 at Pond D. A pair of Canada geese was returning to Pond D with their six 1-week old goslings. The father and 4 of the goslings found their way inside the mesh fence that surrounds the pond. The mother and 2 of the goslings became stuck outside the fence. The mother frantically walked back and forth along an 8 foot (2.43 m) section of the fence trying to find an opening. The father was watching her. After a few minutes the mother began making a nervous medium-pitched "ruh" sound repeatedly. The sound was only loud enough to be heard by the father and other goslings. Eventually the mother found a hole where the mud had sunk. She and the two goslings ducked under the fence and joined the rest of the family. This vocalization was heard on again on 5/26/2011 at Pond B under similar circumstances. This time a father goose was temporarily stuck outside the fence, while his mate and their 2 goslings were inside the fence watching and waiting for him.

Young goslings identify with their family unit from an early age. Families with young goslings can mix and everyone knows who is who. On 5/7/2011 a day old gosling temporarily followed the wrong group of goslings at Pond D. In less than a minute the gosling realized its mistake and rejoined its family. Young goslings can run and swim very fast. Occasionally one strays too far from the family group. When that happens, the mother makes one quiet soft medium-pitched honk. The young gosling recognizes its parent's call, turns around, and runs or swims back to the family. The same call is made to summon all the goslings to move closer to the parents in response to a potential danger.

Once in a while goslings become separated from their parents. Or, a gosling looks around and can't spot their parents in a multi-family setting. When this happens, a gosling makes a soft high-pitched "peep" sound that is repeated twice. The sound can only be heard over a distance of 6 to 8 feet (1.82 to 2.43 m). Mallard ducklings make an almost identical sound under similar circumstances.

Just like a gosling can recognize its parent's call in a crowd from an early age, a parent can recognize its gosling's call. On 5/6/2011 5 goslings at Pond D had made it through the fence and into the pond. At the time the goslings were still small enough to fit through the 3 x 6 inch (7.62 x 15.24 cm) mesh grid. The two parents were briefly detained outside the fence. Since the parents were larger and couldn't fly, they had to take a longer route to enter the pond. The goslings waited and watched their parents. Two of the goslings peeped to indicate their separation anxiety.

On another occasion a family with 5 goslings was strolling along the bank of Pond B. One gosling was particularly energetic. All of a sudden it stopped and looked around, but all it could see were mallard ducks. So it "peep"-ed. The parents were only 4 feet (1.22 m) away, but they came over to reassure their gosling. In both cases the goslings were less than 2 weeks old.

On 6/21/2011 the family with 2 goslings was visiting Pond B. The family became separated temporarily on opposite sides of the fence. One adult and 1 gosling were outside the fence. One adult and 1 gosling were stuck inside the fence. Both sets of geese ran along the fence searching for an opening. The adults called back and forth to each other making the nervous "ruh" sound. Both goslings peeped to express their separation anxiety. The goslings were 5 weeks old at the time.

Undoubtedly Canada geese also have alarm calls and distress calls. Fortunately there were no occasions when the Canada geese at the four ponds or golf course needed to make alarm calls or distress calls.

The quantity and volume of Canada geese vocalizations increases in the fall and winter. This occurs in part because there are more geese in the area since Ponds B and C and the golf course are major stopover points for migratory flocks. This behavior also reflects a sense of urgency about changing weather conditions that is not heard during spring and summer.

It was noted that all the flocks were much more restless and vocal on cold windy days and just prior to a major storm. For example on 1/21/2011 over 100 Canada geese were spread out on and around Pond C which was mostly frozen. Some geese were resting, others were walking around. Vocalizations originated from random individuals and small groups of geese that were scattered about. These vocalizations did not appear to be directed toward an individual or sub-flock. Instead they seemed to reflect commentary on the weather and a strong determination to survive. In particular one young quartet of 4 geese put on quite a show. By moving their heads and necks in all sorts of different positions, up and down, side to side, coiled and not coiled, the geese were able to make a whole spectrum of sounds, from low slow raspy honks to fast high-pitched honks. In essence the quartet used their necks like wind instruments to create a variety of unique music that is not heard the rest of the year.

Aggression, Territorial and Defensive Postures

For the most part Canada geese are peaceful and tranquil birds. Canada geese share the ponds and surrounding areas they inhabit with a variety of other wildlife. However, Canada geese will fight to the death to protect their families.[45] During the night one weekend in May 2007 some drunken vandals broke tree branches, uprooted bushes, and smashed signs along the street leading to Pond A. The next morning I was concerned about what I would find when I reached the pond. The goslings were only 3 to 4-weeks old at the time. There were several large goose feathers on the slope near the street, upturned sod, and evidence of a struggle. I knelt down to examine the feathers and ground. As I did the father goose, mother goose, and the four goslings swam out from behind a bush in the center of the pond. Everyone in the goose family was fine. Apparently the vandals had met their match.

Neither the local flock nor the migratory flocks of Canada geese displayed any territorial behaviors toward each other or any other birds in the neighborhood. None of the Canada geese engaged in any activity that indicated that they considered a particular stretch of lawn or a bank of a pond their personal property. On the contrary, the Canada geese shared the territory in and around the four ponds freely.

Two observations are particularly telling about the non-aggressive nature of Canada geese. Both occurred during the harshest part of winter when the geese were in survival mode. On multiple occasions between mid-December and mid-February people stopped by Pond C to offer visiting flocks bread cubes. When a bread cube landed by one goose, other geese made no effort to retrieve it. They simply let that goose eat the bread cube. When a bread cube landed equidistance between two geese, they both paused for a while, then one of the geese finally ate it. Once when it was particularly cold and windy a couple seagulls (*Larus argentatus*) were in the area. The Canada geese did not fuss at the seagulls when they snatched a few bread cubes either. This polite behavior contrasts with that of mallard ducks who are very aggressive about grabbing bread cubes during winter, even from their own mates.

There was another interesting observation related to bread cubes. When songbirds are fed bread cubes they peck away at the cubes, eating a small bite at a time. Two or more songbirds often eat from the same cube. A single Canada goose, on the other hand, picked up an entire bread cube and swallowed it. If the bread cube was larger than the width of its bill, the goose ran into a problem. The outer rim of a Canada goose's bill is not sharp enough to cut through bread; nor does a goose have the ability to peck a bread cube apart. The geese quickly figured out a solution: dunking the bread in water or snow first.

Canada geese use three postures to communicate their displeasure. When slightly annoyed Canada geese open their bill, stick their tongue out, and make a brief quiet hissing sound. Sometimes Canada geese hiss standing still, other times they walk around. Usually Canada geese only hiss 2 or 3 times. (See Photograph 3.8.)

Photograph 3.8 Take a Hint! (R) The father goose at Pond D sticks its tongue out and hisses at a pesky photographer on 4/4/2012, four days after his mate began nesting. This was a warning not to get too close to the nest. After a couple hisses the father goose had made his point. From then on he just watched the photographer. Only male geese were observed hissing; they only did so during nesting time and when they were raising their young. (L) Goslings learn by observing and imitating their parents. A 4-day old gosling imitates its father by sticking its tongue out and hissing, although the hissing did not seem to be directed toward anyone in particular. © D. Herrmann

If the intended recipient is oblivious to this message or the infraction is more serious, a Canada goose will charge the offending party. To do so a Canada goose lowers its head and neck parallel to the ground and runs full speed ahead with its bill open. Upon seeing this behavior the offending party usually backs off and there isn't any bodily contact.

If by chance the offending party is still being obtuse, the goose will latch onto some feathers on the lower back or tail and not let go. The goose doesn't pull any feathers out. But, by holding onto the offender's feathers the goose renders it immobile. This act never fails to get the message across.

These three behaviors were only observed when the Canada geese were nesting or raising their young. When I was checking the nesting progress of the pair at Pond D, the male hissed at me during the first week or two. After that he simply monitored my whereabouts.

During Spring 2011 two pairs returned to nest at Pond C, rather than the usual one. One male constantly charged the other pair if he felt they got too close to his mate. He started this behavior 3/15/2011. Before that date and after nesting time was over, the two pairs got along fine. But during March and April the one male charged away. He continued to chase the other pair until they were in the water. Sometimes he even chased them after they were in the water.

After these episodes the other couple always swam in unison close together with their necks parallel to the water. After particularly nasty encounters one of the geese briefly laid its neck across the back of its mate. Perhaps this behavior was an act of affection and consolation. (See Photograph 3.9.)

A few mild charging episodes were observed at Pond B. In these instances an adult goose mildly charged a mallard duck (*Anas platyrhynchos*) when it felt the duck came too close to its goslings. The mallard duck got the message and moved out of the way immediately.

Like other birds goslings learn by imitating their parents. One afternoon a gosling watched one if its parents charge a mallard duck. The gosling turned around and charged another mallard duck. At the time the gosling was less than 2 weeks old and not quite 1/3 the size of the mallard duck. The duck seemed more surprised than startled, but it politely indulged the gosling's youthful exuberance.

Photograph 3.9 It's OK. This bonded pair of Canada geese swims close together with their necks low in an act of affection and mutual consolation. For a brief period of time one goose laid its neck across the other goose's neck. This posture usually occurs during pair bonding. However, this episode took place after the pair was chased into Pond C by an aggressive male Canada goose in early spring. © D. Herrmann

3.3 Family Life and Daily Activities

Canada geese family life and daily activities are not covered in much detail elsewhere. In contrast this chapter provides detailed information about:

- Pair bonding
- Spring migration
- Parenting skills
- Gosling development
- Daily routine
- Visiting golf courses

Pair Bonding

Canada geese pair bonding takes place at the wintering grounds.[41] Canada geese first take a mate in their second or third year, and they mate for life.[41, 45, 95, 157, 203] That is quite a commitment given that a Canada goose can live up to 23 years in the wild.[18, 95, 203] The bond between a pair is very strong. If a Canada goose becomes ill or injured during the fall or spring migration, its mate will stay behind with it even though that involves difficult living conditions.[41] If a Canada goose loses a mate, it may or may not find a new mate.[41, 95, 157, 203]

This behavior was observed at Pond C during the fall migration in 2008. Between late October and early November several migratory flocks of Canada geese landed at Pond C to rest for a while before continuing their southern journey. A pair from one of the early flocks stayed behind. One goose rested at the edge of the pond or grazed in a small area. The other more active goose also swam, but stayed near the goose on land. This pair stayed behind for a month, while other flocks of Canada geese came and went, before it finally departed with a migratory flock of mallard ducks.

While the bond between a Canada goose pair is strong, they do not engage in allo-preening or allo-feeding. The long flexible neck of a Canada goose makes allo-preening unnecessary and allo-feeding difficult, if not impossible. In general, birds who raise precocial chicks do not engage in allo-feeding. Canada geese pairs stay close together when walking, flying, swimming, or napping. A Canada goose pair will often imitate each other's behavior by preening, upending, or stretching at the same time. Canada geese pairs also perform a greeting ceremony upon being reunited after a period of separation, if only for a short while. At this time the two geese face each other, lower and raise their heads and necks like they are bowing, and honk softly. (See Photograph 3.10.)

Although other birds mate for life, Canada geese have one interesting trait that has not been reported often in connection with other species—they repeat their courtship behavior each winter,[41] as if they are renewing their vows. That is a rather sophisticated concept. Not many humans do that; they simply observe anniversaries with greater or lesser intensity year after year—assuming they are still married.

Photograph 3.10 Togetherness. A bonded pair of Canada geese shares an interesting find among the silt at the edge of Pond C. Bonded pairs stay close together and go through their daily activities side-by-side. If they become separated, Canada geese pairs engage in a greeting ceremony upon being reunited. © D. Herrmann

Spring Migration

By the time spring migration arrives, Canada geese have already established pair bonds. The literature consistently indicates that Canada geese return to the nesting site where they hatched. That premise seems plausible on the surface, but not when examined in detail. Whose nesting site does a pair return to: the (new) mother's nesting site or the (new) father's nesting site? If the pair hatched at different nesting sites, they cannot return to both nesting sites. As will be seen later, a different pattern of behavior was observed in relation to returning to and selecting nesting sites.

In the spring the first sighting of Canada geese usually occurs during the first week of April. In 2010 the first sighting of Canada geese at Pond A occurred March 9th—three weeks earlier than usual. This sighting involved two adults; they were gone the next day. Pond A was the last pond to thaw after the winter freeze. On March 9th 2 feet (61 cm) of snow still covered the side of the pond that faced north. On March 20th and 21st two Canada geese each were sighted at Ponds A, C, and D—10 days earlier than usual. Throughout April and the first week in May the geese continued to rotate among the four waterways and the golf course. One day the geese were in one location; the next day they were somewhere else. Some, but not all, of the movement from pond to pond can be attributed to non-breeding adult geese, especially those that were too young to have a mate. The rest of the movement can be assigned to nesting area selection. A similar spring migration pattern was noted in 2011 and 2012.

Twice, in 2007 and in 2009, the first sighting of Canada geese returning from spring migration took place when a pair landed on top of a building near Pond A. Perhaps the view from there approximated what the geese would see if they were flying around. Then again, the building the geese landed on was only 5 years old. In the past the geese would have landed on the ground the building now occupied and walked from there to Pond A.

One Canada goose returning from spring migration in 2010 was not so fortunate. This goose lived about 4 miles (5.32 km) north of the four ponds. Originally it and its mate lived in a small pond. The pond was reduced to a quarter of its size to make room for a parking lot. At that point the pond was little more than a drainage ditch, but the pair of Canada geese continued to live there. In the spring it was not unusual to see one of the geese standing on one foot on the roof of a car in the new parking lot. For some reason the geese always seemed to be facing north. A few years later the drainage ditch was filled in. Only one goose returned from spring migration that year. It roamed around the parking lot for a while then left. A week or so later the goose was spotted standing on one foot in a 2 x 3 feet (61 x 91 cm) puddle, again facing north, 0.5 mile (0.66 km) south of where its pond had been. A group of trees had been cleared during road construction, creating the puddle. The goose claimed the puddle as its own and stayed there all summer.

Road construction continued in the fall and winter and the puddle was covered by a new exit ramp. That spring the Canada goose did not have a home to return to. Its pond had been destroyed to make a parking lot, leaving the goose with a small drainage ditch. When the goose's drainage ditch was filled in, it moved to a small puddle. Now the puddle was gone. All of these changes took place within a 5-year period. A few weeks after the goose would have returned from the 2010 spring migration, the immediate area was searched to learn where it had moved. The goose, a 36 inch (0.91 m) tall bird, was found run over on the new road near where its puddle had been. Another victim of habitat destruction. Not only was this goose lost, but so were all of its descendants. The destruction of urban and suburban habitat is as deadly as the destruction of wild habitat. Urban and suburban wildlife literally have no place to go once their habitat is destroyed.

Parenting Skills

Regardless of the local migration among the four ponds, the Canada geese eventually settle down and raise families. Canada geese only raise one brood a year. However, not every Canada goose pair in the local area raises a family every year. Out of a flock of 48-some geese, only 4 pairs (or less than 20%) nest each year on average. Perhaps a pair only nests every 2 or 3 years. Perhaps a pair only raises 3 or 4 broods during their life time. Perhaps a pair does not nest in their senior years. Through this cooperative arrangement, whatever it is, the flock keeps their population balanced with the natural resources that are available to support it. This arrangement ensures that the entire flock is not grounded for parenting duties at the same time, and hence vulnerable. A similar nesting trend was observed for mallard ducks, as discussed in the next chapter. This practice contrasts with humans who tend to over populate an area until war, famine, disease, or natural disasters correct the situation.

It is not clear that the various government bodies that are entrusted with the responsibility of developing and implementing waterfowl management plans are aware that 25% or less of the adult Canada geese in a flock actually raise a family each year. Typical waterfowl management plans assume that if the population of a species is X, it is acceptable to shoot Y% of the birds per year. This approach does not take into account the fact that 25% or less of the adult ducks or geese in a flock actually produce a brood each year. This may explain why the populations of various species fluctuate widely from year to year. For more discussion of this issue see Table 3.3.

Table 3.3 An Invitation to Get Involved: North American Waterfowl Management Plan.

I. The Issue

The North American Waterfowl Management Plan (NAWMP) is on a regular update cycle. The most recent plan was issued in 2012. The process of assembling stakeholder committees, panels of experts, and so forth for the next revision began in earnest during 2013, and the first of seven revised objectives by species was issued late 2014 as an Addendum to the NAWMP. A progress assessment is ongoing about the Plan's effectiveness and feedback is being sought from stakeholders.
Historically hunters, not bird enthusiasts, have dominated the committees, as is evident from the following quotes:[169]

- "Hunters have traditionally supported the Plan's mission and will be the primary stakeholders in Plan activities." (p.4.)
- "Plan partners should find means to encourage participation in waterfowl hunting and secure hunter investment in future Plan activities." (p.4.)

The main focus of the Plan is to set species-level population objectives to "ensure population viability and provide for regulated harvest." (p.3.) To be blunt, the purpose of the Plan is to make sure that there are enough waterfowl alive for hunters to shoot. However, species prioritization has nothing to do with conservation, but rather the "relative values stakeholders [i.e., hunters] place on a species".[168, p.53] Translation: between 1980 and 1999 mallard ducks represented 35% of the total ducks shot; hence, their species prioritization is high.[168, p.57] In comparison, wood ducks only represented 9.1% of the total ducks shot, so their species prioritization is moderate.[169]

There are approximately 2 million registered hunters in North America; most of them from the U.S. Also, U.S. hunters travel to Canada and Mexico to hunt, which provides significant tourism dollars.

II. Who Should Get Involved

It is time for the several million bird enthusiasts worldwide, professional and amateur, who believe that birds have more intrinsic value than simply something to be shot, to get involved in the development, comment/review, and approval of the NAWMP. At present only one stakeholder perspective is being heard.

In fact the Plan solicits joint ventures with other conservation-minded individuals and organizations. For example, a Science Support Team is being assembled to address known deficiencies:[169]

- "... a lack of basic knowledge of population dynamics for some waterfowl species hinders development of conservation strategies." (p.3.)
- "The amount of baseline life history information available varies considerably for individual waterfowl species and by geographic region." (p.13.)

III. Contact Points

The NAWMP is issued jointly by Canada, the United States, and Mexico. The contact information for each country is below.

<u>1. Canada</u>

NAWMP Implementation Office
Wildlife Conservation Branch
Canadian Wildlife Service
Environment Canada
Place Vincent Massey, 3rd floor
351 St. Joseph Blvd.
Gatineau, Quebec K1A 0H3 Canada
(t) 1.819.997.1841
(f) 1.819.994.4445
(e) nawmp@ec.gc.ca
(w) www.nawmp.ca

<u>2. United States</u>

Division of Bird Habitat Conservation
U.S. Fish & Wildlife Service
4401 N. Fairfax Drive, Stop 4075
Arlington, VA 22203 USA
(t) 1.703.358.1784
(f) 1.703.358.2282
(e) dbhc@fws.gov
(w) www.birdhabitat.fws.gov

Table 3.3 contd....

Table 3.3 contd.

III. Contact Points
3. Mexico Secretaria de Medio Ambiente y Recursos Naturales Direcion General de Vida Silvestre Avenida Revolucion 1425, Nivel 1 Colonia Tlacopac, San Angel Mexico D.F. C.P. 01040 (t) 52.55.56243309 (f) 52.55.56243642 (e) felipe.ramirez@semarnat.gob.mx (w) www.semarnat.gob.mx

When people think of bird nests and spring they usually think of a nest in a tree, birds hatching and staying in the nest until they can fly, and the parents flying food back and forth to their young. Canada geese families go through an entirely different experience. To start with Canada geese nests are normally on the ground near a waterway.[83]

The mother selects the nest site and builds the nest. Canada geese nests are 16 to 20 inches (40 to 50 cm) in diameter.[160] They consist of a shallow depression that is lined with down and dried vegetation from the immediate area.[83, 160] It takes about 2 days to build a nest. In 2011 the mother goose at Pond D began building her nest on the morning of 4/7. She began sitting on the nest the morning of 4/9.

Canada geese nests are concealed from view. One year it was not clear whether or not the Canada geese had nested at Pond A. Then one day the mother raised her long neck like a periscope and scanned the area for danger. Once she lowered her head and neck, the nest site was again invisible.

In 2011 the mother goose at Pond D picked a dry somewhat rocky location next to a large black drainage hose. Her body feathers blended in with the dry soil and rocks perfectly, while her black head and neck blended in with the black hose. Unless a person knew the mother goose was there or had binoculars, she would not have been seen. Although both the mother goose and the nest were exposed, they were not noticeable.

While I know where the geese nest, I have never ventured close enough to count the number of eggs. The intent was to observe natural behaviors, not disrupt them. Various sources state that the number of Canada geese eggs laid ranges from 3 to 8, 4 to 6, 4 to 7, or 4 to 8 per clutch, with an incubation period of 24 to 30 or 25 to 28 days.[18, 41, 95, 141, 157, 203] The white eggs are 2.5 by 3.5 inches (6.3 by 8.9 cm).[141]

Mother geese were never observed leaving the nest during the incubation period. In 2011 the mother goose at Pond D reinforced the nest after heavy rainstorms by stretching her neck and pulling in new vegetation. She consistently faced the direction the most noise, movement, and activity were coming from. Sometimes that was toward the pond, other days it was not. When everything was quiet the mother goose tucked her head under her wing and napped.

Soon-to-be father geese engage in some interesting behavior during the incubation period. This behavior was particularly noticeable at Ponds C and D during 2011. In both cases the father geese stayed at the pond where their mate and nest were located. However, the father geese stayed on the opposite side of the pond. The father geese monitored humans, dogs, squirrels, turtles, and other birds who wandered into the vicinity, as they went about their grazing and swimming routine. At times it seemed like the father goose was going out of his way to make himself conspicuous, to draw attention away from the mother goose and nest. Toward the end of the incubation period the father geese appeared a bit anxious or nervous; this was probably related to the vulnerability of the soon-to-be hatchlings during their first 24 hours, prior to leaving the nest.

On the afternoon of 5/5/2011 the father goose splashed all around during a very energetic and prolonged bathing/preening session at the southern end of Pond D. The very next day the mother goose at Pond D was spotted standing over her nest at the northern end of Pond D. Two very tiny newly hatched goslings were barely visible running around in the nest. The father's energetic bathing was: (1) intended to distract predators from the nest at this critical time, and (2) an anticipatory behavior. By 3 p.m. on 5/7/2011, 29 days after the mother began sitting on the nest, six tiny young goslings were walking, running, and swimming with their parents. (See Photograph 3.11.)

The pair of Canada geese at Pond C was not so lucky. The mother goose began sitting on the nest on 4/18/2011, but abandoned the nest after 5/8/2011. The reason is unclear. 2011 was the first year that a Canada geese pair nested at Pond C. Although Pond C is the largest pond and a very popular stopover point during fall migration, the local flock traditionally avoids it as a nesting site. A large snapping turtle (*Chelydra serpentina*) with a 14 inch (35.5 cm) long shell and an 8 inch (20.3 cm) long tail lives in the pond; it might be part of the reason.

Once the hatchlings have left the nest both parents begin to molt their primary flight feathers. This biannual molt is timed to a life cycle event. Both parents are grounded and flightless while their goslings develop and mature. The parents rarely fly again until the goslings themselves are able to fly. Instead, Canada geese parents protect and defend their new family on land and water, never leaving them alone during this 8 to 10-week period.

Photograph 3.11 Nest Time. Canada geese mothers build their nests on the ground from vegetation and their down. It takes about two days to build a nest. Canada geese mothers rarely leave the nest once they start incubating their eggs. Canada geese mothers change the direction they face, to monitor what is going on around them, but that is about all. Once the goslings start to hatch a Canada goose mother stands over the nest. Here the mother stands over two goslings that have just hatched. This photograph was taken on 5/6/2011 23 hours before this new family of six goslings left the nest for their first swim in Pond D. The goslings can be seen swimming in Photographs 3.16 (top left) and 3.17 (bottom right). © D. Herrmann

The timing of a Canada goose's summer molt is determined by whether or not it is raising a family. Adults that are raising a family begin re-growing their new wing and tail feathers the same time their goslings do. That way the family as a whole is ready for flight at the same time. Adults that are not raising a family generally begin their molting and regrowth cycle a month later than those that are.

First the old feathers on the wings and tail are molted. Sometimes 3 or 4 feathers fall out a day and the geese look pretty ragged. Next the geese begin growing new feathers. These appear as narrow light blue cylindrical keratin sheaths that are about 0.25 inch (0.63 cm) wide and 3 inches (7.62 cm) long. At first just the blue sheaths are visible. Then the tip of a feather emerges slowly. Once the complete feather is out the keratin sheath dissolves. The process is similar for most birds and is the source of pet bird dander. (See Photograph 3.12.)

Photograph 3.12 Molting and Renewal. During their summer molt Canada geese look pretty ragged, as wing and tail feathers drop out. A few days later small light blue cylindrical keratin sheaths appear where the feathers used to be. Slowly a new healthy feather emerges from the end of each sheath. Goslings and their parents grow new feathers at the same time, so the whole family will be ready for flight simultaneously. Adults that are not raising a family molt about a month later. © D. Herrmann

Canada geese parents are aware of their mobility limitations, but they find creative ways to improvise when the need arises. One such case is particularly enlightening. On 5/26/2011 a family with 5 young goslings was grazing on the northwest side of Pond D. When they were finished the goslings ducked under the fence and went for a swim. The parents could not fit under the fence. Rather than walk all the way around the fence, the parents walked up on top of a dirt mound. One at a time the parents jumped off the dirt mound, flapped their wings, and coasted over the fence, before plopping into the pond near their goslings. The Canada geese parents understood that they could not fly. Yet, the Canada geese parents knew, between the height of the dirt mound and the flapping of their wings, they could achieve enough lift for a sufficient amount of time to coast over the fence.

Canada geese engage in flying attacks, like pet birds. Instead of holding onto a perch or the side of a cage Canada geese stand up in water and flap their wings vigorously. Canada geese engage in this energetic flapping routine once or twice a week while they are raising their young. This behavior is much more vigorous and of longer durations than simple bathing. This behavior keeps their wing muscles strong while they are grounded for parenting duties. This behavior implies that Canada geese understand the need to keep their flight muscles in shape.

Canada geese have superb parenting skills. Parenting duties are shared by both parents.[45] Rather than teaching their young about the tooth fairy and Santa Claus, Canada geese parents teach their young skills that are necessary for life. Canada geese parents do not show favoritism toward the youngest or oldest gosling; all goslings are treated the same. Canada geese parents are patient and allow goslings to learn new skills when the goslings are ready and comfortable doing so.

Gosling Development

Goslings do not have 20 years to grow up and get an education before they start out on their own. Goslings do not even have 1 year. The accelerated rate with which goslings acquire the knowledge and skills they need to be self-sufficient is truly amazing, as is their rapid growth spurt from being a hatchling to becoming a young adult. Watching a tiny gosling grow into a young adult goose over an 8-week period is like watching a bud appear and the flower unfold. It is beautiful and miraculous at the same time.

Canada goslings learn by observing and imitating their parents. Goslings may not get each behavior exactly right the first time, but after sufficient practice they internalize the correct movements and action. Humans on average spend 20 years learning to become responsible for themselves, or 25% of their average life span. In contrast, Canada geese spend 3% of their maximum life span learning to become responsible for themselves, and they face more dangerous and persistent threats to their survival.

Like many birds, goslings peck their way out of their shell by perforating it around the middle until the shell splits in half. Between 2005 and 2010 the largest number of goslings to hatch in one family was five; that occurred in 2009 at Pond B. For the first time a family of 6 goslings hatched in 2011 at Pond D. In addition two families of 5 goslings hatched in 2011, one at Pond D and one at Pond B, and one family of 3 goslings hatched at the golf course water trap. For the first time 7 goslings hatched in one family on 4/29/2012 at Pond D, 29 days after the mother began nesting. Earlier that week 1 gosling hatched near Pond D on 4/25/2012, 3 goslings hatched 5/2/2012 at the golf course water trap, and 5 goslings hatched 5/13/2012 at Pond B. Newly hatched goslings are usually spotted during the first week of May.

Canada goslings leave the nest when they are only a day or two old.[41, 157] As soon as all the goslings have hatched, the parents lead them to water for their first swim. From then on goslings walk, swim, and feed themselves, unlike altricial songbird hatchlings.[41, 83, 157] Newly hatched goslings are tiny; they are only about 3 inches (7.6 cm) tall—the size of their egg shell—and are covered in yellowish brown down. At first goslings are the same height as the grass they eat. Newly hatched goslings have a short neck and a narrow body. The legs and feet of newly hatched goslings account for about half their height and are out of proportion to the rest of their body. This allows goslings to walk, run, and swim fast. Young goslings can run very fast. But they frequently topple over, especially when going downhill. When that happens, the goslings simply get up and keep on going. The first day or two goslings tire quickly. So they sit down every now and then to rest. At first young goslings do not seem to have any sense of fear. Yet they understand that they are supposed to stay close to each other and their parents. Even during nap time young goslings stay close together. On 5/13/2011 the six 1-week old goslings at Pond D snuggled up under their mother's left wing when it was nap time. It was a cool damp day and the goslings welcomed a little extra warmth. (See Photographs 3.13 and 3.14.)

Photograph 3.13 A Mother's Touch. Newly hatched goslings spend less than 24 hours in the nest. Even so during their first week goslings frequently snuggle under their mother's wings for warmth or a nap, like this 1-day old gosling at Pond D on 4/25/2012. © D. Herrmann

Spring wouldn't be spring in North America without seeing at least one Canada goose family walking or swimming with their recently hatched goslings lined up in single file. While this formation is very cute and endearing for humans to observe, it is also quite practical. Generally the father goose occupies the forward guard position, whether swimming or walking.[45] The goslings, regardless of how many there are, line up in single file behind him. The mother goose generally occupies the rear guard position.[45] This allows both parents to scan for hazards and predators and, if necessary, assume flanking positions to protect their family. This behavior reminds me of the Secret Service escorts that accompany the President, Vice President, and their families. Hmm ... Canada geese have been living in North America much longer than the Secret Service has been in existence. Perhaps the Secret Service appropriated their security escort policies and procedures from Canada geese?

The sight of goslings lined up with their parents is a glimpse of everything that is good about spring: new life, harmony with nature, and hope for the future. The goslings are quite content to queue

Photograph 3.14 Napping in a Clump. Newly hatched goslings identify with each other and their parents. During the first few weeks, goslings nap together in a clump. (Top) The goslings are 1-day old. (Bottom) The goslings are 5-days old. These goslings are from the 2012 Pond D family. © D. Herrmann

up behind their parents. The goslings do not have to be told to queue up. Unlike children at a day care center, the goslings do not have to be tied together with bungee cords for fear that one will run off and get lost or hurt. No goslings, from the day they hatch, have an awareness of self and the fact that they belong to a specific family unit. Multiple families with young goslings can mix and share the same pond, yet the goslings—no matter how young—know which family group they belong to.

For the first few days one of the six newly hatched goslings at Pond D didn't quite get the queuing protocol right. It insisted on lining up behind the mother. When it was time to swim or walk the other 5 goslings lined up behind the father. The mother waited patiently for the sixth gosling to get in line.

When she realized that wasn't going to happen, she swam (or walked) behind the other 5 goslings. And, the sixth gosling lined up behind her.

Another aspect of this queuing behavior is interesting as well. Often the lead adult goose will stop, then change directions by pivoting 90° or 180° while swimming. In this instance the goslings also stop and quickly re-line up behind the lead adult, heading in the new direction. The same thing happens on land. The lead adult goose may be walking in a certain direction, perhaps while grazing. Then he stops and changes course. The goslings automatically adjust their course as well. There isn't any discussion of the matter, the goslings simply do it. Unlike human children, there isn't any fussing or fighting or nagging about who is first in line, who was first in line, or whose turn it is to be first in line. The goslings do not care who is first or last in line. A goose family can be swimming east in a pond, stop, pivot, and start swimming west. The gosling that was first in line is now last in line and vice versa. But none of the goslings take notice or care. (See Photographs 3.15 through 3.17.)

By the time they are 3 to 4 weeks old goslings have doubled in size and are 6 to 8 inches (15.24 to 20.32 cm) tall. To fuel their rapid growth spurt goslings eat constantly, or so it seems. Their down takes on a light brown tinge, but their head remains yellow. A gosling's body begins to fill out and take the shape of a goose, while their neck grows slightly. At this point a gosling is taller than a wood duck (*Aix sponsa*) and about the same height as a mallard duck. Goslings continue to stay close to each other and their parents, regardless of whether they are on land or water. Often goslings will nap together in a clump, as if they were in a nest. The family of 5 goslings from Pond D was observed sleeping in a clump on 5/10/2011 and 5/13/2011. When goslings are taking a serious nap they often

Photograph 3.15 A Reflection of Things To Come. A 1-week old gosling looks at its reflection in Pond D and that of its father who stands guard higher up on the bank. © D. Herrmann

Photograph 3.16 1 to 2 Week Old Goslings #1. Newly hatched goslings are very tiny, only about 3 inches (7.62 cm) tall. Their legs account for nearly half of their height! Newly hatched goslings begin walking, swimming and feeding themselves immediately. (Top left) The 2011 Pond D family takes their first swim after leaving the nest when they are 1-day old. © D. Herrmann

stretch a leg or two out behind then and lay flat—something adult geese never do. And by now the goslings have learned the location of all the fence openings and shortcuts to each pond.

Young Canada goslings are not generally aware that they have wings or even what their wings are for. That realization does not come until they are 3 to 4 weeks old. Unlike young songbirds who have an hour or two to master the art of flying after their first wobbly flight down from the nest, Canada geese learn about wings and flying gradually. The wings of young goslings are little more than down covered stubs since they don't have any flight feathers. At first goslings simply preen around their wings, without taking much notice of them.

Between 3 and 4 weeks of age goslings are able to isolate sensory feedback from their wings. By that time goslings will have observed their parents stretching their wings while preening and bathing. Suddenly one day a gosling will first flap it's down covered wing stubs for balance when running downhill, while bathing, or stretching after a nap. Afterward, occasional wing flapping or stretching becomes part of their daily routine. These activities pique a gosling's curiosity about what its wings are for and are a prerequisite for learning how to fly. In 2011 the family of 5 goslings at Pond D and

Photograph 3.17 1 to 2 Week Old Goslings #2. Young goslings grow rapidly and are constantly on the move exploring their new world. (Bottom right) The 2011 Pond D family takes their first swim after leaving the nest when they are 1-day old. © D. Herrmann

the family of 5 goslings at Pond B both flapped their (soon-to-be) wings for the first time when they were 22 days old. In 2012 the goslings at Pond D also flapped their (soon-to-be) wings for the first time at 22 days of age. (See Photographs 3.18 and 3.19.)

Canada geese accept helpers into the family on occasion. This usually doesn't happen until the goslings are 3 to 4 weeks old. The helper, a gosling from the previous year, provides extra protection for the goslings as they grow and develop. The helper generally walks or swims in the middle alongside the goslings or in the rear with the mother. The helper gosling is old enough to be on its own, but not old enough to form a pair bond. It still identifies with its parents and helps them protect their new family, like an older brother or sister.

By the time goslings are 5 weeks old they are about half the size of an adult. At this point their down is all light brown and much thicker, almost to the point of appearing shaggy. A gosling's tail feathers begin to develop and its ears are visible under the down. Goslings continue to follow their parents and each other, but they venture further out on their own.

For example, on the afternoon of 5/28/2011 one of the five goslings at Pond B found a partially eaten apple. The gosling picked up the apple and began running with it. The other four goslings

Photograph 3.18 3 to 4 Week Old Goslings #1. By the third week goslings begin to isolate sensory feedback from their wings. Goslings first flap their (soon to be) wings at this age, even though they do not have any flight feathers. This is not an attempt to fly, since at this age goslings have not yet seen their parents fly. Rather the goslings stretch their wings for balance when standing up or walking quickly. From top left, these photos were taken at Pond D on 5/20/2012, Pond D on 5/28/2012, and Pond B on 5/26/2011. © D. Herrmann

chased the gosling with the apple until they caught it and one of them was able to steal the apple. The goslings continued to chase and steal the apple from each other, like a game of football, until there wasn't any apple left. No aggression was displayed; it was more like a contest. The parents watched, but did not interfere.

A few days later one of the five goslings at Pond B found a peanut in the shell that had been put out for the squirrels. A similar scenario unfolded. But this game didn't last as long. The gosling quickly learned how to open the peanut by arranging the peanut and closing its bill just right—even though peanuts are not a normal part of a Canada goose's diet.

At about 6 weeks of age Canada goslings begin to molt their down and start to grow a complete set of juvenile feathers. The feathers on their underside and underneath their tail fill out and turn white. The short tail feathers turn black. The crop area expands. Day by day the feather definition on

Photograph 3.19 3 to 4 Week Old Goslings #2. By the fourth week yellow hatchling down has given way to a light brown down, except on the head. Goslings are about a third the height of an adult. Goslings take several naps throughout the day because of their rapid growth spurt. Goslings often stretch one or both legs out behind them when napping, since their legs are still out of proportion to their body. Adult Canada geese do not do this. © D. Herrmann

their back and wings becomes more pronounced. At first small black spots appear on their head and neck. These spots become larger and larger until they coalesce. The head and neck become black and the white ear patches emerge. A gosling's legs have to get used to this sudden increase in size and weight. Goslings sit down and rest frequently while their leg muscles grow.

This juvenile set of feathers closely resembles the parent's plumage. The main difference is that the light beige crop and breast feathers are a bit darker. Several days before and after this molt goslings are not very active; they mostly eat and rest. It is not unusual to see the goslings napping mid-morning and mid-afternoon as they conserve the energy needed to produce a new set of feathers.

Canada goslings do not grow in a linear fashion. After Canada goslings molt and grow their juvenile feathers, their neck and legs expand to the correct proportion in relation to their body and they begin to look like three-quarter size adults. However, the neck and legs do not grow at the same rate—the legs grow faster. All of a sudden a young Canada gosling cannot reach the ground with its head and bill, which makes it difficult to graze. A gosling will try to graze—after all it used to work—and stumble. Realizing that something has changed, the gosling tries a new approach. The gosling walks forward two or three steps and sits down to eat; repeating this process over and over. After a few days its neck and legs are synchronized again and the gosling can resume eating while standing up. This rite of passage must be the Canada geese's equivalent of the awkward teenage years. (See Photographs 3.20 and 3.21.)

Between their 7th and 8th week Canada goslings begin to be curious about onlookers and others in their neighborhood. By now goslings have developed a good sense of who is ok and who they need to stay clear of. Goslings learn to stretch out their fully feathered wings while standing on one foot. Goslings begin to spend more time and are more serious about preening, since they now have real feathers to tend to. Goslings are finally able to rest while floating, with their new long neck curled flat against their body in a half awake half asleep state of contentment.

The brand new set of feathers increases the goslings' curiosity about their wings. At this point it is not uncommon to see goslings run down a hill toward the pond with their wings outstretched. The goslings do not actually flap their wings, but they are imitating a land takeoff. Sometimes two or more goslings will engage in this behavior at the same time. One parent stands guard on the top of the hill and the other parent stands guard on the bottom of the hill while this behavior is occurring. About this time young Canada goslings also first attempt to upend and feed on aquatic plants in deep water. In 2011 the Pond B family of 5 goslings first ran down the hill in preparation for a practice flight on June 20th, 7 weeks after they hatched.

A little later the goslings are ready for their first flying lesson. The parents line up the goslings on a flat stretch of ground near the pond. One parent is behind the goslings, the other parent is in front of the goslings. The parent in front runs toward the goslings, flaps its wings, flies about 18 inches (46 cm) over the goslings' heads, and lands behind them, near the other parent. The other parent repeats the same actions. Then it is the goslings' turn. One by one each gosling attempts to fly over the other goslings, while the parents stand guard. After each gosling has made a couple successful passes, the family returns to the pond for a short swim before continuing their normal daily routine. In 2010 the first flying lesson was observed on June 26th, 8 weeks after the goslings hatched.

Not long afterwards the Canada geese family gathers on the steep side of the pond. The parents demonstrate how to run down the hill, fly a short distance at a low altitude, and make a water landing in the center of the pond. Then they encourage the goslings to try to do the same thing. One by one the goslings run down the hill. They have no trouble becoming airborne, but their first water landings are a bit bumpy. However, after a few practice runs the goslings have mastered the technique. Young Canada goslings must also learn how to takeoff from water, especially in response to an imminent danger. Water takeoff lessons were only observed at Pond C; perhaps because it is larger. On 6/27/2012 goslings practiced bathing, turning upside down in the water, and attempting to walk on the water's surface in preparation for a water takeoff. By the end of the 8th week a gosling's feathers match the

Photograph 3.20 5 to 6 Week Old Goslings #1. Big changes take place in a gosling's appearance during the fifth and sixth weeks. Goslings, now half as tall as their parents, suddenly appear very shaggy. Their ears are visible for a brief period of time. Some tail feathers begin to grow, as do white feathers under the tail. © D. Herrmann

coloring of an adult goose, except for the white chin strap that is a little darker. Goslings also learn how to engage in energetic baths at this time. (See Photographs 3.22 and 3.23.)

Table 3.4 summarizes gosling development stages as they relate to physical traits and learning behaviors.

Daily Routine

Throughout the day Canada geese engage in many common behaviors, such as:

- Upending
- Preening
- Bathing
- Stretching
- Napping

Photograph 3.21 5 to 6 Week Old Goslings #2. During the fifth and sixth week goslings display a new sense of adventure and curiosity. Goslings can be seen exploring new things in their environment, items that were previously ignored. This exploration may be done as individuals or as a group. Here goslings play chase with (top) a newly found apple and (bottom) a peanut. © D. Herrmann

- Participating in flock security measures
- Standing on flat platforms

Canada geese can dive under water in order to munch aquatic plant life or engage in some serious bathing. Unlike smaller geese and ducks, Canada geese rarely go completely under water. Instead they move from a horizontal position floating on the water to a vertical position, with their head, neck, and breast underwater and their tail and the rest of their body above water, pointing straight up. This posture is referred to as upending.[18, 83, 95] While a Canada goose is upending their feet are busy treading water in order to maintain their upright position. Upending is usually a quick up/down movement. However, Canada geese can stay underwater about 10 seconds. If a Canada goose needs to stay upended for a period of time, it blows bubbles out its nostrils to exhale. Canada geese resurface almost at the exact point of entry. Since they are plucking plants from the bottom of the pond, this

Photograph 3.22 7 to 8 Week Old Goslings #1. During the seventh week goslings develop light blue keratin sheaths where their flight feathers will grow in. They also develop an off white chin strap running from ear to ear. At this age goslings begin to engage in adult behaviors like stretching and preening. © D. Herrmann

would imply that their vision under water is rather good, even in water that is not crystal clear. Canada geese appear to spot a particular plant of interest while floating. Then all of a sudden they upend to retrieve it. Often two or more geese will upend at the same time, indicating that a patch of delectable aquatic plants—an underwater salad bar—has been found. Perhaps the flock was searching together for such a find. (See Photograph 3.24.)

Canada geese spend a lot of time preening their feathers. Preening is a meticulous process that involves stroking, arranging, and conditioning all of their 3,000-plus feathers. Often this process involves adding a minute amount of oil from their preening (or uropygial) gland to each feather. Oil from the preening gland facilitates Vitamin D absorption when it is irradiated by the sun.[158] Canada geese start the preening process by stroking the feathers that are easiest to reach—those on their neck and breast. This is followed by preening the feathers on their back and tail. Canada geese often rub their head and neck on their back while preening. Next Canada geese preen the feathers on their underside. Lastly Canada geese preen the top and underside of each wing, first while the wing is retracted, then while it is extended. Sometimes this is accomplished while a goose is standing on one leg. Preening seems to have a calming effect. As a result it is not unusual to see a Canada goose's outer white eyelids toward the end of a preening session.

Photograph 3.23 7 to 8 Week Old Goslings #2. By the eighth week goslings look similar to adult geese, except they are slightly smaller and their feathers are a bit darker. During this final growth phase goslings take a lot of naps. © D. Herrmann

Canada geese rely on their feathers for flight and regulating their body temperature throughout the changing seasons and weather patterns. This latter feature is particularly important for water birds like Canada geese that spend part of every day in the water, regardless if the air temperature is 30°F or 100°F. Consequently, Canada geese take exceptionally good care of their feathers. (See Photographs 3.25 and 3.26.)

Canada geese generally bathe on a daily basis, weather permitting. Canada geese manage to keep their feathers clean—almost immaculate—even though some of the waterways they swim in are muddy or strewn with globs of algae. Canada geese follow a standard procedure for bathing. First, they dunk their head and neck underwater, while throwing water on their back. Canada geese do this twice in quick succession. Next, Canada geese stand up straight and flap their wings rapidly 2 or 3 times. Afterward Canada geese resume a floating position. From there they lean forward and wiggle their tail to shake off any excess water. Normally Canada geese go through this bathing sequence once. Every now and then a Canada goose repeats this bathing sequence twice. Canada geese do not engage in dust baths.

This is the normal year round bathing sequence. A more intense bathing sequence was observed for a few days each year during early spring, just after the ponds thawed and the temperatures began to warm up. In this instance Canada geese went completely under water before flapping their wings.

Table 3.4 Gosling Development Stages.

Weeks Old	Size	Physical Traits and Learning Behaviors
0–2	3–5 inches, 7.62–12.7 cm	• Covered in yellowish light brown down • Walk, swim, and feed themselves within 24 hours of hatching • Short neck, narrow body • Legs and feet are out of proportion to rest of body, which allows them to walk, run, and swim fast • Stay close to parents and each other on land or in water • Nap together, sometimes snuggle under mother's feathers
3–4	6–8 inches, 15.24–20.32 cm	• Mostly light brown with a yellow head • Body starts to fill out, neck begins to grow • Taller than a wood duck, about same height as a mallard • Stay close to parents and each other on land or in water • Nap together, sometimes in a clump • Often stretch 1 or 2 legs out behind them when napping • First flap wing stubs for balance when standing up or walking quickly. • Know/remember where fence openings are
5–6	50% adult size	• Down light brown, becomes thicker and shaggy • Ears are noticeable • Tail begins to develop • Continue to follow parents • Feet and legs grow faster than neck; have to eat sitting down for a few days • Take more naps • Molt baby down, grow juvenile feathers similar to adults, but a little darker
7–8	75% adult size	• Curious about onlookers • Start to stretch out wings while standing on one foot • Spend more time and are more serious about preening • Able to rest while floating, with their neck curled flat against body • First attempt upending in deep water • First flying lessons • Learn water landings • Feathers match adult coloring

Some really exuberant geese rolled over sideways onto their back several times or put their head under water and did a somersault, then splashed up a storm. A few elite bathers even performed a porpoise (*Phocoena phocoena*) routine where they went completely underwater for about 5 seconds before resurfacing, often 2 or 3 times in a row. It almost seemed like this was a contest to see who could go the furthest underwater the fastest. (See Photographs 3.27 and 3.28.)

Members of the local flock and migratory flocks participated in this extra energetic bathing ritual, usually at Pond C. Given the exuberance and level of energy involved, it is likely that more than just bathing was going on. This ritual seemed to express a celebration that spring was here and the flock had survived another winter.

One goose who stopped off at Pond C on 2/15/2011 as part of a migratory flock of 96 geese had a dark brown streak along its left side. The streak was right at the water line when the goose was swimming. Obviously this goose had encountered a tar ball while wintering in the Gulf.

Canada geese frequently engage in a stretching routine before and after a nap. Canada geese methodically stretch one wing at a time while standing on alternating feet. They slowly raise the foot corresponding to the wing being stretched before fully retracting it. Afterward Canada geese wiggle their tail feathers and roll their head and neck around on their back. (See Photographs 3.29 through 3.31.)

Photograph 3.24 Upending. Canada geese upend to retrieve plants from the bottom of ponds. Canada geese spot the plants while they are swimming. This means that Canada geese can see the bottom of a pond clearly from about 2 feet (61 cm) above water. When a Canada goose upends its head, neck, and front half are underwater; while its back half and tail point up in the air. To maintain this position the legs and feet are busy treading water. Canada geese can stay upended for about 10 seconds. However, Canada geese usually upend quickly in a direct up/down movement. Bonded pairs often upend at the same time. © D. Herrmann

Canada geese take several short naps throughout the day, like other birds. Being water birds, Canada geese are not timid about walking or resting on the ground. Canada geese can nap while standing on one leg, sitting on the ground, or floating in a pond. Sometimes Canada geese nap on one leg while standing in a pond. If a Canada goose is taking a short casual nap it tucks its head and neck against its body, but continues to face forward. This seems to be a sort of standby alert position, where the legs and feet get a rest. While in this position Canada geese will often relax their wing muscles and hold their wings loosely about their body. On the other hand, if a Canada goose is taking a serious nap it turns its head around and tucks it under its wing. Canada geese tuck their head under either wing as a matter of individual preference, while floating, sitting down, or standing on one leg. If a Canada goose is napping while standing on one leg, it will tuck its head under the opposite wing for balance. (See Photograph 3.32.)

Often several geese will nap at the same time while floating. One group of 20 geese was napping while floating on 7/2/2011 at Pond C. Every 3 or 4 minutes the geese, in unison, all changed the direction they were facing. The geese took turns napping or being a security lookout, as was evident by which geese did and did not have white outer eyelids visible at any given moment.

At night Canada geese sleep along the perimeter of a pond facing the water. The geese are evenly spaced, with just enough room to accommodate their wingspans for sudden flight or lunging into the water should an emergency arise. Bonded pairs and family members stay close together overnight.

Canada geese adopt several strategies to stay cool during sweltering heat. Nonessential activity is minimized. Sometimes Canada geese settle in tall grass to avoid the heat. Other times geese congregate in the shade under a tree or bush near the edge of a pond. Another practice is to congregate under a tree or bush that overhangs a pond, while staying in the water. If on land, Canada geese hold their wings loosely about them to allow excess body heat to escape. Canada geese have been observed plucking out a down feather or two on occasion to reduce body heat. (See Photograph 3.33.)

Photograph 3.25 Preening is Serious Business #1. Canada geese take preening very seriously because of the importance of feathers to their overall health and well-being. Canada geese do not pop in the shower for 5 minutes, then run out the door so they won't be late to work or school. A single preening session may last up to 45 minutes. During this time each of their 3000-plus feathers is methodically examined, cleaned, and arranged. While doing so Canada geese assume an incredible variety of positions and postures. © D. Herrmann

On cloudy days Canada geese face a variety of different directions when they nap. On windy days, particularly cold windy days, Canada geese face into the wind. If they didn't, the wind would blow their feathers upright and they would lose body heat.

On 12/5/2010 a migratory flock of 76 geese were on the golf course. The geese were all sitting down close together facing into the wind. The flock was positioned at the bottom of a sand hill, using the bank as a wind block. The flock waited for the wind to die down before continuing their journey on 12/8/2010.

On 1/9/2011 a migratory flock of 82 Canada geese were resting at Pond C, which was mostly frozen. The flock faced into the wind. A sudden gust of wind caught two geese by surprise. The wind spun 1 goose, who was sitting on the ice, half way around. Another goose had just stood up to stretch its wings. The wind blew this goose about 5 feet (1.5 m) backward across the ice.

Photograph 3.26 Preening is Serious Business #2. Geese were often seen preening between 10:30 a.m. and 11:30 a.m. For the most part the Canada geese preened while standing on land. Once in a while the geese stood in shallow water near the shore to preen. In those instances the geese incorporated a bill-full or two of water into their preening routine. Sometimes Canada geese preened while sitting down. Bonded pairs frequently preened at the same time. © D. Herrmann

Canada geese participate in cooperative flock security measures. Whenever a flock is grazing, swimming, or napping 1 or more geese, depending on the size of the flock, assume sentry duties and warn everyone of impending dangers. Canada geese hurry to the center of a pond to avoid land or air predators. They may swim or fly there, using which ever method is fastest. Sometimes sudden

Photograph 3.27 Bath Time! #1. Weather permitting bathing is part of a Canada goose's daily routine. The normal procedure involves leaning forward, dunking the head and neck under water, returning to an upright position, and throwing some water on the back and wings. Afterward a goose stands up and flaps its wings quickly 2 or 3 times. © D. Herrmann

flight is the best response to a land-based threat and the flock becomes airborne in an instant. Often other adults will help protect young goslings, since neither the goslings nor their parents can fly.

On the afternoon of 5/26/2011 a family with 5 goslings at Pond B was chased by a dog. The family immediately retreated to the center of the pond. Two other adult geese, who were not breeding and could have flown away, joined the family. The goslings clustered together. The 4 adults formed a circle around them. For the rest of the afternoon, even after the dog was gone, the two other adults stayed with the family. Two adults swam in front of the goslings, and two adults swam behind them. From that date on until the goslings could fly, three adults were always with the goslings. One adult swam (or walked) in front. One adult swam (or walked) behind. And one adult swam (or walked) in the middle alongside the goslings. This was the only family of goslings at Pond B during 2011.

Wood ducks, as their name implies, frequently sit on the branches of trees and bushes near ponds. A few mallard ducks sit on low branches that overhang a pond once in a while. Because of their size, weight, long neck, and large feet, Canada geese do not sit on tree branches. However, Canada geese do like to sit on flat surfaces, such as the top of a: building, car, culvert, rock pile, and so forth. Any flat stable surface will suffice. The height is irrelevant. Canada geese have been spotted standing on a piece of lumber that fell into Pond D, as well as on top of a 16-story building.

Photograph 3.28 Bath Time! #2. On rare occasions a Canada goose or two engages in an extra enthusiastic bath. A Canada goose may even turn upside down and/or go completely under water, like the two geese in the bottom left photograph. There may be a psychological stimulus for this extra enthusiasm. One time a large flock was migrating north for spring and encountered unusually mild weather at Pond C. Three members of the flock participated in an enthusiastic bath. On another occasion a male goose engaged in a prolonged energetic bath the day before his goslings hatched. He may have been drawing attention to himself and away from the nest at this vulnerable time. The top left and bottom right photographs were taken at Pond D 26 hours before Photograph 3.11. © D. Herrmann

One, sometimes two, Canada geese landed on the top of a 16-story building near Pond B. Once there the geese chattered away loudly, as if they wanted everyone to know they were there. Then one at a time the geese jumped off the building and coasted down for a while, before actually flying —like it was an amusement park ride. Sometimes the geese jumped off the building 2 or 3 times in a row. This happened most often during April and early May. Landing on top of a 16-story building, then jumping off it repeatedly serves no purpose. Hence, this behavior can only be considered a play activity. (See Photograph 3.34.)

Photograph 3.29 Stretching #1. Canada geese have three different stretching modes. The most common is to stand on one foot while stretching the opposite leg and wing. © D. Herrmann

Visiting Golf Courses

Many people wonder why Canada geese are frequent visitors to golf courses. If this issue is examined from an historical perspective, it is easy to understand this tendency.

The game of golf is a standard component of western societies as a common recreational activity and a place to conduct informal business. Today many high schools and universities have golf teams. Scotland is given credit for developing the game of golf in the 15th century. However, the terms golf ball (1637), golf course or links (1890), and golf cart (1899) did not enter the lexicon until later, as the game spread to other parts of the globe.[179] The term golf widow appeared in 1927,[179] so one can assume that the game had achieved some degree of prominence by then.

Canada geese were at home in North America long before the surge in the popularity of the game of golf. Two events occurred in North America at about the same time: (1) the dramatic decline in Canada geese populations due to over-hunting and habitat destruction; and (2) the surge in the popularity of the game of golf. It is likely that much of the land used to build all the new golf courses was originally inhabited by Canada geese, especially courses with natural water traps. Canada geese frequent open country which has water resources nearby—the essence of a golf course. Canada geese

Photograph 3.30 Stretching #2. Canada geese also stand up tall on both feet and flap their wings while stretching. This allows them to stretch their neck, back, legs, wings, and tail all at once. © D. Herrmann

do not frequent deserts, mountain tops, or forests. The greens at golf courses are well maintained and usually sport a special low, dense type of grass that is native to Scotland. Canada geese are herbivores. Grass is on the top of their dinner menu.

However, there are safety concerns about Canada geese visiting golf courses. Golf course landscapes are very lush and insect free. After all, members and professionals would not take kindly to being bit by a mosquito or stung by a bee just as they are teeing off or making a critical put. This artificial lushness has been achieved through the extensive use of fertilizers, herbicides, and insecticides.

A 36-hole golf course near Washington, D.C. was recently sold to make room for a new housing development. Pre-construction environmental testing found elevated concentrations of pesticide residues in the golf tee boxes and putting greens.[205] The residues were determined to be typical

Photograph 3.31 Stretching #3. A third stretching mode occurs after a long nap. In this case a Canada goose stretches both wings upward quickly while it is still seated on the ground. Afterward, it stands up. Mallard ducks also do this. © D. Herrmann

agrochemicals that are commonly used on golf courses.[205] Remediation required stripping and removing the surface soils.[205]

These typical agrochemical residues were deemed unsafe for humans to garden, walk around, and play in on their lawns. Given that, how safe can it be for Canada geese to consume golf course grass or drink water from a water trap, especially if the toxin's effect is cumulative? Accordingly, it is (past) time for golf courses to "go green."

The golf course between Ponds A, B, C, and D was reduced in size by 50% between 2004 and 2006 to make room for housing and an indoor tennis center. Since the water trap was not affected, the geese did not mind. The local flock continues to visit the golf course and the migratory flocks continue to use the golf course as a major stopover point.

On 11/13/2010 a humorous incident was observed at the golf course. A flock of 64 migratory Canada geese were there. They arrived the day before. The flock was spread out. Some geese were in the water trap, others were grazing on the fairway or greens. No one had used the course for several weeks because of cold weather.

Even so the groundskeeper had a tizzy about the geese's presence. He tried turning the sprinklers on them. He waited until a group of geese collected on a particular green, then he turned the sprinkler on full blast. At one point he had sprinklers going on six of the nine greens—in mid-November. Canada geese are water birds. Rather than being afraid of the water they either ignored or welcomed it. One enterprising goose took an impromptu shower by spreading his wings and walking around in the spray.

When the sprinkler didn't cause the geese to leave, the groundskeeper hopped in a yellow golf cart and raced toward the geese at full speed. The geese simply stepped aside as if nothing were happening and continued about their business. Over and over again the groundskeeper charged the geese with no results—until he went up a hill too fast and his cart fell over!

Using water to scare water birds isn't exactly logical. Nor is one person charging sixty-four 36 inch (0.91 m) tall Canada geese that are spread out. Talk about having a walnut-sized brain, no logical reasoning ability, and behavior that is driven by raw emotions.

Photograph 3.32 Nap Time. Canada geese take several short naps, usually 15 minutes or less, throughout the day. Canada geese seem to just stop and nap wherever they are. They may assume one of four different postures for napping: sitting on the ground, standing on one leg on land, standing on one leg in shallow water, or floating. A goose's head faces forward for a casual nap. For a more serious nap, a goose tucks its head under one of its wings. The white outer eyelids are visible when a goose is asleep. Even then, a goose opens its eyes briefly every couple of minutes to make sure everything is alright. Bonded pairs often nap near each other. © D. Herrmann

The geese simply ignored this not so bright fellow and stayed put through the morning of 11/15/2010. (See Photograph 3.35.)

3.4 Seasonal Behaviors

The local flock of Canada geese engages in the three seasonal behaviors listed below year after year. The first two behaviors have not been reported before.

- Nesting dispersal
- Reconvening the flock
- Fall migration in a V formation

Photograph 3.33 Staying Cool. On extremely hot humid days Canada geese and goslings seek shade to stay cool: under a tree, in the water, or in tall grass. Canada geese relax their wing muscles and occasionally pluck a few down feathers to reduce heat retention. © D. Herrmann

Nesting Dispersal

A consistent pattern was observed from spring 2005 through spring 2010. Each year the flock returned in March or early April. The breeding pairs dispersed among the waterways. One pair each returned to Ponds A, B, and D and the golf course water trap to nest. The other non-breeding pairs and unattached adults rotated between the larger two ponds, B and C, and the golf course. Goslings were spotted the first week in May. Each pair and the goslings stayed put in and around the pond where the goslings hatched until they could fly.

No Canada geese nested at Pond A in 2006. The fence around the perimeter of Pond A had been removed earlier that year. A Canada geese pair scouted out the area for a few days, but decided that

Photograph 3.34 More Tall Flat Perches Please. Canada geese do not sit on tree branches because of their size and large feet. That does not stop them from standing on tall flat surfaces. Here we see Canada geese standing on the roof of a 5-story apartment building, a rock pile, the top of a 16-story building, and a piece of lumber that was thrown into Pond D. Canada geese were also spotted sitting on the roof of a car. Juvenile Canada geese jumped off the 16-story building 2 or 3 times in a row, while chattering loudly, before coasting into Pond B. This behavior serves no purpose; hence, it can only be considered a play activity. © D. Herrmann

Pond A was not suitable as a nesting site that year. In the past the fence provided protection for the nest and goslings from loose dogs, lawnmower crews, and others who tried to get too close. This behavior indicates that Canada geese pairs are aware of changes to their immediate habitat. This behavior also indicates that Canada geese pairs reassess the safety and security of a nesting site on a yearly basis, rather than just automatically nesting in the same location year after year.

In 2008 no goslings hatched at Pond A. On April 25th, less than a week before the goslings would have hatched, the nest area was flooded after a severe thunderstorm. The parents remained in the area for two weeks, searching the area where the nest had been and looking for their goslings. This expression of loss was consistent with the 2-week mourning period observed in pet birds.

Photograph 3.35 Canada Geese Open. The pros examine the fourth hole, while the spectators watch from the fairway. Canada geese's affinity for golf courses is easy to understand. Their preferred habitat is open grasslands near waterways and their favorite entree is grass—the essence of a golf course. Golf courses pose a health risk to Canada geese because the artificial lushness is achieved through extensive use of fertilizers, herbicides, and insecticides. Residues accumulate at a level that is considered unsafe for human lawns and gardens.[205] Isn't it time for golf courses to go "green"? © D. Herrmann

(See Chapters 7–11.) After that, the pair flew away to Pond C. In subsequent years (2009 and 2010), the Canada geese pair moved their nesting site to the other side of Pond A to avoid potential flooding. This behavior indicates that the Canada geese pair retained a memory of the previous year's nesting experience, and were able to apply the lessons learned from that experience in the future.

In three of the four years that goslings hatched at Pond A, four goslings hatched. Only three goslings hatched in 2009. Four goslings hatched in 2010, but one was taken by a red-tailed hawk (*Buteo jamaicensis*) at three weeks of age. Except for 2010, all the goslings that hatched at Pond A survived into young adulthood.

These observations are inconsistent with the theory that Canada geese return to the site where they hatched to raise their families. These observations are inconsistent with the position put forth by a few sources that the pair returns to nest where the female hatched. Instead, each year only one pair nested at each location.

There were two minor variations to this behavior pattern in 2010. For a couple hours one afternoon in early April a second pair of Canada geese returned to Pond A. Even though the original pair of Canada geese are content to share Pond A with a few mallard ducks, they chased the other pair of Canada geese away. Later on, after the goslings had hatched, on June 2nd to be precise, three adult geese joined the family at Pond A. The parents made no attempt to chase them away. They stayed for a day and a half. Perhaps these were the three goslings that had been raised at Pond A the year before.

Large flocks of Canada geese that summer near large rural waterways may indeed return to the same nesting site en masse year after year. A different nesting pattern may be occurring in the observation area because the four co-located ponds are near a metropolitan area. Each year the local flock returns together. However, the breeding pairs of Canada geese disperse among the four waterways to nest. Is this nesting dispersal behavior due to individual preference, or is it a new metropolitan safety strategy of the flock—if one nesting site experiences a calamity, the other three nesting sites/populations survive?

Table 3.5 summarizes the Canada geese nesting dispersal behavior from 2005 through 2010.

The nesting dispersal behavior of the local flock was consistent from 2005 through 2010. In 2011 an entirely different pattern was observed.

A pair of Canada geese was first spotted at Pond A on 2/26/2011. They swam around for a while, then grazed on the northern lawn. Similar behavior was observed every day from 2/26/2011 through 3/3/2011, even though Pond A was flooded. The pair of Canada geese visited Pond A briefly on the afternoons of 3/5/2011, 3/18/2011, 3/21/2011, and 3/25/2011. A series of thunderstorms followed and Pond A remained at or above flood stage. A pair of Canada geese was not seen at Pond A again until 4/7/2011, 4/14/2011, 4/18/2011, and 4/24/2011. By this time the Pond A pair should have already nested.

The original Pond D was bulldozed into oblivion on 9/17/2010, along with all its plant and aquatic life, to make room for townhouses. A "new" Pond D was created 30 feet (9 m) east of the old location. Instead of being round, the new Pond D is long and narrow. The "pond" has filled up from rainwater and construction runoff. Grass was planted in the area immediately surrounding the pond. However, the remaining 40 acres have been dug up as construction crews continue working.

A pair of Canada geese was first spotted exploring the new Pond D on 3/1/2011. They stood on a mound of dry upturned dirt and looked toward the new Pond D. The construction crew was still working on the retaining wall. The 2 geese left and did not return until 3/13/2011. From then on the pair of geese were seen every day, either swimming in the pond or grazing on the lawn, inside and outside the fence. The female goose began building her nest 4/7/2011. She began sitting on the nest 4/9/2011. Six goslings hatched 5/7/2011. (See earlier Photograph 3.11.)

One pair of Canada geese returned to nest at Pond B on 3/4/2011. Their 5 goslings hatched 4/29/2011. In 2010 this pair lost a young gosling, possibly to a snapping turtle. In 2011 they kept their goslings out of the water for the first week. During the second week the goslings were allowed in the pond, but only in very shallow areas. This behavior indicates that the Canada geese parents understand the size of a snapping turtle and the depth at which it trolls.

For the first time Canada geese returned to nest at Pond C in 2011. Two pairs claimed this nest site. One male was very aggressive toward the other pair. Eventually the two pairs of geese ended up

Table 3.5 Canada Geese Nesting Dispersal Behaviors (2005–2010), Using Pond A as an Example.

Event	2005	2006	2007	2008	2009	2010
1st sighting of adults at Pond A	4/5	4/6	4/2	4/7	5/6	3/9 3/20 3/21 4/5
1st sighting of goslings at Pond A	5/1	-	5/1	-	5/6	5/14
# of goslings hatched at Pond A	4	0	4	0	3	4
Flew away to reconvene with flock at Pond C	7/1	4/7	7/1	5/9	6/12	6/27 1st departure 6/28 1st return 7/10 2nd departure 7/12 2nd return 7/14 3rd and final departure
Mid-August return to nesting site at Pond A for a day	8/20	-	8/18	-	8/15	-
Flock left for fall migration	11/19	11/25	11/23	11/27	12/9	11/15
Notes		1		2	3	4, 5

Notes:

1) No adults nested this year. The fence around the perimeter of pond had been removed. The fence, intended to prevent humans from drowning, protected goslings from lawnmowers, unleashed dogs, and others who tried to get too close.
2) The nest area was flooded 4/25 after a severe storm, less than a week before the goslings would have hatched. The parents remained at the pond for 2 weeks, looking for their eggs/hatchlings.
3) The parents were not seen this year until the goslings hatched. New construction near the nest site may have caused them to be more secretive and depart earlier than usual.
4) When geese first returned to Pond A the south side was still covered with snow. Two adult geese came and left several times before finally staying. This year, for the first time, the Pond A goose family left and returned multiple times after the goslings were grown; rather than the previous single departure. Each departure and return coincided with a severe thunderstorm and flooding of Pond A.
5) The local flock left on 11/15 with a larger flock from up north that spent a couple days here at a stopover point. One goose from the local flock stayed behind with the mallard ducks at Pond B. It did not leave until 12/3/2010. A second lone goose stayed behind at Pond C after a flock of 13 migrated through on 11/28/2010. It did not leave until 12/1/2010.
6) The geese did not nest at Pond A in 2011, 2012, or 2013 since it was in flood stage. The geese did not nest at Pond A in 2014 or 2015 since it was in drought stage.
7) A record of 10 goslings hatched in the Speckleneck family on 5/14/2015 at Pond D. The mother nested on the retaining wall, like she began doing in 2012. The Pond A family had 2 goslings at Pond D in 2015. For the first time, no goslings hatched at Ponds A, B, C, or the golf course during 2015.

swimming, grazing, and napping on opposite sides of the pond. One female goose built a nest and began sitting on it 4/18/2011. But the nest site was abandoned on 5/7/2011. These two pairs nested at Pond C again in 2012. This year they were luckier. Both pairs produced four goslings, all of which grew into young adults. (These are the goslings heard on Disk 1 Track 3.)

At first it was thought that one of the pairs at Pond C might be the pair that abandoned Pond A as their usual nest site. But, on 4/29/2011 two adults and 5 newly hatched goslings appeared at Pond D. The original Pond D pair was still nesting. These goslings hatched somewhere else, possibly near the creek, and walked to Pond D immediately after hatching.

A fourth family of 3 goslings hatched at the golf course water trap on 5/17/2011.

Then the mystery began. In the previous six years the goslings and their parents stayed at the pond where they hatched until the goslings could fly. That was not the case in 2011 or 2012. The Pond B family of 5 goslings stayed put. They moved around from the southern part of the pond to the middle part of the pond, or from the eastern lawn to the western lawn. But that was all.

In contrast the other 3 families took their young goslings on day hikes to and from Pond A, Pond D, and the golf course water trap. The straight line distance between Pond A and Pond D is about 0.75

mile (1 km) north to south. The straight line distance between Pond D and the golf course water trap is about 0.50 mile (0.66 km) east to west. That is a long distance for tiny goslings to walk. Worse, there is a 6-lane busy street between Pond D and the golf course water trap, and a 4-lane busy street between Pond A and Pond D.

Table 3.6 chronicles the day hikes the three families with goslings made during their first month. These little goslings were constantly on the move. May 10th was a particularly busy day. Some days the goslings could not be located at any of the ponds or the golf course. That meant they were en route or visiting another (unknown) location. Frequently, when the goslings arrived at their new location they took a big drink of water then dropped to the ground for a serious nap. It is not clear that the parents understood how strenuous these hikes were for their young families. The family of 5 goslings at Pond B hatched the same day (4/29/2011) the family of 5 goslings appeared at Pond D. However, the Pond D goslings were skinnier because of their constant day hikes.

After June 3rd the families of 2, 5, and 6 goslings abandoned Ponds A and D and moved together to the golf course water trap. The three families of goslings stayed there until the morning of June 19th when they appeared for the first time at Pond B. That meant the 19 geese had crossed another busy 4-lane street. An adult helper was with them. The three families stayed close together and swam around in the north part of Pond B. There was a 19 day difference in the age and development stage

Table 3.6 Day Hikes Goslings Took Among the Ponds During Their First Month (5/7/2011–6/3/2011).

Date	Pond A	Pond D	Golf course water trap
May 7		• families of 5 and 6 goslings	
May 8	• family of 5 goslings	• family of 6 goslings	
May 9	• families of 5 and 6 goslings		
May 10 –10 a.m. –12:30 p.m. –4:30 p.m.	 • family of 5 goslings • families of 5 and 6 goslings • family of 5 goslings	 • family of 6 goslings	
May 11	• family of 5 goslings	• family of 6 goslings	
May 12	• family of 5 goslings		
May 14		• families of 5 and 6 goslings	
May 17			• families of 3, 5, and 6 goslings
May 20		• families of 5 and 6 goslings	
May 21	• families of 5 and 6 goslings		
May 24	• family of 5 goslings		
May 25		• family of 5 goslings	
May 26	• family of 6 goslings	• families of 3 and 5 goslings	
May 30	• family of 5 goslings		
May 31		• families of 2, 5, and 6 goslings	
June 1		• families of 2 and 5 goslings	
June 2	• families of 2 and 5 goslings		
June 3		• family of 5 goslings	

Notes:

1—During this same period the Pond B family of 5 goslings stayed put.

2—On 5/30/2011 one gosling from the family of 3 was run-over when crossing between Pond D and the golf course water trap.

3—None of the 3 families of goslings were located at the four ponds or golf course on: 5/13, 5/15, 5/16, 5/18, 5/19, 5/22, 5/23, or 5/27-5/29.

of the goslings. The Pond B Canada goose family with their 5 goslings was in the north part of Pond B too that day. But, they did not mix with the other three families of geese. Later, the three families with goslings moved back to the golf course water trap.

Three of the four pairs of Canada geese who raised goslings in 2011 engaged in the day hiking behavior. Eventually the three families banded together as a subflock. The hiking behavior mimicked the daily rotation among the 4 ponds and golf course water trap that adult Canada geese engage in when they are not breeding or after their goslings are grown and can fly. Why did these three pairs engage in this unusual behavior? Were they new parents who were not used to being grounded for 8 weeks of parenting duties? Did the "movement" of Pond D and unease about the ongoing construction around it contribute to this unusual behavior? Did the fact that Pond A was flooded the entire time a pair normally would have nested there play a role? Was another factor or combination of factors the cause?

On the morning of 5/30/2011 one gosling did not make it across the street between Pond D and the golf course. It was found lying on the sidewalk just south of a mulberry tree. The family may have stopped there to eat some of berries which were ripe and falling off the tree. It was near the place where two Saturdays earlier the family of 5 goslings was spotted on the lawn behind Pond D.

The young gosling, only 13 days old, still had its yellow and brown feathers. It was lying on its left side, eyes closed, with its black legs stretched out behind it, like goslings do when taking a serious nap. Only this gosling was lying on a cement sidewalk alone. It was one from the family of 3 who hatched at the golf course water trap. This gosling was hit on Memorial Day when no one goes to work, traffic is very light, and there is plenty of time and room to slow down and stop—if a driver is paying attention.

On the sidewalk on the other side of the street, where the geese walk down a hill through thick vegetation to get to the golf course, there were several goose droppings. There had never been any droppings there before. The parents and other two goslings must have waited there, calling out to the gosling who was hit.

Instead of 23 years this little gosling only lived 13 days. It never learned to fly, went on migration, found a life-long mate, or raised a family. This particular corner is deadly. During this same 12-month period 3 deer (*Odocoileus virginianus*), 2 squirrels, and 1 raccoon (*Procyon lotor*) were also killed; all on weekends. None of this had anything to do with survival of the fittest; rather human stupidity and callousness were the root cause.

The red-winged blackbirds (*Agelaius phoeniceus*) were also affected by the changes at Ponds A and D. In the past, Ponds A and D had a thick curtain of cattails. The red-winged blackbirds nested in the cattails. Neither pond had any cattails during 2011 or 2012, due to the fact that Pond D "moved" and Pond A was in a perpetual flood state. The red-winged blackbirds returned to Pond D on 4/17/2011, but they did not nest there. No red-winged blackbirds were seen at Ponds A or D in 2012.

In 2012 another pattern of behavior was observed. The Pond A pair nested somewhere secluded near Pond D since Pond A was still unusable. Their solitary gosling hatched on 4/25/2012. The gosling's initial movements were wobbly and tentative. In previous years this pair had produced 4 or 5 healthy goslings. What caused the change? Did the Pond A pair dine on contaminated food while wintering in the Gulf? Stories about fish that were deformed as a result of the BP Gulf oil spill began appearing in the news at that time. Or, was there another cause? This gosling was not seen again until 4/28/2012 and 4/29/2012 when it was spotted at Pond A with its parents. It was sturdier by then.

Seven goslings hatched at Pond D Sunday morning, 4/29/2012. They were a bundle of energy from the beginning, running and swimming to keep up with their parents and siblings.

Monday, 4/30/2012, both families and the 8 goslings were at Pond A. Tuesday, 5/1/2012, the family of 7 goslings returned to Pond D, but the other family was nowhere in sight. Wednesday, 5/2/2012, both families were at Pond D. The 8 goslings were running and swimming together. Only one set of parents monitored the goslings at a time. Thursday, 5/3/2012, no goslings were spotted anywhere.

Friday, 5/4/2012, 8 goslings and one set of parents were at Pond D. It was the pair that nested at Pond D. The male from that pair could be identified since he has several small white feathers on the front of his neck. He was given the name Mr. Speckleneck in 2011 when he and his mate first nested at the new Pond D. The solitary gosling from the other family could be identified because he was a little larger than the other 7 goslings. This new combined family of 8 goslings alternated between Pond A and Pond D for several days. The Pond A parents did not return. Perhaps they decided it was best for their solitary gosling to grow up in a more normal group environment.

Then, on the evening of Friday, 5/11/2012, the Pond D parents were observed escorting 11 goslings. The 3 newcomers were slightly smaller. They had hatched at the golf course water trap. It is unclear why their parents decided to let them merge into the Pond D family.

None of the goslings were seen again until 5/16/2012 when they all reappeared at Pond D. From then on this new expanded family made Pond D their home, except for occasional day hikes. The Pond D pair accepted and cared for all the newcomers as if they were their own goslings. All of the goslings were comfortable with this arrangement too, even though there was some difference in size. (See Photograph 3.36.)

On 5/30/2012 the Pond D pair and 11 goslings made a day hike to Pond B. They spent the night there. The Pond D pair and 10 goslings returned to Pond D on the evening of 5/31/2012. The missing 11th gosling was found swimming and napping by itself at Pond B on 6/2/2012. It stayed near a group of mallard and wood ducks. Did this gosling get left behind? Did this gosling choose to stay at Pond B? Or, did this gosling accidentally become separated from its expanded family and find its way back to Pond B? For some reason this gosling did not associate with nor was it adopted by the Pond B pair and their 5 goslings.

Photograph 3.36 An Expanded Family. In 2012 the Pond D pair accepted four new goslings that hatched elsewhere into their family. The newcomers were over a week old when they joined the Pond D family. They were not orphans, yet the Pond D pair adopted the newcomers. How and why the four goslings were able to un-imprint from their original biological parents and re-imprint into Mr. Speckleneck's family remains a mystery. These four goslings remained part of the Pond D family from then on. Here we see Mr. Speckleneck leading his expanded family of 11 goslings for a swim at Pond D on 5/11/2012. Note that the 5th, 7th, 10th, and 11th goslings are larger (older) than Mr. Speckleneck's original family. © D. Herrmann

Most likely the latter cause was the reason for the gosling's separation. Canada geese parents are not known for abandoning their young. Goslings are not known for running off on their own. On 6/3/2012 the Pond D pair returned to Pond B and the 11 goslings were reunited.

The new pattern of behavior that was observed in 2012 raises some interesting questions. For several decades now it has been common knowledge that upon hatching precocial ducklings and goslings imprint on their parents and siblings based on some combination of sight and sound. This imprinting is accomplished to such a fine degree that individual goslings can identify their parents and siblings in a crowd. In fact, this is what was observed consistently from 2005 through 2011.

- How then did the Pond A pair convince or encourage their solitary gosling to join Mr. Speckleneck's family of 7 goslings?
- How did the pair from the golf course water trap convince or encourage their 3 goslings to join Mr. Speckleneck's family of 7 goslings and the other solitary gosling?

The Pond A pair's solitary gosling was 8 days old when it changed families. The 3 goslings from the golf course water trap were 9 days old when they changed families.

- How were the 4 goslings, who were a little over a week old, able to un-imprint from their original biological parents and re-imprint into a new family?

According to what is known and assumed about imprinting, this should not have been possible. These goslings were not orphans, yet they were adopted.

Reconvening the Flock

This brings us to the fourth row in Table 3.5 and a behavior that is not discussed elsewhere. From 2005 through 2010 the individual families of Canada geese reconvened as a flock at Pond C after the goslings were nearly full grown and able to fly. Somehow each year all of the pairs, their goslings, and the unattached adult geese came together on the same day and the flock was reunited. This behavior was the counterpart of the nesting dispersal behavior. How they all knew what day to gather is unknown.

Once reunited, a scan of the flock indicates that all the goslings are at essentially the same stage of development and individual families are discernable. Generally this group behavior takes place during the first week of July, when the goslings are about eight weeks old. During the first eight weeks of life goslings learn basic survival skills from their parents. From now on goslings learn social skills and other norms of behavior from the flock, and that they are members of a flock, not just a nuclear family. The red-winged blackbirds and swallows (*Iridopraene bicolor*) generally left Ponds A and D the same week as the Canada geese.

In a typical year the Canada geese family will be at Pond A in the morning, then depart to reunite with the flock midday on the designated date. If the Canada geese family cannot be located at Pond A and it is near July 1st, a check of Pond C confirms their presence there and that of all the other geese families. From then on the flock, in subflocks of various sizes, alternates between Ponds B, C, and D, and the golf course throughout the summer and fall. The two holes nearest the water trap are a favorite hangout. In mid-August the Pond A Canada geese family usually returns to Pond A for a day or two. Perhaps the parents want to remind the goslings that Pond A is their traditional nesting site. Then they, like the rest of the flock, begin preparations for the fall migration. The Pond A pair with their solitary grown gosling returned to Pond A for final visits on 8/26/2012 and 8/29/2012. Mr. Speckleneck brought his family back to Pond D for a day on 8/2/2012.

In 2010 a different behavior pattern was observed. On June 27th the Pond A Canada geese family was observed between 8 a.m. and 9 a.m. By noon they had reconvened with the flock at Pond C, the central meeting place. On June 28th the flock of 42 Canada geese was resting beside Pond C between 8 a.m. and 9 a.m. Between 2:30 and 3:30 p.m. a serious thunderstorm passed through the area. Between

5:30 and 6:30 p.m. a check was made to see how the storm affected Pond A. It was twice as wide and 1.5 times as long. And, amazingly the three goslings and their mother were standing midway up the north side of the pond's bank. The geese walked up and down the slope eating the moist grass. They swam and bathed for a while in areas where the pond had expanded and drank the new cool clean water. The next morning the father goose joined them. This was the first time the Pond A Canada geese family returned after reuniting with the flock, other than their brief visits in mid-August. The geese family stayed at Pond A through the morning of July 10th—an additional thirteen days.

Another severe thunderstorm with heavy rain passed through the area mid-morning on July 10th. By 12:30 p.m. the Pond A Canada geese family had left again. They were spotted at Pond C the next morning. Between 2 p.m. and 3 p.m. on July 12th, the Canada geese family returned to Pond A a second time. They stayed until the morning of July 14th when they left for good. The Pond A Canada geese family was spotted at Pond B the next morning.

What prompted the Pond A Canada geese family to return these two times, especially the first time when they stayed an additional thirteen days? Was it just the severe weather, or were there other reasons at work as well? If it was just the weather, why did the other geese families stay at Ponds B and C? This behavior indicates individual preferences, choices, decision-making, and reasoning by Canada geese pairs and families.

In 2011 the flock reconvened at Pond C on July 5th. In 2012 the flock reconvened at Pond C on July 20th.

Fall Migration in a V Formation

During the fall Canada geese migrate southward in flocks, which can number 12 or more geese. Various family groups come together to form a flock, which functions as an extended family and community. Goslings that hatched during the spring stay with their parents throughout the fall migration, winter, and the spring migration back home.[95, 157, 203] Unattached adults, those too young to have a mate, and those who have lost a mate also affiliate with a flock. Flock sizes during the fall migration tend to be larger and more noticeable than those during the spring migration.[41] The departure date for the fall migration varies from mid-September through late November and depends on seasonal weather patterns and the latitude at which the Canada geese spend their spring and summer.

Understanding the need to migrate—the impending change of seasons and scarcity of food—demonstrates a combination of genetic intelligence and applied intelligence. This understanding represents a shared memory going back hundreds of generations and hundreds of years. So does understanding that there is another place to live for three to four months each year to mitigate these circumstances. This understanding requires knowledge and awareness that there is a spring/summer/fall home and a different winter home, as well as how and when to travel between the two.

Canada geese tend to follow similar migration routes and have traditional resting or stopover points along the way.[95, 203] Canada geese use a combination of genetic intelligence and applied intelligence during migration: shared memory of landmarks, the sun's position, the position of the stars, and the earth's magnetic fields as navigational aids.[41, 136] Migratory routes are updated as needed to reflect changes in the habitat overflown.[95] For example, Canada geese have been observed flying higher than usual during hunting season and/or altering their routes to avoid areas where hunting is permitted.[41] Hence, genetic intelligence is passed from one generation to the next year after year. But, the knowledge passed is not static or frozen in time—it can be updated and refined as needed.

North American Canada geese spend their winters along the Gulf coast and as far south as Mexico.[157] In the past Canada geese populations suffered due to over-hunting and from the destruction of their spring/summer/fall habitat where they nest. Now the destruction of Canada geese winter habitat, particularly wetlands, is of concern.[41] Canada geese's winter habitat is equally important as their spring/summer/fall habitat. Courtship behavior and pair bonding takes place at the winter habitat.[41] It remains to be seen what short-term and long-term detrimental effects the 2010 BP Gulf oil spill will have on Canada geese, mallard ducks, hummingbirds, monarch butterflies, and other species who winter in the area.

The famous Canada geese V formation used during migration is a classic example of cooperative living.[141] First the flock assembles gradually. This process may take a couple days or a couple weeks depending on geography. The flock may have spent the summer together at the same large waterway. Or, the flock may have been dispersed among several smaller waterways in the same general locale. The latter is the case in the local area. Flock members rotate among Ponds A, B, C, and D according to a random pattern throughout the summer, along with occasional visits to the golf course. In preparation for the fall migration Canada geese molt their primary flight feathers and replace them with new strong feathers.[95, 157] This is an example of biological intelligence at work.

Next, the flock begins training for the long flight ahead. After all, one cannot spend the summer basking in the sun by the pond. Then wake up one day and decide to fly 1,000 miles (1,330 km). During late August and early September small groups of Canada geese, usually 2 to 6 individuals, start flying laps around the immediate area. In 2010 the first practice group was observed on August 27th. Later several practice groups were observed on September 28th. The group forms half of the V formation and circles overhead at an altitude 2/3 that of the migration standard. The group circles two or three times before landing and honks like they will during migration. Generally the group repeats their practice laps three or four times a day; sometimes with the same group members, other times there are less or more geese in the practice group. Seven was the highest number of geese observed in a practice group.

One practice group of four was very punctual. One year they were observed doing practice laps promptly at 6:50 a.m., 9:50 a.m., 3 p.m., and 6:15 p.m. each day. Toward the end of practice time, when the sun came up later and set earlier, the times of the first and last practice laps were adjusted by about ten minutes.

At this time of year it is also common to see the flock swimming in formation across Pond C in preparation for migration. Sometimes the flock forms a V; other times the flock just forms one side of a V. Individual geese are evenly spaced at angles to each other, just like they will be when flying. In 2010 several geese were observed swimming in formation on August 29th.

Suddenly the days get shorter. The temperatures get cooler. The angle of the sun changes. Food is less readily available. And, it is time to head south. One morning instead of seeing a practice group doing laps at 7 a.m., the entire flock is airborne in a V formation, honking, and flying in a southern direction. The local flock generally departs during the last two weeks of November.

Pond C is a main stopover point for flocks of Canada geese that are migrating from points farther north. This is probably due to the fact that Pond C is the largest of the four ponds in this area and it is a bit more secluded. Frequently two or three flocks that are passing through the area will arrive at the same time. Depending on weather patterns, flocks from up north may arrive at Pond C before the local flock has started their migration. On occasion migratory flocks also use Ponds A and B as temporary stopover points.

The weather along the east coast of the U.S. was atypical during the fall of 2009. Alternating periods of below normal and above normal temperatures were experienced. This contributed to unusual migratory patterns. Flocks of Canada geese were observed using Pond A as a stopover point on October 18th and Pond B on October 20th—a month earlier than usual.

Frequently multiple flocks landed at Pond B, Pond C or the golf course at the same time during fall migration. The visiting flocks varied from 10% smaller to 10% larger in size than members of the local flock. Sometimes members of the migratory flocks of Canada geese mingled with each other and/or the local flock; other times the flocks stayed apart. Likewise, some of the migratory flocks of Canada geese that stopped at Pond B mingled with the local mallard ducks. Other flocks of Canada geese did not appear to be familiar with mallard ducks; they stayed in the middle section of the pond away from the ducks.

Some of the interactions among flocks were quite interesting. On 11/28/2010 a migratory flock of 6 geese visited Pond B. Perhaps this was a pair and their four goslings that were making the journey for the first time. A lone goose from the local flock was still at Pond B. The local goose called out to the visitors and flew over to see them. One of the visiting geese followed the local goose into the pond. The two of them swam around together for a while. This behavior might have indicated the beginning of a new pair bond. But alas, despite the local goose's hospitality, the migratory flock left the next day without it.

On 12/31/2010 multiple migratory flocks of Canada geese stopped at the golf course. Initially the flocks stayed apart. Gradually members of different flocks began mingling and there was some quiet bantering. The geese lined up single file in groups of 10 or more birds whenever they walked up and down a hill or to and from the partially frozen water trap. (See Photograph 3.37.)

During 2010 the earliest migratory flocks visited stopover points in the area November 12–15. The last migratory flocks visited these stopover points December 31st. Unlike earlier flocks, they encountered light snow and partially frozen ponds.

Twenty-three flocks, with a total of 609 migratory Canada geese, passed through the area during fall 2010. The largest flock size observed was 85 Canada geese. The smallest flock consisted of 6 Canada geese. And, the average flock size was 26 geese.

The typical stopover duration each year was 1.5 days. Usually a flock arrived late afternoon. They spent the next day resting and grazing, or perhaps doing a little swimming. Then the flock left early the following morning. The geese stayed longer if the weather was severe, particularly if it was windy. This allowed the flock to conserve energy. However, the flocks always departed 12 hours prior to a snow storm.

Photograph 3.37 The Line Up. From the day they hatch onward young goslings line up behind their parents, on land and in water. Lining up in single file does not end when goslings become young adults. Canada geese can be seen walking single file throughout their lives, especially during the summer molt when they are flightless. This behavior reinforces a flock identity, since the flock functions like an extended family. Lining up single file also mimics the V flying formation for which Canada geese are famous. © D. Herrmann

Table 3.7 summarizes the visits migratory flocks of Canada geese made to stopover points in the area during the 2010 fall migration.

Migration isn't always as simple as south in the fall and north in the spring, especially in this time of global climate change. Migratory flocks of Canada geese that had flown south for the winter temporarily returned to the mid-Atlantic states more than a dozen times in early 2011. These flocks came back north for a few days when the southeastern states were uncharacteristically blanketed

Table 3.7 Dates Migratory Flocks Visited Local Stopover Points During 2010 Fall Migration.

Month	Dates	Number of Geese	Stopover location	Notes
November	12–15	64 + 12	golf course	• Geese arrived Friday afternoon and left Monday morning. They are 10% smaller than the local flock. • Local flock left with them, except for a lone goose that stayed behind at Pond B. • 64 geese were on the ground at dusk on Friday, when another flock of 12 landed and joined them. They seemed curious and interacted with each other.
	20	11	Pond C	• Geese swam from NW corner to SE corner and back. Awhile later, after some brief bantering, they all took off flying south.
	21	8 16	Pond B Pond C	• Geese at Pond B stayed in the middle section in the sunshine, apart from the local ducks. • Geese at Pond C napped in the sun in the SE corner, then swam to the NW corner. 1 young male mallard duck stayed near them.
	28	6 13 5	Pond B Pond C flying overhead	• Geese at Pond B stayed on the lawn on the west side, across from the local ducks. • The lone goose called out and flew over to socialize with them. 1 goose swam with it for a while. • Geese at Pond C rested in the NW corner. • 1 goose from this flock stayed behind for 2 days at Pond C. • 5 geese flew between the golf course and Pond B.
	30	26	golf course	• This flock grazed near the water trap.
December	3	11	Pond B	• This flock of geese swam freely with the local duck population.
	4–7	8 76	Pond B golf course	• The geese at Pond B stayed on the lawn west and south of the ducks on the 1st day. • The geese at Pond B swam with the local ducks on the 2nd day. • The geese at the golf course are 10% larger than the local flock. • Both groups stayed over a 2nd day because of very windy conditions (+30 mph/45 kph).
	21	58	golf course	• Grazing between patches of snow.
	24–25	85	golf course	• Heavy snow in southeastern states.
	28	28	golf course	• States north and south of here covered in snow.
	29	7	golf course	• Flew in from south.
	30	16 + 22	golf course	• Two flocks.
	31	21 + 42 + 42 + 12 20	golf course water-trap	• Multiple flocks spread out across course and partially thawed water trap. Flock of 12 was 10% smaller than others.

with snow. Also, the 2010 BP gulf oil spill may have decreased the amount of food available at the Canada geese's wintering grounds. The geese had to find food somewhere. They couldn't fly farther south; so they headed north to a known stopover point in hopes of finding food. This new mid-winter partial migration behavior highlights the need to document and preserve waterfowl stopover points.

In the fall Canada geese store body fat to fuel their long migratory journey. That fat is depleted by the time the geese reach their destination. When Canada geese and other waterfowl have to make a sudden partial journey back north, in response to freak snowstorms in the south, there isn't time to accumulate fat reserves. The Canada geese and others must leave immediately or face possible starvation.

When flocks of geese arrived under these circumstances, they were often exhausted. This state was particularly evident when flocks arrived at Pond C on 1/13/2011 and 1/19/2011. The first thing the geese did after landing was sleep. Usually they rested in the northwest corner of the pond. Later, the geese began to eat. After the geese had a chance to rest and refresh themselves, they were very talkative. On extremely cold and windy days, the geese minimized their activity level to conserve energy.

An interesting behavior was observed midday on 1/10/2011. A migratory flock of 82 Canada geese had returned north to Pond C the previous evening. Pond C was about half frozen at the time. The flock was spread out. Some geese were swimming. Some geese were walking around on the ice. Others were upending or resting.

A flock of 18 Canada geese flew over Pond C. The flock of 82 called out to this new flock using the "honk-lite" vocalization. The new flock turned around and landed to join the flock of 82 at Pond C. Once they landed the newcomers mingled with the larger flock and there was a lot of chatter back and forth.

Both flocks had temporarily returned north on a sudden partial mid-winter migration. Both flocks were in survival mode. Yet, the flock of 82 who were already on the ground at Pond C invited the flock of 18 to join them and share the limited resources. How many humans would do the same thing under similar circumstances?

During these mid-winter visits some flocks of geese stayed put at whichever pond they landed. Other flocks traveled back and forth between Pond C and the golf course, Pond B and the golf course, or Ponds B and C in search of food. One pond might be more thawed than the others. More grass could be found under the snow in one location than in the others. Usually the flock stayed put, while groups of 8 to 12 geese traveled back and forth. Pond C was the favorite place to spend the night.

Each year the Canada geese needed a few days to get used to walking and landing on an ice covered pond. Canada geese use their wings, tail, and nails to slow down when landing on ice and regain their balance if they slip. Some geese slept sitting on the ice, just like they do on the ground. Other geese took a nap while standing on one foot on the ice. Often several geese walked close together to break the thin ice at the edge of the pond—and expose more food sources. Geese did not panic when the ice broke; they just swam or hopped out. The geese seemed to be intrigued when they saw their own reflection in the ice. (See Photograph 3.38 and earlier Photograph 1.2.)

A total of 22 mid-winter visits were made by 35 migratory flocks to stopover points in the area during 2011. The longest visit was for 6 days: January 19th through 24th. It is possible that some of the flocks were repeat visitors. Table 3.8 summarizes the mid-winter visits migratory flocks of Canada geese made to stopover points the area during 2011.

Flying in a V formation is a cooperative effort that benefits all flock members. Generally the older and stronger members of the flock fly in the forward positions.[136] These members of the flock have made the journey many times and are the most knowledgeable about the route, stopover points, and possible alternative courses of action should unusual situations be encountered. The younger members of the flock, those making the journey for the first time, and the elderly fill out the back edges of the V. This behavior is a variation of goslings lining up behind their parents.

Photograph 3.38 Ice and Snow. (Top) The weight of multiple geese walking close together is often enough to buckle thin ice near the shore. Here we see a group of geese taking advantage of a new opening in the ice to bathe. (Bottom) When multiple flocks arrived at Pond C at the same time, sometimes they mingled; other times the flocks remained separate. No hostility or aggression was ever displayed. The flocks seemed to understand that Pond C was a temporary stopover point and they were all just passing through. © D. Herrmann

Geese flying in the forward positions, especially the center of the V, expend more energy while creating a slipstream of reduced air pressure and forward suction that minimizes the effort required by the rest of the group.[41, 136] These elite flyers rotate positions during flight in order to rest a bit,[41, 136] similar to professional bicycle racing teams. The V formation also promotes better visibility since each goose can see what is happening ahead and beside them, as well as the birds in front of them.[136] Thus the cooperative effort inherent in flying in a V formation helps ensure that the flock as a whole reaches its destination safely. In time the younger members of the flock will assume the positions of

Table 3.8 2011 Mid-winter Partial Migration North in Response to Snowstorms in the Southeast U.S.

Month	Dates	Number of Geese	Stopover location	Notes
January	1	12	golf course	
	3	44 12	Pond C flying overhead	• 1/4 thawed • Flock of 12 may be same group that was here on 1/1
	4–6	82	Pond C	• Many slept on ice, facing sun
	7	19	Pond C	• There were no geese at Pond C on Jan. 8th
	9	82	Pond C	• Very cold and windy, geese stayed put
	10	82 18	Pond C flying overhead	• 18 flying overhead turned around and joined the flock of 82 at Pond C after hearing them honking
	11	100	Pond C	• Geese grazed at Pond B in groups of 12–16, then flew back to Pond C to swim or rest • There were no geese at Pond C on Jan. 12th
	13	55	Pond C	• Geese seemed very tired, most rested. • Very cold and windy.
	14	16	Pond C	• There were no geese at Pond C Jan. 15–18.
	19-24	100	Pond C	• Very active and talkative after they had rested • Different groups vocalized at different times • Some pulled up dormant grass and ate the roots • Some flew laps between Ponds B and C • When temperature dropped into teens became quiet and still
	25	18	Pond C	• There were no geese at Pond C Jan. 27–Feb. 13, since it was frozen over completely
February	14–16	96	Pond C	• 2/3 thawed • Lots of energetic bathing
	17	12 + 9	Pond C	• Two flocks
	18	36	Pond C	
	19	72	Pond C	• Very vocal on windy days
	20	26 26 6	golf course Pond C Pond B	• Grazing on lawn • Resting on lawn near southern part
	21	26 51	golf course Pond C	• 10 flew over from Pond B
	22	30	golf course	• Grazing on snow
	23	30 + 12 100	golf course Pond C	 • Very energetic bathing, talkative
	24	8 18	Pond B Pond C	• Grazed in same area as they did during fall migration • Calmer today
	26	2 30	Pond A Pond C	
	27	2 2 6 72	flying overhead Pond A Pond B Pond C	• Beginning of spring migration

the elite flyers, and the elite flyers will assume the positions of the elderly, as part of the unbroken chain of life. The flying formations of air forces from around the world are copied from the fall migration patterns of geese.

Migration is a normal and predictable bird behavior.[41, 95, 107, 136, 203] Birds do not encrypt their flight plans or use stealth technology to avoid detection by radar. On the contrary, flocks of Canada geese are visible on radar and from air traffic control towers. Instead of hoping for another miracle, like the landing of US Airways flight 1549 in the Hudson river on January 15, 2009, military and commercial flight patterns should be adjusted on a seasonal basis to accommodate bird migration patterns. This is common sense and there is plenty of scientific evidence to support this proposition. The bird world has been conducting serious studies of bird migration patterns since the mid-19th century, long before airplanes existed. However, only recently the bird community and military and civilian aviation communities began sharing information. We are all responsible for ensuring the safety of migrating birds. If the birds get where they are going safely, so will the rest of us who share the skies. For more discussion of this issue see Table 3.9.

Table 3.9 Humans and Bird Migration: Toward Peaceful Coexistence.

Dr. LeShem, of the Tel Aviv University Department of Zoology, is a leading figure in the effort to reduce bird related airplane crashes. Israel has one of the highest densities of native bird populations per square mile of any country. In addition, some 500 million birds migrate through Israel each year. His research led to the formation of the International Center for the Study of Bird Migration. By studying the factors below Dr. LeShem developed a Bird Plagued Zone map which determines when, where, and at which altitude(s) aircraft flights are forbidden during fall and spring migration:[107]

- Migration routes, altitudes, and times by species
- Daily, seasonal, and annual fluctuations in migration patterns
- Flight speed and other characteristics
- Daily and seasonal progress rates
- Biological factors affecting migration
- Meteorological factors affecting migration

In the 14 years since the Bird Plagued Zone recommendations were implemented in Israel—the major bird migration route between Europe, Asia, and Africa—serious bird/aircraft collisions decreased by 76%.[107] Dr. LeShem's research has been shared with the Bird Strike Committee Europe (BSCE) and the International Bird Strike Committee (IBSC).[103, 104, 105, 106]

In most countries bird strike committees are staffed by volunteers. The birding community tends to be under-represented on these committees. Anyone interested in participating should contact their local committee. Information about the Bird Strike Committee USA can be found on the website: www.birdstrike.org.

There is no reason why governments of other countries cannot or should not implement equivalent Bird Plagued Zone recommendations in their jurisdictions. It is a win-win proposition. Birds' lives are saved. People's lives are saved. The environmental damage that accompanies airplane crashes is avoided, as is the cost of replacing downed aircraft. These statistics from the Federal Aviation Administration about bird strikes in the U.S. between 1990 and 2008 underscore the logic of this position:[56]

- Between 1990 and 2008 a total of 87,416 bird strikes were experienced
  - o 62% of bird strikes occurred during the day
  - o 92% of the strikes were at or below 3,000 feet (914 m) above ground level
  - o 51% of the strikes occurred between July and October
- Of the 38,474 known birds involved in strikes, the majority were:
  - o Gulls 7,470
  - o Pigeons/doves 5,590
  - o Raptors 5,070
  - o Waterfowl 3,175
  - o Shorebirds 2,604
  - o Sparrows 2,358
  - o Starlings/mynas 2,115
- Total aircraft downtime from all bird strikes was 393,521 hours
- Total economic loss from all bird strikes was $308,313,001

3.5 Individual Behaviors

On several occasions goslings and adult Canada geese were observed engaging in behaviors that indicated individual reasoning, individual preferences, and individuality. For example:

- Jumping off tall buildings
- Attempting to upend too early
- Hesitating before jumping in the water
- Swimming and walking away from the family
- Imitating the movements of a water snake
- Preferring the company of mallard ducks
- Constructing pond entrances/exits

Jumping Off Tall Buildings

One or two Canada geese repeatedly flew to the top of a 16-story building near Pond B, then jumped off. The geese half coasted, half flew back to the ground. Before jumping off the building the geese chattered loudly to make sure everyone within hearing distance knew they were on top of the building. Only two geese engaged in this play activity. This behavior was observed during April and May of 2011. Since this is nesting time, most likely these two were young geese that had not yet formed a pair bond. This behavior was not observed in 2012. (See Photograph 3.34.)

Attempting to Upend Too Early

Goslings can be seen trying to upend and imitate their parents when they are only a few days old. However, young goslings usually only do so in water that is a few inches (centimeters) deep. As a rule goslings do not upend in deep water until they are 6 to 7 weeks old and their neck has grown to nearly full size.

On one occasion two adult Canada geese were swimming in Pond A with their three 3-week old goslings. Suddenly the parents upended to retrieve some food from the middle of the pond. Two of the goslings just watched. In contrast, the third gosling tried to follow the parents' actions. Goslings, like other birds, learn by watching and imitating their parents. The gosling turned upside down and dove. But it was still covered in down and too small to have much buoyancy. Worse, its wings were still just stubs; it did not yet have the feathers necessary to flap and get upright again. The parents resurfaced just as the gosling went under. Immediately, both parents fixed their gaze on the spot where the gosling went underwater. The other two goslings turned and faced in that direction too. One parent swam one length toward the gosling. More than 10 seconds passed. All of a sudden the gosling splashed and spluttered to the water's surface, about 4 feet (122 cm) from where it went under. The rest of the family swam over to the gosling and they all rested there briefly.

It is interesting to examine this episode in detail. First, the two adult Canada geese upended to retrieve some aquatic vegetation for nourishment; this is a normal behavior. Two of the goslings decided to watch their parents and store the knowledge of this behavior for later when they were older and larger. The third gosling attempted to imitate its parents' behavior, consistent with the manner in which young birds acquire knowledge and survival skills. For some reason one gosling tried to replicate the parents' behavior, while the other two goslings did not. This episode demonstrates individual reasoning patterns among the three young goslings.

Then there is the Canada geese parents' reaction to this incident. The parents realized their gosling was in danger when they resurfaced. Unlike human parents in a similar situation the Canada geese did not panic, become hysterical, suddenly get religious, or call 911 for someone else to come and deal with the crisis. Instead, they stood guard, watched intently, and moved into position to help if the gosling could not help itself. The adult geese could have lifted the gosling to the water's surface on their bills or wings. Instead, they let the gosling learn how to recover from this situation by itself, thus gaining important knowledge and self-confidence.

Lastly, there is the reaction of the Canada geese parents and siblings after the gosling resurfaced. The gosling splashed and spluttered to the top of the water until it was floating upright again. The two parents and two siblings swam over to the gosling immediately. Unlike humans in a similar situation, there were no irrational displays of excessive emotion or berating the little bird for having tried such a stupid thing or spoiling it thereafter because of its miraculous recovery. No, the family as a whole simply came over to the gosling to make sure that it was alright. And, they stayed there for five minutes so the little gosling could rest and catch its breath before they all continued on their way. This observation of individual reasoning and behavior by goslings took place at Pond A during May 2010.

Hesitating Before Jumping in the Water

A second observation of individual reasoning and behavior by young goslings took place during May 2010 at Pond B. In this instance two parents and their four goslings were on the west side of Pond B in the section the geese normally frequent. The goslings were the same size and stage of development as those at Pond A. The father goose finished grazing, jumped in the water, and began to swim toward the east side of Pond B, a distance of about 36 feet (11 m). One by one three of the goslings jumped in to follow him. In typical fashion the goslings lined up to swim behind the father.

The fourth gosling hesitated. The water's edge was 6 to 8 inches (15.24 to 20.32 cm) below the land where the gosling stood. The gosling looked down at the water for a while. The gosling walked 2 feet (61 cm) along the water's edge in one direction, and looked down at the water. Then the gosling walked 2 feet (61 cm) along the water's edge in the other direction. Again the gosling looked down at the water. The gosling returned to its original position and looked down at the water one more time. At last it jumped into the water and the mother did too. The two of them swam quickly to catch up with the rest of the family, who by now were half way across the pond. The total delay was about 20 seconds.

Three of the goslings jumped in the water right behind the father. The fourth gosling hesitated. Possibly it was concerned or afraid of something—it kept looking down into the water. A large snapping turtle, with a 14 inch (35.5 cm) shell, and several 6 to 8 inch (15.24 to 20.32 cm) red-eared slider turtles (*Trachemys scripta elegans*) have been seen in this pond, as well as fish 6 to 8 inches (15.24 to 20.32 cm) long. Did this gosling have a negative encounter with some form of aquatic life previously? Did the other goslings have the same experience, but decide to follow the father anyway? Three goslings exhibited one behavior, while the fourth gosling exhibited a different behavior in response to the identical situation at the same time. This episode indicates individual reasoning patterns among the four young goslings.

It is interesting to note that the mother goose stayed behind and waited patiently until the fourth gosling was ready to jump in the water. She did not nudge, nag, belittle, bribe, or coerce it. Instead, the mother goose simply stood still and waited patiently.

On this trip all six geese made it safely to the other side of the pond. However, the gosling's apprehension was not entirely misplaced. There were only three goslings eleven days later when I checked on this goose family.

Swimming and Walking Away from the Family

A third observation of individual reasoning and behavior by goslings took place daily at Pond A throughout May and June of 2007. Four goslings hatched at that location in 2007. At various times during the day the goslings lined up behind the father, on land or water, and headed in whatever direction he suggested. Before long, one of the four goslings headed in a different direction, usually 90° away from where the group was heading. It walked or paddled on its own, unaware that anything was amiss. As a result this gosling was nicknamed "Number 4."

Before Number 4 got too far off track, the mother made one quiet, soft, short honk. Number 4 would look up, as if surprised, then hurry to rejoin the group. The mother was in a better position to see that Number 4 had gone astray, since she walked or swam behind the goslings. Hence she, and not the father, did the honking. This individual behavior continued the entire eight weeks the goslings were growing up. Even in the week before the goose family flew away, when the goslings were three-quarters the size of the parents, three of the goslings would be grazing with the parents and Number 4 would be a few feet (~1 m) away doing its own thing.

The four goslings were raised by the same parents in the same location at the same time. Three of the goslings exhibited one behavior, while the fourth gosling exhibited a different behavior. This episode indicates individual reasoning patterns among the four young goslings. Three goslings consistently followed the parents. In contrast, Number 4 swam or walked at right angles away from the family. Then Number 4 made a decision to rejoin the group, after hearing its mother's solitary honk. Again, the parents dealt with this situation in a consistently calm manner.

Imitating a Water Snake

During May 2010 at Pond B an adult Canada goose was swimming across a pond. Four small goslings 3 to 4 weeks old were lined up behind him; the mother was swimming behind them. The adult in the rear swam with her head upright as usual, scanning carefully in all directions. The adult in front had his neck flat, stretched out level with the water. His head and eyes were above water; the underside of his neck was in the water. He swung his head and neck back and forth in an S-like fashion, as if he were attempting to locate and deter predators that might be nearby and attack the goslings. Perhaps the adult goose was imitating the movement of a water snake to warn off fish. Fish can misconstrue the movement of a gosling's tiny feet on the water's surface as insects and bite them; thereby pulling a gosling underwater and causing it to drown. The adult goose continued this protective behavior until he and the goslings reached land on the other side of the pond. This protective posture was not needed to protect the adult goose himself, only the goslings. This behavior represented an unusual cooperative effort by the two parents to protect their young, since they employed different but complementary security strategies. About two weeks later a water snake was spotted at the pond, slithering along the surface of the water, making the exact movements the Canada goose father had made. While other Canada geese parents have been observed scanning the water for dangers, this was the only time one was observed making an S-like movement in the water. (See Photograph 3.39.)

Preferring the Company of Ducks

On several occasions from May through November 2010 individual Canada geese stayed behind at Pond C, while the other geese split up into subflocks and temporarily relocated to Pond B or the golf course. The local flock frequently splits into subflocks and rotates among the four co-located waterways after the current year's goslings are grown. What was unique on these occasions was that an individual goose stayed behind and socialized with mallard ducks. If the mallard ducks went swimming, the goose went swimming. If the ducks basked in the sun at the water's edge, the goose basked with them. The ducks did nothing to discourage the goose. Later in the fall a lone Canada goose stayed behind at Pond B with the mallard ducks for a while, after the local flock of Canada geese had left. This behavior, along with the constantly changing composition of subflocks, infers

Photograph 3.39 Swimming. Canada geese swim frequently for business and pleasure. A Canada goose sits upright with its neck straight when conducting serious business. A Canada goose sits lower and curls its neck as it begins to relax. A Canada goose coils its neck against its back when most content. A Canada goose curls its head forward level with the surface of the water when scanning for water-based predators, like the father goose here who is leading his young goslings across Pond D. © D. Herrmann

that Canada geese individually decide whether or not to participate in flock and subflock activities. This behavior also indicates that Canada geese are aware of other birds in their neighborhood and are willing to socialize with them.

Constructing Pond Entrances/Exits

Throughout the spring and summer of 2010 it was noted that the Canada geese family at Pond A always entered and exited the pond at the same 3 points. If they were on the north, east, or south side of the pond the geese always entered or exited the water from the same place, regardless of whether they were making a leisurely entrance or a rapid entrance in response to a perceived danger. The Canada geese family lined up and passed through the entrance/exit in single file: the father goose first, the three goslings, then the mother goose. The geese did not enter or exit the pond from the west side, since that is the narrowest and most rocky point.

Until 2011 Pond A was surrounded by a thick 1 foot (30.48 cm) perimeter of tall reeds, grasses, and cattails, which in some cases were taller than an adult Canada goose. Upon closer examination it was discovered that the Canada geese had constructed a single entrance/exit on each of the three sides of the pond. To facilitate ease of movement—probably with the goslings' safety in mind—the Canada geese parents had flattened a small area of vegetation on each of the three sides of the pond. This activity created an opening (or pond entrance/exit) the width of one adult goose, thereby eliminating the need for the Canada geese family to randomly push their way through dense vegetation every time they entered or exited the pond.

The entrance/exit was located at the midpoint on each side of the pond. The entrances/exits provided a clear view of what was in the water ahead, for safety purposes. This feature would not have been available if the geese had randomly pushed their way through the vegetation each time they entered or exited the pond. Creating a pond entrance/exit made it easier and faster for the goslings to enter and exit the pond, which was important in emergency situations. The entrances/exits also avoided disturbing the red-winged blackbirds that were nesting nearby in the cattails. (See Photograph 3.40.)

Ponds B, C, and D have plenty of open access; hence it was not necessary to construct any entrances/exits there.

This behavior is significant for several reasons. Construction of pond entrances/exits is a pseudo civil engineering feat equivalent to making tools that will be used repeatedly. This behavior implies an understanding of the need for the entrances/exits and that the entrances/exits will be used in the future. This behavior infers an understanding of the advantages of having (versus not having) standard pond entrances/exits. And, compared to simple construction of a tool that is used once by one individual, construction of the pond entrances/exits benefitted the entire Canada goose family throughout the spring and summer.

3.6 Discussion

Now let's see what these behaviors tell us about Canada geese intelligence, behavior, and individuality. In particular, let's examine the:

- Concepts of community, family and individuality among Canada geese
- Learning, reasoning, and resourcefulness of Canada geese
- Ideas and inventions appropriated by humans

Community, Family, and Individuality

Canada geese experience life as a community of families. The nuclear family is the primary social group in Canada geese society. This is a reflection of the facts that Canada geese mate for life and have superb parenting skills. The patience Canada geese parents demonstrated while their goslings learned new skills and acquired confidence is truly impressive.

From the time a tiny gosling hatches it identifies with its family unit. The gosling can identify its parents and its fellow goslings in a multi-family environment. There is more to this behavior than just imprinting. First, the gosling understands that it is part of a family, and it knows who all the other family members are—not just the parents. In fact, for the first couple of weeks goslings can be spotted sleeping together in a clump. And, the gosling can tell the mother and father apart, even though humans cannot. Second, goslings do not follow their parents around like mindless robots. On the contrary, goslings observe their parents behavior and pay close attention to details. In doing so goslings learn essential life skills and acknowledge that their parents have knowledge they do not. Third, goslings trust their parents to protect them until they are able to care for themselves.

Both the parents and the goslings have vocalizations that express anxiety when they are accidentally separated from their family members. For the adults this is the nervous "ruh" sound. For the goslings this is the "peep" sound. The goslings can identify their parents' vocalizations in

Photograph 3.40 Construction of Pond Entrances and Exits. The Canada geese pair at Pond A constructed entrances/exits on three sides of the pond in 2010. (Top) One entrance/exit follows rocks into the pond on the north side. (Bottom) Another entrance/exit cuts through dense vegetation on the south side of the pond. The entrances/exits were created by flattening vegetation the width of one adult goose. The Canada goose family used these entrances/exits, located at the midpoint of each side of the pond, whenever they entered or exited the pond. The construction and use of these entrances/exits saved the Canada goose family time and enhanced the safety of the goslings since the parents could see what was ahead. © D. Herrmann

a multi-bird setting. Likewise the parents can identify their goslings' vocalizations in a multi-bird setting. If the goslings and adults did not associate themselves with a family unit, there would be no need for these special vocalizations.

In 2008 when Pond A flooded, that Canada geese pair lost their nest and eggs a few days before the goslings would have hatched. The pair stayed in the area searching for their nest, eggs, and goslings for 2-weeks, consistent with the 2-week mourning period observed in pet birds. (See Chapters 7 through 11.) This expression of loss indicated that the pair: (1) identified with the goslings they had anticipated and lost, and (2) knew the timeframe in which the eggs should have hatched. To lose an entire clutch is traumatic since Canada geese only have one family per year.

In 2011 the Pond D father goose stayed on the opposite side of the pond from his mate and their nest, making himself very visible. The afternoon the goslings began hatching the father goose went out of his way to make himself conspicuous by engaging in a prolonged energetic splashing/bathing session. The father goose purposely drew attention to himself and away from the nest at this critical time. In doing so the father goose put himself at risk. The day before a red-tailed hawk snatched a baby rabbit from a nest less than 10-feet (3 m) away. The father goose would not have done so if he did not anticipate the goslings hatching and assume responsibility for their safety and that of his mate.

Goslings stay with their parents long after they have learned to fly. Goslings travel with their parents on their first fall migration south and their first spring migration north. Again, the sense of identifying with their family and the trust they place in their parents are at work.

Grown goslings do not take a mate or raise a family of their own until they are two or three years old. In the interim some young adult geese serve as helpers to their parents. These young geese do not ferry food back and forth to the nest in the typical role of a helper. Instead, they serve as an extra security guard protecting the young goslings. During 2011 the Pond B family with 5 goslings had a helper. In fact on 5/26/2011 when a dog chased the goslings, two adult helpers assisted the parents in providing a security perimeter.

Flocks represent and function as extended families. The local flock migrates south together in the fall and north together in the spring. Breeding pairs disperse among the four waterways to nest. The non-breeding adults, who may number twenty or more geese, hang out together in the interim, taking care not to interfere or draw attention to the nesting pairs. Then, the flock reconvenes after the current year's goslings are able to fly. Throughout the remainder of the summer and fall temporary subflocks of various sizes rotate among the four waterways. All adults participate in cooperative flock security measures whenever needed. When migratory flocks from farther north arrive, the local flock interacts with them yet maintains its identity.

During 2011 three families with young goslings banded together as a subflock. From June 3rd onward the family of 6 goslings that hatched on 5/7/2011 at Pond D, the family of 5 goslings that hatched on 4/29/2011 near Pond D, and the family of 2 goslings that hatched on 5/17/2011 at the golf course water trap banded together, even though there was a 19-day difference in the age of the goslings. They also acquired a helper at that time. There were social and security benefits to this multi-family arrangement.

During 2012 Mr. Speckleneck and his mate accepted 4 goslings that hatched elsewhere into their family even though they were: (1) not orphans, and (2) a bit older and larger than their own goslings. Mr. Speckleneck and his mate treated these 4 goslings just like their own. The expanded family of 11 goslings stayed together from then on. How and why the 4 goslings were able to un-imprint from their biological parents and re-imprint into Mr. Speckleneck's family is unknown.

At the same time Canada geese exhibited individual reasoning, individual preferences, and individuality. In these instances one, or perhaps two, Canada geese responded to a situation differently or chose to participate in a different activity than the other geese present. Both adults and goslings expressed individual reasoning, individual preferences, and individuality. Neither the family nor the flock made any attempt to enforce conformity, except in one circumstance when gosling safety was an issue.

Goslings demonstrated individual reasoning, individual preferences, and individuality when:

- One 3-week old gosling attempted to imitate its parents and upend in deep water before it was old enough to do so. The other two goslings in this family just watched their parents. (Pond A, May 2010)

- One 4-week old gosling hesitated for about 20 seconds before jumping into the water. The other three goslings present jumped in immediately and followed the father goose across the pond. (Pond B, May 2010)
- One gosling repeatedly swam or walked at right angles away from the rest of the family. The other three goslings in the family did not; they followed the parents. (Pond A, May–June 2007)
- One gosling swam behind the mother for the first few days. The other 5 goslings swam behind the father, as usual. (Pond D, May 2011)

Adult Canada geese demonstrated individual reasoning, individual preferences, and individuality when:

- One adult male goose made an "S" movement with his neck, imitating a water snake, to keep water-based predators away from his young family as they crossed the pond. (Pond B, May 2010)
- A pair of Canada geese constructed standard entrances/exits at the midpoints on three sides of Pond A, to make it faster and safer for their goslings to enter and exit the pond. (Pond A, May–June 2010)
- A Canada geese pair expressed loss, similar to the 2-week mourning period observed in pet birds, when their nest was flooded a few days before the goslings would have hatched. (Pond A, April 2008)
- One or two Canada geese repeatedly flew up to the top of a 16-story building and jumped off, half coasting, half flying back to the ground as a play activity. (Pond B, April–May 2011)
- In 2011, unlike the previous 6 years, 3 families took their young goslings on day hikes from Pond A to Pond D to the golf course water trap and back. This behavior mimicked subflocks of adult Canada geese rotating between the four waterways—except that the goslings had to walk between the ponds. (May–June 2011)
- In 2010 the Pond A family returned to the pond for a few days twice after the goslings were grown and able to fly. In the previous 5 years the Pond A family only returned once, for a brief visit in August, as if they wanted to remind the goslings where they had hatched. (Pond A 2010)
- Individual Canada geese preferred to associate with one or more mallard ducks at Pond B or Pond C, while the other Canada geese had gone elsewhere. (2005–2012)
- Subflocks of varying sizes gathered at Pond B, Pond C, or the golf course on different days throughout the summer and fall. (2005–2012)

In summary, individual Canada geese identify with a family, and in turn each family identifies with a flock. A Canada geese flock is an egalitarian society of shared responsibilities. Individual reasoning, individual preferences, and individuality are apparent, and ultimately benefit the flock as a whole.

Learning, Reasoning, and Resourcefulness

The Canada geese consistently demonstrated a high degree of applied intelligence. In particular, Canada geese displayed ongoing learning and reasoning capabilities, self-reliant resourcefulness, and the ability to apply new insights to situations in the future. Learning, reasoning, and resourcefulness were evinced by individual Canada geese, pairs of Canada geese, families of Canada geese, and entire flocks at various points in time. Some learning, reasoning, and resourcefulness episodes were applied to a single experience. Other learning, reasoning, and resourcefulness episodes were applied to multiple events over time. Some of the more notable examples are chronicled below.

One day when 5 young goslings finished grazing they ducked under the fence and went for a swim. The Canada geese parents were too big to fit under the fence. Instead, the Canada geese parents improvised a creative solution rather than walk all the way around the pond, to the other side of the fence where there was an opening. The Canada geese parents walked up on top of a dirt mound near the fence. One at a time the parents jumped off the dirt mound, flapped their wings, and coasted over the fence, before plopping into the pond near their goslings. The Canada geese parents understood that they could not fly at the time—they were grounded for parenting duties. Yet, the Canada geese parents knew, between the height of the dirt mound and the flapping of their wings, they could achieve enough lift for a sufficient amount of time to coast over the fence. This resourceful solution was faster for the parents. More importantly it minimized the amount of time the parents were separated from the goslings. (Pond D, May 2011)

In 2006 a pair of Canada geese returned to Pond A in April as usual. However, the pair did not nest there. A few weeks earlier the fence that surrounded the pond had been removed. In the past the fence protected the nest and goslings from loose dogs, lawnmower crews, and others who tried to get too close. The Canada geese pair recognized the change to the immediate habitat, reassessed the safety and security of the nesting site, and rejected the nesting site that year. (Pond A, April 2006)

On 4/25/2008 the nest area of Pond A was flooded a few days before the goslings would have hatched. In 2009 and 2010 the Canada geese pair remembered their loss and moved their nest site to the other side of the pond to avoid potential flooding. (Pond A, April 2009, April 2010)

When the Canada geese pair returned to Pond A in 2011 and 2012 it was flooded. Pond A remained at flood stage for most of April and May. The pair took note of this, deemed Pond A unsuitable for nesting both years, and nested elsewhere. (Pond A, 2011 and 2012)

At first young goslings are so tiny they can hop through the mesh fence. By their second week this process becomes a tight squeeze. From then on goslings learn how to quickly find openings in and under the fences that surround the ponds. Not only do young goslings learn how to find these openings—they remember where the openings are for future reference. Should an emergency arise, the goslings dart toward the fence openings. (Ponds A, B, and D, 2005–2012)

Canada geese hunker down and conserve energy on extremely cold and windy days, much like other animals. However, Canada geese face directly into the wind. If they did not, the wind would blow their feathers upright and the geese would lose body heat. (Pond B, Pond C, and golf course, winters 2005–2012)

Canada geese seek shade and minimize activity on extremely hot and humid days. Canada geese seek shade under a tree or bush on land or in water. Sometimes Canada geese seek relief in tall grass, which shields them from the sun. (Summers, 2005–2012)

In 2010 the Pond B goose family lost one gosling at an early age, most likely to a snapping turtle. In 2011 and 2012 the Pond B goose family kept their goslings out of the water their first week. During their second week the goslings were allowed in the water, but only in a very shallow area—somewhere the snapping turtle would not go. (Pond B, 2011–2012)

Seasonal migration is a prime example of the genetic intelligence and applied intelligence of individual Canada geese, families of Canada geese, and entire flocks. Seasonal migration reflects learning and reasoning skills, collective memory, and resourcefulness. For example:

- Understanding the need to migrate and the optimum time to do so to minimize exposure to adverse weather. This infers understanding and anticipating current weather conditions in the present location, as well as future weather conditions along the route.
- Fine tuning the skills necessary to successfully navigate a journey of 1,000 miles (1,330 km) or more, without the aid of maps, a compass, street signs, GPS, or air traffic controllers.
- Knowing where viable stopover points are along the journey, and how to find last minute alternatives should a traditional stopover point disappear due to drought, flooding, or urban sprawl.
- Recognizing that there is one place to live during spring, summer, and fall, and a different place to live during winter.

Each of these sophisticated insights requires the constant collection, retention, analysis, and updating of a vast quantity of information.

Throughout the repeated partial 2011 mid-winter migrations north Canada geese displayed considerable resourcefulness under trying conditions. Canada geese wintering grounds in the southeast were uncharacteristically blanketed in snow. The Canada geese had to find food somewhere or face starvation. The geese could not fly farther south. Territory to the west was probably unknown. So, through a process of elimination the geese returned north to known stopover points for a few days. The weather was severe, but at least some food was available.

During this time the Canada geese adjusted their diet to include items they would not normally eat. The geese ate dry dormant clumps of grass, sometimes even the roots. Often the dry clumps of grass had to be pulled up from under snow. The geese accepted bread cubes from people; they also learned how to moisten the bread cubes to make them more palatable. A few geese ate milo, millet, and peanuts that fell from bird feeders.

Trying times bring out the best in some people, and the worst in others. Canada geese flocks were in survival mode during these partial mid-winter migrations north. Yet they called out to other flocks that were flying overhead, inviting them to share the limited resources. Flocks of Canada geese interacted with each other under these difficult circumstances and there was no aggression or hostility. Even though they were hungry, Canada geese made no attempt to steal bread cubes from each other. The Canada geese did not chase seagulls away when the gulls snatched a few bread cubes. Even in survival mode Canada geese exhibited a high degree of social intelligence.

Ideas and Inventions Appropriated by Humans

A complete assessment of Canada geese intelligence must also look at the other side of the coin: what have Canada geese and other waterfowl taught humans? (See Table 3.10.)

To start with, humans never would have considered flying simply from watching insects fly. No, it was the flight of birds that intrigued and inspired us, particularly the long flights made by migratory birds like Canada geese. After decades of experimentation and crashes, humans finally learned aerodynamic principles and became airborne.

On early airplanes the landing gear was always down. Before long humans learned to raise and lower the landing gear to reduce the drag coefficient, like birds do when they are flying.

After a few more decades humans grasped the efficiency of flying in a V formation and the slipstream effect it created. This practice has since been copied by modern air forces around the world. As has the practice of suddenly scrambling military aircraft as an evasive maneuver, similar to mallard duck behavior.

Earlier humans noticed that Canada geese and other waterfowl could swim across ponds without sinking even though they were large birds. Soon humans began building canoes, kayaks, and other boats that were shaped like the body of a duck or goose when it is swimming. Then humans noticed that waterfowl use their webbed feet to navigate and propel themselves forward and paddles were invented. Over time paddles became rudders and propellers. The same principle was applied to create the fins used by scuba divers.

The melodic resonant honking of Canada geese gave rise to car horns, fire alarms, and the sirens used by emergency responders; hence the origin of the phrase "honk the horn".

Humans noticed two other interesting features about the webbed feet of waterfowl. First, the spectacular water landings waterfowl make and how waterfowl use their webbed feet as water brakes. This observation gave rise to water skis and helicopters that can land on water. Second, the fact that large waterfowl like Canada geese can walk along muddy shores and in snow and sand without sinking. Humans learned the principle of distributing weight from this observation and snowshoes (among other things) were invented.

Table 3.10 Ideas and Inventions Humans Appropriated from Canada Geese and Mallard Ducks.

Ideas and Inventions	Observations that Led to the Ideas and Inventions
Flying	• Sight of biannual long distance migratory flocks of ducks and geese
Raising and lowering aircraft landing gear when landing and taking off	• Ducks, geese, and other birds holding their legs and feet straight, parallel to their tail when flying
Flying in a V formation, taking advantage of slipstream effect	• Sight of biannual long distance migratory flocks of ducks and geese flying in a V formation
Boats, kayaks, canoes, helicopters and other planes that can land on water	• Ducks and geese swimming with their body above water
Oars, rudders, propellers, scuba fins	• Webbed feet of ducks and geese and how they use them to swim
Sudden scrambling of military aircraft as an evasive maneuver	• Sudden scrambling of a flock of mallard ducks in every which direction, with their flashing blue speculums acting as a distraction
Car horns, fire alarms, sirens, emergency message systems	• Geese honking and their ability to project sound over long distances
Water skis	• Ducks and geese making water landings with webbed feet
Snowshoes	• Webbed feet of ducks and geese which distribute weight and prevent them from sinking in mud, sand, and snow
Down coats, vests, and quilts	• Ability of ducks and geese to stay warm and dry in cold temperatures and cold water
Periscope	• Long flexible neck of Canada geese which is used to fish, upend, and scan in all directions
Secret Service VIP security escort procedures	• Young goslings lined up single file. Father goose in forward guard position, mother goose in rear guard position. Parents able to scan for hazards and predators in all directions, and assume flanking positions if necessary
Baseball cap	• Duck bill

Humans noticed that Canada geese and other waterfowl can tolerate cold weather and cold water without becoming chilled or wet. Humans then began to appropriate goose down for use in coats, vests, and quilts because of its superior insulation properties.

The long slender flexible neck of a Canada goose can scan in any direction, above and below water. Undoubtedly this capability was noticed by seafarers, and ultimately gave rise to the development of periscopes which were invented in 1879.[179]

The VIP security escort procedures of the Secret Service and other similar organizations are modeled after that used by Canada geese parents.

And, of course there wouldn't be any baseball caps without duck bills. Then again, ducks do not use their bills for shade...

Without this knowledge human travel by air and water, and the resulting commerce and progress, would have been rather limited.

And, what ideas and inventions have Canada geese deemed worthy enough to appropriate from humans? None.

4
Mallard Ducks
Anas platyrhynchos

4.1 Background

Many of the sources consulted use phrases like "most common" or "most well-known" in reference to mallard ducks. In spite of the claims that mallard ducks are common or well known, it was difficult to find more than a few brief paragraphs written about them in any book. Why is it that so little has been written about such a supposedly well known bird? Or conversely, how can a bird be well known when so little has been written about it? Perhaps it is because humans tend to take for granted that which is common, until it is gone—like the Carolina parakeet (*Conuropsis carolinensis*) and the passenger pigeon (*Ectopistes migratorius*).[43] This chapter is one attempt to remedy that situation.

The word duck appeared in the English language sometime before the 12th century,[179] about the same time as the word goose. The word duck is used to refer both to the species (e.g., mallard ducks) and a female duck, similar to the usage of the words goose and gander. A male duck is referred to as a drake. The term mallard duck originated during the 14th century, appearing in English and French (mallart) at about the same time;[179] although no one remembers why the name mallard was chosen. Mallard ducks are also referred to as Wild ducks.[203] They were given their scientific name *Anas platyrhynchos* in 1758.[203]

Mallard ducks are believed to have evolved in the late Pleistocene period in the general area of Siberia.[203] In *The Secret History of the Mongol Queens* Weatherford notes that mallard feathers were used to adorn the top of the tall headdress (boqta) worn by Mongol Queens in the 13th to 15th centuries. Most likely the mallards' iridescent blue speculum feathers were used for this purpose.

Mallard ducks are widely distributed among the temperate zone of the northern hemisphere.[41, 136, 157, 203] Mallard ducks favor fresh water habitats that are surrounded by lush vegetation, such as marshes, wooded swamps, ponds, lakes, bays, rivers, and more recently city parks.[29, 41, 95] They have been introduced in Australia and New Zealand.[203] Mallard ducks are believed to be the ancestor of most domestic ducks.[95, 136, 137, 179, 203] Domestic ducks arose in India approximately 3200 BC and in Egypt approximately 1500 BC.[137]

The population of mallard ducks in the northern Great Plains area of North America is much lower than it was prior to the arrival of European immigrants due to over-hunting and habitat destruction.[95] As of Spring 2010 there were an estimated 8.4 million mallard ducks living in North America.[210] However, the breeding population estimate was 17% less than in 2009.[210] These estimates were prepared before the 2010 BP gulf oil spill.

Now let's examine Mallard duck behavior. Behaviors are presented as they occurred in Sections 4.2 through 4.6. The analysis, comparison, and interpretation of these behaviors are reserved for Section 4.7 Discussion. These observations took place at the locations and during the timeframes described in Annex C.

4.2 Characteristics

Now let's examine characteristics that are common to all mallard ducks, such as their:

- Diet
- Physical Features
- Flight
- Vocalizations

Diet

Mallard ducks are primarily herbivores with some 90% of their diet consisting of plant matter.[157] Plants, particularly aquatic plants, seeds, stems, and roots are the mainstay of their menu.[95, 157] On occasion mallard ducks also eat aquatic insects, acorns, small crustaceans, mollusks, tadpoles, earth worms or small fish; however, these other entrees rarely compose more than 10% of their diet.[95, 141, 157] The consumption of these alternate entrees increases in winter when vegetation is less plentiful. Mallard ducks were also observed eating grain (millet and milo) and peanuts that had been put out for songbirds and squirrels (*Sciuridae sciurini*) during the winter. Newly hatched ducklings are too small to upend and feed on aquatic plants, except in very shallow water. Instead ducklings consume aquatic insects when they are in the water.[95, 141] This source of protein helps fuel their rapid growth spurt.

Physical Features

Through biological intelligence the bill of a mallard duck is specially designed to accommodate its diet. The inner rim of a mallard duck's bill is lined with thin comb-like membranes called lamellae that function like a sieve.[41, 83] A mallard duck can upend and take a bite of aquatic plants. As the duck resurfaces the excess water and sediment drain out of its bill through the lamellae. A mallard duck's bill is sensitive to touch and temperature.[83] In particular the surface inside the front of the upper and lower mandibles is sensitive.[83] This is the area most involved in dabbling. The surface inside the front of the upper and lower mandibles of a mallard duck has 27 touch receptors per mm^2 compared to 23 on the tip of a human index finger.[83]

Mallard ducks frequently feed at the edge of a pond, where the water is shallow. They insert their bill into the soft mud, up to their nares, and retrieve small crustaceans or roots. Small groups of mallard ducks were observed feeding like this for 10 to 15 minutes at a time. The ducks poked their bills in the mud, while walking slowly. The ducks raised their heads slightly after each poke, until their bill was slightly above water. Then they made a faint slurping noise as they expelled the excess water out the sides of their bill with their tongue and swallowed whatever had been retrieved. Once in a while the ducks shook their head side to side to speed up the process of expelling the excess water. Sometimes a duck would pick up a bill full of water, then slowly sprinkle it on its feathers for preening purposes. After four successful pokes the ducks stopped and pulled some leaves off a nearby plant or branch. Then the ducks repeated this process. It was interesting to note that although the ducks were sticking almost their entire bill in the mud repeatedly, once they were upright there wasn't any mud on their bill or feathers. (See Photographs 4.1 through 4.3.)

One duck lifted up a snail that was 1.5 inch (3.81 cm) in diameter—too big to swallow. The duck tried unsuccessfully to crunch the snail with his bill. Next the duck dropped the snail back in

Photograph 4.1 Dabbling. Mallard ducks are considered to be dabbling ducks since they poke around in the silt at the edge of ponds looking for a tasty morsel. Mallard ducks use a combination of their bill and their tongue, which extends beyond the bill, to dabble. They expel excess water through the lamellae that line the perimeter of their bill like a sieve. Mallard ducklings have their first serious dabbling lessons when they are 6 to 7 weeks old. © D. Herrmann

the water and tried to pry it open with the sharp tip of its bill—still no luck. Eventually the snail was granted a reprieve. A few days later a smaller snail was not so fortunate. The duck was able to pull the snail out of its shell and swallow it.

Mallard ducks often cooperate when fishing. Two or three ducks walked in tandem to skim minnow that were swimming near the water's surface. Mallard ducks also stirred fish 2–3.5 inches (5.08–8.89 cm) long out of their hiding place near the bank of a pond by poking around to startle the fish. A mallard duck grabbed a fish horizontally across its bill. Then the mallard duck stood up, rotated the fish parallel to its bill, and swallowed it head first, like a heron. If the fish happened to be more than 4 inches (10.16 cm) long, it was released.

Ducks have between 375 and 400 taste buds.[18, 137] This is among the highest number of taste buds for birds. Most birds have less than 100 taste buds. For example, urban pigeons (*Columbinae*) have between 37 and 75 taste buds.[18] Perhaps ducks have a high number of taste buds to ensure that they don't ingest something dangerous when poking around under water. None of a duck's taste buds are on their tongue.[18, 137] Instead, 30 to 70 taste buds are under a duck's tongue; the rest are on the roof of their bill.[18, 137] Mallard ducks have well developed olfactory lobes and hence a good sense of smell, which comes in handy since they spend a lot of time on the ground.[137]

Mallard ducks are about 24 inches (61 cm) in length when full grown.[137, 203] Mallard ducks have a short stocky tail and a wide wingspan, which ranges from 30–40 inches (76.2–101.6 cm).[137, 203] The average weight of an adult mallard duck is 2.5 lb (1.12 kg).[127, 203] The head of a mallard duck is

Photograph 4.2 Lamellae. The inner rim of a mallard duck's bill is lined with thin comb-like membranes, called lamellae, which function like a sieve. After dabbling excess water and sediment drain out of the bill through the lamellae. Here the lamellae appear as thin black lines just inside the bill. Compare this with Photograph 3.1. © D. Herrmann

the same size as that of a Canada goose. But the neck is much shorter and the bill is wider and more rounded. Female mallard ducks have an orange bill. Male mallard ducks have a yellow bill. There is a dark spot on the pointed tip of a mallard's bill. This spot is black for males and dark brown for females.[203] This sharp point is used for slicing vegetation. A male mallard has two dark lines running the length of the bottom of its bill. A female mallard has faint black streaks running the length of the top of its bill; the exact pattern of the streaks is unique to each bird. The nares are about an inch (2.54 cm) from the base of the bill and parallel to it. As a result mallard ducks breathe air that flows horizontally across their bill, a distinct advantage when flying long distances.

A mallard duck's legs are short. The legs and webbed feet are orange. The bright orange feet of a mallard duck are visible at a distance, especially when they are standing on bare earth or the snow and ice covered bank of a pond. The orange legs and feet of a male mallard are slightly brighter than those of a female, especially when they are in breeding plumage. The eyes of a mallard duck are soft brown and 0.75 inch (1.9 cm) in diameter. Mallard ducks can live up to 29 years.[18, 41]

During breeding season the males (or drakes) adopt fancy dress to impress the ladies. The male's brightly colored breeding plumage draws attention away from the females when they are nesting and the ducklings when they are vulnerable. All adult males in the flock acquire breeding plumage, whether or not they raise a brood. The local male mallard ducks sport breeding plumage from late September or early October through early June. In 2011 the male mallard ducks at Pond B began losing their breeding plumage June 6th, 7 weeks after the first brood of ducklings hatched.

Photograph 4.3 Drinking. Mallard ducks drink clean surface water while swimming or standing at the edge of a pond. They ignore deeper water where sediment and debris have settled. To take a drink a mallard duck lowers its head and neck and fills the lower bill with water. To swallow a mallard duck raises its head and neck slowly. Mallard ducks often take two or three drinks in a row. © D. Herrmann

The head feathers take on a metallic green sheen which is noticeable at a distance. Close up the color of the head feathers varies from metallic green to dark purple to teal, depending on the angle and intensity of the sunlight. At the bottom of the head there is a bright white collar about 0.75 inch (1.9 cm) wide. The breast or bib is a stunning rust brown. The rest of the body and tail are various shades of light gray or white. Under magnification the white side and bottom feathers of an adult male mallard appear white with a tiny medium brown dot in the center. The wings sport a metallic blue patch or speculum about 1.5 inches (3.8 cm) wide and 3.5 inches (8.89 cm) long that is outlined in white. The speculum appears violet when reflected in direct sunlight and serves as a recognition signal.[158] Male mallards have 2 small black curled feathers on the center of their back near the tail. These curled black feathers, reminiscent of Superman's curls, are about an inch (2.54 cm) in diameter and located about an inch (2.54 cm) from where the tail feathers begin.

After breeding season the coloring of male mallard ducks resembles that of females, but is a bit darker. The coloring of female mallard ducks is as modest as the males' breeding plumage is showy. The front and side feathers of a female mallard duck are a beautiful medium brown, outlined in dark brown rounded U's that are about an inch (2.54 cm) long. The lower back has the same color pattern, but the U's are smaller being only 0.5–0.75 inch (1.27–1.9 cm) long. This coloring pattern helps female ducks to blend into the scenery, especially when nesting.[83] The underside feather pattern of female mallards is the reverse of the topside: a white or light ivory background with light brown U's outlining their feathers. This, of course, can only be seen when they upend or are in flight. (See Photographs 4.4 and 4.5.)

Female mallard ducks also have a metallic blue speculum. The metallic blue speculum is identical in male and female mallard ducks. Hence it is not meant to attract a potential partner. Instead the speculum serves a safety function. The metallic sheen from the speculum is quite distracting when a duck is flying, particularly in direct sunlight or when several ducks are flying close together. The effect is similar (but on a larger scale) to watching red-winged blackbirds (*Agelaius phoeniceus*) flying directly toward you. A flashing blue speculum is distracting to humans, who cannot see UV light. Hence, a flashing blue speculum must be even more distracting to other birds, especially predators, who can see UV light. The flashing blue speculum is also believed to be a visual communication signal to members of a flock.[83, 158] Both males and females lose their speculum temporarily during their annual summer molt.

The feathers of a mallard duck are specially designed to keep them warm during the chill of winter, cool in the heat of summer, and dry when swimming. Flexible down combined with the interlocking system of feather barbs and barbules provides an extremely efficient biological system for regulating and retaining or discharging body heat year round. For example, mallard ducks were observed swimming in Ponds A and B on 12/9/2009 and in Pond B on 1/4/2010 and 1/20/2010 when most of the surface was frozen and the wind chill was in the single digits. The ducks swam around in the small area available, hopped up onto the shore, and shook off the excess water, as if this were nothing unusual. Seeing this for the first time gave a whole new meaning to the phrase "like water off a duck's back."

Webbed feet are a specialized biological tool that is perfect for the life style and habitat of water birds like mallard ducks. Webbed feet make swimming much more efficient and allow mallard ducks to navigate, pivot, and change directions easily. Although a mallard duck's legs and feet are bright orange, the webbing between their toes is translucent with a slight orange tinge. The webbing is thin, about as thick as two sheets of wax paper. Since the webbing is very flexible, mallard ducks can wiggle their toes in a variety of shapes and directions. When swimming mallard ducks clench their toes together on the forward stroke and spread their toes on the backstroke. When swimming rapidly or long distances mallard ducks extend their legs all the way back until they are parallel with their body. When floating in one place their legs are down. Mallard ducks three front toes are each about 1.5 inches (3.8 cm) long. Their back toe or hallux is about 0.375 inch (.95 cm) long and 0.5 inch (1.27 cm) higher up on their leg. (See Photograph 4.6.)

Photograph 4.4 Adult Mallard. (Top) An adult female mallard has a brown speckled body and orange mottled bill. (Bottom) An adult male mallard in breeding plumage has a yellow bill, green head, white neck collar, rust breast, and black curls just above the tail. Males in non-breeding plumage look similar to females, except for the yellow bill. © D. Herrmann

Webbed feet make water takeoffs and water landings possible. Mallard ducks can land and takeoff from both land and water, rather quickly if need be. Similar to Canada geese (*Branta canadensis*), a water takeoff consists of stretching the head and body up tall while leaning forward, flapping fully outstretched wings, and what appears to be walking on the water. Actually the legs have been extended and the webbed feet are pressing down hard and fast on the water until a mallard duck can flap its wings for lift. By the third "step" the ducks are airborne. A similar process of running along the ground is used to takeoff from land. To land on water or land mallard ducks gradually slow the wing flapping, tilt to the left, tilt to the right, tilt upright and descend to the intended location. The water landings of mallard ducks are more splashy and sudden than graceful. And they frequently employ their webbed feet as water skis or water brakes to slow down.[41]

Photograph 4.5 Juvenile Mallard. The feather coloring of a juvenile mallard duck is similar to that of a female or male in non-breeding plumage. The bill color of a juvenile matches that of an adult. Hence, this is a male juvenile mallard duck. © D. Herrmann

Another biological advantage to webbed feet has not been reported in relation to ducks or geese. Perhaps because this observation requires poking around in the mud. In addition to the advantages noted above, webbed feet perform the same function as snowshoes, by distributing weight over a larger area. This allows mallard ducks and other waterfowl to walk along the muddy edges of ponds without sinking. Given their weight, if ducks and geese just had three forward toes and no webbing, they would sink in the mud, at least a few inches (centimeters). Wading birds who lack webbed feet are much lighter and have much longer legs. Webbed feet offer the same weight distribution advantage to ducks and geese who find themselves walking along sandbars or in the snow. (See Photograph 3.7.)

Flight

Mallard ducks are water birds, but they are also strong quiet flyers. Mallard ducks generally don't chatter much when flying. They are seen but rarely heard, especially when flying in formation. During migration mallard ducks fly up to 40 mph (53 kph) with 5 wing beats per second and cruise at an altitude of up to 21,000 feet (6,400 m),[12, 18, 136, 137] much higher than Canada geese. Webbed feet come in handy during long migratory flights by acting as an air foil and contributing to horizontal stability during flight. However, it takes a lot of concentration and effort for ducks (and other birds) to hold their feet out straight behind them for long periods of time, as anyone who has studied dance knows.

Vocalizations

Male and female mallard ducks have distinct vocalizations. The vocalizations of a male mallard duck are heard less frequently and described variously as soft reedy notes,[29] a feeble high pitched call,[41] and a quiet "yeeb".[157] Male mallard ducks were heard to make a quiet "yeeb" sound which could only be heard from a distance of 1–2 feet (30.48–61 cm). Male mallard ducks repeated the "yeeb" sound three times, then paused. The male mallard ducks repeated this sequence three times, for a

Photograph 4.6 Orange Feet. Mallard ducks have bright orange legs and feet. Those of a male are slightly brighter than those of a female, especially during breeding season, as shown in the top photograph. Different size scales cover the legs, toes, and webbing. Three toes, each with a nail, face forward. A smaller toe, that is higher up, is on the back of the leg. Thin flexible translucent webbing links the front toes. This webbing makes it easier for mallard ducks to swim and fly, and prevents them from sinking in sand, mud, and snow. © D. Herrmann

total of nine sounds, before going quiet. Male mallard ducks were only observed making this sound when in small groups. Perhaps this communication is reserved for bonded pairs and nuclear families.

In contrast, the loud quacking sound made by female mallard ducks is well known. The word quack, which also entered the English language in the 14th century,[179] is a phonetic imitation of the sound female ducks make. The volume and rapidity of the quacks is varied to convey different messages. Mallard ducks are able to project their quacking sounds over a distance; this facilitates intraflock communication, especially during migration or when spread out around a pond.

These are the two vocalizations traditionally attributed to mallard ducks. Several other mallard duck vocalizations were heard at Ponds A, B, and C. They are described later in conjunction with the behavior during which they occurred. Mallard duck vocalizations can be heard on Disk 1 Track 1, as discussed in Chapter 12.

Photograph 4.7 Natural Camouflage. The plumage of a female mallard duck is designed to blend in with foliage typically found around the edge of ponds, providing natural camouflage. This feature is particularly useful during nesting time. © D. Herrmann

4.3 Family Life and Daily Activities

Mallard duck family life and daily activities are not covered in much detail elsewhere. In contrast this chapter provides detailed information about:

- Life cycle events
- Flock security measures
- Common daily activities

Life Cycle Events

Mallard ducks form pair bonds between late fall and early winter.[95] Male mallard ducks regain their breeding plumage at this time. Mallard ducks first take a mate in their second or third year. Mallard ducks prefer to choose partners from their own hometown. As a result pair bonds are established at the breeding grounds before the flock leaves for migration. Male and female mallard ducks often face each other and bob their heads up and down asynchronously during pair bonding.[83, 161] Mutual head bobbing is a courtship ritual; it is also used as a greeting ceremony on occasion. Mallard duck pairs do not engage in allo-feeding.

Mallard duck pair bonds generally only last one year.[41, 95] This may explain why mallard drakes can be rather aggressive at times when pursuing a particular female. In late fall it is not uncommon to see two males chasing after a lone female mallard duck, flying around and around at top speed above a pond until all three drop into the pond exhausted.

Like other waterfowl mallard ducks migrate on a biannual basis. The local mallard ducks tend to leave their breeding site later and return sooner than Canada geese. For example, during the winter of 2009/2010 the local Canada geese flock was gone by December 9th following several days of cold weather, snow, and ice. In contrast, the local flock of mallard ducks did not leave until February 5th in response to the double blizzard of 2010 when Pond B froze over completely. This behavior reflects the mallard ducks tendency to not venture too far from their breeding site unless they have to, such as in the dead of winter.[41] North American mallard ducks spend their winter anywhere from the lower Mississippi valley through southern Mexico and Central America.[29, 95, 203] It is interesting to note that the V formation mallard ducks fly in during migration matches the V ripple pattern produced when a single duck swims across a pond.

Not surprisingly mallard ducks nest on the ground near water.[83, 157] The nests are totally hidden by vegetation. The female chooses the location and builds a nest about 8 inches (20.32 cm) in diameter from dried vegetation.[95, 161] She lines the nest with special nest down feathers that are darker than her other down feathers.[83] These down feathers are also used to cover and hide the eggs whenever the mother needs to leave the nest.[83]

By mid-April each year the mallard ducks were spread out and the females were very inconspicuous. The mallard ducks primarily nested at Pond B. In 2011 one family of mallard ducklings hatched at Pond C for the first time. In 2012 mallard ducklings hatched at Pond A and the new Pond D for the first time. Each year the mallard pairs went through the motions of searching for a nesting site. For example during Spring 2011 one day a mallard pair was walking on a putting green at the golf course. The next day 2 pairs of mallard were exploring the golf course water trap. Another pair explored the vegetation around the creek. A pair of mallard ducks that had settled into Pond C moved back to Pond B when 2 pairs of Canada geese began to nest there. In the end the majority of mallard ducks usually nested in a remote area of Pond B that is inaccessible to humans. This is consistent with mallard ducks reputation for being secretive when breeding.[161] The site is shaded and has dense vegetation—the same color as the female ducks, ducklings, and non-breeding males. Mallard ducks generally do not practice nesting dispersal. Instead they generally share a large nesting area. (See Photograph 4.7.)

One egg is laid every 24 hours.[161] The female incubates the eggs and only leaves them briefly once or twice a day; beforehand she covers the eggs.[83, 161] Mallard duck eggs are about 2.5 inches (6.35 cm) long and 1.8 inches (4.57 cm) wide.[157] The eggs range in color from white to buff to an olive buff.[29, 95, 157] The coloring varies by geography. Various sources claim that 8 to 10, 9 to 13, 7 to 10, or 10 to 12 eggs are laid each year.[29, 41, 95, 157, 203] Not wanting to disrupt nesting activities, I did not venture close enough to count the number of eggs. The incubation period begins after the last egg is laid and spans 26 to 30 days.[95, 83, 203] Mallard ducks only have one nest per year; they have shown a willingness to parent orphaned ducklings, even of different species, sizes, and ages.[154]

Mallard ducks have specialized molt and plumage cycles that are tied to life cycle events.[83, 137, 153, 161] Male and female mallard ducks have different summer molting schedules. This way both parents are not grounded and flightless at the same time. As a result one parent can defend the ducklings from ground-based dangers and one parent can defend the ducklings from air-based dangers at all times.[83]

Male mallard ducks go through a complete molt, including flight feathers, in early summer, about a month after the ducklings have hatched.[137] Male mallard ducks lose all their flight feathers and are temporarily grounded.[153] They even lose their black Superman curls one at a time. The coloring of male mallard ducks nonbreeding plumage resembles that of females. Male mallard ducks begin to regain their breeding plumage in fall when new pairs are being formed. During this process male mallard ducks develop light blue cylindrical keratin sheaths like those of Canada geese, only smaller. (See Photographs 4.8 and 4.9.)

In 2011 and 2012 one male mallard duck at Pond C retained his brilliant breeding plumage throughout June, July, August, and into September. He was always accompanied by 5 or 6 other mallard ducks that had the traditional brown speckled pattern. Some were juveniles and one was an adult female. This may have been a late hatching family of his, which caused him to retain his breeding plumage as a protective strategy. This is a plausible explanation, given that the other ducks always followed him around on land or water, like young ducklings do.

Female mallard ducks experience a complete molt late in summer, after the ducklings are independent and on their own.[137, 161] Unlike the males the appearance of female mallard ducks does not change when they molt.

Photograph 4.8 Male Mallard Molt #1. Male and female mallard ducks molt at different times. This ensures that new families and entire flocks are not all grounded at the same time. The molt of male mallard ducks is more dramatic as they gradually transition from breeding to non-breeding plumage. They even lose their two black Superman curls just above their tail. © D. Herrmann

Photograph 4.9 Male Mallard Molt #2. Mallard ducks go through a complete molt each summer and are temporarily grounded when they lose their flight feathers. Before long small blue cylindrical keratin sheaths appear and new feathers gradually emerge from the end of each. © D. Herrmann

Juveniles acquire their first full set of adult plumage in late fall, prior to migration. Male juveniles acquire adult plumage about a month after adult breeding males.

Ducklings leave the nest within 24 hours of hatching and are able to walk, swim, and feed themselves.[41, 95] The mother leads the ducklings to water for their first swim.[95] It has been shown that 2-day old mallard ducklings can identify the vocalizations of their mother without even seeing her.[83, 154] Ducklings are quite tiny when they hatch. When standing they are the same height as house sparrows (*Passer domesticus*). And, like house sparrows they are brown. Instead of walking at a leisurely pace along the ground, young ducklings run quickly. Sometimes young ducklings run too fast and they topple over.

In 2011 the first mallard ducklings were spotted at 5 p.m. on April 21st. The ducklings were very tiny; they probably hatched that morning. The 12 ducklings clustered around the mother and walked very rapidly. One stopped to eat some bird seed, but the mother hurried it along. A day later the ducklings were sitting in a semi-circle on the lawn in front of the mother. They were all facing directly into the sun. The ducklings were shorter than the grass. The ducklings could only be seen if they stretched or stood up. A group of 6 male mallard ducks sat on the lawn in front of the ducklings. Three other male mallard ducks were in the water guarding 4 adult red-eared slider turtles (*Trachemys scripta elegans*) who were nearby.

By the fifth day the ducklings were able to take a quick dip completely under water and resurface. The ducklings did this of their own initiative, since the mother was not diving, upending, or bathing at the time.

The sixth day the ducklings were napping on the lawn with the mother. The mother had her head tucked under her wing. Two of the ducklings were snuggled up with the mother. The other ducklings were 1 foot (30.48 cm) away snuggled up together in a circle, much as they would have been in the nest. When they awoke the family went for a short swim. The mother took the outer position and kept the ducklings in a shallow area. Again 3 or 4 male mallard ducks kept watch on shore and near the mother.

The ducklings had doubled in size and their body was beginning to take the shape of a duck by the time they were 1-week old. Ducklings follow their mother, but they do not line up behind her. Instead, the ducklings spread out in a cluster 1–2 feet (30.48–61 cm) around the mother.[83] If ducklings sense danger, they form a tight circle around the mother, whether they are on land or in the water. One duckling stood up and flapped its (soon-to-be) wings, even though it did not yet have any flight feathers. At this time ducklings begin to isolate sensory feedback from their wings. This creates curiosity about what wings are for, since the ducklings will have seen their parents stretch and preen their wings by then. (See Photographs 4.10 and 4.11.)

The mallard ducklings were not seen for several days. Then on May 7th six mallard ducklings were spotted playing by themselves. The mother duck and the other ducklings were nowhere in sight. The six ducklings splashed around in the shallow water for a while, then they walked around on the lawn. Several adult male mallard ducks were nearby.

Midday on May 8th a second new family of mallard ducklings was spotted at Pond B. The mother and 10 ducklings swam or rested at the pond's edge. This new mallard family was in the middle section of Pond B, not the southern section where the mallard ducks usually reside. Other than a solitary male mallard duck, they were they only ducks in that part of Pond B. This behavior may have reflected a preference of the mother duck.

On May 11th the original family of 12 ducklings was spotted together with the mother. They were plucking food from under shallow water. A couple ducklings splashed and bathed. Then the family swam across the pond to where the other adults were resting. At 3 weeks of age the mallard ducklings were about 1/3 the size of an adult. (See Photograph 4.12.)

A family of 5 Canada goslings and a wood duck (*Aix sponsa*) family were there too. At one point a mallard duckling became separated from its family. It made a couple high pitched "peep" calls. The duckling must have heard a reply, because shortly afterward it turned around and headed toward its family. Later on when all the ducklings were roaming around grazing, one mallard duckling followed the wood duckling family for a while. All of a sudden it looked up and realized its mistake. The mallard duckling changed course and ran over to its family.

The mallard ducklings were seen frequently with their mother, particularly running around on the lawn, when they were tiny and less than 2-weeks old. The ducklings were only spotted a couple times

Photograph 4.10 1-Day Old Mallard Ducklings. This newly hatched family of mallard ducklings follows its mother across the lawn and into Pond B for their first swim on 4/20/2012. Upon hatching mallard ducklings instantly identify with each other and their parents. (Top) Note that the mother is scanning skyward for predators. © D. Herrmann

a week as they grew older and larger. The mother duck seemed to be spending more time and effort to conceal their whereabouts. This was true for both broods of mallard ducklings that hatched at Pond B, as well as the wood duckling families. The bills of some of the mallard ducklings (the young males) from the first brood began turning yellow on May 23rd at 4 weeks of age. (See Photograph 4.13.)

Adult mallard ducks are very quiet when the ducklings are young. This too is an attempt to not draw attention to the ducklings when they are vulnerable. The loud quacking sound that is common in fall is not heard during spring or summer. No quacking is heard at all until the ducklings are 6 to 7 weeks old; even then the quacking is quieter and of a short duration.

At 7-weeks of age mallard ducklings resemble adult females and are half the size of an adult. They still stay close to their mother and fellow ducklings. Likewise, 7-week old wood ducklings resemble an adult female wood duck and are half the size of an adult. (See Photograph 4.14.)

On June 24th, at 8-weeks of age, the mallard ducklings learned how to engage in energetic water baths. At the same time the ducklings took their first short low flights over water. The ducklings

Photograph 4.11 1-Week Old Mallard Ducklings. (Top) Ducklings cluster around their mother while walking, swimming, and napping. (Bottom) One duckling stands up and flaps its (soon-to-be) wings, even though it does not yet have any flight feathers. Ducklings begin to isolate sensory feedback from their wings at this time and experiment to learn what wings are for. © D. Herrmann

started in the water, flew a distance of 4–6 feet (1.22–1.83 m) just above the water's surface, and splash landed in the water. Several ducklings engaged in this exercise at the same time. These practice sessions lasted a few days. Mallard ducklings begin their flying lessons the same time the male mallard ducks grow their new flight feathers. Wood ducklings also learned how to fly starting in the water. This contrasts with Canada goslings who learn to fly starting on land.

On July 1st two ducklings from the second brood were swimming in the narrow southern part of Pond B with their mother. The ducklings were 7-weeks old at the time. Since the ducklings could not yet fly, they had to climb up some large rocks to exit the water. The mother duck went first and

Photograph 4.12 2 to 3-Week Old Mallard Ducklings. Mallard ducklings identify with their mother and fellow ducklings. Mallard ducklings stay together in a cluster for safety and social reasons after they have hatched. © D. Herrmann

climbed up the steep rocky slope, using a combination of her bill and the nails on her feet for traction. The ducklings watched her, then did their best to replicate her actions. (See Photograph 4.15.)

Table 4.1 summarizes the life cycle events of the local mallard ducks and the dates they take place.

Some sources implied that a mallard duck father abandons his family after the eggs are laid. That claim is inconsistent with what was observed, particularly for 2011 and 2012. In 2012 the first mallard ducklings hatched on April 20th. By 5:30 p.m. the mother was walking the 11 new ducklings across the lawn to Pond B for their first swim. The tiny ducklings ran fast and did their best to stay up with the mother's brisk pace. A solitary male mallard followed 3 feet (.91 m) behind the ducklings. Once in the water the mother and ducklings stayed close to the edge of the pond. The solitary male stayed on guard 4 feet (1.22 m) away in the deeper water. He even challenged a Canada goose who

Photograph 4.13 4 to 5-Week Old Mallard Ducklings. During the fourth and fifth week mallard ducklings appear darker, yet they still have some fuzzy down. The orange color of their legs and feet becomes more prominent. © D. Herrmann

made a water landing near the ducklings. Likewise, the fathers stayed with the mothers and mallard ducklings that hatched at Ponds A (4) and D (9) in 2012. (See Photograph 4.16.)

Contrary to what is stated in several publications, no hostility or territoriality was observed among co-located mallard nests or young families. Nor was any hostility or territoriality observed among co-located mallard ducks, wood ducks, black ducks (*Anas rubripes*), and Canada geese while they were each raising their families. The only hostility observed occurred during late fall and early winter when the mallard ducks were forming new breeding pairs. Adult mallard ducks tended to harass that year's juveniles, male and female, for no reason. A juvenile would be minding its own business, swimming or walking, and an adult would peck the juvenile's back near the tail with the sharp point of its bill. If one adult harassed a juvenile, two or three other adults would join in, such that the juvenile had to leave the immediate area. Once in a while a feather was pulled out.

Due to their small size and inability to fly ducklings are vulnerable to a variety of aquatic and land predators, such as hawks, falcons, house cats, snakes, fish, and snapping turtles (*Chelydra serpentina*).[41, 95] Ducklings are dependent on their parents and other adults for protection. Some sources

Photograph 4.14 6 to 7-Week Old Mallard Ducklings. By the sixth week mallard ducklings have a mixture of juvenile and duckling feathers. This color pattern blends into the edge of ponds, providing protection on land. The ducklings don't have any flight feathers, but the tail feathers start to grow. The bill of young males begins to assume a yellow color. Prior to this all the ducklings have a dull mottled orange bill. © D. Herrmann

claim that only 10% of ducklings survive their first year.[41] That fact, plus an annual adult mortality rate of 40% to 64%, explains why mallard ducks have large clutches.[18, 137]

The population of mallard ducks at Pond B remained steady. New ducklings hatched each year. However, not all the pairs produce a brood every year. Also, it is unclear if mallard pairs continue to engage in nesting activities in their senior years. The majority of the adult mallard ducks were part of a pair bond but only 2 or 3 broods hatched each year. Males who were part of a pair that did not produce ducklings could be seen flying around Pond B and to/from other ponds during April and May.

This observation raises some questions. Is the mallard duck flock at Pond B limiting their population to match the available food supply and resources, so that each pair only produces ducklings every third or fourth year? Is the mallard duck flock at Pond B limiting the number of pairs that breed

Photograph 4.15 Learning to Climb Rocks. The edges of ponds can be rocky. It is important to know how to climb up a rocky bank, especially during the summer molt when mallard ducks are grounded. Here a mallard duck at Pond B teaches her ducklings how to climb up a rocky slope by stretching her neck, using her bill for a grip, and getting a good foothold. Note the similarity between this learning sequence and that of Photograph 6.2. © D. Herrmann

each year so that the entire mallard population is not grounded and flightless at the same time as a safety precaution? Or, are other factors at work?

Table 4.2 presents the mallard duckling statistics for 2011 and 2012. In 2011 out of an adult population of 36 mallard ducks, only two broods hatched at Pond B. Out of a total of 22 mallard ducklings that hatched at Pond B, only 12 (or 54.5%) survived their first 2 months. Most likely the losses were due to predation by turtles and hawks. For the first time a brood of 8 mallard ducklings hatched at Pond C on June 3rd. Three survived their first few months. As mentioned earlier their father, in breeding plumage, was seen swimming and walking with them well into August.

The mallard ducks continued to expand their nesting areas in 2012, even though Pond B was still considered to be "home". In 2012 11 mallard ducklings hatched at Pond B on 4/27. A week later there were only 9 ducklings. A second family of 5 mallard ducklings hatched at Pond B on

Table 4.1 Time Line of Life Cycle Events for the Local Mallard Duck Flock.

Time Frame	Life Cycle Event
September–November	• Adult males regain breeding plumage • Pair bonds are formed yearly, starting the second or third year of a duck's life • Ducklings acquire adult plumage • Local flock hosts migratory flocks of mallard ducks that pass through the observation area • Adults harass unattached juveniles on occasion
December–February	• Ducks that visit Ponds A, C, and D during the summer and fall stay at Pond B as temperatures drop • Flock moves from their spring/summer/fall home in the south of the pond to a narrow area north of their breeding grounds that gets more sun, as Pond B begins to freeze • Flock clusters together in this narrow area • A large number of ducks swim constantly to keep the water churning and delay its freezing • Other ducks huddle close together on the ice near the edge of the water • If Pond B freezes over completely, the local flock retreats to a nearby creek for a few days • Local flock hosts any migratory flocks that decide to winter at Pond B
March	• Flock spreads out as Pond B thaws, resume flying • Flock reconvenes near nesting grounds • Individual nest sites are selected by females • Nests are constructed by females • Female lays and incubates eggs
April–June	• Ducklings hatch • Ducklings walk, swim, and feed themselves within 24 hours of hatching • Males associate together, assume flock guard duties, using their bright breeding plumage to distract predators • Ducklings begin to preen • Ducklings continue growing and learning about their habitat, neighbors, and flock
July–August	• Adult males molt breeding plumage • Ducklings acquire juvenile feathers • Ducklings learn how to do water landings, water takeoffs, and upend in deep water • Ducklings learn to fly • Adult females do a complete molt

6/2/2012. For the first time a family of 9 mallard ducklings hatched at the new Pond D on 4/28/2012. Four weeks later, only 2 mallard ducklings were left at Pond D. Also for the first time a family of 4 mallard ducklings hatched at Pond A on 5/30/2012. In each case the fathers stayed with the mallard ducklings that hatched in 2012.

Flock Security Measures

Mallard ducks tend to form large communities with it being common to see flocks of 24 or more birds. This association serves a combination of flock social and security needs. Finding food is easier and more efficient when searching in groups. While the group is searching for food a few individuals can stand guard and sound alarms to warn of impending danger. This is definitely preferable to each individual having to simultaneously look for food and predators.

Photograph 4.16 Male Mallard Parenting Duties. Most publications claim that male mallard ducks abandon their family after the eggs are laid. That was not the case during this 8-year study. Male mallard ducks escorted and protected their ducklings on land and in water at all four ponds, from the day the ducklings hatched until they were on their own. The top photograph was taken at Pond B on 4/27/2012. The bottom photograph was taken at Pond B on 4/29/2011. © D. Herrmann

Flock security is a community effort. If mallard ducks sense a low risk hazard, the flock disperses along the perimeter of a pond, near the edge. The ducks are evenly spaced, leaving just enough room to accommodate each other's wingspans should sudden flight be necessary. This is what happens when a lawnmower crew or someone walking a dog comes close to the flock. If a blue jay (*Cyanocitta cristata*) sounds an alarm or a red-tailed hawk (*Buteo jamaicensis*) or peregrine falcon (*Faleo peregrinus*) circles overhead, the flock instantly goes to the center of the pond and maintains a constant vigil above. The hawk or falcon generally leaves the area upon seeing the flock congregate in the center of the pond.

Table 4.2 Mallard Duckling Statistics for 2011 and 2012.

Hatch Date	Location	Number Hatched	Number survived first 2 months	Duckling survival rate (first 2 months)
21 April 2011	Pond B	12	10	83%
8 May 2011	Pond B	10	2	20%
3 June 2011	Pond C	8	3	37%
Total 2011	All ponds	30	15	50%
27 April 2012	Pond B	11	9	82%
28 April 2012	Pond D	9	2	22%
30 May 2012	Pond A	4	4	100%
2 June 2012	Pond B	5	4	80%
Total 2012	All ponds	29	19	66%

Notes:

1—2011 was the first year mallard ducklings hatched at Pond C.
2—2012 was the first year mallard ducklings hatched at Ponds A and D.
3—The adult population was 36 mallard ducks every year this flock was under observation, until 2015. In 2015 the adult population dropped to six mallard ducks; the reason is unknown. They continued to alternate between Ponds A, B, C, and D; however no mallard ducklings hatched this year. The only ducklings to hatch were 7 wood ducklings that hatched at Pond B on 5/15/2015.

Midday on 3/11/2011 a lone male mallard was in the center of Pond A making a distress call. It was a low pitched shrill quacking sound that conveyed a sense of urgency. A red-tailed hawk was perched on top of a short tree that overhangs the pond. The red-tailed hawk descended rapidly toward the lone duck with its talons drawn. The red-tailed hawk's talons could easily fit around the duck's neck. And, at 2.5 pounds (1.12 kg), the red-tailed hawk could readily carry the duck away. The situation didn't look too promising for the lone duck.

The duck was facing straight west when the red-tailed hawk suddenly lunged toward it. The duck went completely under water for over 10 seconds. The duck resurfaced 15 feet (4.57 m) behind his original position and facing straight east. This scenario was repeated several times. Every time the duck resurfaced 10 seconds later in a new and different location. The red-tailed hawk eventually gave up and left, since hawks cannot hover or swim. In the days since individual and pairs of mallard ducks have continued to swim in Pond A.

Midday 4/25/2011 an adult male mallard was spotted that had a white zig zag pattern along his rust bib leading up to his neck. He also limped a little on the left side. Probably this mallard duck had done battle with a hawk over winter and survived; most likely the white feathers indicated scar tissue that had healed.

If mallard ducks sense a high risk hazard from the ground, such as a group of humans or dogs running toward them, the flock instantly becomes airborne, whether they were on land or water. This sudden scrambling of the flock, combined with multiple flashing reflective blue speculum, confuses predators and makes it difficult for them to target a single individual.[30, 95] This scrambling practice has been copied by modern air forces around the world.

Common Daily Activities

Now let's look at some of the common activities mallard ducks engage in throughout the course of a normal day, like bathing, preening, upending, napping, and sleeping.

Feathers are very important to any bird. Feathers are essential for flight. A strong, healthy, complete set of feathers make flight possible. A bird may be able to fly from yard to yard across town with a rumpled, frayed, or incomplete set of feathers. But it would be impossible to make a 1,500 mile (1,995 km) migratory journey under those circumstances. Feathers regulate body temperature throughout the changing seasons and weather patterns. This feature is particularly important to water birds who spend part of every day in the water, regardless of whether the air temperature is 30°F or 100°F. Furthermore, feathers are part of a bird's identity. A male mallard duck is supposed to have a metallic green head and a reflective blue speculum. In contrast, a male cardinal is supposed to be bright red and have a crest. Feather color and arrangement patterns are not some temporary fashion fad. Photographs of mallard ducks from 150 years ago look just like photographs of mallard ducks today. Consequently, birds take very good care of their feathers. In fact, unkempt listless feathers are considered to be symptoms of emotional stress, physical stress, malnutrition, or disease.

Given the opportunity mallard ducks will bathe on a daily basis. How else could male mallard ducks keep their white side and underside feathers so clean and bright? To bathe mallard ducks dunk their head and neck underwater, while throwing water on their back. Mallard ducks do this twice in quick succession. Next they stand up straight and flap their wings rapidly three times. Then mallard ducks resume a floating position. From there they lean forward and wiggle their tail to shake off any excess water. Most of the time mallard ducks go through this bathing sequence once. On occasion mallard ducks repeat this bathing sequence twice. (See Photograph 4.17.)

That is the normal year round bathing sequence. A second more intense bathing sequence was observed for a few days each year during early spring, just after the ponds thawed and the temperatures began to warm up. In this instance mallard ducks went completely underwater before flapping their wings, instead of just dunking their head and neck.

In addition to daily baths, mallard ducks spend a lot of time preening their feathers. Preening is a meticulous process that involves stroking, arranging, and conditioning all of their 3,000-plus feathers. Often this process involves adding a minute amount of oil from their preening (or uropygial) gland to each feather. Mallard ducks start the preening process by stroking the feathers that are easiest to reach—those on their neck and breast. This is followed by preening the feathers on their back and tail. Mallard ducks often rub their head and neck on their back while preening. Next mallard ducks preen the feathers on their underside—at least as far as their bill can reach. Lastly mallard ducks preen the top and underside of each wing, first while the wing is retracted, then while it is extended. Sometimes this is accomplished while a duck is standing on one leg. Preening seems to have a calming effect. A serious preening session is often followed by a nap. Mallard duck pairs do not engage in allo-preening. (See Photographs 4.18 and 4.19.)

As part of their daily search for food mallard ducks frequently upend.[83] Pairs of mallard ducks often upend at the same time. While a mallard duck is upending their feet are busy treading water in order to maintain their upside down position. A mallard duck's feet are so busy paddling that it is possible to see a circular ripple of water around them. A mallard duck can stay upended for more than 10 seconds. If a mallard duck needs to stay upended for a period of time, it blows bubbles out its nostrils to exhale. Mallard ducks can, but rarely go completely underwater in search of food like hooded mergansers (*Lophodytes cucullatus*) do. (See Photograph 4.20.)

Most birds require 12 hours of sleep a day. This generally translates into the need for several naps throughout the day. Mallard ducks are no exception. Mallard ducks have three options for a short nap. They can nap while standing on one leg, sitting on the ground, or floating in a pond. Sometimes they even stand on one leg in a pond while napping. If a mallard duck is taking a short casual nap it tucks its head and neck against its back, but continues to face forward. This seems to be a sort of standby alert position. On the other hand, if a mallard duck is settling in for a serious nap it turns its head around and tucks its head under its wing. There doesn't appear to be any preference about where a mallard duck tucks its head. About half the birds tucked their heads under their left wing. The others

Photograph 4.17 Bathing. Mallard ducks take bathing seriously and will bathe daily given the opportunity. Standard procedure is to dunk the head and neck underwater twice, throw some water on the back, stand-up tall and flap the wings three times, then lean forward and wiggle the tail. On occasion mallard ducks engage in extra energetic baths and go completely underwater. Often if one mallard duck starts to bathe, others near it will too. © D. Herrmann

tucked their head under their right wing. If a mallard duck is napping while standing on 1 leg, it will tuck its head under the opposite wing in order to make it easier to balance.

Mallard ducks like to rest near the edge of a pond. Large groups of ducks congregate on the banks of Pond B and sit on the ground. Often mallard ducks engage in a stretching routine before sitting down. Mallard ducks methodically stretch one wing at a time while standing on alternating feet, wiggle their tail feathers, and roll their head around. They briefly fluff and shuffle their wing feathers once more, then sit down to rest. (See Photograph 4.21.)

Individual mallard ducks frequently hide themselves among the reeds at the edge of a pond when it is nap time. In this position the ducks are half on land, half in the water, and nearly invisible. More than once, thinking no ducks were present, I walked too close to the reeds and was surprised when a duck suddenly jumped in the water or flew away. Mallard ducks tend to assume this position at Pond A more than at the other ponds. Pond A has a couple large bushes near the shore. Several branches overhang the pond. On occasion one or two mallard ducks have been spotted sitting on the lowest branches, just above the water.

Photograph 4.18 Preening is Serious Business #3. Mallard ducks take extremely good care of their feathers. They are not trying to impress someone or put on airs pretending to be something they are not. Rather, good preening is essential to their overall health and well-being. Daily preening sessions, which may last 20 minutes or more, involve cleaning and arranging all of their 3000-plus feathers from head to tail. © D. Herrmann

Male mallard ducks have white outer eyelids. Female mallard ducks, ducklings, and juveniles have light beige outer eyelids. When a duck is asleep its outer eyelids are visible. In fact, these outer eyelids are very noticeable. They are a security feature that hides the fact that the duck is asleep. From a short distance bright white or light beige eyelids contrast with dark head feathers and give the impression that the duck is awake. In other words these bright outer eyelids are intended to deter predators while a duck is asleep. They are a type of reverse camouflage. While napping mallard ducks open their eyes briefly every 60 to 90 seconds to survey the scene.

Current weather patterns determine the direction mallard ducks face when they are napping. On bright sunny days mallard ducks nap facing directly into the sun. As the sun moves, so do the ducks. On windy days mallard ducks nap facing directly into the wind. If they didn't the wind would blow their feathers upright and they would lose body heat. On cloudy days mallard ducks nap facing every which direction. This napping alignment, based on the weather pattern, is consistent regardless of whether the ducks are napping on land or in water. Pairs generally nap near each other with just enough room between them to accommodate their wingspans should they need to take flight suddenly.

Photograph 4.19 Preening is Serious Business #4. Preening involves stretching and bending in all sorts of unusual positions. Preening has a calming or soothing effect. As a result mallard ducks often close their eyes near the end of a preening session. © D. Herrmann

These napping alignment norms hold true across 98% of the ducks present. There are always a few individuals who prefer to do their own thing, even if only for a few minutes. (See Photograph 4.22.)

Mallard ducks often stretch after a nap. To do so a mallard duck stretches its head and neck out straight, parallel to the ground. At the same time the mallard duck stretches both wings up in an M pattern. The duck holds this position for a while, before standing up. This stretching routine is occasionally used as a greeting ceremony in the fall or early spring when mallard ducks are part of a pair bond. This mallard duck stretching behavior is similar to that illustrated in Photograph 3.31 for Canada geese.

In the evening at sunset when it is time to settle in for the night, the flock spreads out along the perimeter of the pond. Half of the ducks are on one side of the pond, half of the ducks are on the other side. The ducks are in the water huddled close together in groups of 2 or 3 facing the center of the pond. Their heads are tucked in. A few ducks cluster under the low branches that overhang the pond. The ducks generally stay put until just before sunrise when they awake.

Photograph 4.20 Upending. Mallard ducks upend as part of their daily search for food. The head, neck, and breast go under water and the rest of their body points straight up in the air. While pursuing a fish, snail, or some aquatic plant life, the feet of a mallard duck are busy treading water to keep them upright. Mallard ducks can stay upended for about 10 seconds. If a mallard duck needs to stay upended for a period of time, it blows air bubbles out its nares to exhale. Bonded pairs often upend at the same time. © D. Herrmann

Songbirds and other birds that live in trees have a somewhat regular daily schedule, being most active and consuming their largest meals at sunrise and sunset. In contrast mallard ducks, being water birds and ground birds, do not have a fixed daily routine. Mallard ducks walk, swim, fly, and nap whenever and wherever they decide to. Mallard ducks eat when they are hungry and don't when they are not, unless some particularly interesting morsel presents itself. The only fixed part of a mallard duck's daily routine is the fact that the flock reconvenes prior to sunset in order to settle in for the night. Pond B was the chosen rendezvous point.

Photograph 4.21 Stretching. Mallard ducks stretch their wings one at a time periodically throughout the day. This may be done while standing on land, standing in water, or sitting down. Stretching a wing involves stretching the corresponding leg at the same time and adjusting one's balance. © D. Herrmann

4.4 Seasonal Behaviors

Mallard ducks exhibited some consistent behaviors year after year that were tied to the changing seasons. For example the:

- Fall restlessness and preparation
- Winter cooperative group survival tactics
- Spring return to a more relaxed life style

Fall Restlessness and Preparation

Mallard ducks become restless mid-fall when the days are noticeably shorter and the temperatures are cooler. The level of activity, chatter and movement increases across the flock, all of which convey a sense of urgency. It is common to see several mallard ducks swimming in formation at this time, either as a V or partial V. There is a lot more wing flapping while in the water as well. This is not part of bathing—there is no head or neck dunking. Rather, this behavior is focused on exercise and conditioning the wing muscles. Often 1 to 3 ducks fly a distance of 10 feet (3 m) while being only 1–2 feet (30.48–61 cm) above water. Then they land in the water. This feat is repeated 3 or 4 times in a row. (See Photograph 4.23.)

Photograph 4.22 Napping. Mallard ducks nap standing on one leg on land or in water, sitting on the ground, or floating in a pond. Sometimes mallard ducks nap with their heads facing forward. Other times they tuck their heads under their wing. Mallard ducks open their eyes briefly every 60–90 seconds while napping to survey the situation. Their off-white outer eyelids are visible when the mallard ducks eyes are closed. This type of natural camouflage creates the impression that the ducks are awake when in fact they are asleep. When mallard ducks nap on the ground, it is always near water. Bonded pairs nap near each other. © D. Herrmann

There are changes in dietary behavior too. Mallard ducks who ignored leaves when they were fresh and had just fallen from a tree now pull up these partially decomposed leaves from the bottom of a pond and eat them. Mallard ducks consume more fish and snails as the temperatures drop. Mallard ducks were observed eating 2 to 3 fish during one feeding in the fall, compared to the norm of 1 fish every week or so during summer. This increased activity level leaves less time for socializing. Before long mallard ducks have grown their winter down and their feathers appear to be much fuller.

Most mallard ducks migrate south during the fall and return to their nesting grounds in the spring. Table 4.3 lists the dates migratory flocks from farther north passed through the area in Fall 2010. Ponds B and C are regular stopover points for flocks of migrating mallard ducks. The migratory flocks of mallard ducks were all the same size as the local flock of mallard ducks.

Photograph 4.23 Wing Flapping. Mallard ducks engage in wing flapping exercises throughout the year. Wing flapping occurs more often and is most noticeable in the fall when mallard ducks are strengthening their muscles for migration. This wing flapping is more energetic than normal stretching or bathing. © D. Herrmann

Occasionally a pair or two of American black ducks were spotted in the midst of the mallard ducks. In these instances the black ducks mingled with the mallard ducks, but stayed together as a pair. Male and female black ducks look alike. The appearance of black ducks is similar to female mallard ducks. Black ducks are slightly larger and a bit darker than female mallard ducks.[29] Also, black ducks have greenish yellow bills, unlike the orange bill of a female mallard duck.[29]

Migration isn't always as simple as south in the fall and north in the spring, especially in this time of global climate change. As shown in Table 4.3 flocks that had flown south for the winter returned to the mid-Atlantic states five times mid-winter. These flocks came back north for a few days when the southeastern states were uncharacteristically blanketed in a foot (30.48 cm) or more of snow. Also,

Table 4.3 Dates Migratory Flocks of Mallard Ducks Passed Through the Area (Fall 2010–Winter 2011).

Date	Number of Ducks	Location	Notes
I. Normal Fall Migration South (2010)			
Nov 21	7	Pond C	
Nov 27	9	Pond C	
Nov 28	9	Pond C	
Dec 2	42	Pond C	
Dec 4	18	Pond C	
Dec 7	24	Pond C	
Dec 8	24 12	Pond B Pond C	
Dec 9	12 24	Pond B	Two flocks were seen. The flock of 12 may have moved over from Pond C, which froze.
Dec 11	24	Pond B	
Dec 12	24, plus 2 black ducks	Pond C	
Dec 13	24, plus 2 black ducks	Pond B	Might have been flock that was at Pond C the day before.
Dec 16	24	Pond B	
II. Partial Migration North in Response to Snowstorms in the Southeast (Winter 2011)			
Jan 3	20	Pond B	
Jan 19	28	Pond C	
Jan 21	12	Pond C	
Jan 27	36	Pond B	
Feb 18	36	Pond C	

less food may have been available in their wintering grounds due to the 2010 BP gulf oil spill. The ducks had to find food somewhere. They couldn't fly farther south, so they returned north to a known stopover point in hopes of finding food. This new mid-season partial migration behavior highlights the need to document and preserve waterfowl stopover points.

In the fall mallard ducks store body fat to fuel their long migratory journey. That fat is depleted once they complete their journey. When mallard ducks and other waterfowl have to make a sudden partial journey back north in response to freak snowstorms in the south, they don't have time to accumulate fat reserves. The mallard ducks and others must leave immediately or face possible starvation. When flocks of mallard ducks arrived under these circumstances, they were often exhausted. The first thing the ducks did after landing was sleep. Later, the ducks began to eat.

The average duration of a stopover was 1.5 days during the normal Fall 2010 migration. Generally the mallard ducks would land mid to late afternoon, rest and eat for a day, and they were on their way early the following morning. If a flock was detained due to inclement weather, the mallard ducks could be seen flying short practice flights overhead to stay in shape.

Winter Cooperative Group Survival Tactics

The local flock of 36-plus mallard ducks does not migrate south in the fall, unless Pond B freezes over completely in response to heavy snow and extremely cold temperatures. Instead the local flock stays put and applies the energy they would spend on migrating toward surviving winter here. The local flock socializes and mingles with migrating flocks as they pass through Pond B. No hostility or aggression was displayed toward the visitors. In fact one flock of 20-plus mallard ducks joined the locals for the winter during 2010. This flock arrived in early December and stayed until early March.

The flock was much more vocal on windy days and just prior to storms. At this time of year the flock makes a quiet chattering or clicking sound at regular intervals. Males and females participate. The clicking sound lasts for 15 to 20 seconds. The flock is quiet for a couple minutes. Then they repeat the clicking sound. This sound may be a roll call or some sort of security message. The chatter intensifies whenever anyone, large or small, flies over the flock or a blue jay makes an alarm call. Once spring arrives the flock quits making this sound.

A small group of mallard ducks visited Pond C regularly during the summer and fall. As Pond C began to freeze over the ducks moved into the small area that was still open. At this point the ducks flew from one side of Pond C to the other, since they could no longer swim. It took 4 to 5 days for the ducks to master the art of landing on ice and snow. At first the ducks skidded about 1 foot (30.48 cm) before they stopped. To slow the skidding the ducks sat down. Occasionally the ice buckled. When that happened the ducks simply swam over to the edge of the ice and hopped out of the water nonchalantly. When Pond C froze over completely these mallard joined the others at Pond B.

As Pond B began to freeze over the mallard ducks clustered in a small area of the pond north of their nesting grounds. This part of the pond receives the most sun during winter. When it was cold and windy, the flock stayed in the water near the shore. Individual ducks snuggled under bushes and the embankment (near where the turtles hibernate) for warmth. On extremely cold days the mallard ducks were quiet, still, and huddled close together to save energy. Sometimes they rested by standing on one foot or sitting on the ice. At first there was some aggression due to the sudden crowded conditions. After 2 or 3 weeks the mallard ducks adjusted to their new situation and were more relaxed, especially on sunny days. However, the adults continued to harass the juveniles.

At times the small area the mallard clustered in was only 30 feet (9.14 m) in diameter. Yet it accommodated 64 mallard ducks. The rest of Pond B was frozen over. This area didn't freeze. Through a cooperative winter survival strategy the mallard ducks, by constantly swimming, kept the water churning and delayed its freezing. The mallard ducks understood that their food was under the water and they needed to be able to reach it. The mallard ducks also walked close together around the perimeter of the open area to increase the chance of some of the ice sinking. There was some slipping and sliding when new snow fell on the ice. One female mallard duck who slipped and fell onto her back used her wings to get upright again, much like humans automatically stretching out their arms when they stumble, fall, or slip on ice. (See Photographs 4.24 and 4.25.)

Spring Return to a More Relaxed Lifestyle

The mallard ducks began to fly as the temperatures warmed up and Pond B started to thaw. The small group of mallard ducks even returned to Pond C for a few hours at a time on sunny days. Eventually spring arrived and the mallard duck flock at Pond B returned to their normal spring-summer-fall home and daily routine. So did their neighbors.

A small flock of 10 wood ducks share Pond B with the mallard ducks. Wood ducks are the only native species of perching ducks in the United States; hence their name.[161] They were frequently seen sitting on low branches over Pond B. In spring and summer the male wood ducks sit in the shade on large tree limbs about 20 feet (6.1 m) above the water. Wood ducks nest in tree holes 3 to 60 feet (.91–18.29 m) above ground.[141, 161] They may have been guarding their nest cavity. (See Photograph 4.26.)

Wood ducks almost became extinct in the 19th century due to over-hunting and the clearing of forests.[95, 137,161, 208] A nest box campaign helped spur their recovery.[95, 137,161, 208]

Photograph 4.24 Winter Home #1. The local mallard duck flock clustered in a small area of Pond B north of their nesting grounds for the winter. By constantly swimming the mallard ducks kept the water churning and delayed its from freezing. Ducks that were not swimming stayed nearby on the ice or snow. Their necks are retracted and their feathers are puffed up to maximize heat retention. © D. Herrmann

The wood ducks at Pond B mingle freely with the mallard ducks, even during nesting time. The wood ducks at Pond B are very energetic but not aggressive at all. However, they are wary of people. These wood ducks do not migrate; instead they stay put during winter and stick it out with the mallard ducks.

Photograph 4.25 Winter Home #2. Each winter it takes the mallard ducks a few days to get used to walking on snow and ice again. Mallard ducks use their tail feathers to slow down when landing on ice and their wings for balance if they slip. The nails on their three front toes provide traction. © D. Herrmann

Male wood ducks have striking plumage and a unique head crest.[141] Female wood ducks look like a three-quarter size female mallard, including the blue speculum, but they have a thick white eye circle and a thin yellow eye membrane. At times only the distinctive calls of the wood ducks were heard. The wood ducks' high pitched cheery vocalizations are quite a contrast to those of mallard ducks. A female wood duck often makes this sound as a contact call or part of a greeting ceremony.[153, 161] Wood ducks can be heard in the background on Disk 1 Track 1. (See Photograph 4.27.)

Unlike mallard, wood ducks go completely under water when they bathe. Perhaps that is because of their smaller size. Wood ducks were not observed upending; apparently they do so on rare occasions.[95]

In northern climates wood ducks have 1 brood a year. In warmer climates wood ducks may nest twice a year. During 2011 the first family of 10 wood ducklings was spotted on April 24th at

Photograph 4.26 Male Wood Duck. One to four male wood ducks congregated on a particular tree limb during nesting time each spring. Most likely they were guarding the nest. Only male wood ducks were seen there, not females. The male wood ducks were only seen on this tree limb during nesting time. The tree limb is about 20-feet (6 m) above ground near the edge of Pond B. © D. Herrmann

Pond B. In spite of their tiny size, the ducklings ran very fast. A second family of 8 newly hatched wood ducklings was spotted at Pond B on May 11th. They swam behind the mother in single and double file. The mother swam in a large S-like pattern across the pond to confuse predators. Wood ducklings are very adept at hopping; they were often seen hopping up or down a distance of 8 inches (20.32 cm) or more. That is not surprising since wood ducklings hop down from their nest cavity a day after they hatch.[95, 141, 137] Wood ducklings are 15% smaller than mallard ducklings of the same age. (See Photograph 4.28.)

Photograph 4.27 Female Wood Duck. Female wood ducks have exquisite but delicate coloring. In direct sunlight the speculum of a female wood duck reflects a light turquoise blue. Otherwise, the speculum appears dark. It is believed that the speculum of ducks is used as a visual communication signal. Female wood ducks have a small crest, yellow legs, and brown eyes that are lined with a thin yellow membrane. © D. Herrmann

The last week of May 2011, one then two wood ducklings were observed swimming by themselves across the pond. The mother and other wood ducklings were nowhere in sight. At the time one of the wood ducklings attempted to stand up in the water a couple times and flap its wings stubs, like an adult would do. Later on July 2nd a lone wood duckling was observed traveling around with a pair of Canada geese and their 5 nearly grown goslings. The Canada geese parents did not deter the wood duckling. Both of these instances were unusual. Wood ducks are known to combine broods and foster orphans.[95]

Photograph 4.28 Wood Ducklings. Wood ducklings are very energetic from the time they jump out of their tree nest onward. Like mallard, wood ducklings identify with each other and their mother upon hatching. From top left we see 1-day old, 1-week old, 4-week old, and 7-week old wood ducklings. © D. Herrmann

Wood ducks tend to explore the more remote and inaccessible areas of Pond B and the nearby creek. Usually this is done in small groups of 2 or 3. Females, juveniles, and males in non-breeding plumage readily blend into the scenery when doing so. As a result wood ducks were accidentally flushed from the area on occasion. When that happened, the wood ducks did not always take flight. Often they hurriedly scurried along the top of the water—not actually flying or swimming—for a distance of 8 to 10 feet (2.43 to 3.04 m) while they assessed the seriousness of the threat.

A small flock of 10 hooded mergansers share Pond C with the mallard ducks, Canada geese, and belted kingfishers (*Megaceryle aleyon*). The hooded mergansers do not mingle with the mallard ducks; they keep to themselves. There is no hostility or aggression—the two groups just don't mix. The hooded mergansers are very wary of humans. They tend to stay on the western side of the pond away from people. Hooded mergansers do not upend. Instead they dive, stay under water 10 to 15 seconds, and resurface 8–10 feet (2.43–3 m) from where they went under. During these dives hooded mergansers pursue small fish, tadpoles, crustaceans, and water insects, aided by their superb underwater vision.[29, 95, 137]

This flock of hooded mergansers is only seen during fall and winter. They stay at Pond C during fall and winter as long as the pond isn't entirely frozen. When Pond C freezes over completely, the hooded mergansers retreat to a nearby creek for shelter and food. Table 4.4 lists the dates hooded mergansers were seen at Pond C during winter 2010/2011. After April 3rd the hooded mergansers disappeared to participate in nesting activities. They were not seen again until May 11th when a single female hooded merganser was spotted at Pond C swimming with 2 small ducklings behind her. This new family was not seen again after that date. During spring and summer hooded mergansers prefer more wooded territory;[95] hence they retreat to the creek near Pond C.

These observations raise a question. If a flock of ducks (mallard, wood ducks, black ducks, or hooded mergansers) doesn't migrate for several years will anyone remember the route and stopover points? What will happen if the climate changes suddenly and they have to migrate? The younger ducks will never have made the long journey. Will they know to associate with a migratory flock that is passing through the area?

Table 4.4 Sightings of Hooded Mergansers at Pond C During Winter 2010/2011.

Month	Day	Number of Ducks	Notes
December	2	6	
	3	6	
	4	7	
	7	7	12/9/2010 Pond C frozen
	12	2	Partial thawing; 12/16/2010 Pond C frozen
January	4	1	Partial thawing
	17	3	
	19	3	
	20	3	
	21	3	1/27/2011 Pond C frozen
February	14	4	Partial thawing
	21	4	
	22	4	
	23	6	
	24	6	
	26	6	
March.	1	7	
	2	3	
	3	2	
	4	10	
	7	2	
	8	4	
	9	2	
	22	2	
	24	3	
	26	4	
	29	2	
April	3	4	Disappeared afterward to begin nesting activities

4.5 Individual Behaviors

Mallard ducks are flock birds and socialize in large groups. Be that as it may, members of the flock do express their individuality on occasion. In particular they exhibit evidence of individual decision-making. The observations below relate to adult mallard ducks. The expression of individuality was most notable in regard to three activities:

- Formation of subflocks
- Rotation among ponds and other waterways
- Socialization outside the flock

Formation of Subflocks

Pond B is home for the local mallard duck flock. That is where the majority of mallard ducks can be found on any given day and at any given time. Even at Pond B, however, the dynamics of subflocks are apparent. Some ducks are swimming. Some ducks are walking around. Other ducks are resting while sitting down. A few might be bathing or feeding at the pond's edge. All this activity is going on at the southern end of Pond B where the mallard ducks normally congregate. In addition groups of 1 to 4 mallard ducks are usually in the wooded area between the southern and middle sections, in the middle section, and in the narrow northern end of Pond B. If mallard ducks were simple automatons as Descartes implied, endlessly executing mindless stimulus/response routines, all the ducks would be doing the same thing at the same time in the same place. But the ducks never are, not even when they are home at Pond B.

There are over 36 mallard ducks at Pond B. Each mallard duck makes a decision about what part of the pond to be in (south, wooded area, middle, or north), what activity to participate in (flying, swimming, bathing, feeding, preening, stretching, vocalizing, upending, walking, or resting), and which other ducks or subflock to associate with. The formation of subflocks is temporary and related to a particular daily activity. Each duck decides for itself whether or not to participate in subflock activities. Hence, mallard ducks are gregarious but they are also individuals.

Rotation Among Ponds and Other Waterways

Further evidence of individuality is seen in the random patterns by which mallard ducks rotate among the four ponds. Subflocks of various sizes visit the other ponds on different days and at different times. Sometimes only 1 or 2 mallard ducks visit a particular pond. The next day 10 ducks are at that pond. Then for a day or two no ducks are seen there. Even more interesting is watching a small group of mallard ducks swim in one pond, takeoff and fly to another pond, make a water landing, and start swimming there. For example, at noon on July 14, 2010 a group of five ducks flew from Pond D to join a single duck who was swimming in Pond A. Earlier in April small groups of ducks took a quick dip in the community swimming pool, then flew directly to Pond A. These flights are short and deliberate. The ducks are not roaming around looking for water. On the contrary, they know the exact location of each pond and fly directly to it. Mallard ducks start their day at Pond B. Temporary subflocks form and take outings to one or more of the other ponds or the community swimming pool or the nearby creek. Before dusk, the subflocks return and the flock reconvenes at Pond B.

Why do the mallard ducks rotate among the four ponds, creek, and community swimming pool? A lone male mallard even visited the new Pond D on 5/25/2011. He was seen there 2 days in a row. The ducks do not rotate among the ponds out of necessity. There is plenty of food, water, space, and security at Pond B. It is a recognized fact that the act of engaging in play—an activity that does not contribute to survival, but is merely fun—is a sign of intelligence. Given the frequency and precision with which these mallard ducks rotate among the four ponds, the swimming pool, and the nearby creek, and on occasion the golf course water trap, they must be a rather smart flock.

Table 4.5 summarizes the rotation of mallard ducks among the four ponds during 2010. The ponds were surveyed at different times on different days to capture the broadest spectrum of behavior. Not every location was surveyed every day. A dash in a cell means that no birds were seen at that location. A blank cell means that location was not surveyed that day. An asterisk indicates a migratory flock that was resting at one of the ponds as a stopover point.

Socialization Outside the Flock

Mallard ducks are aware of other life forms around them, and not just predators, given their inclination to socialize outside the flock. A single Canada goose stayed behind on occasion to socialize with the mallard ducks at Pond C. This behavior implies that the goose was interested in socializing with the small group of mallard ducks. At the same time this behavior implies that the small group of mallard ducks, usually 2 to 4 ducks, was interested in socializing with the lone goose and accepted it. When this occurs the ducks and goose walk around together, sit and rest in the same area, and even swim together. The goose honks and the ducks quack. They function just like a subflock of the same species. This pattern of individuals socializing outside the flock took place at Pond C year after year.

Groups of mallard ducks and Canada geese mix freely from August onward at Pond B. This is after both flocks have diminished parenting duties. This pattern of group socialization is particularly interesting because it occurs at Pond B and in the southern area that the mallard ducks consider home. The mallard ducks could view the Canada geese as rivals—after all they eat similar food—and run them off. Instead, the mallard ducks welcome the Canada geese as neighbors. House sparrows also walk freely among the mallard ducks at Pond B. House sparrows have been spotted foraging within inches (centimeters) of young ducklings that were only days old. The adult ducks did not express any concern, panic, or hostility. Once in a while tree squirrels roam into the ducks' home turf too. The ducks acknowledge the tree squirrel's presence, but again they express no concern.

A third pattern of socialization outside the flock was observed at the golf course. On a few occasions each year a single mallard duck from Pond B visited the Canada geese at the golf course. This behavior is the reverse of the lone Canada goose socializing with the mallard ducks at Pond C. The mallard duck swam with the geese in the golf course water trap or sat on the edge of the water trap with them. There were usually 10 to 14 Canada geese there at the time. On the afternoon of 9/26/2010 the mallard duck even participated in flying laps around the golf course with the Canada geese. But the mallard duck drew the line when it came to walking up and down the fairway or on the putting greens. Mallard ducks just don't do that. Since mallard ducks are smaller than Canada geese, this pattern of individual socialization implied more than just curiosity. It also implied a degree of trust.

One to four mallard ducks at a time visit Pond A during May and June when the Canada goose pair is raising their young. In this instance the mallard ducks and Canada geese do not really socialize. Instead the two groups acknowledge each other and share the pond. There are no displays of hostility or territorial disputes. The ducks seem to understand that this is the Canada geese's family time and do not intrude.

There are reports of mallard ducks becoming tame and tolerant of people. Most often these reports pertain to ducks living in city parks who experience frequent contact with humans. Perhaps the most famous is Quackers, a male mallard duck that lives in New York City's Central Park.[172]

The local flock of mallard ducks has limited contact with humans (except for lawnmower crews), even though they live near a metropolitan area. The ducks rarely encounter humans at Pond C, never at Pond D, and only on occasion at Pond A. The ducks at Pond B see small groups of humans, usually mothers with young pre-school children, two or three times a week. Because of the noise and sudden movements of small children, this flock of ducks is not very interested in or trusting of humans. If a

Table 4.5 Mallard Duck Rotation among the Ponds during 2010, Including Visits by Migratory Flocks.

Month	Day	Time	Pond A	Pond B	Pond C	Pond D and elsewhere
January	4	3 PM	-	24	-	-
	20	2 PM	-	24	-	-
March	1–7	1 PM	-		4	-
	24	4 PM				2 at creek
April	6	9 AM	1		-	-
	11	11 AM	-		1	
	20	11 AM	-		1	
	22	1 PM	-	6		
May	4	5 PM	-		2	
	6	1 PM	-		2	
	8	6 PM	2		2	
	14	6 PM	2		2	
	15	12 PM	2			-
	20	1 PM	-		20	
	30	12 PM	-	16		
June	11	12 PM	-	24		
	13	12 PM	-	24		
	27	2 PM	-	24	-	
July	8	10 AM	-	24		
	11	9 AM	1			
	12	3 PM	-	24	3	
	13	4 PM	-	1 flew from creek to Pond B		
	14	12 PM	5 flew from Pond D to join 1 at Pond A		-	-
	15	10 AM	1	20	2	
	16	12 PM	-	24	3	
August	11–22	-	Pond A four times normal size due to flooding. 4, 6, then 10 ducks visited			
	26	12 PM	-	36	4	
	27	8 AM	2	36	6	
	28	7 AM	14			
	29	8 AM	14		-	
	30	8 AM	10			
	31	8 AM	12		6	
September	1	7 AM	21			
	2	7 AM	10		4	

Table 4.5 contd....

Table 4.5 contd.

Month	Day	Time	Pond A	Pond B	Pond C	Pond D and elsewhere
September	3	2 PM	1			
	17	10 AM	1	30	4	
	21	6 PM	-	36	2	-
	22	10 AM	-		2	-
	24	9 AM	-		2	-
	25	10 AM	-	36	2	-
	26	4 PM	-	36	1	1 in golf course water trap with 30 Canada geese
	28	6 PM	-		4	-
October	1	6 PM	-	24		-
	2	1 PM	2			-
	7	4 PM	8	34	2	-
	8	5 PM	-	36	2	-
	12	5 PM	-	30	6	-
	13	4 PM	-	30	2	-
	16	3 PM	-	18	5	-
	22	4 PM	-	24	5	-
	23	3 PM	-	24	3	-
	28	5 PM	-	36		-
	30	2 PM	-	36	1	-
	31	3 PM	-	36	1	-
November	2	1 PM	-	24	1	-
	5	1 PM	-	24	6	-
	6	1 PM	-	24	6	-
	10	1 PM	-	36	3	-
	11	4 PM	-	36		-
	12	2 PM	-	36	7	-
	13	2 PM	-	36	5	-
	14	3 PM	-	36	5	-
	15	3 PM	-	48	3	-
	19	1 PM	-	48	7	-
	20	2 PM	-	42	3	-
	21	2 PM	-	42	8	-
	24	3 PM	-	42	1	-
	25	3 PM	-	42	3	-
	27	10 AM	-	42	3 + 9 *	-
	28	12 PM	-	42	2 + 9 *	-
	30	12 PM	-	42	3	-

Table 4.5 contd....

Table 4.5 contd.

Month	Day	Time	Pond A	Pond B	Pond C	Pond D and elsewhere
December	2	1 PM	-	42	3 + 42 *	-
	3	2 PM	-	42	3	-
	4	12 PM	-	42	3 + 18 *	-
	5	12 PM	-	42	4	-
	7	1 PM	-	42	24*	-
	8	2 PM	-	42 + 24*	3 + 9*	-
	9	1 PM	-	42 + 24* + 12*	-	-
	10	2 PM	-	42 + 24*	-	-
	11	1 PM	-	42 + 24*	-	-
	12	11 AM	-	42	24*	-
	13	2 PM	-	42 + 24*	4	-
	16–31	11 AM	-	42 + 24*	-	-

Notes:

1—Ponds were surveyed at different times on different days to capture the broadest spectrum of behavior.
2—Not every location was surveyed every day. A dash means no birds were seen at this location. A blank means that site was not surveyed.
3—An asterisk indicates a migratory flock resting at a stopover point.

large group of ducks has gathered in the area of Pond B that the mallard ducks frequent, they simply ignore the humans. Small groups of ducks elsewhere just fly away. More than once, if I stood openly watching a particular bird or small group of birds for more than a few minutes, the ducks would fly away even if I was at a respectable distance from them. Then again, if you are a mallard duck living at Pond B and you have more than thirty other mallard ducks to socialize with, not to mention the Canada geese, why would you be interested in socializing with a human?

Behavior rarely follows rigid rules. There are always exceptions and variations. On the evening of 9/24/2010 I made a routine check of Pond C. A lone goose had been seen there a few days in a row earlier in the week. The goose wasn't in its usual place so I went to the pond's edge for a closer look. Two mallard ducks were swimming in the pond at the time. To my surprise the two ducks swam right to me. They hopped out of the pond and walked toward me. The ducks appeared to be young and were only about 18 inches (45.72 cm) in length. Both had brown juvenile plumage. One duck had a blue speculum. The other duck was just starting to grow its speculum. The white feathers which outline the speculum were present, but there were no blue feathers. One duck had a yellow bill; the other duck's bill was orange. One duck made the quiet "yeeb" sound attributed to males. The other duck did not make any vocalizations.

I knelt down to the ducks level so I would not appear threatening. The two ducks came and stayed within inches (centimeters) of me. They were curious and walked all the way around me. One duck took a dandelion leaf from my hand. In one quick precise movement the duck snatched the entire leaf, the tip of its bill touching the end of my fingers without biting them. The ducks grazed around my shoes, but didn't touch them. The ducks seemed intrigued when I spoke softly and one responded with its own vocalization. If I moved a short distance in one direction the ducks followed me. This close interaction continued for over 15 minutes, until the ducks were startled by a loud noise. At that point the ducks retreated to the pond.

I had visited Pond C numerous times when ducks were present over a period of several years; yet the ducks had never expressed any interest in interaction. What was special about this visit?

Were these ducks young enough that their curiosity still outweighed their fear? These ducks had been socializing with the lone Canada goose. Did that experience contribute to their acceptance of and interest in me? Or, were other factors at work?

I decided to explore this situation further and went back to Pond C the next morning (9/25/2010). The two ducks and the lone goose were there. Again the two young ducks came right to me. They walked around me as before and took another leaf. This activity piqued the Canada goose's interest, so it came ashore too. The goose came within 3 feet (0.91 m) of me and watched the ducks interacting with me. If I spoke the male duck made a few quiet vocalizations. If I moved around the ducks stayed with me. This interaction continued for 20 minutes.

I returned to Pond C the following day (9/26/2010). The goose was gone and there was only one young mallard duck. The duck was quite a distance from me. But he swam around the perimeter of the pond until he reached me. At that point he hopped out of the water and walked up to me. It was the duck with the full blue speculum and yellow bill who made the quiet "yeeb" vocalizations. The young male duck went to a lot of effort to come visit me, but he was more cautious than before. Perhaps because he was by himself. I walked slowly along the edge of the pond. The male duck did the same. He poked his bill in the soft mud every now and then for food as the frogs, minnows, and water insects darted away. The male duck continued following me for about 12 minutes, before he turned and went the other direction. A red-eared slider turtle family was basking on driftwood in front of him. The adult turtle was about 8 inches (20.32 cm) in length. The three baby turtles were about 3 inches (7.62 cm) in length. On that same afternoon a lone mallard duck was observed socializing with some Canada geese at the golf course water trap. It may have been the partner of the lone young mallard duck at Pond C.

Between 9/27/2010 and 10/11/2010 I did not see the two young mallard ducks at Pond C, because of heavy rain, lawnmower crews, and people walking their dogs too close to the pond's edge.

On the evening of 10/12/2010 the two young mallard ducks were at Pond C. As soon as they saw me they began running to me—a distance of about 30 feet (9.14 m). I knelt down and the two young mallard ducks were just as curious as before: walking around me, looking at me, and making quiet vocalizations. The two young ducks continued interacting with me for 15 minutes, even though they hadn't seen me for over 2 weeks.

I visited Pond C again on the afternoon of 10/13/2010. I looked around but didn't see the two young mallard ducks. Then all of a sudden I saw them. The two young mallard ducks were clear across the pond on the other side. Yet they swam right to me—a distance of about 200 feet (60.96 m). How did they recognize me that far away? Two other people walked by, but the ducks ignored them. This time the two young ducks stayed in front of me. They seemed to enjoy me talking to them and made some quiet vocalizations in return. One duck engaged in a thorough preening routine. The other duck, after eating some small pink flowers, sat down for a nap and closed his eyes. He was only 1 foot (30.48 cm) from me. This act implied more than curiosity or eagerness to socialize with me; it implied trust. When the duck sat down for his nap he was perfectly aligned with the angle and direction of the sun. His head faced directly toward the sun. Every now and then the duck briefly opened one or both eyes and looked at the sun.

As I continued to visit the young mallard ducks I varied the time of my visit and my outfit. This experiment was intended to confirm whether or not the young ducks recognized me and if so, how. The variation in the time of my visits made no difference to the ducks. They still swam or walked to me without hesitation, but continued to ignore other people. At first I wore khaki pants and a black jacket, with a black camera/shoulder bag. Then I alternated black pants. That change in appearance made no difference to the young ducks. One day I wore a royal blue sweatshirt and sweatpants and a different pair of shoes. After about 10 seconds hesitation the young ducks started swimming toward me. Each time I walked along the same side of the pond and knelt down. It is likely that a combination of some aspect of my appearance and my behavior (walking on the same side of the pond, kneeling down, talking to the ducks, etc.) was used to confirm my identity.

Fall became winter and hats and gloves were added to my outfit. The young ducks paid no attention to the gloves. The first time I wore a hat the young ducks hesitated for a few seconds before coming to visit. After that the ducks ignored my hat. As Dr. Skutch observed:[154]

> "Recognition of individuals is proof bird's minds are capable of fine discriminations and establishing order in complex situations."

Or, to put it another way mallard ducks are capable of recognizing individual people. But, how many people can recognize an individual mallard duck?

The young male mallard always greeted me with 1 to 3 quiet vocalizations when he first saw me. Sometimes he bobbed his head up and down as part of a greeting ceremony. Another time he stretched his head and neck out straight, parallel to the ground, while raising his wings in an M pattern. The visits were purely social since I did not offer the young ducks any food. More than once the young male would visit for 10 to 15 minutes, then pretend to leave. He would get back into the water and swim along the edge 10–12 feet (3–3.66 m). If I followed him, he hopped out of the water again and walked to meet me. It seemed as if he was gauging how interested I was in socializing with him. After all, on most occasions he swam clear across the pond to visit with me.

The young male mallard duck visited for another 10 to 15 minutes before continuing on his way. Frequently he stretched and preened while I watched. On 11/12/2010 after the young duck finished preening, he slowly lowered his head twice and tapped the ground with his bill while looking right at me. Still looking at me he twice made a quick motion like he was going to sit down. I inferred that he was asking me to sit down, so I knelt down. As soon as I did so the young duck sat down, turned his head around, and tucked his bill under his wing for a nap. He was aligned directly with the sun. Even though he was less than 3 feet (.91 m) from me, he closed his eyes, only opening them every now and then.

The young male mallard developed a small dark green spot on one side of his head by 11/10/2010. The green appeared to be underneath his regular head feathers. He was still molting and the rest of his feathers were various shades of brown, like a non-breeding male. In contrast, the adult males had regained their breeding plumage in early October. The green spots on the young male's head expanded slowly. At the same time he continued to molt his front and side juvenile feathers. At one point his feathers were sort of a half juvenile, half adult pattern. By 11/21/2010 he had a complete set of adult breeding plumage.

By mid-November the morning temperatures were much cooler. The local flock of mallard showed signs of restlessness. Several migratory flocks of large and small birds had already passed through the area. This was the young male mallard's first fall/winter. Yet his behavior indicated that he sensed the impending changes.

The young male mallard exhibited three new behaviors on 11/14/2010. That afternoon the young male mallard flew across the pond when he saw me. His journey involved a water take-off and a water landing. This was the first time he flew across the pond to meet me; before he always swam across the pond. Later, a large flock of small dark birds flew from behind me and over the pond. Upon seeing them the young male mallard made a 1-syllable high pitched alarm call to warn me and left. This sound was totally different from his usual quiet vocalizations. He came back after the flock was gone. The young male mallard walked around with me for another 25-minutes. Sometimes he was grazing on land, other times he was in the water. He tapped my shoe once on purpose while grazing. That was the first and only time he did that. Something else startled the young male mallard and he took off flying in a hurry toward the other side of the pond. Then he looped back around and flew only 2-feet (0.6 m) above my head, as if he were trying to encourage me to go with him for safety.

In this one visit the young male mallard flew three times, made an alarm call, and tapped my shoe. In the previous 7 weeks the young male mallard had not exhibited any of these behaviors. Most likely these behaviors were due to: (a) a nervous restlessness about the impending fall/winter, something he had not experienced before, and (b) the fact that he was growing up. In particular his protective behavior toward me was a sign that he was maturing.

The last week of November 2010 was cold and windy. The young male mallard did not socialize as much. Instead there seemed to be a sudden urgency to his search for food. As Canada geese flocks passed through Pond C the young male mallard associated with them, whether they were swimming, grazing, or resting. He also socialized with migratory flocks of mallard ducks that passed through Pond C. On 12/5/2010 it looked as if the young male had found a girl friend from one of the migratory flocks. Alas, she eventually left with her flock without him.

Questions arose about why the young male mallard spent most of his time at Pond C, away from home (Pond B). Did he prefer the peace and quiet? Did he enjoy the company of the Canada geese and migratory flocks? Was he avoiding the harassment adult mallard ducks directed toward juveniles? Or, was there another reason? Some combination of these factors might explain his interest in socializing with me.

Questions arose about whether the young male mallard would comprehend winter and the need to migrate or would he try to stick it out by himself at Pond C. December 9th Pond C froze over and the young male mallard left. He returned briefly on 12/12/2010 when there was some partial thawing. Pond C froze again on 12/13/2010 and he left. The young male mallard spent the rest of the winter at Pond B with the flock in the small open area. Whenever there was any thawing he returned to Pond C for a few hours.

It was possible, but not always easy, to spot the young male mallard in the midst of the crowded flock. His white neck collar was thinner and his bill a duller yellow than most of the adult males. Also, whenever anyone threw bread cubes to the ducks he was the first one on the scene. He even jumped into the air to catch the bread cubes. That, of course, caused the adults to peck him.

April 3, 2011 the young male mallard was spotted back at Pond C. He had survived his first winter. He remembered me and swam over for a visit. The young male mallard performed a greeting ceremony then we went for a walk like before. Sometimes he was grazing on the lawn, other times he was in the water. The entire time he was very vocal. He made a steady "rahn, rahn, rahn" medium-pitched sound. It was lower than a female quacking and louder than his quiet "yeeb". He had not made this sound before. After about 25 minutes a young female mallard swam from the northwest corner of the pond to join him. She was very quiet and reserved. She is a little smaller than most female mallard ducks and her beak was more black than orange. She stayed 15-minutes. Then they both flew away together. Over the winter the young male mallard had grown up and found a buddy. Survival demanded the former and enabled the latter.

4.6 Comparison of Mallard Duck and Canada Geese Behavior

Because they are neighbors and both members of the Anatidae family, it is interesting to compare the similarities and differences in mallard duck and Canada geese behavior.

Mallard ducks and Canada geese share the same temperate zone fresh water habitats that are surrounded by lush vegetation. Within the last two centuries members of both populations have begun living in city parks. Since mallard ducks and Canada geese share the same habitat, it is not surprising that they have similar diets. Mallard ducks mostly eat vegetation, while Canada geese are true herbivores.[157] During autumn a Canada goose will pick up and eat a yellow, orange, or red leaf that has just fallen from a tree and is floating on top of a pond. In contrast, mallard ducks ignore the floating leaves and continue to poke around the bottom and sides of a pond. Both mallard ducks and

Canada geese upend. For Canada geese this is usually a quick up/down movement. In comparison, mallard ducks can stay upended for 10 seconds or more.

Like other waterfowl mallard ducks and Canada geese have webbed feet. The difference between them is the color and size. The legs and feet of a mallard duck are orange. While the legs and feet of a Canada goose are black and slightly larger. Canada geese walk, swim, or fly. Mallard ducks have also been observed hopping on occasion, primarily at the edge of a pond. Instead of walking out of the water, a duck may hop up to the shore. Canada geese jump up or down distances of 6–8 inches (15.24–20.32 cm) when they need to, but less frequently than mallard ducks. Canada geese and mallard ducks spread and contract their toes when they walk on land, just like they do when swimming.

Mallard ducks and Canada geese share a long distance biannual migration. Both are strong quiet flyers. One difference is that mallard ducks tend to leave later in the fall and return earlier in the spring. Mallard ducks and Canada geese share wintering grounds just like they share summer habitats. Both birds build their nests on the ground near water. Even the material used to construct the nests is similar, including the down lining provided by the parents. Male mallard ducks and Canada geese molt their flight feathers to coincide with a life cycle event—the incubation and hatching of their new family.

Except for not being able to protect themselves, mallard ducklings and Canada goslings are rather self-sufficient upon hatching. Ducklings and goslings leave the nest when they are only one to two days old and are able to walk, swim, and feed themselves. Because they need protection and have a lot to learn in a short amount of time, mallard ducklings and Canada goslings follow their parents around everywhere on land or in water. The mother mallard duck generally leads her brood. This behavior contrasts with that of Canada geese where the father leads and the mother follows the goslings. Mallard ducklings and Canada goslings grow rapidly. In about 8 or 9 weeks both are near full size and learn to fly. Natal down gives way to juvenile feathers which are replaced by adult plumage in quick succession.

Both mallard ducks and Canada geese rotate among the four co-located ponds. Canada geese nest at Ponds A, B, and D, as well as the golf course water trap. Pond B is home for the mallard ducks and that is the primary area in which they nest. The mallard ducks know about the other three ponds and visit them frequently. In particular the mallard ducks visit Pond A after a heavy rain. They seem to understand that Pond A will flood and expand after a storm, creating a larger swimming area. If there is a storm in the late afternoon or evening, some mallard ducks will arrive at Pond A the next morning, almost like clockwork. For example, Pond A quadrupled in size in mid-August 2010 after a series of storms. The first morning four ducks were there. The second day six ducks were there. Then ten ducks. The ducks continued to visit all ten days the pond was flooded. The ducks frequently walked along the expanded edges of the pond, plucking newly submerged plants.

Canada geese frequent the golf course water trap throughout spring, summer and fall. In contrast, only 1 or 2 mallard ducks visit the golf course water trap and then only on rare occasions. Small groups of 1 to 4 ducks visit two other water locales that the Canada geese do not: a nearby creek and a community swimming pool. On average three or four times a year 2 to 4 mallard ducks are spotted at a nearby creek or flying back from there to one of the ponds. The creek is shallow and varies in width from 6–24 feet (1.83–7.31 m). Dense vegetation lines the creek. This area is home to a lot of wildlife, especially birds. The ducks are usually near the narrow part of the creek.

The community outdoor swimming pool is filled with water during April. Groups of 2 to 4 mallard ducks visit the pool in the early morning hours almost on a daily basis. They swim and walk around the deck area which has a lot of plants. The mallard ducks continue visiting the pool for 15 to 20 minutes a day during good weather until the pool opens (for people) in May. After that they stop coming. The pool is closest to Pond A. The mallard ducks generally head in that direction when leaving the pool. The mallard ducks were only seen visiting the pool in the evening once, a Friday evening during 2006.

Pond B is home for the mallard ducks; however, they share it with Canada geese. During May and June when the Canada geese and mallard ducks are raising their young, the ducks and geese primarily inhabit different parts of the pond. The Canada geese reside in the middle section of Pond B, while the mallard ducks reside in the southern section of Pond B. Come August, after parenting duties have diminished for both, the mallard ducks and Canada geese mix freely in and around Pond B. They can be seen resting or walking together on the slopes of the pond. The ducks and geese swim in different groups, but share the same space. If there is a row in this mixed environment, ducks only fuss with ducks and geese only fuss with geese, and it is usually over quickly.

One scene occurred on the afternoon of 9/26/2010 that was particularly amusing. A low thin fence, about 24 inches (61 cm) high, surrounds Pond B. The intent is to prevent young children from drowning. Sections of the fence have become loose and bent over time. One section near where the ducks like to rest is bent, creating a triangle shaped opening. The opening is just big enough for one duck to walk through at a time. A Canada goose can fit through the opening if they bend their head and neck down. In this instance a group of 14 ducks and geese that were outside the opening decided to return inside to the resting area. Because the opening is small, the ducks and geese lined up single file and began to walk through the opening one at a time. Some of the ducks in the middle of the line decided this was taking too long and started fussing at the ducks ahead of them. The ducks in front turned around to fuss at the ducks behind them. Soon all the ducks were fussing at each other and the line stopped moving. The Canada geese decided to sit out this round and simply flew over the fence. Upon seeing this, the ducks quit fussing and did the same thing.

A notable physical difference between mallard ducks and Canada geese (other than the different necks) is their feathers. Male mallard ducks sport very showy and colorful feathers during breeding season. In particular their bright green metallic head feathers stand out. Male Canada geese on the other hand have the same color feathers all year; they do not have special breeding plumage. Nor is there any difference between the coloring of male and female Canada geese. The coloring of female mallard ducks, their ducklings and non-breeding males allows them to blend into the habitat in which they live. This example of biological intelligence is particularly useful when nesting.

There are similarities and differences between mallard duck and Canada geese bills. Male mallard ducks have yellow bills, while female mallard ducks have orange mottled bills. In contrast, male and female Canada geese have black bills. Both mallard ducks and Canada geese have nostrils that are parallel to their bills to facilitate breathing while flying. Both have bills that are lined with lamellae. Both have a sharp point at the tip of their bill which is used to slice vegetation. The bill of a mallard duck is a bit longer, wider and rounder on the tip than that of a Canada goose.

The life span of mallard ducks and Canada geese is different. The maximum life span of mallard ducks is 29 years, compared to 23 years for Canada geese.[18, 41, 137] In other words mallard ducks can live 26% longer than Canada geese.

Average life spans are hard to compute, given abnormalities like hunting and oil spills that are not part of nature and over which birds have no control. The average annual adult mallard duck survival rate is estimated to be 52%.[137] The top three human-related causes of bird deaths annually (excluding oil spills and car crashes) in the United States are:[18]

- Hunting: 120 million per year
- House cat predation: 118 million per year
- Insecticide, herbicide, and pesticide poisoning: 72 million per year

Canada geese and mallard ducks have different temperaments. Within any flock (as in any society) there is a spectrum of temperaments. Individual birds may be more timid and wary or more aggressive than the so-called norm. However, as a whole Canada geese are quieter and more reserved than mallard ducks who are boisterous and vocal. Mallard ducks are very aggressive when competing for food in the winter, especially when people throw them bread cubes. The mallard that grabs a bread cube first is pecked unmercifully until it can fly or swim away. Often it seemed as if the ducks spent more energy competing for a single bread cube than it could possibly provide. Perhaps hunger overruled logic. About 8 to 10 of the ducks did not bother to compete. They may have decided it was more efficient to save their energy. Mallard duck adults harass juveniles during the fall and winter when new pair bonds are being formed. In contrast, Canada geese never exhibited aggression toward their young.

Canada geese and mallard ducks have different grazing patterns. Adult Canada geese roam up and down the slopes of a pond. When munching vegetation they may walk 30 feet (9.14 m) or more from the edge of a pond or all around a golf course. In contrast, mallard ducks usually stay within 8–10 feet (2.44–3.05 m) of the pond's edge, preferring not to venture too far from the water. Mallard ducks tend to rely more on water for safety. This may explain why Canada geese frequent golf courses and mallard ducks generally do not. (See Photographs 4.29 and 4.30.)

Another difference between mallard ducks and Canada geese is the cruising altitude during migration. Mallard ducks migrate at altitudes up to 21,000 feet (6,400 m), while Canada geese migrate at an altitude of 3,300 feet (1,005 m), occasionally going as high as 12,000 feet (3,657 m).[41, 136, 137] That means that mallard ducks on average are flying more than six times higher than Canada geese,

Photograph 4.29 Swimming #1. Mallard ducks swim for a variety of reasons, not just to get from point A to point B. Why else would they regularly visit a community swimming pool that offers no food or rotate among the four co-located ponds? © D. Herrmann

Photograph 4.30 Swimming #2. Mallard ducks float, swim at a leisurely pace, or speed across the water depending on what they decide to do. They may (or may not) stop to dabble, bathe, or upend. Should the need arise, mallard ducks can takeoff as easily and quickly from water as they can from land. © D. Herrmann

even though they are flying similar routes to similar destinations. The air temperature is colder, the oxygen is thinner, and the ducks have to flap their wings more frequently at higher altitudes.[12] As a result, migration requires extra endurance on the part of mallard ducks. It is interesting to note in passing that around the world groups of migratory birds have assigned themselves different cruising altitudes without prolonged treaty negotiations or relying on hundreds of air traffic controllers.

There are notable differences between mallard ducks and Canada geese when it comes to family life. Mallard duck pair bonds only last a year. They do not mate for life like Canada geese. The same drake and duck could establish pair bonds in subsequent years and may in fact do so, but it is not a given.

Members of a mallard duck flock generally share a large nesting area. For the most part they do not practice nesting dispersal like Canada geese. Instead, the nesting area is considered home and during spring and summer the flock does not travel far from there. The mallard ducks primarily nested at Pond B and in one central area. Individuals and small groups visited the other three ponds. But the largest concentration of ducks was always found at Pond B. Mother mallard ducks assume the majority of parenting duties, especially when it is time to take the new family for an outing. Father mallard ducks' main function is to provide protection; for example by using their bright breeding plumage to draw attention away from the young ducklings.

As a flock mallard ducks are not overly interested in humans. Instead they tolerate humans or acknowledge their presence. In more extreme cases they deliberately fly away from visitors and onlookers. Perhaps because mallard ducks associate in large groups there is not much motivation to socialize with humans. That was the case at Pond B. Away from the group and home, individual ducks and groups of 2 ducks at Pond C did express curiosity about humans. This behavior contrasts with Canada geese, where whole families and groups of ten or more geese will walk right up to a human.

Table 4.6 summarizes the similarities and differences in mallard duck and Canada geese behavior.

Table 4.6 Comparison of Mallard Duck and Canada Geese Behavior.

<table>
<tr><th colspan="2">I. Similar Behaviors</th></tr>
<tr><td colspan="2">• Share the same habitat
• Mallard ducks eat 90% plant matter. Canada geese are herbivores
• Spread and contract their toes when they walk on land, just like they do when swimming
• Face the sun when resting on sunny days, face into the wind on windy days
• Upend to obtain food that is underwater
o Mallard ducks tend to stay upended longer
• Biannual migration
o Mallard ducks leave nesting area later and return sooner
o Mallard ducks migrate at an altitude of up to 21,000 feet (6,400 m) compared to 3,300 to 12,000 feet (1,005 to 3,657 m) for Canada geese
• Nest on ground near water, only one nest a year
• Summer molt of flight feathers timed to life cycle event—hatching of chicks
o Only male mallard molt at this time. Female mallard molt after ducklings are independent
• Precocial hatchlings leave nest when only 1–2 days old, walk, swim, and feed themselves
• Young follow parents when on land or in water
• Young learn to fly at about 8 weeks
• Rotation among the four co-located ponds
o Ducks also visit community swimming pool and nearby creek
o Canada geese frequent golf course water trap</td></tr>
<tr><th colspan="2">II. Different Behaviors</th></tr>
<tr><th>Mallard Ducks</th><th>Canada Geese</th></tr>
<tr><td>• Male mallard has special breeding plumage
• Walk near the edges of ponds
• Sit on branches that overhang ponds
• Do not mate for life
• Do not practice nesting dispersal
• Female has primary parenting duties
• More boisterous and vocal than Canada geese</td><td>• Plumage is same for both sexes year round
• Roam up and down the slopes of ponds (and golf courses)
• Do not sit on branches
• Mate for life
• Practice nesting dispersal
• Parenting duties are shared</td></tr>
</table>

4.7 Discussion

Now let's see what these behaviors tell us about mallard duck intelligence, behavior, and individuality. In particular let's examine:

- Vocalizations
- Common and cooperative behaviors
- Individual and unique behaviors
- Ideas and inventions appropriated by humans

Vocalizations

The literature consulted was adamant that mallard ducks only make two sounds: males "yeeb" and females quack. Male mallard ducks were heard "yeeb"-ing and female mallard ducks were heard quacking. However, a variety of other mallard duck vocalizations were heard as well, which have not been documented before.

The majority of the sources consulted insisted that only female mallard ducks quack. Apparently no one told the flock of mallard ducks at Pond B. Some of the male mallard ducks living there do indeed quack. It is a lower pitched quieter quacking sound than the female mallard ducks make. Also, it is a steady quacking sound with each quack being the same duration. Female mallard ducks frequently vary the pitch and duration of their quacks to convey different messages. Male mallard ducks did not make a quacking sound when the flock was gathered en masse at Pond B—it would have competed with a female duck's quacking. Rather, male mallard ducks made a quacking sound when they were swimming by themselves in Ponds A or C or in the north part of Pond B, particularly on pleasant sunny days during the spring and summer. (See Photograph 4.31.)

Photograph 4.31 Male Mallard Quacking. Male mallard ducks were heard quacking, contrary to claims that only female mallard ducks quack. Male mallard ducks quacked when they were by themselves in Ponds A or C or the northern part of Pond B. Male mallard ducks make a slightly lower pitched steady quacking sound, with each quack having the same duration. © D. Herrmann

A second new vocalization was heard during winter after the flock had assembled in the narrow part of Pond B north of their nesting grounds. This vocalization was a quiet nervous chattering or clicking sound. All of the adult ducks present participated in the chattering. The chattering lasted 15 to 20 seconds. After a couple minutes of silence the chattering was repeated. The volume and rapidity of the chattering increased in response to a perceived danger. This vocalization seems to serve as a roll call and as a standby alert message when the flock is in survival mode. This vocalization is made throughout the day.

Mallard ducks also make an alarm call. This call is used more often when a small group of mallard ducks are visiting other ponds (and hence more vulnerable), than when they are at home in a large group setting at Pond B. The mallard duck alarm call consists of a sudden loud intense high pitched "rhee" sound. The alarm call is made once by the duck that spotted the danger, before it departs for safety.

Mallard ducks have a special vocalization that is reserved for a greeting ceremony upon being reunited with their partner, family, or sub-flock. This sound was only heard during late fall and early spring. This sound was not heard at Pond B. Instead this sound was only heard at Pond C when individuals and pairs of mallard ducks were present. This greeting consists of a medium pitched "rahn" sound that is repeated over and over.

The mallard duck distress call is a shrill variation of the normal male quacking sound. This loud low pitched shrill quacking conveys a sense of danger and urgency. This sound was heard when a lone male mallard was being attacked by a red-tailed hawk at Pond A. He continued making the sound for a few minutes after the red-tailed hawk left, even though no other mallard ducks were present. Presumably female mallard ducks make a similar distress call, although none was heard.

Most likely there are other quiet vocalizations that were not heard. A flock of 30 or more mallard ducks living together would need to communicate many other ideas. For example small groups of ducks would need to communicate messages about taking an outing to one of the other ponds. Also, the local flock and migratory flocks would need to exchange information. These quiet short distance messages would not be heard by humans in the boisterous environment of Pond B.

In addition there must be some quiet short distance messages that are exchanged between a mother duck and her ducklings. On several occasions the ducklings were spread out 2 to 3 feet (0.61–0.91 m) from the mother. As soon as the mother saw me all the ducklings ran or swam right to her, even though I didn't hear a sound. Then again there is no need for this sound to carry more than 2 or 3 feet (0.61 or 0.91 m). As a result it is reasonable to infer that mallard duck communication is more varied and complex than has previously been acknowledged.

Table 4.7 summarizes the categories and nature of the new mallard duck vocalizations that were identified at Ponds A, B and C. These mallard duck vocalizations can be heard on Disk 1 Track 1.

Common and Cooperative Behaviors

Several common mallard duck behaviors were observed. They included:

- Recognition of Pond B as home
- Lack of territoriality
- Lack of hostility toward neighboring birds and migratory flocks
- Awareness of a group identity

All members of the local flock of mallard ducks acknowledge Pond B as home. The majority of the flock is always present in the southern part of Pond B. The flock shares a community nesting ground at Pond B; most likely many members of the flock hatched there. The flock winters at Pond B in an area north of their spring-summer-fall habitat. Even the individuals, pairs, and sub-flocks of mallard ducks who frequent Ponds A, C, and D, the creek, the golf course water trap, and the community swimming pool during the day return home to Pond B as sunset approaches. Yes, Pond B is home for this flock.

As such the members of the local mallard duck flock do not display any aggression or signs of territoriality toward each other, even during nesting time. On the contrary, adults, juveniles, and ducklings from multiple families mingle freely on land and in water.

Likewise members of the local mallard duck flock do not display any aggression or signs of territoriality toward their bird neighbors or migratory flocks that are passing through the area. Wood ducks, black ducks, Canada geese, songbirds, and even tree squirrels roam among the mallard ducks freely in and around Pond B. In winter when the local flock is in survival mode clustered in a small area of Pond B and food is scarce, they make no attempt to chase away migratory flocks or those who have flown north for a few days to escape snowstorms in the southeast. Instead, these visitors are allowed to join the local flock in the small open area of Pond B.

Table 4.7 Mallard Duck Vocalizations.

Sound	Who Makes It	Attributes and Usage
I. Traditional Vocalizations		
"yeeb"	Male	• A soft sound that can only be heard at a distance of 1 to 2 feet (30.48 to 60.96 cm) • Usually repeated 3 times in a row • Only heard when male is in a small group setting • May be reserved for bonded pairs and nuclear families
Quacking	Female	• A medium pitched moderately loud sound that carries over distances • Volume and rapidity of quacks is varied to convey different messages • Only heard in a large group situation • Usually only 1 female in the group quacks at a time
II. New Mallard Vocalizations Heard at Ponds A, B, and C		
Quacking	Male	• A lower pitched steady quacking sound that is quieter than the sound females make • Only heard when a solitary male was swimming by himself in Ponds A or C or in the north part of Pond B, usually on sunny pleasant days
Chattering or clicking	Both	• A quiet chattering or clicking sound that lasts 15 to 20 seconds at a time • All the adult ducks present participate • After a couple minutes of silence, the sound is repeated • Volume and rapidity of the sound intensifies in response to a potential danger • May also serve as a roll call • Heard most often during winter in a large group setting
"rhee"	Both	• A sudden loud intense sound consisting of 1 high pitched syllable • Functions as an alarm call • Only heard in a small group setting • Only heard when hawks, falcons, or other perceived predators flew over Pond C
"rahn"	Male	• Medium pitched steady sound that is repeated over and over • Only heard when a solitary male or pair was swimming in Pond C • May be part of a greeting ceremony upon being reunited
Shrill quacking	Male	• Loud low pitched shrill steady quacking • Functions as a distress call • Heard when a lone male mallard in Pond A was being attacked by a hawk
"peep"	Ducklings	• A soft quick high-pitched sound that is repeated twice • Individual ducklings make this sound when they become separated from their mother and family group • Canada goslings make an almost identical sound under similar circumstances

Note: Mallard duck vocalizations can be heard on Disk 1 Track 1.

Only one act of mallard duck aggression was observed. It only lasted 5 seconds. A mother mallard duck was walking her ducklings from the pond to the grassy lawn. The ducklings were scampering quickly in a cluster. A male wood duck was nearby. A male mallard duck thought the wood duck didn't get out of the way fast enough. He lowered his head and neck, stuck out his tongue, and charged the male wood duck in the same manner as a male Canada goose without touching him. By then the mallard ducklings had passed the wood duck and that was that. At the time the mallard ducklings were only 4 days old.

The local flock of mallard ducks exhibits a group identity. As just discussed they collectively consider Pond B home. Adult mallard ducks recognize individual members of their flock whether they are at Pond B or elsewhere. Even when they are among 1 or more large migratory flocks the locals seem to know who is who. The same is true for the visiting flocks of mallard ducks. Mallard ducklings are aware that they belong to a specific family unit. Ducklings are able to recognize their mother within a group of 4 or more female mallard ducks. Ducklings also identify with their fellow ducklings. If 1 or 2 ducklings wander off, they know which group of ducklings to run or swim back to.

This sense of community allows the local mallard duck flock to engage in several cooperative behaviors, such as:

- Flock security measures
- Protection of young ducklings
- Delaying a small part of Pond B from freezing

The local mallard duck flock participates in cooperative security measures in response to perceived land, water, or air threats. The group is notified of a potential threat by an alarm call, then everyone pursues a coordinated response. This, of course, contrasts to the mass panic and hysteria seen when humans are in a similar situation. The local mallard duck flock assumes three different security postures, depending on the nature of the threat, as shown below. The coordinated response continues until the danger has passed.

- *Low risk land-based threat*: The flock immediately disperses evenly around the perimeter of the pond in the water, a short distance from the edge. There is just enough space between the ducks to accommodate each other's wingspans, should sudden flight become necessary.
- *Medium risk land or air-based threat*: The flock rushes to the center or deepest part of the pond. They may swim or fly there, using which ever method is fastest.
- *High risk land, water, or air-based threat*: The flock becomes airborne instantly using a coordinated scrambling pattern. This sudden movement in every which direction, combined with multiple flashing blue speculum, confuses predators.

Male mallard ducks were observed cooperating to protect ducklings that were less than 3-weeks old. When the ducklings were resting on the lawn, they sat in front of the mother. Three or four male mallard positioned themselves on the other side of the ducklings, forming a sort of protective barrier. The same thing happened when the ducklings were in the water. The mother positioned herself to keep the ducklings in a particular shallow area. She was assisted by two or three male mallard ducks. Other male mallard ducks stood guard on the bank of the pond.

This cooperative protection extended to monitoring the whereabouts of turtles. Whenever the ducklings were taking an extended swim, 2 or 3 male mallard ducks posted themselves about a foot (30.48 cm) from where the red-eared slider turtles were basking. The male mallard ducks faced toward the turtles and formed a semi-circle around the turtles like a security fence. Red-eared slider turtles were not observed challenging the male mallard duck security fence. However, several young ducklings, especially wood ducklings, were lost to snapping turtles each year.

Since the local mallard duck flock does not migrate they rely on cooperative group winter survival strategies. The most impressive of these is the community effort to delay the pond from completely freezing. As Pond B begins to freeze the flock moves from its usual location in the south of Pond B to a small area north of their nesting grounds that receives more sun. Throughout the winter two-thirds of the flock is always swimming, regardless of the temperature. This cooperative constant churning of the water delays this small area of the pond from freezing and covering up their food supply. The one-third that is resting stay huddled close together on the ice near the water. Throughout the day the flock makes an almost constant chattering or clicking sound that serves as a roll call and standby security alert message. By banding together and engaging in cooperative activities the flock survives the harsh winter. If individual or pairs of ducks tried to do this, they would not survive. Apparently all the mallard ducks understand that.

Individual and Unique Behaviors

The local flock of mallard ducks is gregarious, has a community identity, and cooperates to ensure the survival of the flock. At the same time expressions of individuality are frequently seen. Adults and juveniles express individuality. Most expressions of individuality were observed away from the mallard duck flock's home in the southern part of Pond B. Was that because mallard ducks felt freer to express their individuality when they were away from home and the flock? Or, was that because the mallard ducks who frequently visit the other ponds and waterways are more individual by nature?

A lone male mallard dove completely underwater, stayed there for over 10 seconds, then resurfaced 15 feet (4.57 m) behind his original position and facing a different direction to avoid a red-tailed hawk. The lone male mallard was observed doing this at Pond A. Every time a red-tailed hawk dove toward the mallard duck flashing his talons, the mallard duck dove underwater. When a hawk flies over the mallard duck flock at Pond B, the ducks cluster together in the center of the pond. The fact that there are 30 or more ducks clustered closely together discourages the hawk and it goes elsewhere to hunt. In contrast the lone male mallard duck at Pond A was on his own. To sit alone in the center of the pond guaranteed that he would be the red-tailed hawk's lunch.

So the mallard duck devised a new individual protective strategy. While mallard ducks can dive completely underwater, they rarely do so. When mallard ducks dive underwater they don't stay there for over 10 seconds. This mallard duck understood that the red-tailed hawk could not hover, and as a result stayed underwater just long enough to ensure that the red-tailed hawk would have to retreat. In the end this novel strategy was successful. The lone male mallard has appeared at Pond A several times since the red-tailed hawk attack. This behavior implies he has confidence in his individual protective strategy.

Individual mallard ducks hide themselves among the reeds at the edge of a pond when it is nap time. In this position the ducks are half on land, half in water, and nearly invisible. This behavior was observed at Pond A, the northwest corner of Pond C, and the narrow wooded part of Pond B. Usually only 1 to 3 mallard ducks hid in the reeds at a time. This behavior may be an individual preference and an extension of the practice of engaging in different behaviors when away from home. Or, this behavior may be an individual security practice. In other words a mallard duck spends more effort to conceal itself when napping away from the flock. Either way, this behavior reflects individual decision-making.

Adult mallard ducks were observed sitting on low branches over Pond A. Sometimes a solitary male mallard was observed sitting on a branch about a foot (30.48 cm) above the pond's surface. Other times a solitary pair of mallard ducks was seen sitting together in the same bush. Mallard ducks were not observed sitting on branches over Ponds B, C, or D or at the creek. However, mallard ducks have been spotted sitting on fallen branches in and on the bank of Pond C. One day a pair of mallard ducks was seen sitting on a log that was floating in Pond C. Wood ducks frequently sit on branches that overhang Pond B. For some reason the mallard ducks have decided not to engage in similar behavior at Pond B.

Communities are composed of individuals and individuals decide on a daily basis whether they are going to participate in community or individual activities. So it is with the local flock of mallard ducks at Pond B. Throughout the day individual mallard ducks decide:

- What activities to engage in
- When and how long to engage in an activity
- Where to engage in an activity
- With whom to engage in an activity

A mallard duck may decide to swim, walk, fly, stretch, preen, nap, or upend. It decides whether or not to vocalize while engaging in an activity. Each mallard duck decides whether to engage in the activity now or later. Each mallard duck decides whether to engage in the activity in the southern part of Pond B, the middle part of Pond B, the narrow wooded part of Pond B, Pond A, Pond C, Pond D, the creek, or the golf course water trap. Each mallard duck decides whether to pursue this activity on its own or as part of a pair or sub-flock. Each of these decisions is made by an individual mallard duck and reflects individual preferences. A mallard duck may make one decision today and a totally different decision tomorrow. Or, like the two juveniles who spent most of Fall 2010 at Pond C, a duck may make the same decision several days in a row. External factors, like a change in weather or the appearance of the lawnmower crew, may alter these decisions. Regardless, the activity pursued reflects individual decision-making.

The same is true in regard to socializing outside the flock. Mallard ducks mingle with wood ducks, black ducks, and Canada geese at Pond B, but they don't really socialize with them. In contrast, 1 to 3 mallard ducks were frequently seen socializing with 1 or more Canada geese at Pond C or the golf course water trap throughout the summer and fall. One male mallard duck socialized with a flock of migratory Canada geese during early winter. In these instances the mallard ducks swam with the Canada geese, upended when the Canada geese upended, grazed with the Canada geese, and napped when and where the Canada geese did. Often this sub-flock consisted of 1 to 3 mallard ducks and 1 to 3 Canada geese and they were all comfortable with this arrangement.

The same 1 to 3 mallard ducks that socialized with Canada geese at Pond C socialized with individual humans, especially the young male mallard duck. This behavior was not observed at any of the other waterways. The decision to socialize with humans day after day over a period of more than 2 months was clearly an expression of individuality. The initial contact may have been prompted by juvenile curiosity. However, the fact that the contact continued until Pond C froze implies more than curiosity. It implies that this social interaction was important to the ducks and a degree of trust had been established. This becomes evident when you consider the facts that the ducks had to swim clear across the pond (~ 200 feet/61 m) to visit with me and greeted me with extra enthusiasm when they hadn't seen me for a few days. Also, the young male exhibited a protective behavior toward me when a perceived predator flew over Pond C on 11/14/2010. If the social interaction was not important to the young male mallard duck, he would not have done that. Furthermore, there is the young male mallard duck's greeting when he was first spotted back at Pond C in Spring 2011.

When the 2010/2011 winter chill sat in for real a few people began to offer bread cubes to the mallard ducks at Pond B. The quantity and frequency of bread cubes were not sufficient to constitute survival rations. Instead, the bread cubes were more like a Christmas cookie that provided a few extra calories. The mallard ducks scurried over to where the bread cubes landed in the water. Then there was some pushing and shoving until one lucky duck snatched up the bread cube. Once in a while the wood ducks and black ducks participated in this scrimmage. Most of the time they chose to sit it out.

The young male mallard duck from Pond C tried a different tactic to increase his chances of capturing a bread cube. Instead of waiting for the bread cube to land in the water, he jumped 1–2 feet (30.48–60.96 cm) in the air to catch it. More often than not he succeeded in snatching the bread cube. This was rather creative behavior on the young male's part considering this was his first winter. None of the other 36-plus mallard ducks at Pond B engaged in this behavior. He was the only one.

Sometimes individual behaviors are borne out of necessity. Still these behaviors reflect a high degree of individual reasoning, problem solving, and decision-making skills. That is certainly true for the young female mallard duck with 1 leg that was spotted at Pond B on 4/4/2011. There was no sign of injury or that she ever had two legs. Several red-eared slider and snapping turtles reside in the southern part of Pond B. Maybe she lost her leg as a duckling. The young female mallard had not been noticed before. Is she a local? Or, was she a member of a migratory flock and decided to stay at Pond B? How did she survive winter?

The young female mallard ate laying flat on the ground, scooping up food with her bill. She seemed to be very hungry. It is probably very difficult for her to upend, which might explain her hunger. None of the other ducks got in her way, even though they fussed at each other. Apparently the young duck knew she was vulnerable laying on the ground. Every few minutes she stood up on her leg, using her bill as a prop to get up, and looked around and behind her. The young female mallard duck is a strong flyer and swimmer, even though she only has one "oar". In order to survive the young female mallard taught herself all of these skills. Walking, swimming, and upending with one leg are not part of her genetic intelligence. Nor was there a role model she could learn from. The young female mallard duck only had her own resourcefulness to rely on.

Table 4.8 summarizes the individual and unique mallard duck behaviors that were observed.

Throughout this chapter we have seen that mallard ducks are intelligent, creative, and social beings. The flock knew to move to a small area of Pond B that was in the sun during winter because there was a better chance of it not freezing. The mallard ducks sat down and used their tail feathers as a brake when slipping on ice. When an individual mallard duck slipped over on her back, she used her wings to get upright again. The young male knew to dive under water and stay there since a hawk cannot hover. The young female mallard duck taught herself how to walk, swim, and upend with 1 leg. These actions and others are examples of applied intelligence at work.

Mallard ducks exhibit a significant awareness of themselves, their flock, and their surroundings. Mallard ducks are a community of individuals who band together for social and security reasons. Individual ducks are capable and comfortable spending the day together with the flock in a large group setting or by themselves or with one or two other ducks at one of the nearby ponds. Mallard ducks can identify and remember individual people, even with a change of outfit or if they haven't seen them for several weeks or months. Mallard ducks are spunky and frequently engage in play activities, like visiting other ponds, the creek, the golf course water trap, or the community swimming pool. Mallard ducks exhibit a healthy balance of curiosity and caution.

Table 4.8 Individual and Unique Mallard Duck Behaviors.

Activity	Who Did It	Setting and Frequency Observed
Diving under water, changing direction and position to avoid a hawk	Adult male	• Lone male swimming at Pond A • Once on 3/11/2011
Hiding among reeds at edge of pond to nap	Adults	• Individual adults napping at Pond A, the narrow part of Pond B, the northwest corner of Pond C, or the creek • Weekly from spring through fall each year
Sitting on branches over pond	Adults	• Adult male and/or female when visiting Pond A, mostly during 2010
Engaging in individual and sub-flock activities	Adults and juveniles	• All locations • Daily spring through fall each year
Visiting Ponds A, C, and D, the creek, golf course water trap, and swimming pool	Adults and juveniles	• 1 to 4 ducks • Daily spring through fall each year
Socializing with 1 or more Canada geese	Adults and juveniles	• 1 to 3 ducks • Ponds A, B (middle), and C, and golf course water trap • Daily summer through fall after current year's ducklings are on their own
Socializing with humans	Juveniles	• 1 to 3 ducks • Pond C • Fall 2010
Jumping for bread cubes	Juvenile male	• Ponds B and C • Winter 2010/2011
Walking with 1 leg	Young female	• Pond B • Observed infrequently from 4/4/2011 on

Mallard ducks function as a cooperative egalitarian community. Flock security is a community effort, as is surviving winter and protecting young ducklings. Each member of a flock has the same standard of living, based on their own initiative and industry. There are no "worker" ducks or "queen" ducks. Each duck is responsible for itself. If a duck decides to bask in the sun beside the pond all day, no one is going to bring it food. This complex society welcomes a high degree of individuality and resourcefulness, which in the end benefits individuals as well as the flock. A sophisticated pattern of communication is needed in this social environment, especially given the variety of neighbors that mallard ducks share their ponds with, not to mention all the migratory flocks of mallard ducks and other birds that pass through the area twice a year. Unfortunately humans have barely begun to comprehend the scope and complexity of mallard duck communications.

Ideas and Inventions Appropriated by Humans

A complete assessment of mallard duck intelligence must also look at the other side of the coin: what have mallard ducks and other waterfowl taught humans? The thirteen ideas and inventions humans appropriated from ducks and geese were discussed in Chapter 3 and are summarized in Table 4.9.

And, what ideas and inventions have mallard ducks deemed worthy enough to appropriate from humans? None.

Table 4.9 Ideas and Inventions Humans Appropriated from Mallard Ducks and Canada Geese.

Ideas and Inventions	**Observations that Led to the Ideas and Inventions**
Flying	• Sight of biannual long distance migratory flocks of ducks and geese
Raising and lowering aircraft landing gear when landing and taking off	• Ducks, geese, and other birds holding their legs and feet straight, parallel to their tail when flying
Flying in a V formation, taking advantage of slipstream effect	• Sight of biannual long distance migratory flocks of ducks and geese flying in a V formation
Boats, kayaks, canoes, helicopters and other planes that can land on water	• Ducks and geese swimming with their body above water
Oars, rudders, propellers, scuba fins	• Webbed feet of ducks and geese and how they use them to swim
Sudden scrambling of military aircraft as an evasive maneuver	• Sudden scrambling of a flock of mallard ducks in every which direction, with their flashing blue speculums acting as a distraction
Car horns, fire alarms, sirens, emergency message systems	• Geese honking and their ability to project sound over long distances
Water skis	• Ducks and geese making water landings with webbed feet
Snowshoes	• Webbed feet of ducks and geese which distribute weight and prevent them from sinking in mud, sand, and snow
Down coats, vests, and quilts	• Ability of ducks and geese to stay warm and dry in cold temperatures and cold water
Periscope	• Long flexible neck of Canada geese which is used to fish, upend, and scan in all directions
Secret Service VIP security escort procedures	• Young goslings lined up single file. Father goose in forward guard position, mother goose in rear guard position. Parents able to scan for hazards and predators in all directions, and assume flanking positions if necessary.
Baseball cap	• Duck bill

5

Ruby-throated Hummingbirds

Archilochus colubris

5.1 Background

Hummingbirds are only found in the New World.[45, 83] There are 341 species of hummingbirds in the Americas.[41, 45, 136] Sixteen species of hummingbirds breed in the United States.[161] Only the ruby-throated hummingbird nests and spends its summers east of the Rocky Mountains,[41, 161, 172] east of the Mississippi river,[29, 141, 182] or east of the Great Plains and the 100th meridian.[3, 22, 45, 95, 136, 203] Ruby-throated hummingbirds winter along the Gulf coast, from Texas, Mexico, and Central America to Panama. Ruby-throated hummingbirds live in wooded areas with flowers, parks, suburban gardens, orchards, open woods or the edge of woods.[3, 29, 41, 95, 203] Ruby-throated hummingbirds were assigned their Latin name, *Archilochus colubris*, in 1758.[203] The ruby-throated hummingbird's closest relative is the black-chinned hummingbird, *Archilochus alexandri*, which occupies the western half of North America.[95, 182] These two species may have diverged during the glacial period.[95, 182]

Hummingbirds are capable of performing aerial acrobatics that other birds cannot. For example hummingbirds can hover, fly straight up and fly straight down; they can even fly backwards.[29, 30, 41] Hummingbirds often appear to be constantly in motion.[29] To do so their wings beat so fast that they produce a humming sound—hence their name.[29, 41] Ruby-throated hummingbirds are among the smallest birds in North America.[29] Even so, hummingbirds have been described as being "exquisite, unabashed, and strangely fearless".[22]

During the 1800's thousands of hummingbirds were killed for the European fashion industry and jewelry or stuffed as decorative items because of their beautiful iridescent feathers.[41, 45] Hundreds more died while being shipped to Europe for zoos or private collections.[83] At that time it was not known that in addition to nectar ruby-throated hummingbirds need to eat insects for protein.[83] Hummingbird populations are still declining due to destruction of summer and winter habitats, use of insecticides and herbicides, and house cat predation.[41, 182] Seven species of hummingbirds are listed as threatened in the ICBP/IUCN Red Book due to habitat destruction, deforestation, and agricultural encroachment:[166]

- Chestnut billed hummingbird, *Amazilia castaneiventris*, which lives in the eastern Andes in Colombia. This species' population has been "unrecorded" for 30 years
- Glow-throated hummingbird, *Selasphorus ardens*, which lives in western and central Panama
- Mangrove hummingbird, *Amazilia boucardi*, which lives along the Pacific coast of Costa Rica
- Oaxaca or blue-capped hummingbird, *Eupherusa cyanophrys*, which lives in the mountain range of southwest Mexico

- Sapphire-bellied hummingbird, *Lepidopyga lilliae*, which lives along the north coast of Colombia
- Scissor-tailed hummingbird, *Hylonympha macrocerca*, which lives in southwest Mexico
- White-tailed hummingbird, *Eupherusa poliocerca*, which lives in northeast Venezuela

Unlike the ruby-throated hummingbird, which occupies a wide range of territory and migrates, these seven species of hummingbirds occupy specialized or narrow environmental niches; hence they are more vulnerable.

5.2 Characteristics

Now let's look at characteristics that are common to all ruby-throated hummingbirds, such as their:

- Diet
- Torpidity
- Flight
- Physical Features
- Vocalizations

Diet

Ruby-throated hummingbirds have a high metabolic rate because of their almost non-stop activity level. Ruby-throated hummingbirds cannot eat a large quantity of food at once because of their tiny size; nor would they be able to fly if they did so.[182] Hence, they consume many small meals throughout the day, often several meals per hour.[41, 182]

A ruby-throated hummingbird's diet consists primarily of nectar, insects, and spiders.[18, 22, 41, 182, 203] Ruby-throated hummingbirds occasionally rely on sap in early spring, before nectar is available.[3, 203]

Ruby-throated hummingbirds may visit over 1,000 flowers a day while feeding.[41] They seem to prefer cupped or tubular flowers that are red, orange, bright pink, and yellow—those that are on the long wavelength end of the light spectrum.[29, 41, 161, 182] In general the flowers frequented by ruby-throated hummingbirds have no scent, are horizontal or hanging flowers often with re-curved petals, and have a diurnal flowering time.[18] Ruby-throated hummingbirds are able to extract nectar because of their unique ability to hover.[22, 30] Ruby-throated hummingbirds seem to prefer nectar with a high sucrose content over that composed of simpler sugars like glucose or fructose.[182] Nectar provides a ruby-throated hummingbird's fluid and energy or carbohydrate needs. In return, ruby-throated hummingbirds provide superb pollinating services to the flowers they visit.[30, 182]

To meet their protein, vitamin, mineral, and amino acid needs ruby-throated hummingbirds consume spiders and insects, in particular gnats, flies, aphids, and beetles.[182] Ruby-throated hummingbirds catch these entrees while in flight or snatch them from flowers or tree bark.[22, 41, 163, 203] Ruby-throated hummingbirds are adept at catching insects in flight, in part because of the design of their bill. When pursuing an insect a ruby-throated hummingbird may flex its lower bill up to 25%.[163] That action causes the bill to snap shut quickly, like a spring-loaded hinge, often in less than 1/100th of a second.[163] Hatchlings are primarily fed insects because the high protein content supports their rapid growth and development.[41, 203]

Torpidity

Frequent small meals fuel a ruby-throated hummingbird's energy needs throughout the day. During the night ruby-throated hummingbirds adopt a different strategy, in part because they need to rest and in part because many flowers close at night. At nightfall ruby-throated hummingbirds enter a state of torpidity whereby their heart rate, body temperature, and breathing rate drop substantially and they assume an inactive or resting metabolic rate to conserve energy.[30, 41, 182, 203]

At sunrise ruby-throated hummingbirds awake, re-energize, and resume feeding.[41] Before entering a state of torpidity, ruby-throated hummingbirds need to ensure that they are beyond the reach of predators. If ruby-throated hummingbirds were attacked while in a state of torpidity, they would not be able to awake and flee in time. If ruby-throated hummingbirds encounter stressful environmental conditions, like colder than normal temperatures or a shortage of food, they can enter a state of torpidity to save energy until conditions improve.[182]

Flight

The flight of ruby-throated hummingbirds has been described as "rapid and darting".[22] Ruby-throated hummingbirds tend to hover briefly, perhaps even back-up or turn sideways for a quick visual scan, then dart deliberately to their next destination.[83] Occasionally they spread and pump their tail during flight.[41, 182] The wing beats of ruby-throated hummingbirds have been clocked at 55 to 80 beats per second—the fastest rate of any bird.[18, 83, 95, 137, 203] Ruby-throated hummingbirds are believed to be able to fly up to a maximum speed of 60 mph (90 kmph) for short periods of time.[18, 83] These flying feats are possible because of two physical features. The wing muscles of ruby-throated hummingbirds account for 25% to 35% of their total body weight.[41, 203] In addition, the wings of a ruby-throated hummingbird are only attached at the shoulder joint. As a result, ruby-throated hummingbird can rotate their wings 180°, in a figure 8 pattern, and in other unusual modes.[41, 203]

Ruby-throated hummingbirds migrate south for the winter. Many ruby-throated hummingbirds migrate 3,000 miles (4,545 km) or more 1-way. Some ruby-throated hummingbirds island hop across the Caribbean until they reach their final destination.[41] Other ruby-throated hummingbirds take the longer land route around the Gulf of Mexico.[41] Many ruby-throated hummingbirds take a short cut and fly approximately 500 miles (758 km) across the Gulf of Mexico.[3, 12, 41, 45, 83, 137, 182, 203] This route requires flying non-stop for about 20 hours.[12, 41] Flying 500 miles (758 km) in 20 hours represents a sustained rate of 25 mph (38 kmph). Relative to body mass, this 500 mile (758 km) route is the longest distance flown non-stop by any bird.[18]

The same journey is made again when it is time to go north for the summer. Both the north and south migrations are timed precisely to correspond to the availability of food and blooming flowers along the route.[2, 136, 172] In fact, banded ruby-throated hummingbirds have been observed following the same migration path and appearing at the same flowers and feeding stations on time year after year.[12]

Ruby-throated hummingbirds do not migrate in large flocks. Instead individuals or small groups, usually mothers with the young birds that hatched that year, make the journey. Frequently, male ruby-throated hummingbirds begin their migratory journeys 2 weeks before female ruby-throated hummingbirds.[3, 136, 203] The date ruby-throated hummingbirds begin their fall migration and return from their spring migration depends on the latitude of their summer home. Ruby-throated hummingbirds molt annually, just prior to the fall migration.[203]

Physical Features

Hummingbirds are known for their beautiful colors and iridescent feathers.[30, 182] Ruby-throated hummingbirds have a bright metallic green head, neck, back, and wings. Their underside is white or light grey.[22, 29] The males have an iridescent ruby colored throat patch or gorget.[22, 29, 41] The color reflected from the male's throat patch varies from a brilliant ruby color to a dull red to a dark green or black, depending on the angle of the sun's rays.[137, 203] Immature ruby-throated hummingbirds resemble females.[182]

Ruby-throated hummingbirds are tiny, being only 2.75 to 3.5 inches (6.9 to 8.8 cm) in length.[3, 22, 29, 182, 203] Ruby-throated hummingbirds have a wingspan of 3.25 to 4.25 inches (8.2 to 10.7 cm) and weigh between 0.10 to 0.11 ounce (2.8 to 3 gr).[18, 76, 137, 141, 161, 203] Ruby-throated hummingbirds have a long straight slender bill, about 0.75 inch (1.9 cm) in length.[29, 182] The narrow tongue of a hummingbird extends beyond the bill.[41, 203] The tongue has a forked tip, to assist in snatching insects, and grooves along the sides to guide nectar into the digestive system.[22, 41, 203] Ruby-throated hummingbirds have tiny feet; this helps to minimize their inflight weight.[22, 41] Ruby-throated hummingbirds only use their feet for perching; unlike other birds they do not walk or hop.[83] A ruby-throated hummingbird's eyes are large in proportion to the size of their head.[83] In general male ruby-throated hummingbirds are smaller than the females.[83, 182, 203]

Due to their compact size and high metabolic rate, ruby-throated hummingbirds exhibit some impressive physiological traits, for example a:[18]

- Heart rate of 615 beats per minute—the fastest of any bird
- Breathing rate of 250 breaths per minute

The maximum life span of a male ruby-throated hummingbird is generally 7 years.[18, 182] Female ruby-throated hummingbirds have been known to live up to a maximum of 9 years.[18, 137, 182] Because of their tiny size, ruby-throated hummingbirds bathe in the morning dew and/or lawn sprinklers.[182]

Vocalizations

The vocalizations of ruby-throated hummingbirds are acoustically proportioned to match their diminutive size. These vocalizations are rarely heard due to the rapid flight of ruby-throated hummingbirds and the humming sound produced by their wings. Ruby-throated hummingbird vocalizations have been described as a "shrill little squeak", a "mouse-like squeak", "rapid squeaky chirps", and a "soft tchew or tchup".[22, 29, 182, 203] Female ruby-throated hummingbirds have also been heard making a "tic-tic" sound when feeding their young.[182]

5.3 Family Life

For the most part ruby-throated hummingbirds are solitary birds; they do not form lasting pair bonds or large family groups.[41, 136, 203] This behavior may reflect a practical aspect of the life of ruby-throated hummingbirds, rather than a social preference. Nectar and insects occur in small quantities over a widely distributed area. This fact, combined with the need for a ruby-throated hummingbird to eat frequent meals, more or less makes it impractical for large flocks of ruby-throated hummingbirds to travel around together. Never the less, the males perform rather elaborate courtship displays, flying in large upside down arcs similar to the swing of a pendulum.[136, 137]

Female ruby-throated hummingbirds begin building a nest about a week before mating occurs.[41, 95, 161, 182, 203] Frequently new nests are built in the same location year after year.[182] Ruby-throated hummingbird nests are constructed in the fork of a tree branch near water.[22, 141, 161] The nests are constructed of dandelion or thistle plant down, bud scales, pieces of cattail, fern fronds, lichen, and bits of tree bark that are held together and to the tree with spider silk.[22, 29, 41, 76, 141, 161, 182, 203] The materials used to build a ruby-throated hummingbird nest provide both insulation and camouflage, since visually the nest blends into the tree branch.[76] Ruby-throated hummingbird nests are 1.6 inches (4 cm) in diameter and 10 to 20 feet (3 to 6 m) above ground.[29, 41, 95, 141, 161, 203] The female fashions the cup shape of the nest by pressing her feet and body against the plant material.[3, 95, 182]

Two tiny white eggs are laid one day apart in the May/June time frame; they are approximately the size of a green pea.[22, 29, 41, 83, 95, 141, 161, 182, 203] Ruby-throated hummingbird eggs are the smallest bird eggs laid in North America.[30]

Female ruby-throated hummingbirds assume all nesting and family raising responsibilities.[3, 83] The female incubates the eggs for 16 to 17 days.[18, 41, 95, 161] After the chicks hatch, the mother feeds them up to 3 times per hour.[161, 203] Ruby-throated hummingbird hatchlings do not make a begging call (a shrill peep) until they feel the breeze from their mother's wings nearby.[30, 137, 182] This behavior prevents announcing the hatchlings whereabouts to potential predators until the mother is present to protect them. Hatchlings are primarily fed insects, supplemented with some regurgitated nectar.[41, 203]

Ruby-throated hummingbird hatchlings grow feathers by the time they are 9 to 12 days old.[182] Ruby-throated hummingbird hatchlings begin to exercise their wings by the time they are 15 to 20 days old and they fledge at 21 to 24 days of age.[18, 22, 41, 95, 182, 203] Ruby-throated hummingbird hatchlings exercise their wings while perching on the rim of their nest and holding on tight with their feet.[12] This behavior ensures that they don't accidentally fall out of the nest.[12] Young ruby-throated hummingbirds remain dependent on their mother for up to 4 weeks.[182] During this time young ruby-throated hummingbirds develop their wing muscles and learn survival skills, like how to find nectar and insects.[182] In addition, their bill grows to full size. Young male ruby-throated hummingbirds acquire their first full set of adult plumage in the wintering grounds.[161]

Ruby-throated hummingbirds have one to two broods a year, depending on the climate, weather, availability of food, and prior nesting success.[95, 182, 203] The survival rate of ruby-throated hummingbird eggs to hatchlings to young adults is less than 50%.[182] Common predators include birds of prey, jays, wrens, squirrels, snakes, and lizards.[182]

5.4 Individual Behaviors

Now let's look at the characteristics of the ruby-throated hummingbirds that were observed, for example their behavior, daily activities, and life stages. Behaviors are presented as they occurred. The analysis, comparison, and interpretation of these behaviors are reserved for Section 5.5 Discussion. These observations took place at the locations and during the timeframes described in Annex C.

Ruby-throated hummingbirds were observed from a 2nd story balcony that was 4-feet (1.21 m) wide by 12-feet (3.65 m) long. The balcony faced northeast and bordered the woods of a nature preserve. During warm weather many plants were moved out on the balcony, mostly cacti and succulents. The movement outdoors combined with the variation in temperature and humidity, day to night and day to day, along with exposure to direct sunlight stimulated new growth and flowers.

A round cactus with hooked spines produced a large yellow waxy flower with a bright red center. Once in a while there were two yellow flowers at the same time. A columnar cactus 1.5 inch (3.8 cm) in diameter sprouted a series of tiny light purple flowers that circled its tip, like a headband. This cactus only bloomed once a year, usually during the end of May. The hooked cactus bloomed through August, with a new flower every couple of weeks. Every year around June 1st the sansevieria produced a few spikes that were covered with small white flowers about 0.5 inch (1.27 cm) in diameter. These flowers bloomed in succession and had a delightful scent, like lilies, that was noticeable 2 or 3 feet (.6 to .9 m) away.

Each year the plants on the balcony were augmented with some annuals; in particular snapdragons, marigolds, and moss roses. The snapdragons and marigolds were transplanted into pots. The moss roses were placed in hanging baskets on the railing. In effect the balcony became a small garden during the warm months. The spring plant migration occurred mid to late April each year. The fall plant migration took place mid-November, coinciding with the first freeze warning.

This scenario made it possible to observe the ruby-throated hummingbirds up close on a regular basis at different times throughout the day, from sunrise to sunset, year after year. Often the hummingbirds were only 4 to 5-feet (1.21 to 1.52 m) away, which necessitated standing still. If there was any sudden movement, the hummingbirds darted away. Electric lights bothered the hummingbirds, so they were turned off during the early morning and early evening hours. It is unclear whether the incandescent lights produced a reflection that was scary, or the fact that the hummingbirds could see through the sliding glass doors was the problem. If a light was on in the room next to the balcony, that was ok. This close proximity viewing enabled individual ruby-throated hummingbird behaviors to be identified, such as their:

- Feeding habits
- Visiting pattern
- Strategy for dealing with inclement weather
- Preparation for fall migration
- Spring return
- Long distance visual identification

Feeding Habits

The precision of the ruby-throated hummingbirds flight and navigation skills was impressive. The ruby-throated hummingbirds were not foraging around looking for food. Rather, their flight pattern was direct and deliberate. The precise timing of the ruby-throated hummingbirds arrivals and departures made it clear that they were flying a regular route and not an ad hoc one. In fact, one morning a ruby-throated hummingbird arrived before the feeder was refilled. It flew right up to where the feeder should have been and hovered there for a few seconds. Next the ruby-throated hummingbird retreated briefly to a nearby tree that faced the balcony. As soon as the replenished feeder was replaced, the ruby-throated hummingbird returned to the balcony. The ruby-throated hummingbirds arrived each morning shortly after sunrise. Their last visit each evening was just before sunset and the flashing arrival of the lightning bugs.

The ruby-throated hummingbirds zoomed toward the target feeder or flower, then stopped short about 1.5 inches (3.81 cm). The ruby-throated hummingbirds examined the feeder or flower from that close vantage point before feeding. Once a determination was made that everything was alright and the feeder or flower contained food (nectar and/or insects), the ruby-throated hummingbird would proceed to eat. The skill required to hover in the correct position, while inserting its bill in exactly the right place through the bee guard and into the small opening in the feeder that held the nectar, was remarkable. Equivalent precision and timing is required to extract (and not drop) nectar from flowers.

Sometimes a ruby-throated hummingbird would take a drink for about 5 seconds, pause, and take a second drink. Other times it only took 1 drink. When it was time to leave, the ruby-throated hummingbird would fly backward a few inches (centimeters) before darting away. Once or twice a year a ruby-throated hummingbird flew toward the sliding glass doors, instead of the woods, when leaving the balcony. Each time when it was about 6 inches (15.24 cm) away from the glass, the ruby-throated hummingbird stopped suddenly and hovered. Somehow the ruby-throated hummingbird figured out that it was seeing its own reflection and that of the trees. Then, it turned and left toward the woods.

During the first year an experiment was conducted to learn the ruby-throated hummingbirds feeding preferences. Various combinations and configurations of moss roses and hummingbird feeders were hung from the balcony railing. The ruby-throated hummingbirds only visited the moss roses. They ignored the cactus flowers, sansevieria flowers, marigolds, and snapdragons even though they were in the preferred color range.

First a single tube feeder was hung with a moss rose planter on the left side of the balcony. The ruby-throated hummingbirds ignored the planter with the feeder. Instead they visited the other two moss rose planters. Next, a single tube feeder was hung with a moss rose planter on the right side of the balcony. The ruby-throated hummingbirds ignored that planter. Instead they visited the other two moss rose planters. Then a single tube feeder was moved to the center moss rose planter. Again, the ruby-throated hummingbirds bypassed the planter with the feeder. In each case, the tube feeder/hanging planter combination was left in place for one week.

Having exhausted all possible combinations and locations of a single tube feeder and a moss rose planter, a single tube feeder was hung by itself evenly spaced between two moss rose planters. The ruby-throated hummingbirds went right to the feeder and ate from it the very first morning. Apparently something about the combination of a moss rose hanging planter and a single tube feeder did not seem right to the ruby-throated hummingbirds. Perhaps the ruby-throated hummingbirds understood that a plastic tube feeder was not part of a (living) moss rose plant and thought it was a trap. Thereafter, single tube feeders were hung by themselves between the moss rose planters.

Afterwards another experiment was conducted. The intent was to determine whether ruby-throated hummingbirds preferred feeders with single or multiple feeding stations, bee guards, and perches. The ruby-throated hummingbirds sampled the multi-station feeder a couple times, but spent more time at the single station feeder. If a ruby-throated hummingbird stopped to perch it was only for a second or two before it took another drink or departed. The vast majority of the time only one ruby-throated hummingbird fed at a time. So, single tube feeders became the standard configuration.

Visiting Pattern

On a typical visit a single ruby-throated hummingbird would approach from the side of the balcony. It would stop to sample the first feeder or flower, perhaps fly over to sample a second feeder or flower, then dart away. The ruby-throated hummingbird only paused about 2 seconds between or after feeding. The entire visit lasted less than 25 seconds and consisted of 9 steps:

- Arrive at the balcony
- Scan 1st feeder or flower
- Feed
- Pause/hover
- Fly to 2nd feeder or flower
- Scan 2nd feeder or flower
- Feed
- Pause/hover
- Depart

One day a bee was eating at the feeder when a ruby-throated hummingbird arrived, even though the feeder had a bee guard. The ruby-throated hummingbird hovered for a while and waited for the bee to leave. The bee did not leave. The ruby-throated hummingbird apparently became tired of waiting —it takes a lot of energy to hover—and simply stuck its bill through the bee and into the feeder to obtain some nectar. Then the ruby-throated hummingbird continued on its way.

Table 5.1 highlights some interesting statistics that were derived from the ruby-throated hummingbird visits. These statistics are based on the dates and times recorded for ruby-throated

Table 5.1 Summary of Ruby-throated Hummingbird Visits.

• Earliest visit (time of day):	5:50 a.m.		
• Latest visit (time of day):	8:45 p.m.		
• Earliest visit (calendar date):	April 25th		
• Latest visit (calendar date):	September 27th		
• Shortest interval between visits:	5 minutes		
	A.M.	P.M.	Total
% of visits hummingbirds appeared from right side of balcony (east)	48%	17%	35%
% of visits hummingbirds appeared from left side of balcony (west)	33%	59%	44%
% of visits hummingbirds came directly to/from trees across from balcony (north)	19%	24%	21%
Total	59%	41%	100%

hummingbird visits. The ruby-throated hummingbirds were not under 24/7 surveillance the entire 10 years. Undoubtedly, they made visits to the balcony at other times as well. Regardless, the data is sufficient to derive some general trends.

The earliest visit by a ruby-throated hummingbird was observed at 5:50 a.m. That visit occurred in early June. The latest visit by a ruby-throated hummingbird was observed at 8:45 p.m. That visit occurred in late June.

The earliest date ruby-throated hummingbirds were observed returning from spring migration was April 25th. The latest date ruby-throated hummingbirds were observed prior to leaving for fall migration was September 27th.

The shortest time interval between visits by ruby-throated hummingbirds to the balcony, and its feeders and planters, was 5 minutes. This observation was constant whether normal weather or inclement weather was being experienced. In the case of normal weather the sequential visits were probably made by different ruby-throated hummingbirds. In the case of inclement weather, the sequential visits were probably made by the same ruby-throated hummingbird.

During the morning, ruby-throated hummingbirds appeared from the right (or east) side of the balcony 48% of the time. More visits were made in the morning hours (59% of the total visits), even though there were more hours of daylight in the afternoon and evening. This statistic probably reflects the need to replenish energy quickly after exiting a state of torpidity.

During the afternoon and evening, ruby-throated hummingbirds appeared from the left (or west) side of the balcony 59% of the time. This observation may reflect ruby-throated hummingbirds flying a reverse route.

Ruby-throated hummingbirds traveled directly to and from the trees across from the balcony 21% of the time. That was a distance of 10-feet (3 m) or less. These shorter journeys occurred most often when the ruby-throated hummingbirds were nesting, had a fledgling in tow, or were sitting out a storm.

Strategy for Dealing with Inclement Weather

The ruby-throated hummingbirds route and visits remained fairly constant, except for two special instances: thunderstorms and fall migration. If a thunderstorm or heavy rain began when a ruby-throated hummingbird was feeding, it would quickly retreat to the trees nearest the balcony instead of continuing its journey. The ruby-throated hummingbird rested briefly in the trees. Then, every 5 to 10 minutes, the ruby-throated hummingbird made a short journey down from the trees to the feeders on the balcony. The ruby-throated hummingbird understood that the winds were too strong and the weight of rain water on its wings was too heavy to continue its regular journey. Instead of challenging nature or engaging in risky behavior, the ruby-throated hummingbird stayed nourished through short trips to and from the balcony, while sitting out the storm near a ready source of food.

In comparison, when a sudden severe rain storm appears suddenly the majority of humans insist on continuing to drive at or above the speed limit even though visibility is limited and braking distances are increased. One or two souls may pull over to sit out the storm. But most continue on as if it is a contest to see who is in control—man or nature. Usually 10 minutes or so after a storm begins, police sirens can be heard indicating that a wreck has occurred.

Preparation for Fall Migration

Each fall, during the second or third week of September, the ruby-throated hummingbirds began carbohydrate-loading for the long journey ahead. At those times the feeders often needed to be refilled twice a day. A ruby-throated hummingbird would stop at the first feeder, take two drinks, go to the next feeder, and take two or more drinks. Then the hummingbird perched in the nearby trees for a couple minutes before returning to feed again. It was as if each hummingbird was trying to store as much reserve nectar (fuel) in its tiny body as it could and still be able to fly. This pattern continued for about a week. Then all of a sudden all the ruby-throated hummingbirds were gone. Fall, just prior to the southern migration, was the only time of the year when more than one ruby-throated hummingbird was seen feeding at a time.

About the same time the hummingbirds were carbohydrate-loading, a flock of eastern bluebirds *Sialia sialis* would arrive. Fifteen to eighteen bluebirds arrived each fall. The bluebird flock was obviously on its way south for the winter. The bluebirds would land all at once on the ground near the bird seed feeder. They consumed all the available food in a rapid and animated fashion, as if they were hungry from a long flight. Then they retreated to the nearby trees to roost. As soon as the bird food was replenished, the bluebirds indulged themselves again. The bluebirds only stayed a day or two and then they were gone. The bluebirds were only seen during their fall/southern migration, not during their spring/northern migration.

In a lucky year a group of monarch butterflies might also be spotted on their journey south during the fall. Two bushes nearby had plentiful spikes of small white flowers in September. The monarchs would descend on these two bushes for a day or two, before continuing their journey. At times it seemed like the whole bush was covered with monarchs, with more than 20 butterflies per bush. The monarch butterflies were not foraging or randomly looking for food. They did not stumble onto these bushes. On the contrary, the monarch butterfly flock flew directly to this pair of bushes, as if it were a known milestone along their journey.

Spring Return

Six months later the ruby-throated hummingbirds made the long journey back north. The feeders and planters were hung out on the balcony in mid-April to be sure they were out and full when the hummingbirds returned. And return they did, year after year. It was always amazing that first day in April when the ruby-throated hummingbirds zoomed up to the balcony and feeders. Each tiny hummingbird, 3.5 inches (8.8 cm) long and weighing practically nothing, had just completed a journey of over 1,000 miles (1,515 km). It may have spent the winter in Panama, Mexico, or southern Texas. Then, without the aid of the global positioning system (GPS), radar, maps, street signs, or air traffic controllers, each tiny hummingbird made it back to the exact same balcony and feeders.

This feat cannot be simply dismissed as instinct. On the contrary, some rather sophisticated genetic intelligence, applied intelligence, logical reasoning, and memory are involved. The flight plan may or may not stay the same every year, given climatic changes. The departure and arrival destinations and dates may or may not stay the same every year. The navigational landmarks are in a constant state of flux. Lakes and rivers dry up or flood. Trees grow taller or are chopped down. Coastlines erode or become inundated. Buildings are built or torn down. New roads are constructed, and so forth.

A hummingbird, like all migratory birds, carries an historical memory of the migration route. But this memory has to be updated in real time to reflect changing realities. To put it another way, if navigating from Maryland to Mexico and back is so simple, why do humans have to have all the navigational aids they do? If it is so simple to update migration routes in real time, why do airline flights get so scrambled up and down the east coast every time there is a change in the weather pattern? It would appear that the intellect of tiny ruby-throated hummingbirds is better at solving these types of problems than humans.

The possibility of a special observation happened each spring after the ruby-throated hummingbirds returned: the sighting of a fledgling ruby-throated hummingbird. A full grown adult ruby-throated hummingbird is tiny. A fledgling ruby-throated hummingbird is even tinier. Fledgling ruby-throated hummingbirds were spotted five times:

- 10 June 1989 at 11:20 a.m.
- 2 July 1989 at 10:30 a.m.

- 4 May 1990 at 10:30 a.m.
- 5 May 1990 at 6:20 p.m. and 7:25 p.m.

During the two sightings in 1989 the fledgling hummingbird approached from the right (east) side of the balcony. During the three sightings in 1990 the fledgling hummingbird traveled directly to and from the trees across from the balcony. On 10 June 1989 the fledgling was preceded by an adult hummingbird at 11:10 a.m. and another at 11:15 a.m. The fledgling appeared to be trying to follow the adults. But its flight was still a little wobbly and less precise. The first few visits it took two tries before the fledgling was able to insert its bill into the feeder correctly.

Given the maximum life span of a hummingbird, it is likely that two or possibly three generations of ruby-throated hummingbirds visited the tube feeders and planters during the 10 years they were observed at the nature preserve.

Long Distance Visual Identification

Several years later on August 8th a ruby-throated hummingbird was spotted hovering outside a 2nd story sunroom window (in a different location) that faced northeast. The hummingbird hovered for a while, then surveyed the borders of the window with its bill, as if it were trying to get into the sunroom. The ruby-throated hummingbird touched the upper glass and lower screen gently with its bill, then it turned around and left quickly.

The ruby-throated hummingbird had to leave the woods, fly across 12 feet (3.65 m) of lawn, a 12 foot (3.65 m) parking lot, and a 4 lane road in order to reach the 2nd floor sunroom window. What prompted it to do that? There were no hanging flower baskets or tube feeders outside that window or along the path to it. However, there was a cactus indoors that had just started to bloom. The flowers were 1 inch (2.54 cm) in diameter and light coral in color with a dark purple center. One flower opened and closed each day for a week. Had the ruby-throated hummingbird spotted the cactus flowers from a tree that far away? If so, that would imply that, in addition to their other impressive skills, ruby-throated hummingbirds have superb long distance vision.

A ruby-throated hummingbird was spotted at the same window a week later. The following year a ruby-throated hummingbird visited 3 different windows on the 2nd floor, all facing northeast. These visits occurred almost daily from July 12th through July 26th. All of these visits occurred in the morning, between 7:15 a.m. and 10:00 a.m. Inside each window a plant was flowering: a cactus, a geranium, or African violets. This confirms the earlier observation that ruby-throated hummingbirds are attracted to flowers by their color, not their scent.[18]

5.5 Discussion

Now let's see what these behaviors tell us about ruby-throated hummingbird intelligence, behavior, and individuality.

Biological Intelligence

When it comes to biological intelligence ruby-throated hummingbirds have it in spades. Ruby-throated hummingbirds have the fastest heart rate of any bird—615 beats per minute—and one of the fastest breathing rates—250 breaths per minute.[18] Both are autonomic processes. Still, the ability to send, interpret, and act upon an electrical impulse that fast is impressive. The human body doesn't perform any function even near that speed.

Even more impressive is the ruby-throated hummingbirds ability to enter and exit a state of torpidity voluntarily on a daily basis or when stressful environmental conditions are encountered.[30, 41, 182, 203] This process allows ruby-throated hummingbirds to assume a resting metabolic rate to conserve energy.[30, 41, 182, 203] The modern equivalent is switching a laptop computer to hibernate mode. If humans had this ability people would be less likely to freeze to death in winter, die from heat exhaustion, or starve during famines.

Furthermore the long slender bill of a ruby-throated hummingbird is perfectly designed for extracting nectar from flowers, as is their tongue which has long narrow grooves to guide the nectar into the digestive system.[22, 41, 203] The lower bill can snap shut like a spring-loaded hinge, often in less than 1/100th of a second.[163] That capability plus a forked tongue makes catching insects a breeze.[22, 41, 203]

Ruby-throated hummingbirds flap their wings at 55 to 80 beats per second—faster than any bird—and they can fly up to 60 mph (90 kmph).[18, 83, 95, 137, 203] Then there is the ruby-throated hummingbirds 20 hour 500 mile (758 km) nonstop journey across the Gulf of Mexico. All of this is possible because of the superb design of the ruby-throated hummingbirds wings. First, their wing muscles account for 25% to 35% of their total body weight.[41, 203] Second, since a ruby-throated hummingbird's wings are only attached at the shoulder, they can rotate their wings 180°, in a figure 8 pattern, and in other unusual modes.[41, 203]

If ruby-throated hummingbirds were machines they would be described as a fine example of precision engineering.

Rapid Decision-making Ability

The ruby-throated hummingbirds consistently demonstrated an ability to make decisions rapidly. In turn this infers an ability to process a large volume of information quickly and send the appropriate messages to and from the brain with equal speed. This rapid information transfer is most evident during ruby-throated hummingbird flight.

Flying 60 mph (90 kmph), or even 25 mph (38 kmph), requires instantaneous processing of a large volume of visual data. At the same time a ruby-throated hummingbird is sending signals to and from its wings to beat 55 to 80 times a minute. These signals rarely reflect repetitive motion. Instead a whole series of start, stop, hover, turn left, turn right, go up, go down, and back-up commands are involved, all of which have to be synchronized with incoming and changing visual data.

As an example consider the one or two times a year when a ruby-throated hummingbird flew toward the sliding glass doors, instead of the woods, when leaving the balcony. The ruby-throated hummingbird flew the 4 foot (1.21 m) distance quickly. Yet, each time when the ruby-throated hummingbird was about 6 inches (15.24 cm) away from the glass, it stopped suddenly and hovered, instead of crashing. The ruby-throated hummingbird quickly figured out that it was seeing its own reflection, turned around, and headed safely toward the woods. The entire flight, including stopping and turning around, was less than 5 seconds.

Ability to Merge Historical and Contemporary Information

Ruby-throated hummingbirds have a well-developed sense of genetic intelligence as is evinced by their bi-annual migration. Understanding the need to migrate, when to migrate, what route to take, and so forth requires the ongoing combination and synthesis of historical and contemporary information. This ability is particularly significant because ruby-throated hummingbirds do not migrate in large flocks. Instead, individuals or small groups, usually mothers with the young birds that hatched that year, make the 1,000 mile (1,515 km) plus journey. Different ruby-throated hummingbirds take different migratory routes. Some ruby-throated hummingbirds island hop across the Caribbean, some take the longer land route around the Gulf of Mexico, and others fly approximately 500 miles (758 km) non-stop across the Gulf of Mexico.[3, 12, 41, 45, 137, 182, 203] A high degree of self-reliance, resilience, and resourcefulness is needed regardless of which route is taken. Each year historical route and destination information is updated with new information on both the southern and northern journeys. The ruby-throated hummingbirds did not just return year after year—they returned all the way from the Gulf of Mexico to the exact same balcony. That accomplishment bespeaks more than superb navigation skills.

Ability to Anticipate and Plan for Future Events

Ruby-throated hummingbirds demonstrated an ability to anticipate and plan for future events. Ruby-throated hummingbirds started carbohydrate loading before beginning the fall migration. They understood how much to consume so that it was not too much (which would have interfered with their flying ability) or too little (which would not sustain them throughout the journey). Ruby-throated hummingbirds knew how to time their southern and northern journeys with precision to correspond to the availability of nectar and insects along the way. This implies that ruby-throated hummingbirds understand the distance and duration of their journey and the changing weather patterns that will be encountered en route. This understanding reflects a fine tuned ability to anticipate and plan for the future.

Female ruby-throated hummingbirds begin building their nest about a week before mating occurs.[41, 95, 182, 203] This is another example of anticipating and planning for the future. The female ruby-throated hummingbird prepares the nest beforehand, in anticipation of the eggs that will be laid and the chicks that will hatch. Often new nests are built in the same location year after year.[182] This practice implies that female ruby-throated hummingbirds remember from year to year which nesting sites were successful.

Reasoning Ability

The ruby-throated hummingbirds repeatedly demonstrated an agile reasoning ability and creative problem solving skills in response to novel situations. These traits are essential for survival given that for the most part ruby-throated hummingbirds experience life as solitary birds. The ruby-throated hummingbirds flew deliberate flights not only to the balcony, but also directly to a particular feeder or hanging planter. The ruby-throated hummingbirds were not foraging. Instead they had created and stored a spatial memory map of where the feeders and planters were located. They also learned when the feeders were refilled each morning. Recall the incident when a ruby-throated hummingbird arrived before the feeders were full. At the same time the ruby-throated hummingbirds stopped 1.5 inches (3.8 cm) short to scan the feeder or planter each time before feeding. This implies an understanding of the need to check the current status of each potential food source, rather than operating on auto pilot. Concurrently ruby-throated hummingbirds learned how to gently extract nectar from each feeder and planter so that none was spilled.

In addition there is the ruby-throated hummingbirds strategy for dealing with inclement weather. A ruby-throated hummingbird cannot out fly a thunderstorm, yet it needs to eat frequently. So, the ruby-throated hummingbirds reasoned that the best approach was to sit out the storm near a known source of food.

A comical variation of this strategy occurred when a ruby-throated hummingbird got tired of waiting for a bee to leave the feeder. In the end the ruby-throated hummingbird found a way to have some nectar anyway.

The ruby-throated hummingbirds demonstrated an ability to distinguish between an inanimate manmade object and a living plant when they refused to eat from the combined moss rose/tube feeder configuration. Both items were accepted when they were hung separately. But when they were combined the ruby-throated hummingbirds decided something was wrong, through a contemporary assessment of the situation.

Studies have confirmed that hummingbirds remember which flowers they have taken nectar from and do not go back to them again within the same day.[189, 190] That would waste valuable time and energy.[189, 190] However, ruby-throated hummingbirds are able to distinguish between a limited quantity of flower nectar in a specific location and a hummingbird feeder which contains a large quantity of food. The fact that ruby-throated hummingbirds returned to the same tube feeder several times each day, coming from different directions, is evidence of their ability to make this distinction.

Hatchling ruby-throated hummingbirds do not make a begging call until they feel the breeze from their mother's wings.[30, 137, 182] This practice implies that the hatchlings understand the difference between when their mother is and is not present. This practice also implies that the hatchlings understand that there is no point in making a begging call when their mother is not nearby. This, of course, has the fringe benefit of not alerting predators to the whereabouts of an unattended nest. This behavior reflects a combination of genetic and applied intelligence. That is, as the hatchlings get older they: (1) associate the mother with food, and (2) realize she is nearby when they feel the breeze of her wings.

These observations should put to rest, once and for all, the mistaken notion that brain size has anything to do with intelligence.

Ideas and Inventions Appropriated by Humans

A complete assessment of ruby-throated hummingbird intelligence must also look at the other side of the coin: what have ruby-throated hummingbirds taught humans? Two concepts that humans learned from hummingbirds come to mind quickly:

- Hovering mid-air
- Drinking straws

Hummingbirds are the only birds that can hover, fly straight up, fly straight down, and fly backward.[29, 30, 41, 83] That feat was initially noticed by inhabitants of the Americas. After airplanes became a viable technology, humans attempted to duplicate hummingbirds' aerobatic prowess. Eventually helicopters and other hovercraft were developed. Today most metropolitan areas could not function without the helicopters used by medical staff, law enforcement, emergency responders, traffic reporters, news reporters, government and corporate officials.

Most likely the long slender bill of hummingbirds and the sight of them drinking nectar contributed to the development of drinking straws. Originally it was thought that hummingbirds used their bill like a straw to drink nectar. Later it was learned that ruby-throated hummingbirds have small grooves along the sides of their tongue to guide nectar into their digestive system.[22, 41, 203] As a result, extracting nectar involves more of a swallowing process. That explains why a hummingbird's bill does not appear sticky or gooey even though it has been drinking nectar all day. Besides, there would be no need for an extensible tongue if a hummingbird was using its bill as a straw. (See Table 5.2.)

And what ideas have ruby-throated hummingbirds deemed worthy enough to appropriate from humans? None.

Table 5.2 Ideas and Inventions Humans Appropriated from Ruby-throated Hummingbirds.

Ideas and Inventions	Observations that Led to the Ideas and Inventions
Helicopters, hovercraft	• Ruby-throated hummingbirds hovering, flying straight up, straight down, and backwards
Drinking straws	• Ruby-throated hummingbirds drinking nectar from flowers using their long slender bills

6
Cardinal
Cardinalis cardinalis

6.1 Background

It took a while for the scientific community to settle on a Latin name for cardinals. At various points in time a cardinal was referred to as *Cardinalis Virginianus*, *Fringilla cardinalis*, *Loxia cardinalis*, *Cardinalis phoenaceus*, and *Cardinalis grosbeak*, prior to its current name *Cardinalis cardinalis.*[73] A book published in 1737 simply refers to cardinals as the "American red bird".[73] Cardinals are also referred to as Northern Cardinals and Virginia Cardinals.[3, 33, 95] The *cardinalidae* family was defined in 1901.[85, 203]

Cardinals are the official state bird of seven states: Illinois, Indiana, Kentucky, North Carolina, Ohio, Virginia, and West Virginia.[41, 95]

The red-crested cardinal (*Paroaria coronata*), native to Argentina, Brazil, Bolivia, Paraguay, and Uruguay, resembles a cardinal.[85, 203] It has a red crest, red head, red throat and small black eye patch. The red-crested cardinal has a similar shape and size. However, it has a white body, light gray wings and back, and a light beige bill. Both are related to grosbeaks, tanagers, and buntings.

Cardinals only live in the New World.[3, 22, 29, 41, 95, 136, 153] Today cardinals occupy a broad range in North America, east of the Great Plains, from southern Ontario, Nova Scotia and the Dakotas to Texas and the Gulf coast; they also inhabit a few isolated pockets in southern California and Arizona.[3, 29, 41, 95, 136, 153] The San Gabriel River Wildlife Sanctuary is one of the few locations where cardinals may be found in California.[80] In Arizona cardinals mainly reside near Tucson.[80]

The presence of isolated pockets of cardinals in southern Arizona and southern California is hard to explain. Especially since cardinals do not migrate. It is a long distance from Texas, the western edge of cardinals' normal range, to southern Arizona and even further to southern California. The terrain is inhospitable and it is too far for a cardinal to fly in a day. There are no recorded releases in these areas; neither is an ideal location for an intentional release.

One plausible explanation can be derived from studies conducted of small mammals in the southwestern states during the 1920's and 1930's. Stein makes the observation that, "The formation of isolated populations [of small mammals] coincided with the gradual transformation of southern California ... from wet to dry habitats".[159] In other words, the small populations that survived scattered to the few isolated areas that would still support their life form. The cardinals present today in southern California and southern Arizona may have undergone the same experience.

Cardinals were introduced in Hawaii in 1929.[41, 83, 136] Introduction of non-native species in Hawaii is not new. Exotic flora and fauna were traditionally presented to the Hawaiian monarchy as gifts.[159] Unfortunately, westerners who arrived on the islands in the last century have continued and expanded this practice. As early as the 1930's Alexander noted the tendency to experiment with non-native birds in Hawaii, such as the "Kentucky and Brazilian cardinals" which were reported to be thriving.[159] That, of course, meant that the native birds which were displaced retreated to more and more inaccessible regions or simply disappeared.[159]

Originally, cardinals lived in the southeastern United States. Hence, it is rather curious that cardinals are sometimes referred to as northern cardinals. They began expanding their range north in the early 1900's due to moderating temperatures and the addition of berry producing shrubs and bird feeders in new suburban neighborhoods.[3, 29, 83, 95, 137, 153] In fact, a book published in 1918 describes cardinals as "a southern bird becoming more and more common during summer in states north of Virginia, especially Ohio".[22] The first record of wild cardinals breeding in Canada was made in 1901.[3]

Cardinals mainly live in the edge of woods, thickets, brushy swamps, suburban gardens, mixed woodlands, parks, and along streams.[3, 29, 41, 95, 136, 119] The isolated pockets of cardinals that live in southern California and Arizona inhabit brushy to semi-open desert washes, particularly territory where mesquite is common.[95, 119] Cardinals have demonstrated a moderately low sensitivity to habitat fragmentation.[137] This undoubtedly facilitated the ability of cardinals to expand their range northward. Cardinals are year round residents; they do not migrate.[22, 29, 41, 119] Most cardinals spend their entire lives within a few miles (kilometers) of where they hatched.[41, 119]

The fame of cardinals spread with unfortunate consequences after the publication of James Lane Allen's book, "The Kentucky Cardinal".[22] During the 1800's and early 1900's the brilliant red plumage and "agreeable melodious song" of cardinals instantly made them subject to mass capture and export for the European caged bird trade, where they were known as Virginia nightingales.[22, 41, 73, 81, 83] A book published in 1879 states that:[73]

> "This bird is a very old favourite, not only in England but in America, whence are annually exported immense numbers to all parts of Europe, one London dealer receiving in a single consignment 700 pairs."

Seven hundred pairs??? Most people who live in the U.S. today where cardinals live are not privileged to see 10 pairs of cardinals simultaneously during their life time. The most the author, who has been observing birds for over 30 years, has ever seen is five pairs of cardinals at one time. That occurred one year during the dead of winter. There was 10 inches (25.4 cm) of snow on the ground and the wind chill was -30° F. These five pairs of cardinals came to a bird feeder at the same time for a few days. During the rest of the year only one or two pairs of cardinals came to this feeder.

The same book goes on to mention one dealer where 70 cardinals died in two months, as well as the fact that "thousands ... perish miserably through improper feeding ... literally starving to death".[73] Like the ruby-throated hummingbirds, it was learned much too late that cardinals must consume insects for protein.[33, 81] Others cardinals are reported to have died from the heat in open aviaries that offered no shade.[33] The poaching and smuggling of parrots for the pet bird trade today, which is pushing many species to the brink of extinction, is a continuation of practices that began in the 1600's, 1700's and 1800's with cardinals, ruby-throated hummingbirds, and other New World birds.

At that time cardinals were described as traveling in flocks.[22] Today cardinals are only seen as singles or in groups of 2, 4, or rarely 6 birds. It would appear that 136 years later cardinal populations have not yet recovered from the earlier mass capture and export. Since then cardinals have been protected against capture by federal and state laws. However, as late as 1939 it was not uncommon to see painted buntings and cardinals in cages on porches in the Cuban neighborhoods of Key West, Florida.[80]

6.2 Characteristics

Now let's look at characteristics that are common to all cardinals, such as their:

- Diet
- Physical Features
- Temperament
- Vocalization

Diet

Cardinals consume a varied diet of seeds, insects, berries, fruit, buds, and flowers.[29, 41, 95, 119, 136, 153] This factor contributed to the ability of cardinals to expand their range. Cardinals are definitely not fussy eaters. On the contrary, one study recorded cardinals consuming 51 different types of insects and spiders, 33 types of fruit and berries, and 39 types of seeds.[41] Cardinals tend to hop and/or fly, rather than walk around, when they are foraging.[95]

More insects are consumed during breeding season than the rest of the year because of their protein content.[29] Insects are fed to hatchlings for the same reason.[95] Beetles, grasshoppers, caterpillars, ants, flies, spiders, and centipedes are some of the more common cardinal insect entrees.[95] During fall and early winter, when insects start to disappear, berries are an important addition to the diet.[41]

Cardinals are popular visitors to suburban bird feeders; they often visit in pairs or as a family group.[29, 41, 119, 136] Over the years cardinals have shown a preference for gray-striped sunflower seeds, black oil sunflower seeds, safflower seeds, peanuts, cracked corn, and white proso millet.[41, 83, 119] Like all seed eating birds, cardinals have to compare the energy cost of obtaining and opening a seed to the energy value of the contents.[137] That is why, given a choice, cardinals will select sunflower seeds, safflower seeds, and peanuts first over other grains that might be available. Practicality, not taste, is the driving factor, especially during winter when energy needs are higher.

Physical Features

Adult male cardinals are brilliant red. They have a black mask around the eyes, bill, and upper throat. Male cardinals have a red bill and a red crest, which they can raise or lower at will.[3, 22, 29, 41, 73] During winter the wings of a male cardinal take on a slight grayish tinge.[22] There is some variation in the color of the feathers of adult male cardinals, from brilliant red to red to a reddish orange. This variation is due, in part, to the amount of red carotenoid pigments in the food eaten just prior to and during the annual fall molt.[171]

Adult female cardinals are the same size and shape as male cardinals. However, their feathers are a light gray or beige color, with a slight red tinge on the crest, wings, and tail.[3, 22, 29, 41] Adult female cardinals also have black eyes and a red bill.[3, 22, 29, 41] The face mask of a female cardinal is smaller and less pronounced, being gray in color. The tail of a female cardinal may be shorter than that of a male.[22, 41]

Cardinals have a conical or wedge-shaped bill, which makes them adept at cracking open seeds.[3, 41, 119] The upper mandible has sharp edges and is grooved to hold seeds in place.[41] The strong lower mandible has an angled cutting edge at the base to simplify removing seed husks.[41, 119]

Cardinals are 8 to 9 inches (20.32 to 22.86 cm) in length and have a wingspan of 10.25 to 11.0 inches (26.03 to 27.94 cm).[3, 22, 29, 41, 73, 83, 136] Cardinals weigh 1.4 to 1.6 ounces (39.2 to 44.8 gr).[18, 41, 83] Their resting heart rate is 375 beats per minute.[18] Cardinals have a high oxygen requirement because of their high metabolic rate.[137] As a result cardinals breathe 45 times per minute while resting; this rate increases during flight.[137]

The feet and legs of cardinals are a light beige fleshy color. Three toes face forward and one backward in an anisodactyl arrangement.[41]

Cardinals rely on biological thermoregulation to keep them warm during winter and cool in summer. In winter when temperatures drop, cardinals fluff up their feathers to retain heat; this increases the number of insulating air pockets.[137, 153] In summer when temperatures rise, cardinals follow the reverse process by holding their feathers flat against their body.[153] This way no excess body heat is retained.

The maximum life span of wild cardinals is 16 years.[18, 137] The maximum reported life span of cardinals living in captivity is 22 years.[18] While, the estimated average life span of cardinals living in the wild is believed to be 4 to 13 years.[18, 83] The wide range in the estimated average life span is due to the fact that the adult annual survival rate is only 50 to 75%.[153]

Temperament

Consistent with their brilliant plumage, cardinals are not shy birds. On the contrary, cardinals are considered to be bold.[29, 41] Cardinals have been described as being a "haughty aristocrat ... among the feathered tribes", and displaying a "courtly dignity" or "self-conscious superiority".[22] The bright coloring of cardinals makes it impossible for them to blend into the scenery, especially during winter. Their bright red feathers may in fact serve the exact opposite purpose. Red is considered to be a warning sign to stop or stay away in the world of non-tropical birds. That fact may account, in part, for the boldness of cardinals. (See Photograph 6.1.)

Vocalizations

Unlike other songbirds, cardinal vocalizations are heard year round and both sexes readily participate.[22, 29, 44, 83, 136, 137] Cardinals have between 8 to 12 different types of songs in their repertoire.[18, 137] These vocalizations occur in the 3,000 to 4,000 Hz range.[137] They have been described as sounding like a "clip, clip", "cheer, cheer", "purty, purty, purty, purty", "sweet, sweet, sweet, sweet", "cheo, cheo, cheo, cheo", and "strong rich whistle".[22, 29] Cardinals do not vocalize much when they are molting since they are vulnerable.[73]

Some geographical differences in vocalizations have been noted in cardinals. In particular, cardinals living in the isolated pockets in southern California and Arizona sound a bit different than those living in the eastern states and provinces.[153] That would imply that cardinal songs are learned, generation to generation, rather than inherited. Female cardinals tend to have a richer song and learn new songs more quickly than males; often they can match their partner's song exactly.[44]

Cardinal pairs have been noted for the antiphonal duets they sing during incubation and while raising their hatchlings.[5, 44] Rather than just making beautiful music together, the two are exchanging important information to minimize the number of times the location of their nest is revealed.[44] If the female cardinal matches the male's song, that is a message to maintain the status quo and not come near the nest.[44] On the other hand, if the female cardinal responds with a different phrase, that is a message to bring food to the nest.[44]

6.3 Family Life

Cardinals first form pair bonds when they are 1-year old.[153] Cardinals are monogamous and they form life-long pair bonds.[41, 153] During courtship male and female cardinals sing to each other and sway back and forth together.[95] The male cardinal feeds the female as a demonstration of his ability to supply her with food while she incubates their eggs.[41, 95, 137]

The female cardinal builds the nest, starting in late March or early April.[41, 73, 95, 153] It takes 3 to 9 days to build a nest.[41, 83] The nest is made from twigs, leaves, stems, weeds, dry grass, and bark strips.[29, 33, 41, 95, 153] The open cup nest is attached to a fork in a tree branch, 3 to 12 feet (0.91 to 3.65

Photograph 6.1 Not Exactly Shy. Cardinals are very visible and rarely make an attempt to blend into the scenery, other than while nesting. The temperament of cardinals matches their bright bold plumage. Cardinals' bright red feathers contribute to their boldness. The color red is frequently considered to be a warning sign to stop or stay away by non-tropical birds. © D. Herrmann

m) above ground.[3, 83, 95, 119] Cedar, yew, and holly trees are among the preferred locations to build a nest.[33] Cardinals conceal their nests in dense foliage, usually near a stream or other source of water.[29, 33, 41, 95] Cardinals do not reuse their nests.[153] In an effort to protect their nest and young, cardinals are somewhat territorial while nesting.[119, 153]

The female lays 3 to 4 eggs.[3, 29, 41, 83, 95, 153] The eggs are a dull white with brown spots and weigh 6.3 grams each.[3, 18, 22, 29, 33, 41, 73, 81, 83, 95, 153] The female incubates the eggs for 12 to 13 days.[41, 73, 81, 83, 95, 153] During this time the male feeds the female.[41, 95, 137]

The altricial young hatch at the same time.[153] Both parents feed the hatchlings.[33, 41, 44, 83, 95, 153] Cardinals have demonstrated a willingness to feed orphans of other bird species.[73] The mouth of a cardinal hatchling is lined in red;[41] this contrasts with many songbird hatchlings whose mouths are lined in yellow. The hatchlings fledge at 9 to 11 days.[41, 83, 95, 153] Fledglings continue to follow their parents around for a few weeks. (See Photograph 6.2.)

Some interesting observations were made in this regard during September 2013. Two cardinal fledglings began to appear at the bird feeder mid-month. This is rather late in the year for cardinals to hatch and fledge in the north. One fledgling was larger than the other and had a few red feathers; it was probably a male. This fledgling was accompanied by an adult male cardinal. The other fledgling was alone and quite small, probably not more than a few days out of the nest. It looked like the fledgling in Photograph 6.2. Whenever the small fledgling attempted to eat at the bird feeder the same time as the larger fledgling, the adult male cardinal chased it away, often quite aggressively. The adult male cardinal would then feed the larger fledgling, even though it was able to eat on its own. After

Photograph 6.2 Just like Dad. This fledgling cardinal takes it first flight down from the nest at 3:13 p.m. on 6/10/2011 near Pond B. Note the fledgling's tiny crest and black bill. The fledgling followed its father on short flights from rock to rock to the grass and back up onto a tree branch. The father moved 1-step ahead, looked back, and waited patiently for the fledgling to follow. Each time the fledgling landed almost exactly where the father had been. This entire sequence only took 42 seconds. Prior to this afternoon there was no indication that a cardinal nest was nearby. © D. Herrmann

a couple days the smaller solitary cardinal fledgling began hanging around with a flock of juvenile house sparrows (*Passer domesticus*). This way it could eat at the feeder without being harassed by the adult male cardinal. The house sparrows made no attempt to discourage or exclude the young cardinal. Since the fledgling retained its grey feathers, it is most likely a female cardinal.

This particular male cardinal took his parenting duties a little too seriously. On one occasion it chased an adult blue jay (*Cyanocitta cristata*) away and snatched a peanut the blue jay was about to take. The blue jay was more than twice the size of the adult male cardinal. To add insult to injury the adult male cardinal sat there and ate the peanut, feeding half of it to its fledgling. Two blue jays returned and swooped down to take other peanuts. The adult male cardinal charged both of the blue jays. But this time he let them leave with the peanuts. The adult male cardinal shared one more peanut with its fledgling then they both left.

Twelve to fifteen peanuts (in the shell) are placed near the bird feeder each morning year round between 8 and 10 a.m. This usually triggers a race between the cardinals and the blue jays (and any tree squirrels that are nearby). The cardinals, male and female, swoop down and take a peanut. Then the cardinals retreat to a tree and eat the peanut. In contrast, the blue jays zoom down, take a peanut, and store it in a tree branch or shrub. The blue jays repeat this action as fast as they can, while keeping an eye out for predators, until all the peanuts are gone. Some mornings the blue jays grab the peanuts so fast the cardinals don't get any. After all the peanuts have been retrieved, the blue jays take time to eat them.

On warm days, especially when no cardinals are present, the blue jays are more selective. A blue jay may sample—pick up—two or three peanuts before deciding which one to take. It appears that the blue jay is weighing the peanuts to determine which one contains the largest (or heaviest) nuts. After all, the outside shell does not always reflect the size or number of peanuts inside. Shells that look like they contain two nuts may only contain one nut. On the other hand sometimes they contain three nuts. The process of sampling the nuts implies that the blue jays understand this.

It is unclear why the small cardinal fledgling was not accompanied by an adult. During the preceding winter three pairs of cardinals were often seen at the feeder together. Also, it was unusual for the adult male cardinal to be hostile to the small fledgling, even if it was not his, since cardinals are known for feeding orphans.[73]

During December 2013/January 2014 the small cardinal fledgling continued to hang around with the house sparrows and benefitted from their knowledge of winter food sources and cooperative security measures. The cardinal fledgling also learned how to maneuver in tight formation at the ground feeding station and during sudden flight. Twenty or thirty house sparrows may descend on a ground feeding station at once and pack in close together. The flock is able to burst into sudden flight from this position without colliding. These are not behaviors cardinals engage in. Yet, by affiliating with the house sparrows the young cardinal was able to learn these skills.

Sadly the young male cardinal did not survive his first winter. He fledged mid-September. By mid-October that year winter began and average temperatures remained 30°F below normal for the next 4 months. The ground was completely covered by snow and ice for 30 or more consecutive days. Although the feeders were filled 3 or 4 times a day, this young bird did not make it. He was found frozen laying in the snow under his usual night roosting tree at 8 a.m. 2/12/2014 when the air temperature was ⁻14°F. He was 5 months old.

Within ten days two mourning doves (*Zenaidura macroura*), eight house sparrows, and two American robins (*Turdus migratorius*) were also found frozen in the same area. The robins returned six weeks earlier in 2014 because of unusual snow, ice, and cold in the southeast U.S. due to the "polar vortex". Less than half of them survived until nesting time. No birds were found frozen the previous winters when temperatures were within the normal range.

The problem with sudden or prolonged unusually cold weather is that there comes a point where birds cannot consume enough calories to prevent themselves from freezing to death—assuming that food, water, and gravel for maceration are even available under these conditions. Climate change is particularly deadly for birds that do not migrate.

Young cardinals start to sing when they are 3 weeks old; however, it takes about 2 months of practice before fledglings become proficient at vocalizing.[41] Prior to that their main vocalization is that of hatchlings making a begging call.

Fledgling cardinals resemble females, except that their bill is blackish.[3, 153] Young cardinals retain their juvenile plumage until the end of their first annual fall molt; at that time they acquire adult plumage.[41] About a month later the bill of a young cardinal changes to the red adult color.[41] It is interesting to note that female hatchling zebra finches (Chapter 8) also have blackish bills.

Cardinals living in the north have 1 to 2 broods a year.[22, 41, 95, 153] Cardinals living in the south may have up to 4 broods a year, depending on the weather and the availability of food.[41, 136]

6.4 Individual Behaviors

Now let's examine behaviors which distinguished individual cardinals, such as:

- Exploring sliding glass doors
- Exploring a metal slide
- Arriving at the feeder first
- Using tail feathers as a tool

The behaviors are presented as they occurred. The analysis, comparison, and interpretation of these behaviors are reserved for Section 6.5 Discussion. These observations took place at the locations and during the timeframes described in Annex C.

Exploring Sliding Glass Doors

In the fall of 1984 a young male cardinal visited the balcony facing the nature preserve long after the ruby-throated hummingbirds (Chapter 5) were gone and the plants were back inside. The male cardinal was first spotted on the balcony one morning shortly after sunrise. The cardinal had just landed on the railing. He sat there for a while, then hopped down to the center of the balcony. The cardinal walked cautiously toward the sliding glass doors, as if something had piqued his curiosity. The lights were on. He may have seen the plants that were behind the glass doors, or he may have seen his own reflection or that of the trees behind him. Something piqued the cardinal's curiosity. And, he did not seem alarmed at all that he was being watched.

What prompted this young cardinal to leave the bird food and cluster of other birds on that particular fall day and venture forth onto the 2nd floor balcony? Did he know that this was where the bird food came from? Perhaps he was just the inquisitive type and wanted to explore why the plants inside the sliding glass doors were still green and growing, while the plants outside, where he was, had lost their leaves and were dormant. Bees frequently fly outside windows that contain flowering plants. But that behavior occurs during warm weather when the bees are looking for food. Cardinals do not eat plants and it was fall.

Something obviously intrigued this cardinal, since he returned several days in a row. Some safflower seed was put on the balcony to encourage his visits. This particular cardinal was brilliant red and young. This was probably his first winter. In addition to his curiosity, the cardinal had another unique feature. One of his tail feathers was always crooked. Because of this he was named Mr. Rumpletail. It was unclear whether the feather grew in crooked or he groomed it that way as a matter of individual style.

It did not take long for Mr. Rumpletail to understand that glass was a translucent barrier. Mr. Rumpletail would walk right up to the glass, tap the glass with his beak, and tilt his head as if he were contemplating something. When a bird tilts its head people often interpret this as a sign of curiosity or contemplation and in fact it may be. However, birds' eyes do not move as freely as the eyes of humans. Hence, birds need to tilt their heads on occasion to get a closer view of something of interest.

Mr. Rumpletail never attempted to fly into or through the sliding glass doors no matter what season it was, what time of day it was, what angle the sun was at, or how bright or dark the day was. This behavior was rather intelligent on Mr. Rumpletail's part, especially since he was less than a year old when he first visited the balcony. Many birds, indoor pet birds and outdoor wild birds, never figure out glass, often with fatal consequences. Perhaps because birds can see ultraviolet light, glass looks different to birds than it does to humans.

Mr. Rumpletail continued to visit the balcony throughout the winter and in subsequent years, whether or not there were any safflower seeds. Food was not the objective; exploring was. A larger quantity and greater variety of bird food were available out back. Mr. Rumpletail would dine with the other birds then stop by for a short visit of about 5 to 10 minutes. Mr. Rumpletail was the only cardinal and only bird that explored the sliding glass doors on the balcony from 1984 to 1994.

A variation of Mr. Rumpletail's behavior was observed briefly in 2012 and again in 2013. However, these behaviors were not play activities.

During 2012 on a cold windy late March day a male cardinal sat on a brick window ledge looking at the flowering plants inside. The cardinal tried to fly into the room several times. The cardinal even flew around the perimeter of the window while tapping on the screen. The cardinal returned two consecutive days. On the third day the cardinal was accompanied by a female cardinal. They both tried to enter the room with the flowering plants. The cardinals may have been seeking food, or they may have decided that location was warmer since the flowers were blooming.

During August 2013 on an extremely hot dry day a male cardinal sat on the brick window ledge again and tried to enter the room with flowering plants. This cardinal made several vocalizations. It is likely that he was seeking food and water due to the intense heat and dry conditions. The cardinal would have associated a blooming plant with adequate water.

Exploring a Metal Slide

During Spring and Summer of 1985 a wooden jungle gym near the edge of the woods captured Mr. Rumpletail's attention. The jungle gym consisted of chin-up bars, a balance beam, a metal slide, and an assortment of climbable and movable parts. The jungle gym was surrounded by a mixture of wood chips and mulch.

Mr. Rumpletail zeroed in on the metal slide for an in-depth examination, usually mid-morning. Mr. Rumpletail took short flights around the slide, as if he were examining the slide from all different angles. Next Mr. Rumpletail landed on top of the slide. Then he landed on the bottom of the slide. Lastly, Mr. Rumpletail walked around on the ground next to and under the slide, and into and out of the slide's shadow. This pattern of behavior lasted 30 to 40 minutes and continued for a couple weeks. The top of the slide was about 4 feet (1.21 m) off the ground; a height at which cardinals are comfortable perching. The slide was the only part of the jungle gym that was all metal; however, being galvanized steel, it was not shiny. Why was Mr. Rumpletail curious about the slide? From a human perspective at least, it seemed like the balance beam or chin-up bars would have been a more natural place for Mr. Rumpletail to perch and play.

One morning a slight pinging sound was heard coming from outside: 3 short pings and a pause, 3 short pings and a pause. An attempt was made to determine the source of the pinging sound. Surprisingly it was Mr. Rumpletail. He had positioned himself directly underneath the slide, half way between the top and the bottom, facing outward toward the bottom of the slide. Mr. Rumpletail flew straight up about 2.5 feet (0.76 m) until his crest touched the underside of the slide. Then he pinged the slide with his beak once, and coasted back to the ground. Mr. Rumpletail repeated this process three times in a row. Then he rested for a few seconds, before starting all over again. The pings were probably no more than 5 or 6 seconds a part—the time it took for Mr. Rumpletail to fly up to and down from underneath the slide.

Mr. Rumpletail had discovered another manmade limitation. The slide, which was about 16 inches (40.64 cm) wide, represented a manmade ceiling that blocked sunlight and rain, and through which he could not fly. These facts seemed to intrigue Mr. Rumpletail. He would repeat his pinging routine several times. Next Mr. Rumpletail would position himself beside the slide, about a foot over from where he had been. From there he would fly upward unimpeded 2 or 3 feet (0.60 or 0.91 m) above the slide. He looked down at the slide and coasted back to the ground. Mr. Rumpletail ascertained that here he could fly as high as he wanted to; in contrast, when he was under the slide he could not. That was an impressive display of reasoning by Mr. Rumpletail. Having constraints on upward flight is not something birds generally worry about, since the sky does not have a ceiling. Birds may fly amidst trees and dense foliage, but there is always a way up and out.

Mr. Rumpletail continued to ping the slide most every morning. Usually between 10 a.m. and 11 a.m. he would stop by and perform his pinging routine for 20 to 30 minutes, before continuing on with his day. If something startled him, like a lawnmower or children running through the area, he would interrupt his routine and retreat to the nearby trees. Once the disturbance was gone, he would resume his routine. This daily pattern of behavior continued throughout the warm weather and into fall. One day Mr. Rumpletail even sat under the slide (dry) during a brief rain shower.

Mr. Rumpletail continued to explore and play with the slide, performing his pinging routine year after year throughout the 10-years. Toward the end he must have been pretty elderly for a cardinal. After Mr. Rumpletail discovered the slide's manmade flight ceiling, he visited the sliding glass doors less often. No one other than Mr. Rumpletail was observed pinging the slide or tapping the sliding glass doors. Even though other birds saw Mr. Rumpletail doing this on a regular basis, none of them attempted to engage in these two activities. Mr. Rumpletail was the first cardinal to live in that location after that building was built. He discovered the translucent barrier of the sliding glass doors and the slide's manmade flight ceiling.

Arriving at the Feeder First

When Spring 1985 arrived Mr. Rumpletail sang his unique well-rehearsed song from the treetops, as male cardinals do. Soon he was joined by a young female cardinal and they became a bonded pair. She was named Lady Cardinal. Mr. Rumpletail and Lady Cardinal were frequently seen flying around and dining together. Lady Cardinal waited in the trees nearby whenever Mr. Rumpletail visited the metal slide or the sliding glass doors on the balcony. In addition to being curious, Mr. Rumpletail always seemed to be calm and content. He was never timid, nervous, or fearful, even in the thick of winter.

Mr. Rumpletail and Lady Cardinal were not observed building a nest or raising their young, consistent with the reputation of cardinals for well concealed nests. Their nesting area and parenting activities were deeper in the woods and private. By the second year, however, two pairs of cardinals made this area their home. The two pairs of cardinals stayed close to each other throughout the day. When it was time to eat, only two cardinals visited the bird food at a time. Sometimes one pair would eat at a time. On other occasions the two female cardinals would eat while the two male cardinals waited nearby in the trees, on the lookout for predators. Afterward, the four cardinals traded places.

Mr. Rumpletail was never shy about bird food. He was usually the first bird to arrive after the food was replenished. Mr. Rumpletail was not intimidated by other (non-cardinal) birds who were eating or squirrels. He simply flew up to the bird food, landed where he wished to, and ate what he wanted to. He made no attempt to chase the other birds away and minded his own business. If, however, the other birds crowded Lady Cardinal, Mr. Rumpletail would step in.

Mr. Rumpletail and his extended family, as well as other birds in the area, always ate twice as much food the day before a major snowstorm. During the storm the cardinals stayed hidden and warm. Afterward, the cardinals resumed their normal routine. On more than one occasion the cardinals' behavior gave a more accurate prediction of meteorological patterns than the local weather forecast.

This behavior was observed consistently from year to year and at different locations many years later. Whenever fresh bird food was put out, a cardinal or two were the first birds to arrive. Again, the cardinals were neither intimidated by nor aggressive toward the other birds or squirrels present. (See Photograph 6.3.)

Because cardinals do not migrate it was possible to observe Mr. Rumpletail and his family year round from 1985 to 1994. After a few more years, three pairs of cardinals called this area home. Most likely they were three generations of Mr. Rumpletail's extended family. It was a treasure

Photograph 6.3 Bold is Beautiful. Cardinals are usually the first birds to arrive after the food at the bird feeder has been replenished. Cardinals are neither intimidated by nor aggressive toward other birds that arrive after them. (Top) Most birds fly away quickly when a squirrel or chipmunk arrives; not the cardinals. (Bottom) A cardinal dines alongside a red-bellied woodpecker. © D. Herrmann

to observe all six of the cardinals settled in the trees or dining together. Mr. Rumpletail was easy to identify because of his unique tail style. The second male had all the markings and coloring normally associated with a cardinal. The third male cardinal had normal markings, but his feathers had more of an orange-red tinge, than a true red. Hence, it was easy to identify who was who when observing the male cardinals. Telling the female cardinals apart was another matter. Lady Cardinal was slightly larger than the other two female cardinals. However, the two young female cardinals were indistinguishable, at least to humans.

Lady Cardinal did not make it through the last winter; after that Mr. Rumpletail was alone. Mr. Rumpletail hung out with the other four cardinals and played his pinging routine while they were in the trees nearby.

Using Tail Feathers as a Tool

Another observation of individual behavior by cardinals was made by Vornberger.[172] One fall while Vornberger was photographing birds in Central Park a swarm of newly hatched termites were crawling all over a small boulder. Like other songbirds, the beak of a cardinal is not designed for snatching moving insects off a hard surface.[172] So the cardinals present devised a creative solution for partaking of the suddenly plentiful insects. The cardinals used their tail feathers to guide the termites up onto their wings.[172] From there, the cardinals could easily snatch the termites.[172] The cardinals were the only birds present to use their feathers as a tool to catch the termites.[172] This behavior and posture are a variation of that used by some perching birds *Passeriformes* that is referred to as "anting", whereby birds encourage certain species of ants to climb up their feathers in order to remove parasites from them.[158]

6.5 Discussion

Now let's see what these behaviors tell us about cardinal intelligence, behavior, and individuality. In particular, let's examine:

- Curiosity
- Play activities
- Confidence
- Creativity
- Ideas and inventions appropriated by humans

Curiosity

Curiosity and an inquiring mind are considered to be signs of intelligence. Asking basic questions about the world around oneself, such as: "What is this?", "How does this work?", and "Why did that happen?" is how learning takes place and knowledge is acquired. Mr. Rumpletail definitely had an inquiring mind and a high degree of curiosity. He also had a ready ability to distinguish between the natural world and manmade objects. Mr. Rumpletail explored the limitations and constraints posed by manmade objects, particularly limitations and constraints that did not exist in the natural world. Glass, a translucent barrier that Mr. Rumpletail could see through but not pass through, was one such limitation. A metal slide was another.

Mr. Rumpletail did not stumble onto the manmade metal slide by accident or play with it for a day or two and forget about it. Quite the opposite. Mr. Rumpletail studied the slide: top, bottom, and sides, the ground around the slide, and the changing patterns of sunlight and shadows for a couple weeks. After which Mr. Rumpletail discovered that the slide represented an upward flight ceiling. Mr. Rumpletail learned that he could touch the underside of the slide with his bill, but could not fly through it, any more than he could fly through the sliding glass doors. This 16 inch (40.64 cm) wide strip of curved metal was in essence a manmade flight ceiling. Mr. Rumpletail knew that this manmade flight ceiling only existed at this precise location. Right next to the slide or elsewhere, he could fly upward as high as he wanted. Mr. Rumpletail did not make this discovery, declare it interesting, and forget about it. On the contrary, Mr. Rumpletail not only performed his pinging routine throughout that first summer and fall, but he resumed playing it every year thereafter for 10-years. That implied memory as well as knowledge. Mr. Rumpletail remembered what he had learned about the slide and its manmade flight ceiling from year to year. And, he may have been checking the slide and manmade flight ceiling year after year to see if anything had changed.

Play Activities

The act of playing or the ability to play is considered a sign of intelligence in any species.[79] Play constitutes a break from day to day tasks associated with survival. Play represents time spent doing something simply for fun or to engage in a new experience. Play is active and requires creativity, resourcefulness, rapid assimilation of new information, and decision-making. The act of playing enhances intellectual, physical, and social skills.

Mr. Rumpletail took time out of each day, except the harshest days of winter, to play and explore. This is evidence of intelligence and emotions; or put more simply, the ability to engage in an interesting and fun activity that had nothing to do with survival. This behavior raises questions about:

- Why did Mr. Rumpletail engage in exploring?
- What was the object of his explorations?

Mr. Rumpletail was not seeking gold he could carry across the ocean to a queen or king. Mr. Rumpletail was not seeking a discovery through which he could become famous instantly. Nor was he seeking wealth from which to launch a multi-national company. No, Mr. Rumpletail was merely curious and wanted to learn more about the world around him; in this case the manmade objects in the world around him. Isn't that the true sign of intelligence and the hallmark of a real explorer or scientist?

Confidence

The cardinals demonstrated behavior that could be considered bold, consistent with their reputation, throughout the 10-years. On closer examination, this behavior wasn't really bold in the sense that the cardinals were daring or in common parlance trying to make a statement. Rather, the cardinals' behavior evinced a quiet confidence or assurance.

For example, the cardinals were consistently the first to arrive at the bird feeder after it was refilled. The cardinals were neither afraid of nor aggressive toward any other birds or squirrels that came later. The cardinals made no attempt to chase the other birds away or hoard the food. Cardinals are not known to cache. The cardinals simply ate what they wanted or needed, then left.

Cardinals also displayed confidence in their interactions with humans. The cardinals would allow humans to come within 6 feet (1.82 m) of them. That is much closer than most wild songbirds will allow. On one occasion a male cardinal went through an entire preening routine, only looking up occasionally, when he was only 6 feet (1.82 m) away. That behavior implies confidence and self-reliance that the cardinal could, from that distance, deal with whatever happened. (See Photograph 6.4.)

Cardinals consider 6 feet (1.82 m) of separation to be a safe distance. Less than 6 feet (1.82 m) is not considered a safe distance and will cause a cardinal to depart suddenly. (See Photograph 6.5.)

On a few occasions cardinals were observed while a falcon was circling directly overhead. The other birds in the area immediately darted away to a safe location. In contrast, the cardinals stayed where they were out in the open—but monitored the falcon's whereabouts.

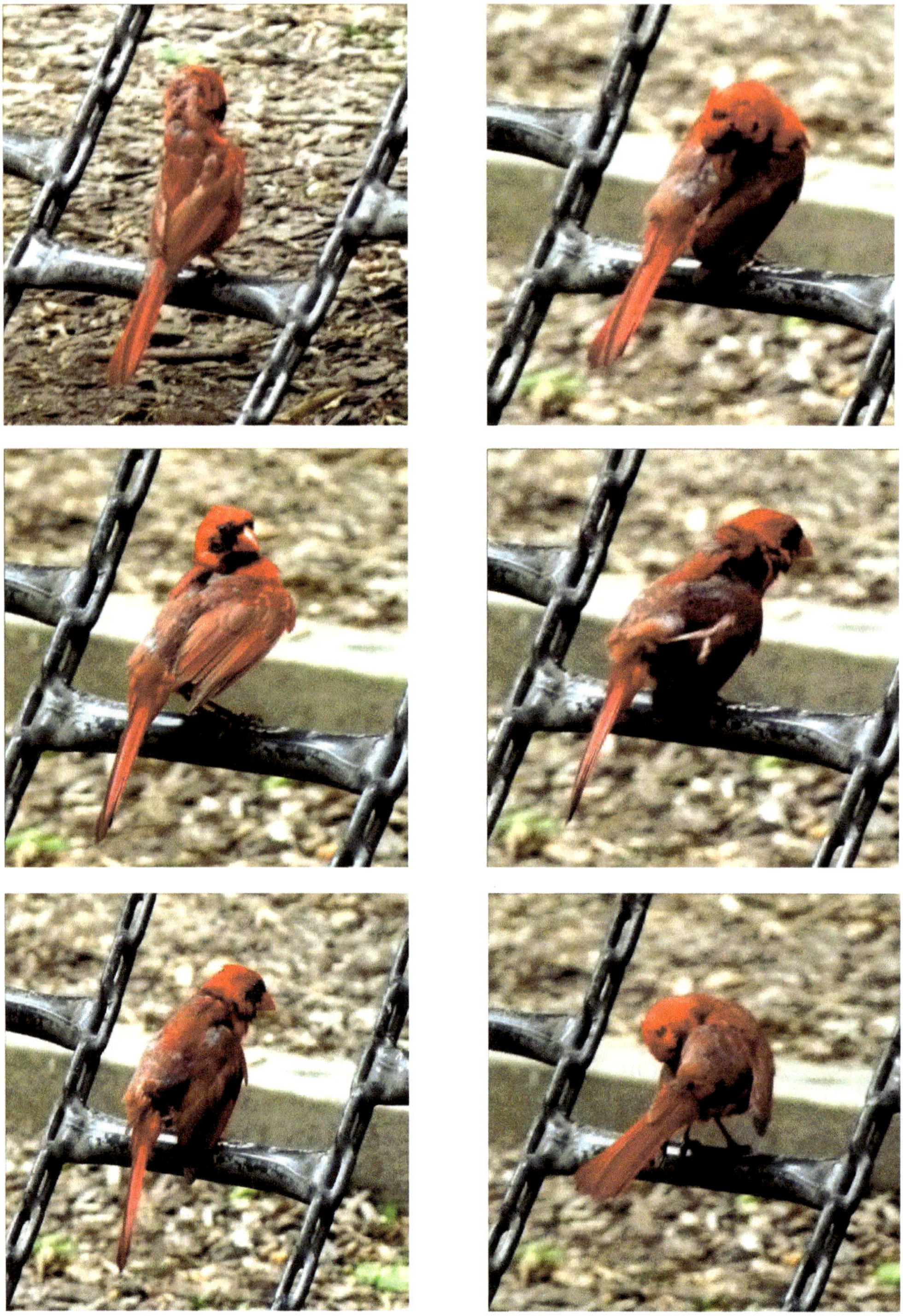

Photograph 6.4 A Quick Touch-Up. A male cardinal stops to preen on 6/22/2011 near Pond B. The cardinal starts with the usual stretching and preening movements. In the fourth photograph the cardinal engaged in "indirect scratching" (placing the leg over the shoulder of the wing). While in this position, the cardinal could not fly. Indirect scratching is less common than direct scratching (placing the leg under the wing, for example to scratch the head). By the fifth photograph the cardinal moves his leg back under his wing. This entire preening sequence only lasted 40 seconds. © D. Herrmann

Photograph 6.5 A Safe Distance. Most cardinals in this study considered 6 feet (1.82 m) to be the minimum safe distance between them and humans, regardless of whether they were on the ground or up in a tree. Whenever a human approached closer than that, even slowly, the cardinal took off in a hurry. © D. Herrmann

Creativity

Cardinals as a species and as individuals have demonstrated creative behavior, especially in response to novel situations.

Most cardinals spend their entire lives within a few miles (kilometers) of where they hatched. Yet over a century or so cardinals were able to expand their range northward from Virginia to Canada. This behavior implies an ongoing creative assessment of new information. For example, cardinals recognized the:

- Changing weather patterns, such as the milder winters north of their normal range
- Availability of new food sources during winter, such as berries on ornamental shrubs and seed offered at suburban bird feeders

At the same time cardinals refined their techniques for staying warm during winters that were much harsher than traditionally experienced in the southeast. Cardinals did not expand their range from Virginia to Canada in one generation. Instead, cardinals expanded their range northward gradually, while constantly acquiring, processing, and storing new information about habitats, predators, climate, and the availability of food and water. In summary, the continual habitat reassessments and creative resourcefulness made it possible for cardinals to successfully double the size of their range in a relatively short period of time. Similarly, the cardinals living in Arizona and California demonstrated resourcefulness in finding an oasis where they could live while the land for hundreds of miles around became a desert.

Cardinals use their vocalizations in creative ways which we are only beginning to understand. Cardinal vocalizations are not limited to spring time and courtship. Instead cardinal vocalizations are heard year round and both sexes readily participate.[22, 29, 44, 136, 137] Cardinals have between 8 to 12 different types of songs in their repertoire.[18, 137] Since cardinals form life-long pair bonds, it stands to reason that they are exchanging a lot more information than just courtship messages. The antiphonal duets sung by cardinal pairs during incubation and later while raising their hatchlings are particularly noteworthy.[5,44] Subtle messages are embedded within the beautiful music about when and when not to bring food to the nest.[44] These duet exchanges are somewhat unique to each pair, suggesting a small private vocabulary.

Furthermore, some geographical differences in vocalizations have been noted in cardinals. In particular the vocalizations of cardinals living in isolated pockets in southern California and Arizona sound a bit different than those living in the eastern half of North America.[153] Eastern and western cardinals have not been separated from each other long enough for them to develop as separate species. However, eastern and western cardinals have developed some different vocalizations. This implies that cardinals' songs are learned, generation to generation, rather than inherited. It is also possible that a different vocabulary is needed to describe the arid habitat of the southwest, than the lush suburban gardens and parks in the east.

Vornberger's observation of cardinals dining on termites reflects real-time creative problem solving. As Vornberger points out, it is rare for termites to hatch in the fall.[172] It is also rare for cardinals to use their tail feathers to guide insects up onto their wings in order to eat them. A few cardinals in a specific location were presented with a new situation and challenge. Instead of letting the opportunity to dine on an abundance of young termites pass them by, the cardinals improvised by using a variation of "anting" behavior. Most likely this small group of cardinals will remember that this new behavior was successful, should they encounter a similar situation in the future.

The presence of a curious inquiring mind, the ability to regularly engage in play activities, the confidence to do and go where they want when they want to, and the creativity to double the size of their range in little over a century, develop specialized communications, and respond successfully to novel opportunities, clearly indicates that cardinals are intelligent individuals.

Ideas and Inventions Appropriated by Humans

To complete our assessment of cardinal intelligence we must examine the ideas and inventions humans appropriated from cardinals. The combination of a cardinal's bright red plumage and their bold confident behavior inspired several common everyday practices. The bright red feathers alone did not do the trick. After all, there are several bright red flowers.

The combination of a cardinal's bright red plumage and their bold confident behavior captures one's attention and respect. This dynamic combination inspired the development of red stop signs and traffic lights. People who want to stand out in a crowd wear red clothing, especially royalty. When teachers grade students papers and editors edit copy they use red ink because it draws the reader's attention. In a similar vein, advertisements frequently feature big bold red lettering. This was particularly true when newspapers were only printed in black and white, with occasional red lettering in an advertisement. On the other extreme overdue bills are often stamped in big bold red letters to intimidate the recipient. (See Table 6.1.)

And what ideas and inventions have cardinals deemed worthy enough to appropriate from humans? None.

Table 6.1 Ideas and Inventions Humans Appropriated from Cardinals.

Ideas and Inventions	Observations that Led to the Ideas and Inventions
• Red stop signs • Wearing red clothing to standout in a crowd or indicate royalty • Using red ink to grade or edit papers • Big red lettering in advertisements and overdue bills	• The bright red plumage of male cardinals and their bold confident behavior

7 Budgerigar

Melopsittacus undulatus

7.1 Background

Budgerigars are native to Australia, in particular the dry interior known as the Outback. Budgerigars inhabit arid and semi-arid regions like dry woodlands, savanna, grasslands, open plains, and scrub lands.[36, 60, 94, 137] Flocks are frequently seen in spinifex, dry mallee, and mulga scrub, acacia trees, and eucalyptus shrubs.[41, 203] Budgerigars are believed to have lived in this area as a distinct species for 5 million years.[203] The Aborigines, who arrived in Australia about 40,000 years ago, referred to these birds as betchegara or betcherrygah.[36, 203] Budgerigars are also referred to as Shell Parakeets, Shell Parrots, Undulated Grass Parakeets, Warbling Grass Parakeets, Canary Parrots, Zebra Parrots, and Budgerygahs.[34, 36, 81, 112, 203]

Florida is the only other place on the planet to have budgerigars living in the wild. A flock of escaped birds was able to establish itself there in the 1950's and has been reported on since then. This is a bit unusual, given the difference in climate between the dry interior of Australia and the warm moist climate of Florida. There is speculation that these budgerigars rely at least in part on food from bird feeders. The Florida budgerigar flock reached a peak of 20,000 individuals between the mid 1960's and mid 1970's.[147, 204] Since then the range and population of the Florida budgerigars have declined dramatically.[147] Today there are only 25 to 99 bonded pairs in two counties.[147, 204]

The primary predators that concern budgerigars are hawks, other raptors, and tree snakes.[123, 124] Two key defenses have been developed to foil these predators. First, budgerigars are very sensitive to sound and movement.[123, 124] Hence, they can often hear predators coming and take evasive action. The hearing range of budgerigars is impressive: a 40 Hz lower limit to a 14,000 Hz upper limit.[18] The lower limit is lower than some owls. Second, when flying budgerigars form large flocks and twist and turn with precision.[41, 137] This action confuses predators and prevents them from targeting a specific bird.[41, 137] Severe droughts are a major environmental challenge for budgerigars because of the loss of water and food supplies.[36, 41, 59] Thousands have been known to die near dry water sources.[36]

Budgerigars were first described in European literature in 1805 by George Shaw.[84] John Gould, a naturalist, brought the first budgerigars to Britain in 1840.[36, 70, 84, 117, 167, 178, 202, 203] Gould is also credited with giving budgerigars their scientific name, *Melopsittacus undulatus*.[203] His brother-in-law, Charles Coxen, raised the first clutch of budgerigars in captivity.[123, 124] The first young budgerigars sold as pets fetched a price of 27 pound sterling each.[139] By 1850 the Antwerp Zoo had a display of budgerigars.[84, 139] From then on the popularity of these birds spread throughout Europe and the rest of the world. In fact a book published in 1909 refers to budgerigars as the "commonest and most freely imported bird in Europe".[34] Fortunately, in 1894 Australia had the foresight to ban any further exports, so that budgerigars could continue to live free in the wild in safe numbers.[115] It took another 100 years for other countries to adopt this practice to protect their native parrot populations. Some countries have yet to do so.

All wild budgerigars are green with yellow heads. This color pattern is often referred to as a normal green or nominate green budgerigar. The other colors of budgerigars seen in pet stores are the result of genetic manipulation by breeders. In fact, after 170 years of captive breeding 70 different color mutations are officially recognized by budgerigar show societies around the world.

Today budgerigars are undisputedly the most popular pet bird in the world and have been for decades. It is estimated that there are 4 million pet budgerigars in the U.K. alone.[143, 167] All the sources consulted used terms like intelligent, resourceful, social, active, and energetic to describe budgerigars, regardless of whether they were describing wild or pet birds. Undoubtedly these traits have contributed to the popularity of budgerigars. Another factor is the ability to learn and repeat human language words and phrases (not necessarily in context). Two rather talkative budgerigars made it into the Guinness Book of World Records. In 1958 a budgerigar named Sparky Williams was entered into the Guinness Book of World Records for being able to say 531 words and 383 sentences.[70] In 1994 a budgerigar named Puck entered the record books for being able to say 1,728 words.[203]

Table 7.1 lists the major milestones in the history of budgerigars (from a human perspective).

Table 7.1 Milestones in the History of Budgerigars (compiled from multiple sources).

Approximate Date	Significant Event
5 million years ago	• Budgerigars began living in the Australian Outback as a distinct species
40,000 years ago	• Aborigines arrived in Australia; referred to these birds as betchegara
1805	• George Shaw first person to describe budgerigars in European literature
1840	• Naturalist John Gould brings first budgerigars to Great Britain
1840's	• Charles Coxen, brother-in-law of John Gould, raises first clutch of budgerigars in captivity • First budgerigars sold as pets for price of 27 pound sterling each
1850	• Antwerp Zoo one of first to display budgerigars
1872	• First budgerigars with yellow mutation hatched in Belgium
1881	• First budgerigars with blue mutation hatched in Holland
1880's–1900	• Budgerigars become very popular throughout Europe
1894	• Australia bans exports of budgerigars
1920's	• Budgerigars introduced as pets in the U.S. and Japan; became popular very quickly
1950's	• Flock of escaped budgerigars begins living in Florida; the first and only flock to establish itself in the wild outside of Australia
1958	• Budgerigar named Sparky Williams entered into Guinness Book of World Records for being able to say 531 words and 383 sentences
mid 1960's–mid 1970's	• Population of wild budgerigars in Florida about 20,000 individuals living along the Gulf coast from Pasco to Sarasota counties
1994	• Budgerigar named Puck entered into Guinness Book of World Records for being able to say 1,728 words
today	• Population of wild budgerigars living in Australia about 5 million • Population of wild budgerigars living in Florida 25 to 99 pairs • Estimated 4 million pet budgerigars in the U.K. • 70 different color mutations recognized by budgerigar show societies around the world

7.2 Characteristics

Now let's look at the characteristics of wild budgerigars, their behavior, and the activities that comprise their daily lives. In particular let's examine their:

- Physical characteristics
- Diet
- Socialization and daily activities
- Family life
- Vocalizations

Physical Characteristics

Budgerigars are small, thin, agile birds measuring 7 to 7.5 inches (17.7 to 19 cm) from head to tail.[84, 117, 203] Their tail accounts for nearly half their length; that is the distinguishing feature of a parakeet. The word parakeet is a general term and does not refer to a specific type of bird. Instead, the word parakeet refers to any slender long-tailed member of the parrot family. Budgerigars weigh 0.8 to 1.1 ounce (22.4 to 30.8 gr), have an 8 inch (20.32 cm) wingspan, and are strong flyers.[203] A budgerigar's legs and feet are pink with small outer scales that become darker and take on a slight blue tinge as the bird ages.[60, 132] A budgerigar's legs are 0.5 inch (1.27 cm) long. They have zygodactyl feet with two toes facing forward and two toes facing backward. The outer toes are 0.5 inch (1.27 cm) long, while the inner toes are 0.375 inch (0.95 cm) long. There are small nails on the end of each toe, which grow continuously.

Wild budgerigars come in one color pattern that allows them to blend into their natural habitat for protection. Budgerigars have a bright yellow face and forehead. Their body is a bright medium green. Budgerigars have small oval shaped purple patches below each ear. These patches fluoresce under ultra violet (UV) light, specifically near UV wave lengths of 300 to 400 nm.[133] Four small round black dots run across the top of a budgerigar's throat, with two dots on each side of their beak. Their tail feather is a dark cobalt blue. The outer tail feathers and wing feathers have yellow fringes. The top and back of an adult budgerigar's head is striped with thin black bars. These stripes continue all the way to the cere on young budgerigars until they are 4 months old.

Budgerigars have a small hooked dull yellow or beige beak, about 0.5 inch (1.27 cm) long. The cere is royal blue for males and light beige to medium brown for females. The cere of a female budgerigar changes color throughout the year. When nesting the cere is brown; the rest of the time it is light beige.[34, 81] The cere of juveniles is light pink until they are six months old.

Budgerigars have round black eyes, 0.125 inch (0.317 cm) in diameter, that are surrounded by a dull white or light yellow outer rim. Young budgerigars have solid black eyes until they are 10 to 12 weeks old.[132] Budgerigars have tetrachromatic vision which means that they have four classes of cone cells and can see all colors, unlike humans.[203] However, like most birds they need full spectrum lighting including UVA and UVB to see everything clearly.

The ear structure of budgerigars is more complex than that of mammals, which accounts for their extended hearing range.[113] Budgerigars have a distinct advantage over humans: hair cells within their ears regenerate if they are damaged, thereby preventing hearing loss and deafness.[113] In addition, unlike most vertebrates budgerigars are not temporarily deafened by extremely loud noises; through biological intelligence their ears are protected from acoustical trauma.[74]

Another feature of the physiology of budgerigars is noteworthy as well. Budgerigars, unlike other parrots, have large salt glands. The salt glands are located between their skull and eyes. This allows budgerigars to expel excess salt from their bodies through their nasal cavities as droplets.[137] This is a handy feature indeed. Budgerigars live in the arid Outback, which is prone to occasional droughts.

If budgerigars cannot find any fresh water sources, they can temporarily drink from brackish inland pools since their salt glands will expel the excess salt.

Diet

Wild budgerigars are granivores and ground feeders. Their diet consists primarily of seeds, in particular ripe seeds found on or near the ground, like seeds from grasses and ground cover.[34, 117, 137] Twenty-one different types of seeds have been found in the crops of wild budgerigars, along with fine grains of gravel.[59] Favorites include the seeds of spinifex, Mitchell grass, porcupine grass, saltbush, and occasionally ripening wheat.[41, 94] This spinifex is of the genus *Triodia* and refers to a group of grasses with spiny seeds and sharp leaves, not the spinifex found in sandbars.[120, 179, 202] These grasses, like budgerigars, had to develop a degree of drought resistance to survive in the arid interior regions.[179] Budgerigars also consume dirt and possibly tree bark to supplement the minerals in their diet.[7]

A study conducted in Australia during the mid-1970's confirmed that budgerigars only ate the seeds of ground vegetation year around, in spite of seasonal movements, changing weather patterns, and varying amounts of daylight.[183] This study reported that budgerigars ate seeds that varied in length from .002–0.1 inch (0.5–2.5 mm) and in weight from 0.3–1.3 mg.[183] Bonded pairs, individuals, and juveniles within a flock ate the same diet.[183]

Socialization and Daily Activities

The average life span of wild budgerigars is 7 to 8 years, with a few individuals making it to 10 or more years.[84] Wild budgerigars lead an active nomadic life. Their daily movement from place to place is tied to the availability of food and water, while their flight is described as being straight and rapid.[34] Flock membership is not fixed and fluctuates over time. There is no hierarchy or structure to flocks, other than the fact that bonded pairs stay together. Flock size varies widely. Flocks of 10 to 100 budgerigars are common; however, flocks of up to 1,000 or more may congregate.[59, 60] In particular, budgerigars gather in large flocks for protection against predators when drinking water.[41] Because of their nomadic lifestyle, budgerigars do not exhibit any territorial behaviors. Likewise, due to the dynamic nature of their flocks, budgerigars are very social. This fact has positive ramifications for pet budgerigars, in that they are less likely to develop the negative behaviors often associated with larger parrots.[113] A few sources implied that there are differences in behavior and sociability between males and females in captivity. However, no such differences have been noted in the wild; nor were any differences noted among the 28 budgerigars discussed in this chapter.

Like many land birds budgerigars are most active early in the morning and late in the afternoon.[60] Individual birds awake 10 to 20 minutes before sunrise. They begin calling to each other and engage in preening and stretching activities. Some birds jump back and forth between adjacent branches.[59] It was noted that budgerigars discussed in this chapter woke up each morning at the same time in relation to sunrise year round, regardless of whether it was sunny, cloudy, or overcast. Other than the zebra finches (*Poephila guttata castanotis*) in Chapter 8 who are also native to Australia, no other indoor birds discussed in this book exhibited this trait.

Sunrise signals that it is time for breakfast and the flock leaves their overnight community roost. They take off flying toward various areas that should offer food this time of the year. Energetic flying is usually accompanied by equally energetic chatter. By early morning the flock arrives at their intended destination. They may be joined by other flocks, as a large number of budgerigars forage together for their first meal of the day.

By mid-morning it is time for some water. So, the flock flies to the nearest water source; this may be miles (kilometers) away from the feeding area.[59] Multiple flocks may congregate near the water source at the same time.

Midday it is time to rest and digest the morning meal. Budgerigars seek a high safe place for the flock to roost.[123, 124] Some budgerigars chatter softly while they are resting.[123, 124] Budgerigars are known to sit among the leaves of gum trees for shade during the midday heat.[34]

Mid-afternoon the flock enjoys another water break. Late afternoon it is time for the last meal of grass seeds. This meal must provide enough energy to last throughout the night and until morning. At sunset the flock leaves the feeding area. They fly toward trees that will provide a high safe place for the flock to rest overnight.[59, 123, 124] During flight many birds engage in energetic chatter. The chatter continues for a while after the flock has settled into its roost. As the sky gets darker the chatter slowly fades and it is time to sleep. Tomorrow this routine will be repeated.

Budgerigars alternate feet throughout the night while they sleep; they stand on one foot while resting the other. Budgerigars retain a degree of alertness while they are asleep, thus maintaining a constant vigilance for natural and man-made hazards. Four states of alertness within budgerigar sleep patterns have been identified:[13]

- Wakefulness
- Drowsiness
- Slow wave sleep
- Paradoxical sleep

Paradoxical sleep is characterized by slight motor and ocular movements.[13] Like a periodic safety test, paradoxical sleep interrupts slow wave sleep for 3 to 9 seconds at a time.[13]

Table 7.2 summarizes the typical daily routine of wild budgerigars.

Table 7.2 Typical Daily Routine of Wild Budgerigars (compiled from multiple sources).

Time of Day	Activities
Before sunrise	• Individuals begin calling softly, stretching, and preening; may jump between branches
Sunrise	• Flock leaves overnight community roost • Flock flies to feeding areas • Energetic chatter while flying
Early morning	• Flocks gather in large numbers to forage for 1st large meal of the day
Mid-morning	• Flock flies to nearest water source for a drink
Mid-day	• Flock seeks a high safe place to rest • Flock rests at community roost • Individuals may chatter while resting
Mid-afternoon	• Flock flies to nearest water source for a drink
Late afternoon	• Flocks gather in large numbers to forage for 2nd large meal of the day
Sunset	• Flock leaves feeding area • Flock seeks a high safe place to rest overnight • Individuals may chatter while flying and settling in for the night

Family Life

Bonded pairs perform many of their daily activities together.[113] In addition, bonded pairs engage in beak touching, allo-preening, allo-feeding, repetitive strutting, and head bobbing.[34, 94, 113] Most budgerigars chew wood. This activity helps to keep their beaks trimmed; it is often a precursor to nest building.[94] The breeding cycle of budgerigars is linked to the availability of food and water. For males and females mate selection is influenced somewhat by the UV fluorescent properties of purple cheek patches and lower back feathers.[113, 133]

Budgerigars are monogamous colony breeders and mate for life.[41] Nests may be carved out of trees, stumps, or even fence posts; they are lined with wood dust and shavings.[36] Four to eight 0.5 inch (1.27 cm) long white oval eggs are laid, one every other day.[34, 36, 41, 59, 81, 84] The eggs are incubated by the mother for 18 days.[59, 81, 178] The eggs hatch in the order in which they were laid.[81] The altricial young are fed by the mother, who in turn is fed by the father. The mother feeds the youngest and smallest chick first.[41] Shortly after hatching budgerigars begin to make a begging call when they are hungry.

Budgerigar hatchlings grow down quickly and their eyes open when they are 10 days old.[84] At this point their eyes are solid black. Budgerigar hatchlings begin to grow their first complete set of feathers when they are 21 days old. A week later budgerigar hatchlings begin to stretch their wings and wiggle about. Most budgerigar hatchlings take their first wobbly flight down from the nest when they are 30 to 35 days old.[84] Budgerigar fledglings continue to follow their parents around and are fed by them while they are improving their flying skills and learning to avoid predators.

By the time they are six weeks old juvenile budgerigars are usually weaned and start feeding themselves. At this time they also begin to acquire a vocabulary and learn social skills and the norms of behavior from the flock. At the same time juvenile budgerigars are maturing physically. They develop a white iris around their black eyes. They lose their baby stripes as their forehead becomes a solid yellow. Their cere assumes the color of an adult and is no longer the light pink of a juvenile. By six months of age a budgerigar has become a young adult.

Table 7.3 summarizes the developmental stages of young budgerigars.

Table 7.3 Developmental Stages of Young Budgerigars.

Stage	Time frame	Events
Egg	Days 1–18	• Eggs are incubated by mother
Nestling	Day 1 Day 10 Day 21 Days 27–29	• Altricial hatchlings are fed by mother • Hatchlings make a begging call • Down starts to appear • Eyes open • Feathers begin to grow • Start to stretch wings and move about
Fledgling	Day 30–35 Day 30–41	• Take first flight from nest • Continue to follow and be fed by parents
Juvenile	6 weeks 3 Months 4 Months	• Become weaned, begin to feed themselves • Flying skills improve • Start to acquire a vocabulary • Develop white outer rim around black eyes • Learn social skills from flock • Lose stripes on forehead
Young Adult	6 Months	• Cere assumes adult color, no longer light pink of juvenile • Bird is now a young adult

Vocalizations

Young budgerigars gradually learn communication skills from their parents and the flock. When young budgerigars are reared in isolation or experience hearing loss, they exhibit abnormal vocalizations.[88, 113] Wild budgerigars have been described as "constantly warbling and chattering in a most amiable manner".[36] Nine functional categories of wild budgerigar calls have been identified to date:[59, 130]

- Normal chatter
- Prior to the group taking flight
- Alarm/warning
- Conflict

- Roosting chatter
- Courtship warbling
- Distance call
- Nestling/fledgling begging for food
- Nesting female soliciting food from male

Given that captive budgerigars have been known to learn and repeat hundreds of human words and phrases, these nine types of calls may be more complex than they appear on the surface, especially to human ears. While the talkative pet budgerigars mentioned earlier did not necessarily use human words or phrases in context, the ability to hear, discern differences, and repeat hundreds of words and phrases implies advanced communication abilities, not to mention a good memory.

Next let's look at the characteristics of indoor budgerigars, their behavior, daily activities, and stages of life, in comparison to their wild cousins. Sections 7.3 through 7.6 describe the behavior of individual budgerigars. Behaviors common to most or all budgerigars are described with the first budgerigar, Uzziel, in Section 7.3. The behaviors are presented as they occurred. The analysis, comparison, and interpretation of these behaviors are reserved for Section 7.7 Discussion. These observations took place at the locations and during the timeframes described in Annex C.

7.3 Uzziel

Uzziel arrived on January 9, 2000. Uzziel was a petite light blue budgerigar with a white head. One small feather on the covert of her left wing was slightly crooked. No matter how many times she preened, the feather remained crooked until she molted and a new feather grew in straight. The striped baby feathers on Uzziel's head were gone, but her cere was still the indeterminate color of a juvenile, but leaning toward beige. We estimated that she was close to six months old. Eventually her cere turned solid beige. For some reason Uzziel never had a leg band.

Pet bird leg bands serve a purpose, but also have some serious drawbacks.[138] Legitimate domestic breeders put a uniquely numbered band on each chick they raise. Breeders record the band number and hatch day of each member of a clutch, and the identity of the parents. This information proves that the bird was legally bred in a particular country and is not a wild caught bird that was smuggled in illegally. Leg bands also provide traceability in case there is an outbreak of a communicable disease. Infected birds can be traced to a specific breeder and clutch.

On the other hand, leg bands occasionally pose a hazard to young birds, particularly the small ones like budgerigars. Blue I was one such bird. In his case the diameter of the band was not wide enough. As a result the band became embedded in his leg as he grew from a nestling into a juvenile. By the time Blue I came home from the pet store the band was already stuck on his leg. The veterinarian worked slowly and cautiously for a half hour to remove the band. He treated the sensitive skin with a topical ointment and antibiotic and wrapped Blue I's leg with gauze. The veterinarian predicted correctly that the bandage would not last very long. Blue I pulled the bandage apart thread by thread on the way home. But the damage had already been done. Blue I could not spread out his toes. Obviously the blood supply to them had been cut off at an early age. Blue I kept his toes clenched together and used them as a prop for balance when he walked or perched.

Upon arriving home from the pet store I opened the travel cage and slowly stuck my hand in, with my index finger extended like a perch. Uzziel looked at my finger for a minute. Then she took a firm grasp of my finger with her beak to make sure it was a solid surface before stepping up. This "firmness testing" is often mistaken for a bite. Birds use their beaks to touch and explore their world. When confronted with a new object, birds—particularly the parrot family—will test the object's firmness before stepping on it. After all, birds cannot perch on something that is not solid. This was the first of three times during her life that Uzziel's beak and my finger had a close encounter.

I lifted Uzziel out of the travel cage slowly and sat down on a chair, while talking to her softly. Amazingly, Uzziel made no attempt to fly away. Rather, her dark inquisitive eyes scanned me and everything in the room. As she did her little feet gripped my finger tighter and tighter. Uzziel had just endured a half hour trip home in a travel cage, with her view obstructed by a towel, only to emerge in a room where everything was strange. In short, her whole world had been turned upside down without any warning or explanation. After 10 or 15 minutes Uzziel closed her eyes tightly. No, she was not sleeping. Her eyes were just closed, as she absorbed the fact that she was now in a totally new environment. Uzziel was breathing rapidly and clinging to my finger for dear life. And there she stayed for two hours.

At last her grip and breathing relaxed and Uzziel opened her eyes. I walked slowly over to her new cage, with Uzziel still clinging to my finger. The large cage door was open. I showed Uzziel where the water dish, dry food dish, and millet spray were located. She looked up at me, then hopped off my finger for an energetic lunch. I had lunch too seated next to Uzziel. Little did I know this act established what would become a life-long pattern. Breakfast and dinner we each did our own thing. But, lunch was different. Uzziel would always have lunch "with me", no matter what time it was. Afterward Uzziel would sing a little song while sitting on her food dish.

The previous new bird to join the aviary was Greenie II, a normal green budgerigar. Greenie II died June 1, 1997, but not of natural causes. At 6:30 p.m. the Sunday before his death Greenie II was found laying on the floor of his cage and breathing with difficulty. His paper was splattered with drops of blood. In my absence someone had broken into the aviary and ransacked it, apparently searching for something. She must have terrified Greenie II so much so that he flew back and forth against the sides of his cage in panic, injuring himself severely. It is horrifying to think what must have transpired to cause this calm gentle bird to behave like this.

Greenie II hung on valiantly for a few days. The last evening Greenie II slowly but methodically played with each of his toys one more time and loaded up on millet spray, his favorite snack. Because of the nature of his death and the fact that Greenie II was only 5 years old at the time, I decided not to get another budgerigar until after what would have been his normal life span. His cage was cleaned and stayed just like it was, in tribute to this gentle soul.

Greenie II was a sweet bird. He loved to fly laps around the room, taking off and landing on his cage top each time. Greenie II had a funny but gentle way of expressing his displeasure with something. He would simply lay his beak on my finger, as if to say "I don't like this." But he never made any attempt to bite.

Uzziel had shown that she was willing to explore a trust-based relationship by sitting on my finger for two hours her first day here. The very next day we began a routine that was to continue for months. In addition to breakfast and a clean paper in the morning, lunch together, and evening flying/playing time, Uzziel would hop onto her training perch and I would walk her around slowly from room to room. Uzziel loved to look out the windows and see the plants and trees up close. I held her high. I held her low. I held her at arm's height. Uzziel learned that she could trust me to keep her safe. (See Photograph 7.1.)

Early on in addition to learning trust, Uzziel began to express her likes and dislikes. She tolerated occasional pats, but made it clear that she did not like them. In her publications and seminars Liz Wilson often reminds people that "pet birds are not dogs with feathers".[113] Uzziel agreed. Instead of pats she preferred me talking to her and being nearby. If I were working at the desk or sitting on a chair, Uzziel wanted her cage to be right next to me so she could "participate" as part of the flock.

During spring and fall when it wasn't too hot nor too cold, Uzziel loved being by the open window so she could see and sing along with the birds outside. You could hear the happy chatter all day long on those days, even from the sidewalk outside. That first winter (2001) Uzziel seemed a bit sad when the birds left. So a songbird CD was purchased for her. The first time the CD was played Uzziel went crazy and burst into song. The CD was played two or three times a week until spring

Photograph 7.1 Building Trust. Uzziel sits on top of her brass baby cage a few days after her arrival. When it was time for a walk Uzziel hopped onto her training perch. She was not forced to "step-up". Instead the perch was simply held inside her cage. If and when she was ready, Uzziel hopped onto the perch. If she wasn't interested another attempt was made later. After her walk Uzziel was allowed to hop onto the top of her cage. That way Uzziel could decide when she wanted to go back inside. Letting pet birds make decisions about their daily activities is an essential part of building a trust-based relationship. It also allows pet birds to grow emotionally and intellectually. © D. Herrmann

when the birds returned. This became an annual ritual. Uzziel, a budgerigar native to the Australian grasslands, made friends with the finches and other songbirds native to Maryland. The rose-colored house finches (*Carpodacus mexicanus*) in particular would come and sit on the railing and sing to her year after year. Often she was more excited about singing with the finches than other budgerigars.

Today it is common knowledge that a variety of toys of different colors, shapes, and composition are needed not just for amusement, but for mental stimulation and challenge as part of a bird's healthy emotional, intellectual, and physical development.[99] There are several categories of pet bird toys, such as:[99]

- Swings
- Peck or push/pull toys
- Musical toys

- Reflection toys
- Foot toys
- Climbing toys
- Moving/action toys
- Puzzle toys
- Chewing toys
- Preening toys
- Shredding toys

We also now know that pet birds should have at least five or six toys and that new toys should be rotated in and out of their cage every six weeks or so to keep them interested and challenged.[99] After all, none of us still play with the toys we had when we were one year old. From the beginning Uzziel made her toy preferences known. Swings, bells, balls with bells, mirrors, and anything cotton was in. Ladders were out.

Uzziel's eyes were solid black until late middle age. Afterward, some white was visible. Most publications state that this change takes place during a budgerigar's first few months of life. However, all of the budgerigars discussed in this chapter retained solid black eyes until they were at least a year old.

Budgerigars blink their eyes about 40 times a minute, so fast that it is hardly noticeable.[99]

Budgerigars are also known to wink on occasion. Budgerigars do not wink at each other. Rather, budgerigars wink to acknowledge the presence of another living entity. For example, if a human winks most budgerigars will wink back. This behavior demonstrates that budgerigars can: (1) distinguish between themselves and other living entities, and (2) discern when another living entity is looking directly at them. This winking behavior was also observed in Meyer's parrots *Poicephalus meyeri* and red-bellied parrots *Poicephalus rufiventris* (Chapter 9), Indian ringneck parakeets *Psittacula krameri manillensis* (Chapter 10), and mitred conures *Aratinga mitrata mitrata* (Chapter 11). Sometimes budgerigars wink and nod their heads once or twice. The larger parrots also combine winks with bowing on occasion. A few sources infer that winking may also be a sign of deference.[99] It is unlikely that this winking behavior occurs in the wild.

Uzziel spent her first several months doing the silly things young budgerigars do as they explore and learn about the world around them through play and experimentation. This is a time when everything is new and everything is exciting. Uzziel had the advantage of exploring a predator-free world. It was not without hazards though: mirrors, windows, and cacti, all of which had to be covered when she was out.

Inevitably a young budgerigar will hang upside down, either from a perch, swing, or cage top. A young budgerigar will stand upright one moment, then hang upside down the next. While hanging upside down young budgerigars slowly let go with one foot, then the second foot, as they half tumble, half fly to the cage bottom. Their first physics lesson: the laws of gravity. Uzziel thought this exercise was tremendous fun, like an amusement park ride. She would go from perch to swing to cage top, letting go each time and enjoying the controlled crash. Sometimes Uzziel would hang upside down on her swing and flap her wings, so that it would rock back and forth, before she let go.

Young budgerigars enjoy playing with each other and often engage in comical antics, like stepping on each other's tails, so that forward movement is impaired. Young budgerigars have to get used to their long tail feathers. At first they seem to forget their tail is there, until it gets stuck between the cage bars or under a perch or toy. Often young budgerigars will pull their own tail (once), as if it were in the way. In fact, ShortTail pulled his tail feather out a couple times when he was very young; hence his name. But young budgerigars soon learn that it is more fun to pull another bird's tail; better yet to step on it so the other bird cannot go anywhere.

Young budgerigars do lateral somersaults and lay on their backs with their feet up in the air. At this stage young budgerigars have learned that they can assume this posture, but they haven't decided if or why they should assume this posture. This is as amusing to them as it is to watch, especially when two or three birds roll over at the same time. By the time they are a year old and agile flyers, budgerigars no longer lay on their backs.

Table 7.4 summarizes the common play activities young indoor budgerigars engage in as they learn about themselves and their world. Many of these activities are common throughout the parrot family when in captivity. It is unknown whether or not young wild budgerigars engage in similar activities.

Table 7.4 Common Play Activities Young Indoor Budgerigars Use to Learn About Themselves and Their Environment.

Common Play Activities
Hanging upside down • Young budgerigars hang upside down from a perch, swing, and/or cage top • At first they hang upside down by both feet • Next they let go with one foot, then the other • They learn about the laws of gravity as they half tumble, half fly to the cage bottom • Young budgerigars engage in this activity until they are about 1 year old
Tail stepping • Young budgerigars have to get use to their long tails • Often they pull their own tail, as if to get it out of the way • Next they pull another budgerigar's tail in jest • Then they learn to step on another budgerigar's tail for fun, so that bird cannot move forward • This activity continues until budgerigars are 6 to 7 months old
Laying on their backs • Young budgerigars do lateral somersaults and lay on their back with their feet up in the air • They lay there for 1 to 2 minutes before getting up • If one budgerigar does this, other birds in the same aviary will participate too • Young budgerigars engage in this activity until they are 6 to 8 months old
Punching toys, toys punching back • Young budgerigars chew on hanging toys and punch them • The harder they punch the toys, the harder the toys swing back at them • At first the young budgerigars don't understand and punch the toys harder and harder • By the time they are 6 months old, young budgerigars have learned another physics lesson
Food, fingers, and toys • Young budgerigars are extremely curious and explore their world with their beaks • At first food, fingers, and toys are all considered the same thing • Young budgerigars play with food, especially a new item, as if it were a toy • Young budgerigars chew toys as if they were food • Young budgerigars sample fingers holding treats, as if they are part of the treat; they even try to eat their paper • Eventually young budgerigars sort all this out

Swings and mirrors were Uzziel's favorite toys, not just as a juvenile but throughout her life. Her favorite was a wooden swing with a bell on top. She would play with the swing for hours, ringing the bell or putting it on her head like a hat. At one time Uzziel had two swings in her cage, both wooden. One had a jingle bell; the other was a bell with a clapper. She would hop from swing to swing like a one bird bell choir, ringing the jingle bell and wearing the other bell like a hat.

At least twice a week one swing would end up upside down, evidence of a busy and happy day, with Uzziel trying to perch on the thin wire. Young budgerigars are immensely curious about their world. This curiosity coupled with their limitless energy leads to some intriguing discoveries as they explore and customize their environment, while challenging themselves mentally and physically. Pet birds often play with toys in a manner never intended or envisioned by the toy designer. In this case, Uzziel preferred the challenge of trying to perch on the thin wire part of a swing to the ease of balancing on the much wider wooden dowel.

Years later Uzziel was in a four bird aviary when acrylic swings with cement perches and rope tassels were in vogue. One of the swings broke. With four birds, four swings are needed, especially before bedtime since many birds like to sleep on their swing. The only spare swing was the wooden one with the bell from Uzziel's baby/quarantine cage. The wooden swing was put into the aviary. Uzziel rushed excitedly over to the swing, like a long lost friend. Uzziel played and played with that swing and would not let the other birds near it for several days. It was amazing that she recognized and remembered the swing. It had been almost four years since she had seen the swing and now she and the swing were in a different cage.

Anyone who has had a budgerigar knows that mirrors are one of their favorite toys, even in multi-bird aviaries. Uzziel was no exception. A popular style is oblong, 2 inches (5 cm) high by 3 inches (7.62 cm) long, with two small metal wires, and multiple moveable colorful beads strung across the wires. Uzziel loved to talk to the bird in the mirror while energetically sliding the beads back and forth, like a Chinese merchant using an abacus, so that the clicking sound of the beads could be heard. She had one of these mirrors in every cage she lived in, from her baby/quarantine cage onward, whether she was on her own or in a multi-bird aviary. This mirror like the wooden swing, remained a favorite friend throughout her life.

When Uzziel was about five years old and living in a multi-bird aviary, a new type of toy mirror appeared which she took to immediately, as did the other birds. The mirror consisted of a narrow u-shaped bracket which attached to the side of the cage and a 2 inch (5 cm) round double-sided mirror which rotated 360 degrees. Uzziel talked to the bird in the mirror while pecking the mirror. The harder she pecked the faster the mirror would spin around. Sometimes Uzziel pecked so hard that she fell through the mirror, which only added to the fun.

It is amusing to watch a budgerigar play with the "bird" in a mirror. At times these play sessions, which are motivated by social needs not vanity, are quite animated. Some people have ridiculed the act of a budgerigar playing with a mirror as an indication of the lack of intelligence. However, is a budgerigar playing with a mirror really any different than a human watching images on a television or movie screen and bonding with and/or reacting emotionally to fictional characters and plots?

In her later years Uzziel figured out that the bracket was just the right size to support her laying down. She would talk to the bird in the mirror for a while, then take a short nap, secure in knowing that the other bird would wake her up if there were any danger. These two mirrors seem to comfort Uzziel after she lost her buddy Blue Boy. She did not go near the mirrors during the 2-week mourning period. Uzziel was so lost and sad during this time, it was not certain she was going to pull through. Gradually Uzziel began to play with the mirrors again, as if to acknowledge that Blue Boy was gone, but these two old friends were still here.

Indoor birds go through about an approximately 2-week mourning period after losing their buddy. The signs are consistent: a profound sense of sadness and loss, reduced appetite, reduced activity level, searching for and calling out to the lost bird. After the mourning period, indoor birds slowly return to their normal routine. Some eventually bond with a new bird, others choose to remain single. This behavior was observed repeatedly over 30-years for all of the species of birds discussed in Chapters 7–11. Mourning behavior should not be treated as an illness requiring medication. Instead, the bird should be given emotional support while it learns to deal with its loss. Often the process of being rehomed causes a pet bird to go through a mourning period, as it grieves for the loss of its former buddies and caretaker. This is more likely to occur if a bird spent many years or even decades in the same home.

Neither altricial nor precocial birds hatch knowing how to fly. Instead they must learn how to fly, and swim in some cases, just like people have to learn how to walk. The first flight of a wild fledgling is rather precarious since hungry predators usually loom nearby. The first flight of a pet fledgling is endearing as they wobble to the left, wobble to the right, over shoot their intended landing place, and decide where they are going ten seconds after they arrive there. Sometimes they just drop to the floor tired. Some birds cannot see a white or light beige wall and will fly right into it. (It is best to check this out in a small room, before letting a pet bird out to fly in a large room. That way a bird cannot build up enough velocity to hurt themselves.) Other pet birds have trouble with the concept of ceilings and continually bump into them. After all, the sky does not have a ceiling; ceilings are a manmade limitation.

Uzziel's flight feathers had been clipped at the pet store. She first attempted to fly when most, but not all, of her flight feathers had grown back. After she balanced her left and right momentum, Uzziel landed on an African violet, a soft forgiving place to land. Next it was off to the sansevieria. Uzziel overshot the height at first, but succeeded landing there the second time. Feeling adventurous Uzziel made a U-turn the length of the room. As she began to tire Uzziel coasted toward the floor. At the last minute Uzziel saw a perch a made a bee-line for it. At least to Uzziel it looked like a perch. In actuality it was a barbell with free weights. Uzziel, weighing less than an ounce (28 gr), perched very properly in the middle of a 30 pound (13.44 kg) set of weights. (See Photograph 7.2.)

Like the regular walks on Uzziel's training perch, a special routine was established for Saturday mornings which continues to this day. After the normal paper change and fresh food and water, I held a millet spray while Uzziel ate it from my hand. At first Uzziel gobbled up the millet like it was a race, probably in response to some lingering memory from the pet store about eating quickly before the other birds came. After a while she ate the millet spray slowly and methodically, which became a hallmark of her characteristic neatness. Uzziel was such a neat eater that she would push the empty millet and oat hulls gently to one side of her food dish. At this time a check was made to see if Uzziel needed a new cuttlebone or mineral block, or if any toys or perches needed special cleaning or replacing. Then it was off to the swimming pool.

At first Uzziel swam in a small dish on the bottom of her cage. She would splash away, summer or winter, until all the water was gone. Later she moved up to the swimming pools that hang on the side of a cage, like a private spa annex. This type of pool has an opaque brightly colored bottom, clear top and sides, and a single entrance/exit. There is plenty of room to splash and walk around. Unless two birds try to go in at once, then they get stuck like too many people in a telephone booth. The concept of getting stuck seems to puzzle budgerigars and they wiggle around, splashing, long tails in the way, until one bird finally gets free. Usually this scene attracts the curiosity of other birds in the aviary and they fly over to the entrance of the swimming pool to see what is going on. One of the original stuck birds may get free, but now a new pair is stuck as this hilarious scene repeats itself over and over. (See Photograph 7.3.)

Uzziel was neither an adventurous eater, nor a picky eater. A basic seed mixture was combined with a jar of specialty seeds, a different flavor each month; some natural pellets and dried greens were sprinkled on top. Given that iodine deficiency is common in indoor budgerigars, a mineral supplement was added to her water. Uzziel explored new foods to decide what she did or did not like. Apples, baby spinach leaves, green leaf lettuce, and especially shredded carrots were greeted with enthusiasm. So much so that some days her beak was stained with carrot juice.

All the indoor birds in the aviary receive filtered water. It stands to reason that if so many parts per million (ppm) of this or that chemical commonly found in tap water are harmful to humans weighing over 100 lb. (45 kg), these chemicals must likewise be harmful to birds whose weight is measured in grams.

In a group setting birds will often sing themselves to sleep as night approaches. This seems to have a calming effect on the flock and also serves as an evening roll call. During her first year, Uzziel talked in her sleep. It sounded like a muted version of her happy chatter during the previous day. This behavior was similar to the roosting chatter of wild budgerigars. Around the first anniversary of her homecoming Uzziel stopped talking in her sleep. This pattern of behavior did not return until nine years later. About a month before her death Uzziel started talking in her sleep again. So did Daisy.

Photograph 7.2 A Perch is a Perch #1. After 170 years of captive breeding, pet budgerigars are considered to be domesticated. When they are out flying pet budgerigars seek places to land that resemble the perches in their aviary. That may include a variety of items such as curtain rods, lamp shades, and bicycle handlebars or spokes. (Top) Uzziel decides to perch on a barbell and free weights. (Bottom) Twelve years later Uzziel II selects the crossbar underneath a plant stand as an impromptu perch. © D. Herrmann

Photograph 7.3 Budgerigar Splash About. In the wild bathing is a lively community event as several flocks of budgerigars gather in and around a water hole at the same time. Indoor budgerigars are just as enthusiastic. (Top left) Lilac II does some serious splashing, while Yellow Birdie and Yellowhead watch and wait their turn. (From top right) Sunflower II, Blue II, and ShortTail take a quick plunge. © D. Herrmann

When Uzziel joined the aviary standard budgerigar perches were 0.375 inch (0.952 cm) in diameter smooth wooden dowels. The only difference was the length. Mid-decade all sorts of recommendations came out stating that pet birds needed (not a luxury, but a valid medical need) a variety of perches of different textures, shapes, composition, and diameters.[99] Think about it for a minute. Have you ever seen a deciduous tree where all the branches are exactly 0.375 inch (0.952 cm) in diameter and smooth? Or, to bring the analogy closer to home, can you imagine wearing the same pair and size of shoes all day every day your entire life? Of course birds need a variety of perches of various textures, shapes, composition, and diameters to exercise different leg and foot muscles, relax different stress points in their feet, and keep their nails trim.[99] The only question is why it took us scientifically advanced humans over a century to figure this out.

Uzziel started out with 0.375 inch (0.952 cm) wooden dowels, but by mid-life she had a collection of wooden, cholla, pumice, calcium, sand, and rope perches, not because of conspicuous consumption or mindless materialism, but out of necessity. In particular Uzziel played with the cardboard perch covers that are coated with sand. These covers fit snugly over a smooth wooden perch. Each day Uzziel carefully selected a couple pieces of sand from the perch. Uzziel then spent several days gently prying the sand cover loose from the perch, which caused it to slide around. Uzziel engaged in this behavior from the time she was a juvenile through her senior years.

Uzziel's baby/quarantine cage was 30 inches (76.2 cm) long by 14 inches (35.56 cm) wide by 30 inches (76.2 cm) high, brass, with 0.5 inch (1.27 cm) bar spacing and a translucent amber plastic bottom and tray. When Uzziel was around ten months old, reports began being published about dangers associated with brass bird cages. Brass is an alloy of zinc and copper. Zinc, like lead, is toxic to birds. Chrome, galvanized, and brass cages, dishes, and toys wear out over time, causing birds to ingest minute particles of lead and/or zinc which are toxic. Experts recommend testing cages, dishes, and toys with a magnet.[87] If the item is attracted to a magnet, it contains lead and/or zinc and is harmful to birds.[87]

Whenever a new bird joins the aviary they spend their first four to six weeks in a baby/quarantine cage 24 inches (61 cm) long by 18 inches (45.7 cm) wide by 24 inches (61 cm) high. This gives them a chance to get used to their new environment gradually, while in the 30-day quarantine period. The cages birds are kept in at most pet stores are quite small and often contain several birds. Moving from that setting to a large aviary in a new home all at once can be overwhelming. After a few weeks in the baby/quarantine cage, when the new birds are comfortable with their new environment and ready for more room, they are moved to a large aviary.

Uzziel had become an agile flyer by then and was ready for a full size aviary. Her new home was a free standing white powder coated steel aviary, 37.5 inches (95 cm) long by 27.5 inches (70 cm) wide by 63 inches (1.6 m) high. Her favorite toys, perches, and swings were moved into the new aviary, some new items were added, and she was ready to go. Uzziel loved her new home. She could fly side to side and top to bottom. Because the aviary was on casters, it could be rolled from room to room and by the window for sun.

People tend to think of cages as something to confine a pet bird, like an aquarium contains fish. Naturally, pet birds have a different view. Their cage is their home, their private retreat, a place of safety that keeps scary humans and other pets out. Birds know where their cage is located and that their food, water, and toys are inside. That is why after flying around outside their cage for a while, pet birds will always return home. Even young birds find their way home the first time they are let out to fly. This behavior was observed consistently, among not only all 28 budgerigars discussed in this chapter, but also all the birds discussed in Chapters 9–11.

When Uzziel came home most books discouraged keeping two or more pet birds together. The books warned in dire terms that one bird would dominate the other, often to the point of preventing the other bird from eating. By the first anniversary of Uzziel's homecoming, both of these statements were being refuted. The behavior of dominating the food dish was due to the fact that at that time pet birds, particularly budgerigars, were kept in cages that were way too small. Overcrowding and fear that there would not be enough food to go around created the aggressive behavior, not a lack of social needs.

Asher was a small slate blue male budgerigar, about one year old, who came from the same pet store as Uzziel. Asher did not have a leg band either. He spent the first few weeks in a baby/quarantine cage, across the room from Uzziel. She ignored him. Then Asher's cage was placed next to Uzziel's. She ignored him. Eventually Asher moved into Uzziel's flight cage. It was not love at first sight. Over time they learned to play together, but they were never a bonded pair. Asher did not last very long. His liver ruptured before his first homecoming anniversary, a quick but painful death. Budgerigars, like Asher, that have genetic problems usually do not survive to their second year.

When Uzziel was 4.5 years old a crosstown move was made. Uzziel took all the commotion in stride, in spite of having to spend almost twenty hours in her small travel cage. Uzziel was placed in her travel cage, the same one she came home in from the pet store, at 10 p.m. the night before. The movers were scheduled to arrive at 8 a.m. and there would be too much commotion then. Uzziel's travel cage was placed away from the major traffic and partially covered with a towel. Because she trusted me, Uzziel seemed puzzled but did not panic, resist, or complain. At first light, around 5 a.m.

the next morning, Uzziel let out one solitary squawk. Other than that, Uzziel was quiet the rest of the day, watching what was going on with interest.

After the movers left, it was time to take Uzziel to the new location. This involved a walk outside and a ride on the subway. It was a beautiful sunny summer day. Uzziel loved being outside and chattered away with all the wild birds. Being her first subway ride, Uzziel thought the chimes that sounded every time the doors opened and closed were a toy, so she sang back to them. The other passengers, well let's just say that they had some free entertainment that ride.

When Uzziel was placed back in her aviary, the first thing she did was to take a long bath and preen her feathers. True to form Uzziel was always fastidious about her appearance; that took precedence over flying, eating, or playing with toys.

An application was submitted to adopt a budgerigar, a new buddy for Uzziel. In the end four zebra finches (Chapter 8) and two blue budgerigars were adopted. They all arrived November 15, 2004.

The medium blue male budgerigar was named Blue Boy. He had an extra-long tail and was slightly larger than most budgerigars. Possibly he had some English budgerigar genes. Blue Boy was middle-aged, but his exact age was unknown since he had been found flying loose outside. The powder blue female budgerigar was named Precious. She was small like Uzziel, very energetic and inquisitive. Precious only had one leg, but that did not slow her down or dampen her spirit. She was also an adult. Her exact age was unknown, as was whether she hatched with one leg or lost a leg in an accident. Instead of being nervous or anxious about being in a strange place, both Blue Boy and Precious seemed relieved and content to finally be in a stable environment again. Over the years it has been noticed that it is reassuring for a new bird to find out that other birds are living in their new home. In contrast to her reaction to Asher's arrival, Uzziel was excited about the arrival of Blue Boy and Precious and chattered away happily.

When Blue Boy arrived he was as happy and chipper as you would expect for a budgerigar. But that night, when everything was quiet, Blue Boy made a wheezing sound like he was having trouble breathing. So it was off to the veterinarian his first morning here. At the last minute a decision was made to take Precious along for a check-up too (not knowing anything about her medical history), and Uzziel for a nail trim. The veterinarian prescribed some antibiotics for Blue Boy, which cleared up his respiratory infection right away. Precious passed her physical with flying colors.

It takes two people to trim a bird's beak or nails properly. One person, usually a certified veterinarian technician (CVT), holds the bird while the veterinarian does the trim. Both are delicate procedures. If the beak is not trimmed properly, a bird will not be able to eat, preen, or climb. If the nails are not trimmed properly, they will bleed.

Uzziel, being the independent bird she was, resisted the nail trim. While the technician was able to hold her still, Uzziel squeezed her little feet together like a fist, making it nearly impossible to hold on to individual toes. The veterinarian was only able to trim two toes on one foot and gave up after that. As the technician put Uzziel back in her travel cage she uncharacteristically bit him.

The veterinarian expressed concern that Uzziel was so tiny, while being almost five years old. I was shocked when, as we were leaving and without giving any reason or indication of what was wrong, the veterinarian said it did not matter that only two out of eight nails had been trimmed because Uzziel was not long for this world. In a strange and sad twist of fate the veterinarian died four months later. Uzziel, on the other hand, lived over five more years. This was Uzziel's one and only visit to a veterinarian's office. She was never ill and never needed another nail trim.

Precious and Blue Boy went through the standard procedure for introducing a new bird into the aviary: 30 days in a separate quarantine cage way on the other side of the room, then gradually moving their temporary cage closer, until it was next to Uzziel. This process allows all the birds to see and hear each other and get acquainted slowly. This process also builds curiosity about the other bird(s). This was certainly true in Uzziel's case. She was delighted to have new budgerigars in the aviary.

When it was time to fly Uzziel was let out to fly first. When she was back home, Precious and Blue Boy were let out to fly. Before long Uzziel figured out that Blue Boy was a boy and she popped her tail up—the budgerigar equivalent of flirting and saying she was interested. One morning Uzziel was standing on her perch with her head hanging low, like she was pining because she was in an aviary by herself, apart from Precious and Blue Boy. It was time to let them all out to fly and move in together in Uzziel's large aviary.

With Uzziel and Blue Boy it was love at first sight. From that day on they were inseparable. Precious and Blue Boy had not been a bonded pair. Rather, they were two birds that needed a new home at the same time and arrived here together. Precious had been more of a temporary roommate to Blue Boy. Instead of rejection, resentment, or jealousy, Precious accepted the situation and was okay with Uzziel and Blue Boy being happy together. Blue Boy liked Uzziel. Uzziel liked Blue Boy. No need to interfere.

Then again, maybe Blue Boy wasn't Precious' type. Like humans, birds have particular traits they are looking for when they choose a mate. A budgerigar is not smitten by another budgerigar just because they are of the opposite sex. Uzziel's reaction to Asher demonstrates that fact. This individual mate selection behavior was captured visually in Chapter 3 of "The March of the Penguins" documentary.[188]

Blue Boy and Uzziel did everything together: swimming, flying, playing with toys, and munching millet sprays. To show affection birds preen each other and feed each other. These behaviors are known as allo-preening and allo-feeding. Blue Boy and Uzziel engaged in these behaviors often. Blue Boy was very protective of Uzziel. If she wanted to play with a certain toy, sit on a particular perch, or eat out of a specific food dish, he would shoo the other birds away. Uzziel had the funniest way of letting her preferences be known, without being aggressive. She would simply climb or fly next to the perch, swing, or toy she wanted and stare at the bird currently using the item. Most of the time, her stare alone would do the trick. When that didn't work, Blue Boy would gallantly come to the rescue. Uzziel and I now had several shorts periods of interaction throughout the day, instead of a long period of interaction together in the evening. In fact, this new routine worked out best. If I spent too much time with Uzziel, Blue Boy tried to shoo me away. Blue Boy was a very strong flyer, which is probably how he ended up loose outside.

Over the years Uzziel and Blue Boy shared their aviary with several other birds. The number of homeless birds continued to increase. We continued to adopt, as long as there were adequate room and adequate time to take proper care of each bird. We moved the birds around so that everyone could get acquainted with everyone else. On occasion we had to temporarily isolate a bird that was too aggressive. As the new birds came and settled in, Uzziel always took note, in a mixture of curiosity and eagerness—new neighbors, more birds to play with. It did not matter if they were other budgerigars, tiny zebra finches, or big birds; she was happy to see them. Uzziel never got jealous or acted up, like human children often do when a new baby brother or sister arrives. Uzziel played with all the new birds, but she and Blue Boy remained a bonded pair throughout.

Table 7.5 summarizes the activities common to bonded pairs of indoor budgerigars. These activities occur whether or not the pair builds a nest and raises a family. These activities approximate those of bonded pairs of wild budgerigars.

Uzziel continued to call out to the rose-colored house finches year after year, especially upon their return each spring. Uzziel had a unique song or call she would make when talking to the finches. Whenever she saw or heard them, she would get excited and drop whatever she was doing. It is unclear why Uzziel was attached to finches. She did not have the same reaction to other outdoor birds or the zebra finches indoors. Maybe it was because the first birds she sang to (and heard a reply from) as a juvenile were rose-colored house finches.

During Uzziel's middle years the colorful plastic water and food dishes were replaced with stainless steel dishes. The stainless dishes were not as attractive, but they did not harbor bacteria like the plastic dishes and they could be sterilized in the dishwasher. At first 10 ounce (300 ml) dishes

Table 7.5 Common Activities of Bonded Pairs of Budgerigars.

Common Bonded Pair Activities
Allo-preening • Bonded pairs preen each other in the hard to reach areas like the head, face, and neck; particularly around the eyes and cere
Allo-feeding • Bonded pairs feed each other a partially digested mash • This is similar to the mash a mother budgerigar feeds nestlings • Primarily the male feeds the female, as proof he will take care of her while she takes care of the nestlings
Head bobbing • Males bob their head up and down when they are excited and trying to impress a female • Some females bob their head up and down in return • Head bobbing often precedes allo-feeding
Strutting • Males strut back and forth rapidly while trying to impress a female • Often they walk right up to their chosen female, backward about 8 inches (20 cm), then toward the female again • Occasionally a female mirrors the male's strutting behavior • Strutting often precedes allo-feeding
Beak touching • Budgerigars often tap their beaks together once or twice in conjunction with strutting and head bobbing
Chewing paper and wooden toys • Some female budgerigars chew a small hole in their paper, the male may join in this activity • It is more common to see a bonded pair chewing a wooden toy together. The female selects which toy to chew and the male joins her • This activity simulates chewing a nest cavity in a tree
Male chasing others away • When a male is strutting back and forth and bobbing his head up and down to impress a female, he will chase away other males that get too close to his partner
Sitting together • Bonded pairs frequently sit together on the same perch, especially after eating or when it is nap time or time to settle in for the night

were tried. While two birds could easily eat out of one dish at the same time, the 10 ounce (300 ml) dishes were too big. A switch was made to 5 ounce (150 ml) stainless dishes. They were the perfect size in terms diameter and depth for one bird.

Stainless dishes have an added bonus over plastic dishes. Metal hoops hold the dishes. While the dishes are being cleaned and replenished, the birds can perch and play on the empty hoops that hold the dishes. Birds of all ages seem fascinated that they can occupy the same space the dishes usually do, and will frequently hang upside down in the middle of the hoop. The hoops soon became a new favorite toy and hanging upside down in the middle of the hoops a new favorite past-time. Some larger birds even throw a dish on the floor just so they can play with the hoop. (See later Photograph 10.15.)

Since one 5 ounce (150 ml) dish provided food or water for one bird, several dishes were installed in different locations and heights to accommodate a multi-bird aviary. Often the birds would fly or hop from dish to dish to see if all the dishes had the same contents. To stimulate their curiosity, on even days the dish contents were the same; on odd days the dish contents were varied. Uzziel was intrigued by the transition from plastic to stainless dishes. She was the first to see another bird (her reflection) in the bottom of the dish. Uzziel even tried to find that bird on the outside and bottom of the dish.

Around that time reports began to be published about the sensitivity of birds to pollution, carbon monoxide, car exhaust, paint fumes, smoke, paraffin from burning candles, out-gassing of plastics, other assorted natural and manmade chemicals, and various dusts.[99] The natural environment for birds is in the wild where the unlimited air is clean or at least cleaner than confined air in a metropolitan area. After reading reports about the dramatic increase in childhood asthma rates in metropolitan areas, a portable HEPA carbon zeolite air purifier was purchased for the aviary. It is run a few hours every morning and evening, so the birds can breathe clean fresh air.

Another team studied conure wasting disease, which was linked to severe calcium deficiency. As the research team put the pieces of the puzzle together it was determined that all birds, not just conures, need a certain amount of sunshine, specifically UVA and UVB, in order to metabolize vitamin D_3, which is a prerequisite for calcium absorption. In other words, conure wasting disease is equivalent to rickets in humans.[6] It makes sense when you remember that birds are supposed to be outside in direct or at least indirect sunshine all day, particularly tropical birds. In addition to the nutritional benefit sun lamps provide an emotional benefit as well. Unlike humans, birds can see ultraviolet light. The sun lamps help the birds to see better. In particular they can see the true color of plants and feathers, which helps them to identify each other. This in turn improves their emotional health.

Over the years it was noticed that the birds in the aviary seemed to slow down during the late fall and winter months. Then, with the arrival of spring they became energized again. This behavior was attributed to the winter blahs, to which humans are also subject. After reading these reports full spectrum sun lamps (20% UVA, 5% UVB, color temperature (CCT) 5500K which is equivalent to sunny daylight at noon, and a color rendering index (CRI) of 93 out of 100) were installed in the aviary to compensate for the sunshine the birds missed from November through March. The lamps are on four to five hours a day to make up for the shorter days. What a difference—all the birds are their usual lively selves year round now. Uzziel, who was always fastidious about her appearance, seemed to enjoy preening even more when the full spectrum lights were on.

We lost Blue Boy in the fall of 2007. Blue Boy protected Uzziel, but was never aggressive. Blue Boy raced to be one of the first to eat the organic baby spinach leaves every morning. Blue Boy had a favorite swing for sleeping and was usually the first or second bird into the swimming pool. He ate millet sprays out of my hand, but his primary interest was Uzziel.

One of Blue Boy's endearing traits was giving speeches. Blue Boy would bob his head up and down and chatter away as if he were saying something profound, while walking back and forth. This routine was different than his courtship strutting. Sometimes he would even twist his head left and right while bobbing it up and down. It reminded me of several politicians. During State of the Union speeches, political conventions, and rallies Blue Boy had a knack for picking winners. If the speech (politician?) was really good, Blue Boy would simultaneously go into his "speech" routine also. Sometimes he would get really carried away and flap his wings for added effect in addition to the head bobbing and strutting back and forth. In contrast, if the speech was replete with rambling repetitive run-on sentences and empty platitudes, Blue Boy would just sit quietly or fall asleep.

Blue Boy was never ill, other than the respiratory infection he had when he arrived which cleared up quickly. Because Blue Boy was found flying loose, we had no way of knowing his age other than that he was an adult. Blue Boy never gave a sign that anything was wrong. Birds in general are very adept at hiding an illness. This safety strategy is designed to prevent attracting the attention of predators.[178] His activity level, appetite, coloring, and droppings were always normal.

On Blue Boy's last morning he was a bit hesitant about going to the food dish. Blue Boy hopped on a perch near the food dish and waited there a while before hopping up to eat breakfast. That was unusual for him. Later that evening Blue Boy was laying on the floor. He was still alive, but it was obvious the end was near. I placed a small dish of water and millet spray next to him and patted him. He calmly took a drink of water, looked up at me and winked to acknowledge my presence. Ten minutes later this gentle soul was gone.

As mentioned earlier, Blue Boy's death was very hard on Uzziel. She was in her eighth year and the average life expectancy for budgerigars is 7 to 8 years. Uzziel was slowing down a bit, but she had the same spunky spirit.

Birds have five distinct stages during their lives: nestling, juvenile, young adult, middle-aged, and senior. The transition from nestling to juvenile is obvious. The feathers grow in. The bird grows nearly full-size and learns how to fly and be responsible for itself. This transition occurs early in life, when a bird is only a few weeks old. During the transition from juvenile to young adult subtle color markings on the feathers, cere, and eyes take place. Birds spend 70–80% of their lives in the middle-aged stage, but there are few if any visible changes. Birds do not get wrinkles, grey hair, or midriff bulges. They do not feel compelled to dye their hair or have face lifts or tummy tucks. Instead a budgerigar looks the same when they are seven years old as they did when they were two years old. The transition from adult to senior bird is more obvious, as this little package of perpetual energy begins to slow down a bit.

As is common in senior birds, when Uzziel molted not all the feathers grew back. This process is similar to hair thinning in humans. Senior birds have trouble staying warm if they are missing feathers on their body. Fortunately, Uzziel did not have this problem. But she was missing a couple wing feathers which made it difficult to get around. Uzziel, always an agile flyer, would take off, fly horizontal for a foot (30.48 cm) or so, then coast to the floor. She would climb right back up and try flying again and again, always with the same result. Each time Uzziel would have a puzzled expression, as if to say "Hmm ... this used to work." Since she was in an aviary, Uzziel was coasting down almost 4 feet (1.21 m). There was a concern that Uzziel might hurt herself falling from that height onto a metal tray. So bubble-wrap was placed under the paper to soften her landing.

Bubble-wrap was first introduced into an aviary in 1990. At that time Greenie I was 10 years old. She was missing a few feathers on front where she had molted, but the feathers had not grown back. As a result Greenie I had trouble staying warm during winter. At night Greenie I would shiver, which caused her cage bars to rattle. Also, once in a while Greenie I would lose her balance while sleeping, and fall to the cage bottom. Bubble-wrap was placed on the bottom of her cage, with an extra inch (2.54 cm) sticking out on one side that curled up like a sleeping bag. Bubble-wrap provides extra cushioning on the cage bottom, which senior parrots with limited balance need in case they slip off their perch. In Greenie I's case, bubble-wrap also provided the additional insulation and warmth she needed during nap time. Greenie I quickly learned to take her naps on the floor curled up in the bubble-wrap. The insulation provided by the bubble-wrap kept her warm. Since there was an opening on one side, Greenie I could make a quick exit if she needed to. Greenie I began retreating to the bubble-wrap for a nap every day after breakfast, like it was her favorite quilt.

Before long, Uzziel was climbing more than flying, having come to terms with her new situation. That involved a lot of climbing, given the dimensions of the aviary. Sometimes Uzziel would be climbing around the aviary, as if in a hurry to go somewhere. Then all of a sudden Uzziel would stop and look puzzled, as if she forgot where she was going. The other birds in the aviary were less than half Uzziel's age. The other birds were not aggressive toward Uzziel. But, they did not understand why Uzziel could not keep up with them when they were all playing together.

Table 7.6 summarizes the life stages of indoor budgerigars.

It was time for Uzziel to "sell the big house" and move into a more appropriately sized "little old lady condo". It was a bittersweet moment—sad about having to take her out of the aviary, away from the other birds, but happy and grateful that she had lived this long. All her favorite things (wooden swing with bell, oblong mirror with beads, round spinning mirror, sand perch covers, cotton string toys, cuttlebone, and millet spray, along with the bubble-wrap) were moved into a 24 inch (61 cm) long by 18 inch (46 cm) wide by 24 inch (61 cm) high cage. Her new cage was placed near the aviary, so she could still see and hear her friends.

Table 7.6 Life Stages of Indoor Budgerigars.

Characteristics of Life Stages
Juvenile (6 weeks–6 months) • Happy, fearless, curious, trusting • Likes being with a group of birds • Explores its new world with energy and enthusiasm • Engages in comical play activities by themselves and with other birds • Open to new food and toys
Young Adult (6 months–2 ½ years) • Continues to engage in common play activities until it is about 1 year old • Confident, but more cautious around new people • Still has limitless energy and curiosity • Open to new food and toys
Middle-aged (2 ½–7 ½ years) • Energy level decreases a little • Food and toy preferences established; will consider new items, but not as eagerly as before • Prefers a consistent daily routine • Usually part of a bonded pair • Has a preferred group of birds it likes to socialize with
Senior (7 ½–10+years) • Energy level decreases noticeably; while they still fly, walk, and climb around they do so less often and slower, and are content to stay in one place for a longer period of time • Takes more and longer naps throughout the day • Eats smaller meals and more often • Doesn't chatter as much as before • Flying strength wanes a little • Not all feathers grow back when molting; need to make sure bird is warm during cold weather
Acknowledge Impending Death (last 24 hours) • Play with all their favorite toys one last time, even ones they haven't touched in months • Have one last large meal of their favorite treat, usually a millet spray • Sing a farewell song; often this is the same song budgerigar nestlings sing when they are hungry • Snuggle up in the corner of their cage for security, like they are back in the nest • Often more energetic than usual the day before their death • Mate often feeds their partner, grooms them and sits next to them to keep them warm, even on a low perch or cage bottom
Mourning Period (~ 2 weeks) • Exhibits a profound sense of sadness and loss, hangs their head low • Reduced appetite • Reduced activity level • Bird calls out to and searches for lost mate

Then it was time to move Uzziel. I picked her up gently, held her a while and patted her, which she tolerated. As I sat Uzziel in her new cage, she took a firm grasp of my finger with her beak before stepping forward into her new home. This was the second time she did this, the first being eight years earlier on the afternoon of her arrival. Again, it was not a bite (no pinching or puncturing). Uzziel accepted this transition, like all the changes in her life, without complaining or self-pity.

Uzziel settled into her new home with enthusiasm. She became quite adept at climbing up and down and around the sides of her cage. Uzziel could jump from the top perch to the bottom tray and keep on going, with a little help from the bubble-wrap. The first few days were spent exploring, as Uzziel learned where the perches, toys, cuttlebone, millet spray, food and water dishes were now located.

Next, Uzziel developed a new daily routine. Nights were spent sleeping on the high perch, with her right side leaning on the oblong mirror. She would awake at 7:30 a.m., climb down to the front of her cage and have a big drink of water. Then she would clean and carefully arrange all her feathers. As her balance on one foot became a bit precarious, Uzziel devised an ingenious alternative: she would hang high on the side of her cage in a vertical position. She could hang from one or both feet and preen energetically as long as she wanted while gravity did the work. Uzziel thought this new approach was great fun and would slide her head back and forth in glee. She could also stretch out her wings one at a time from this position.

Breakfast was promptly at 8 a.m. and shredded carrots got top billing. In contrast to her juvenile days, breakfast was a leisurely affair with each seed or pellet being chosen carefully. Some days this would be followed with a morsel or two of gravel, other days not. Next it was off to the round rotating mirror for a quick hello to that "bird" before settling in for a nap. The set time for her second breakfast was 10:30 a.m. This meal was followed by an hour or so of play time and exploring. Uzziel might visit the cuttlebone, run around on the floor or climb up to the other high perch, just to make sure it was still there. Then it was off to the oblong mirror. Uzziel slid the beads back and forth so energetically that the clicking sound could be heard across the room. As the beads slowly stopped clicking, Uzziel would settle in for another nap. Her favorite wooden swing with the bell was there, but because her balance was precarious, Uzziel did not play with it. She tried once, but slipped off. At this point in her life Uzziel needed solid steady perches.

Uzziel would still have lunch with me, usually between 1:00 and 2:00 p.m. After she was finished, Uzziel would always sing a little song while sitting on her food dish. In the morning or in the evening when the other birds were having their daily songfests, Uzziel would join in, adding a note or two here and there. But after lunch she sang her own song.

Birds generally sleep 12 hours a day, because of their high metabolic rates. After lunch Uzziel settled in for a serious nap, some days on her round mirror, other days on the high perch with the oblong mirror. Perhaps knowing that the other "bird" was there gave her a sense of security. Uzziel did not venture forth again until around 8:30 p.m., in spite of the other birds' chatter. All of a sudden she would wake up, do some quick preening and climb down for dinner. Uzziel usually ate a big dinner, chattered a bit, then played with the round mirror. After a short nap, she would have her second dinner, a smaller fare, at 10:30 p.m. Then it was off to the high perch and oblong mirror for the night. Without consulting a clock or watch, Uzziel stuck to this routine year round.

When she was up, Uzziel was happy and energetic. She enjoyed attention, 5 to 10 minutes at a time, but still did not want to be held or patted. She preferred for me just to talk to her. She also enjoyed hearing the other birds singing and scampering around. Every once in a while she would pop her head up like she heard a familiar sound. Then she would call out two or three times for Blue Boy, but there was no answer. Bonded pairs have unique calls they make to each other.

January 9, 2009, was the ninth anniversary of Uzziel's homecoming. Uzziel was at least nine years old. She had made the transition from juvenile to young adult to middle-aged to senior bird with aplomb and was still going strong. Along the way Uzziel had adjusted to new situations and changes in her life and environment as easily and gracefully as climbing from one rung of a toy ladder to the next. To celebrate Uzziel's backyard birds CD was played several times. Uzziel received an extra millet spray, extra shredded carrots, a misting, and lots of attention.

Uzziel continued happily in her routine. August 2009 Uzziel began chewing a small hole in her paper about 1 inch (2.54 cm) in diameter. Other birds, from zebra finches to Indian ringneck parakeets, have shredded their papers with enthusiasm, particularly around nesting time. But Uzziel had never shredded her paper before. It is unclear what prompted this behavior at this stage in her life.

That fall Uzziel had a night attack. About 2 a.m. Uzziel started making a panic call, a sound she had never made before. She raced up the side of her cage, jumped off, and repeated this procedure three or four times, while making a panic call. Then Uzziel hung on the side of her cage, her whole body shook two or three times. Her eyes expressed total panic and she was breathing rapidly. I talked to her slowly, trying to calm her and assure her everything was alright. But, I had no idea what was happening. Uzziel had never had "night fright" attacks, which some birds do. Uzziel was always the one that stayed calm when the other birds panicked during thunderstorms. I kept talking slowly, trying to soothe her. After ten minutes or so, Uzziel finally calmed down and went back to sleep. In the morning Uzziel was just fine and acted as if nothing had happened. This scenario was repeated two more times, about a month a part. Each time Uzziel was fine in the morning, as if nothing had happened. The fourth and last time Uzziel had an attack was at 4 p.m. in the afternoon. There was still no explanation, but she was fine afterwards and thankfully the attacks did not reoccur.

As time went on Uzziel began taking her naps on her paper. Her cage was up high and the tray and base provided a safe shelter. She laid horizontally on the floor, using her beak to prop up her head on the tray's edge. A parrot's head is the heaviest part of its body, because of the beak. Sleeping this way allowed her to save some energy and stretch her feet out flat; a smart adjustment on her part. At first Uzziel would only sleep on the floor when she thought I was not watching. Later she did not mind if I saw her. I started to put her millet spray on the floor, so she could walk over and have a snack without having to climb all the way up the side of her cage.

About the same time Uzziel made another ingenious adjustment. When Uzziel slept by the oblong mirror on the high perch, she had to hold on with both feet. Generally birds sleep on one foot and rest the other foot, alternating throughout the night. At this stage of her life Uzziel could not balance on one foot. This was another reason for the naps on the floor. Using both feet took too much energy, and did not allow her to enter a deep sleep. Uzziel, in one of her creative solutions to life's challenges, figured out how to put her right foot on the wire that held the mirror's beads. She kept her left foot on the perch. Her nails would catch on the wire, holding her in place, and gravity would do the rest. As she fell into a deep sleep, her little head would hang to the left contentedly, as if she were totally relaxed.

January 9, 2010 was the tenth anniversary of Uzziel's homecoming. Uzziel, the little tiny puff of feathers the veterinarian had written off in 2005 was now at least ten years old.

Members of the parrot family are not programmed for snow. The first few times a budgerigar sees snow falling, particularly a heavy snow with big fluffy flakes, its eyes get real big and it stares at the snow in amazement and disbelief, the "this is not supposed to be happening" type of disbelief. Members of the parrot family react the same way when seeing large trees bending back and forth on very windy days. Most pet birds become nonchalant about snow after their third or fourth winter. A few pet birds get wide-eyed every time they see snow fall or trees bending in the wind. Uzziel, having been through many winters, hardly noticed the heavy snow during the winter of 2009/2010. In contrast, the younger birds found every snow flake to be a source of wonderment and a bit scary.

Uzziel, being the little trooper she was, stuck to her routine throughout the snowstorm and multi-day power outage that followed. There was a concern that Uzziel would try to sleep through the night and be nearly starved the next morning, since without electric lights it got dark at 5 p.m. But about 8:30 and 10:30 each night during the outage Uzziel started rumbling around trying to find her food. So I shined the flash light over her cage until she had finished eating and was back asleep. Uzziel was always the first one up for breakfast during the storm and multi-day power outage.

Uzziel stuck to her routine during the power outage and she continued her routine after the power was restored. The first few days afterward, her naps were a bit longer and the play times a bit shorter. Uzziel continued to sing her little song after lunch.

Monday, February 15th Uzziel spent more time than usual napping on her paper. I had to stop by several times before I could change her paper without waking her. Because Uzziel was in a horizontal position so long, her droppings were stuck together around the vent area. This situation can be dangerous for a bird. If a bird's digestive system becomes blocked, it can starve to death in a short period of time due to its high metabolic rate. I watched Uzziel closely. She was eating enough and by late afternoon when she started climbing up the side of her cage the droppings fell off.

Tuesday Uzziel seemed better. She was climbing, playing with her toys, and preening while hanging from the side of her cage as usual. Uzziel had a good appetite and sang her little song after lunch. But her droppings were stuck around the vent area again, until gravity solved the problem. Uzziel was very energetic in the evening.

Wednesday morning Uzziel was up and about. She drank her water and ate her breakfast with gusto. But, she seemed to be in a hurry to go somewhere. Uzziel walked from one side of the cage to the other and climbed up and down without staying in any one place very long. After a while Uzziel was hurrying toward the corners of her cage and trying to snuggle up there, first one corner, then another. She had never done this in the past. Unfortunately, I had observed this behavior before. Often when a pet bird is ill or shortly before its death, it will attempt to snuggle up in the corner of its cage. The close quarters seem to remind the bird of when it was a hatchling in the nest, linking the beginning of its life with the end.

Before long, Uzziel could only walk back and forth; she could no longer climb up the sides of her cage. I sat Uzziel's cage on the floor, took off the top, and picked her up. I held Uzziel in my hand, patted her, and talked to her softly. Although Uzziel had never liked being held she did not resist. She briefly grasped my finger gently with her beak, like a final handshake. I continued to pat Uzziel and talk to her, as she lay in my hand. Then I noticed something glistening. At first I thought sunlight had caught her eye. Then I remembered the sky was dark, cloudy and ominous. I looked closer and realized a tear was welling up in Uzziel's eye. Uzziel was crying, without making a sound. Uzziel shed one tear, Uzziel who had never been afraid of anything.

Fifteen minutes later Uzziel's whole body shook with a tremor, as if every electrical impulse had been triggered at the same time. Then Uzziel was gone. Her indomitable spirit was as strong as ever, but her little body had worn out. Uzziel spent her first two hours here in my hand, clinging to my finger. Likewise, Uzziel spent her last two hours in my hand, resting before the long flight home. When a bird dies, the last thing it does is to fully stretch out its wings, swing its feet back, and flap its wings like it is taking flight once more.

Uzziel was with us from Sunday, January 9, 2000, 12:12 p.m. until Wednesday, February 17, 2010, 3 p.m., a total of ten years and thirty-nine days (or 3,691 days).

Consistently over a 10-year period Uzziel was intelligent and creative. Uzziel could be happy, sad, content, lonely, inquisitive, puzzled, or playful. She was always trusting and patient with overly energetic noisy young birds. Uzziel never complained or felt sorry for herself, even as she faced many changes and challenges late in life. Uzziel was never aggressive, fussy, or timid, nor did she develop the negative behaviors large parrots often do. In summary, Uzziel was a pretty amazing little bird.

7.4 Precious

Precious arrived with Blue Boy on November 15, 2004. Precious is a beautiful petite powder blue budgerigar with a bright white head and silver white wings. Precious' disposition is consistently pleasant. Precious is a sweet, happy, feisty, energetic, and inquisitive bird, even though she only has one leg (the right one). Precious has never been aggressive, fearful, or timid toward humans or other birds. Precious was an adult female when she arrived, probably 5 years old. Her cere was medium brown and protruded a bit—one sign of a middle-aged female budgerigar. It is common to see budgerigars rub their beak back and forth on a perch. That action helps to keep their beak and cere clean and trim. That activity is easy for a budgerigar with two legs to perform. It is not so easy for Precious to perform. As a result her cere becomes slightly overgrown between polishing. Precious' life story and how she dealt with her disability is remarkable and demonstrates that many bird behaviors are individually learned and unique to individual birds.

Precious made it clear from the beginning that she did not view herself as disabled or expect special treatment. Nor did she indicate that she felt sorry for herself because she was not like the other budgerigars. Precious was aware that she was different than two-legged budgerigars, in part because she sat much lower on a perch and "walked" differently than they did. However, that did not hold Precious back at all; quite the contrary. Precious did, or at least tried to do, everything the other budgerigars did. And, as will be illustrated later, Precious even figured out how to do a few things that the other budgerigars could not. Precious is the epitome of the "use what you've got" philosophy, often turning her disability into an advantage.

As for special accommodations, they were simple, straightforward, and the other birds enjoyed them too. A few key perches were wider than usual for budgerigars to make it easier for Precious to balance. Precious sat on the narrower perches, but the wider ones were easier for her to rest on. Bubble-wrap was placed under her paper to ensure a soft landing whenever she went down to the tray. Instead of a thin wire swing, swings with acrylic arches were provided so she could lean against the left side of the swing. This was Precious' favorite place to sleep for several years. Round stainless steel dishes were replaced with colorful oblong plastic dishes with perches. To eat out of a round stainless dish Precious had to lay across the dish, then turn her head around to eat. She could eat from a stainless dish this way, but it certainly wasn't convenient or comfortable. This maneuver did not work for a round stainless water dish; hence the oblong plastic dishes with perches on which she could balance. Lastly a 3 x 6 inch (7.6 x 15.2 cm) wire mesh "loft" was added to the back wall of the aviary, complete with its own toys: a cotton string toy with colored beads, a small manu mineral log with millet, and a tiny mirror with a bell.

Budgerigars, like other birds, alternate resting their right and left legs throughout the day and night. Precious learned to lay on the loft to rest her right leg. She also learned how to lay on her side, stretch out her wings one at a time, and scratch her head with her right foot while on the loft. Stretching and scratching are easy for a two-legged budgerigar to do while standing on a perch, but impossible for Precious to do. Instead, she taught herself how to use the loft and the laws of gravity to her advantage.

The other birds thought the loft was fun because they could run across it or hang upside down from it. Before long, two lofts were installed in Precious' aviary: one for her to rest and preen on, and one for play. When Precious finishes resting and preening, she joins in the fun, playing with the beads, hanging upside down with one foot, or chewing on the mineral log. After the latter she has a small brown dusty spot on her head. Precious can spin around almost 360° when suspended from the loft. The other birds, who are hanging on with two feet, cannot. (See Photograph 7.4.)

Precious is a precision flyer. When she was put in her aviary she flew from side to side and top to bottom several times with uncanny accuracy, even though she had not been in the aviary before. Whenever Precious wants to go somewhere (to a perch, dish, toy, or swing), she flies to that precise spot. In contrast, the other budgerigars fly to the side of the aviary nearest the item, then climb from there. Precious always reaches the item faster with her more direct route—a distinct advantage if the item is a new treat. Most birds hold their legs and feet straight back, facing their tail, when they fly. In contrast, Precious holds her leg folded forward toward her beak when she flies. This way her center of gravity is co-located with her beak, making it easier to balance her flight.

Because she only has one leg Precious cannot walk by alternating feet like other birds do. Nor can she grasp a perch by putting two toes in front and two toes in back. Instead Precious had to devise on her own an entirely new method of locomotion. To do so Precious clenches all four toes together and curls her leg under the middle of her body. She uses this posture to balance on a perch and/or hop from point to point. It is unclear how long it took Precious to learn or when she learned to do this. Precious was already quite adept at maneuvering like this when she arrived. It is amazing to watch her land on a perch with1-leg or hop the length of a smooth 0.5 inch (1.27 cm) wide by 36 inches (91 cm) long perch without falling off. The development of this method of locomotion and the level of expertise involved can only be explained by the creative problem solving, ingenuity, and persistence of the individual bird involved. Precious could not have learned this behavior by observing adult

Photograph 7.4 Precious' Loft. (Top) Precious rests her solitary leg on a small wire loft on the back wall of the aviary. Precious often takes short naps while resting on the loft. Precious learned to lay on her side while on the loft in order to scratch her head and stretch her wings. (Bottom) Precious is joined by her buddy Dandelion. Two-legged birds enjoy the loft too, because they can run along the top or hang upside down from the bottom. © D. Herrmann

budgerigars. All she had to work with were her own mental and physical resources. Consequently, this behavior cannot be attributed to genetic programming or "unconscious sleepwalking". Rather, this behavior is akin to the skill acquired by a professional trapeze artist after years of training and practice—except that Precious taught herself. (See Photograph 7.5.)

Members of the parrot family enjoy climbing around the perimeter of their cage or aviary. It is not uncommon to see birds climbing around and around their cage, from top to bottom and side to side; often very rapidly. In this instance a parrot's beak functions as a "third foot". As a parrot climbs two of their three digits are attached to the side of their cage. The third digit is used to move forward

Photograph 7.5 Perching on One Leg. Through ingenuity and persistence Precious taught herself how to balance, perch, walk, and climb using her solitary leg. To balance on a perch Precious clenches all four toes together and curls her leg under the middle of her body, as shown here. Precious also uses this posture to hop from point to point, similar to a pogo stick. This is unusual since budgerigars don't hop; they walk or run. Most birds hold their legs and feet straight, facing back toward their tail, when they fly. In contrast, Precious holds her single leg folded forward toward her beak when she flies. This way her center of gravity is co-located with her beak, making it easier to maintain balance during flight. Otherwise, there would be more weight on her right side. © D. Herrmann

in whatever direction they are going. Climbing presented another challenge for Precious, since she only has two digits: a beak and a leg. But that did not prevent her from trying. On predominantly horizontal bars that are spaced 0.5 inch (1.27 cm) or less apart, Precious can climb reasonably well, although not as fast as the other budgerigars. Vertical bars, where the horizontal crossbars are 6 inches (15 cm) or more apart, present a real challenge. Precious tries to climb up a vertical bar by alternating between her foot and her beak. It is a slow process. When that doesn't work, or she begins to lose her grip, Precious dangles from her beak and swings from side to side in order to get enough momentum to propel her foot to the next hold, much like people do when swinging over a wall in an obstacle course. This way Precious can partly overcome the 2 inch (5 cm) gap between her beak and her foot. Usually this approach works. Once in a while Precious slips; in which case she simply flies to the intended destination. Climbing is another example where Precious, through creative problem solving, ingenuity, and persistence, developed and taught herself a unique skill.

In addition to hopping, climbing, and flying, Precious created a fourth means of getting from Point A to Point B. The aviary Precious lives in is 37.5 inches (95 cm) long by 27.5 inches (70 cm) wide by 63 inches (1.6 m) high. For stability halfway up on all four sides there is a solid 1 inch (2.54 cm) square horizontal ledge. On her own Precious figured out that she could lay on her left side facing outward, grasp the vertical bars with her right leg, and scoot along the ledge, and rather quickly. By doing this Precious can scoot under dishes, toys, and perches—places the other birds cannot go. In fact one of Precious' favorite pastimes is to lay on the ledge, between the aviary bars and a string of four brightly colored whiffle balls, each with a rattle inside. She positions her head and beak right next to the whiffle balls. Holding the top ball, she shakes all four balls until they rattle. Precious does this for 10 to 15 minutes at a time. Sometimes she shakes the whiffle balls so hard that the string comes apart and the balls fall on the tray. Precious resumes play as soon as the toy is put back together and inside the aviary. The other budgerigars have to sit on the perch below the string of whiffle balls and shake the bottom ball.

Precious can scoot the entire length of the aviary and turn corners without falling. Precious scoots along the ledge frequently and seems to enjoy doing so; probably because she is the only bird that can do this. The other budgerigars cannot scoot because their left leg is in the way. A couple birds tried (once) and fell off the ledge. Since there is no equivalent to wild budgerigars scooting along a smooth ledge on their side, this unique behavior cannot be attributed to genetic programming either. (See Photograph 7.6.)

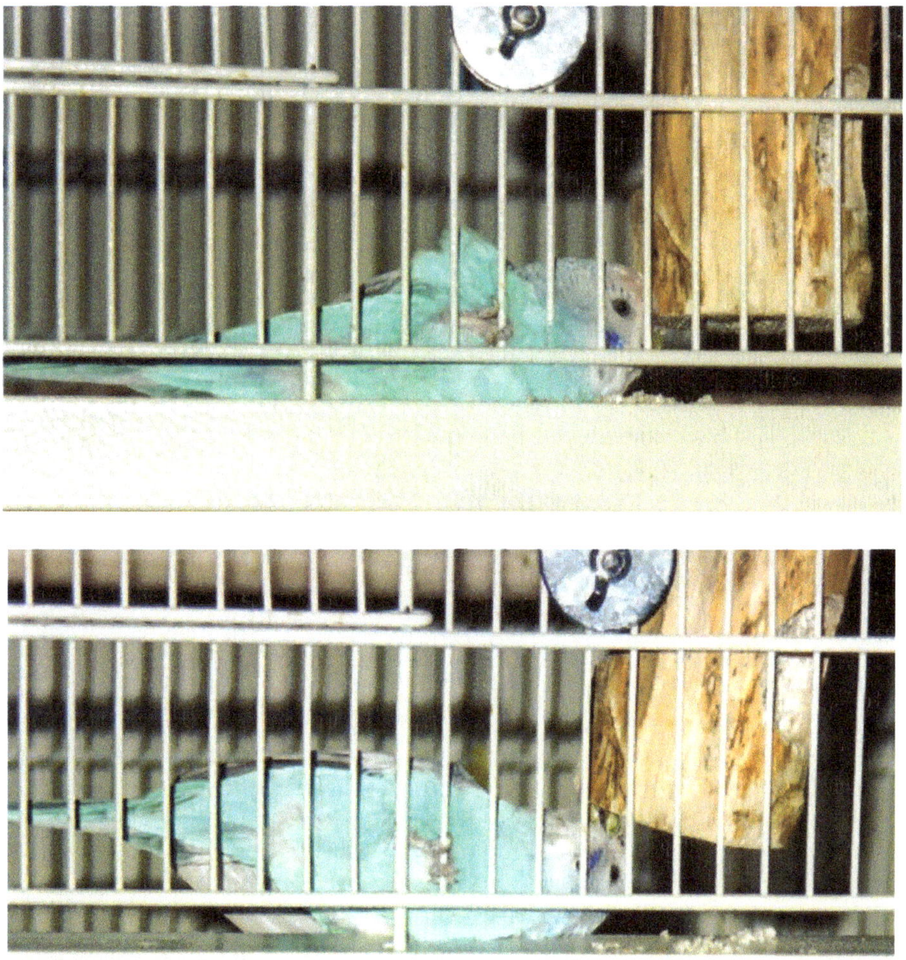

Photograph 7.6 Scooting Along the Ledge. Precious figured out a novel way of getting from point A to point B: scooting along the ledge. Precious lays horizontally on the ledge and grasps the vertical bars with her right leg to propel herself forward, as shown here. Precious can scoot under dishes, toys, and perches—places the two-legged budgerigars cannot go—and rather quickly. Here Precious scoots over to chew the kabob toy. There is no equivalent to this scooting behavior in the wild. © D. Herrmann

Precious is aware that she is different than the other budgerigars because of the need for different means of locomotion. But Precious has never given any indication that she considers herself disabled. Equally informative is the reaction of the other budgerigars to Precious' situation. Obviously the other budgerigars can see that Precious sits about 0.5 inch (1.27 cm) lower on a perch, hops instead of walking, and climbs differently than they do. They also know that Precious can scoot along the ledge and they cannot. However, the other budgerigars never give any indication that they consider Precious disabled or display any aggression toward her, as wild birds occasionally do toward a sick or injured bird. In fact, two males considered her petite height attractive and pursued her in a romantic fashion. Regardless of whether the budgerigars are playing, swimming, eating, singing, flying, or engaging in another social activity, Precious is a full participant. If the budgerigars are rough housing or playing tag, Precious holds her own, even with birds half her age or younger.

Humans who have a disability in one area often have enhanced cognitive, perceptual, and/or sensory abilities in other areas. So it is with Precious. Because she has had to solve several problems other budgerigars did not have to deal with, Precious is by nature very inquisitive. Whenever a new toy, a new swing, a new perch, or a new kind of food or treat is introduced into her aviary, Precious is the first one to check it out. She finds new items and new experiences, like moving the aviary to the other side of the room or closer to the window, exciting. Precious' behavior is the opposite of birds who are change phobic.

Whenever a new toy is introduced, Precious is the first to experiment and figure out how to play with it. For example, one of Precious' favorite toys early on was called a bird bungee cord. It consisted of a coiled plastic cord (similar to a telephone cord), two square plastic discs strung over the cord, and a round plastic ball with spikes on the bottom. Each piece was a different bright color. The bungee toy hung on the side of the cage above a perch. Precious figured out how to bounce up and down while holding onto one of the spikes on the bottom of the ball with her beak. She would do this for 10 to 15 minutes at a time, on a daily basis when the toy was new. None of the other birds in her aviary or the other budgerigar aviaries figured out how to do this. Instead they focused on how to move the square discs up and down the bungee cord. It would be interesting to learn which method of play the toy designer envisioned.

Precious is the first to figure out acrylic toys that have movable parts that are supposed to be spun, shaken, or moved from here to there. Likewise Precious is the first to attack shredding toys. In particular she likes kabob toys made from the stem of dried yucca plants. The first one was 8 inches (20 cm) tall and 2 inches (5 cm) in diameter. Two 1.5 inch (3.8 cm) round openings had been started on each side of the kabob to encourage chewing. At first the other budgerigars were intimidated by the size of the kabob toy. After a few days they would sit on the top of the kabob or try to push it around. In contrast, Precious understood from the beginning that this was a shredding toy and began playing with it immediately.

Precious scooted along the edge until she was underneath the bottom of the kabob, then began carving out the soft fibrous material with her beak. Next she carved out pieces through the openings on the sides and top. The other birds watched Precious for a while, then they joined in. Precious was the first to figure out that the openings on each side of the kabob connected to each other, once the center was hollowed out. The other budgerigars may have seen this, but Precious was the first to try to climb through the holes from one side to the other. Unfortunately she got stuck. Her head, half her body and her wings made it out the other side. But the hole was a snug fit and there wasn't anything for her foot to push against at that point. Precious did not panic or make any alarm or distress calls even though she was trapped. She simply kept trying to move forward or backward. The kabob had to be cut in two with pruning shears to free her. (The outer layer is rigid.) Precious did not panic then either. She simply flew to a perch after being freed.

A few weeks later a new larger kabob was put in Precious' aviary. This one was 3 inches (7.6 cm) in diameter with 2.5 inch (6.3 cm) openings on the sides. How would Precious react? Would she remember being stuck and avoid the kabob? Or, would she remember the fun she had shredding the other kabob and play with the new one anyway? What would the reaction of the other birds be—after all they saw Precious get stuck and what it took to free her? The answer was quick in coming. Within a minute after the new larger kabob was hung in her aviary, Precious was busy carving out the soft middle. After watching Precious for a while, the other budgerigars joined her. No one got stuck this time, or in any future kabobs, because the holes are larger. Interestingly, this large size kabob is marketed for cockatoos and macaws. (See Photograph 7.7.)

Photograph 7.7 Chewing Kabob Toys. Carving out the soft fibrous material inside kabob toys is a favorite play activity for many pet birds. (Top) One day Precious decided to climb through the holes and got stuck. Interestingly, Precious did not panic—even when the kabob had to be cut in two with pruning shears to free her! (Bottom) A few weeks later when a new larger kabob was placed in Precious' aviary she resumed play immediately, without hesitation. Slate II accompanies her, illustrating that budgerigar play is often a group activity. © D. Herrmann

A partial globe was another favorite toy. It consisted of two circular wires about 6 inches (15 cm) in diameter. One wire ran horizontally, while the other ran vertically. Both wires were covered with woven palm leaves alternating with colorful wooden blocks. This toy was hung from the top center of all three budgerigar aviaries, where everyone could see it. Precious was the first one, and for several weeks the only one, to explore the globe toy. She hung from the wires by her right foot and wiggled around among the dried palm leaves and wooden blocks. It didn't take Precious long to figure out that this was another place where she could rest her foot. She proceeded to chew all the dried palm leaves off the bottom half of the globe. That caused the four wooden blocks to slide together, creating a small platform in the middle where the wires intersected. Precious could lay on the platform and rest her leg. The platform was just the right size for one budgerigar and the globe was the highest item in the aviary; that made it a special retreat. Precious slept there overnight on occasion.

The other birds in Precious' aviary watched her customizing and resting on the globe for about 3 weeks before venturing forth. To rest on the platform they have to stand flat footed on the wooden blocks. And, the narrow wire is too thin for a good grip. Since neither position provides a steady balance, they don't visit the globe very often. While the other birds may sit on the blocks occasionally, Dandelion figured out how to use the sharp edge of the blocks to scratch his head. In the second aviary the budgerigars visited the globe two or three times over a period of several months. To date the toy still looks brand new. In the third aviary the budgerigars treated the globe like a shredding toy. After watching Precious, they removed all the dried palm leaves from the top and bottom of the globe. Only the wooden blocks and the thin wires remain. These budgerigars have not returned to the globe since they finished their shredding task. In this instance the same toy was installed in all three budgerigar aviaries and three different results were observed. Precious was the catalyst for playing with this toy and she alone figured out how to customize it to fit her unique situation.

Another toy that caught Precious' fancy consisted of three plastic pegs. The pegs are 0.125 inch (0.31 cm) in diameter and 2 inches (5 cm) long. The pegs are arrayed on a triangular base which is attached to the side of the aviary. Each side of the triangle is 2 inches (5 cm) long. The individual pegs can be pushed and pulled through the triangle. All the pieces are different colors. This toy was installed in all three budgerigar aviaries, with one peg on top and the other two pegs on the bottom. Nearly all the budgerigars pushed and pulled the pegs back and forth through the triangle and/or tried to perch on the narrow pegs. Yellowhead even enjoyed napping on the toy in the sun.

Precious developed a different strategy for playing with this toy. She pushes the top peg outside the aviary, then lays across the two lower pegs as if they were a perch. The toy is located in the top third of the aviary, which gives her a good view. And, the two pegs are just the right size for one budgerigar. This peg toy became Precious' favorite place for an afternoon nap. Understanding that the two lower pegs could be used as a perch—if the top peg were pushed out of the way—indicates that Precious understood the spatial relationship of the three pegs before and after they were moved inside and outside the aviary and how to align all three pegs correctly. (See Photograph 7.8.)

Birds, squirrels, rabbits, and such enjoy bathing and are meticulous about keeping their feathers or fur clean. Early one morning a rabbit was observed scraping the dew off the grass with its front feet, then using the moisture to wash its face and body. Squirrels keep their underside bright white, even though they spend the day digging in the soil and climbing through vegetation. In contrast, humans have to be taught to bathe. In fact, some adults have to be told of the necessity of daily personal hygiene.[128]

Every budgerigar jumped at a chance to bathe. None of them were taught or told to bathe. Even budgerigars that were housed singly when they were very young or in quarantine understood the need to bathe and that swimming was fun. Most budgerigars rush over to the "swimming pool" as soon as it is installed. A few budgerigars wait a minute or two. Most budgerigars jump right in and start splashing around, usually by walking low in the water while flapping their wings. A few budgerigars sit at the entrance for a while and/or take a drink of water first to make sure it is water

Photograph 7.8 Mastering the 3-peg Toy. The three pegs on this toy can be pushed outside or pulled inside an aviary. (Top) Precious learned that the two bottom pegs were just the right size for her to rest on. At first Precious pushed the upper peg out of the way. Eventually she broke the peg to get it out of the way permanently. (Bottom) In contrast, two-legged budgerigars stand on one peg and play with the other pegs, as Yellowhead demonstrates. © D. Herrmann

before jumping in. Swimming time stimulates a lot of sudden energy and chatter in the budgerigar aviaries. The budgerigars are given the opportunity to bathe at least once a week. Generally the group participates for about 30 minutes.

Precious is as enthusiastic about swimming as any budgerigar. If not the "first one in", she is the second. Since she can't walk around in the swimming pool, Precious jumps from the entrance and does a "belly flop" into the middle of the pool. From there she hops and splashes around. Precious is the only budgerigar to put her head completely under water. (The water is an inch (2.54 cm) deep.) That is the only way she can rinse her back. Perhaps she is fearless and just having fun. Or, perhaps a combination of factors is at work. Regardless, whenever Precious comes out of the swimming pool she is completely soaked. (See Photograph 7.9.)

Photograph 7.9 Precious Swimming. Precious does some serious swimming, while Slate II and Dandelion watch. Precious hops and splashes from side to side using her solitary leg. She is the only budgerigar to put her head completely underwater. © D. Herrmann

When it comes to diet, Precious has a good appetite. She likes everything (seed mixtures, extra oat groats, millet sprays, shredded carrots, apples, raw shelled green peas, baby organic spinach, and natural pellets), but never overeats. She particularly enjoys spinach leaves and millet sprays if I hold them for her; after which she sings a short song. Precious is also enthusiastic about chewing up cuttlebones and mineral blocks. Her favorite mineral block is light pink and 1 x 1 x 2 inches (2.54 x 2.54 x 5 cm) in size. To chew it, Precious lays on top of the mineral block while holding onto a bar with her foot. Precious has a small pink spot on her white head after a serious chewing session.

About a year after she arrived Precious suddenly started dumping, one beakful at a time, all the food in her seed/pellet dish on the floor. Her sudden change in behavior was puzzling, since she had never been a fussy eater. Perhaps she was tired of eating "the same old thing" every day, so the brand was changed. Precious continued to dump the seed mixture on the floor. Another brand and another brand were purchased to no avail. Precious ate her fruits, vegetables, and millet spray. But no matter what brand of seed mixture she was served, it ended up on the floor.

Precious was watched carefully one weekend. She did not rush to dump the food on the floor, as if she were looking for something in particular or throwing a tantrum like a young child. Instead, Precious was very methodical and steady and seemed to enjoy what she was doing. Perhaps the food or brand of food was not the problem. It was decided to try an experiment.

Two identical oblong plastic food dishes were placed on opposite sides of Precious' aviary. One contained the usual seed mixture and pellets. The other dish was filled with various dry materials on different days: corn cob bedding, shredded paper, cotton balls, etc. Precious left the seed dish alone, but proceeded to empty the other dish, one beakful at a time, regardless of what dry material it contained. Precious' new behavior had nothing to do with throwing food on the floor. Rather, she was "carving" out a nest in the oblong dish, similar to what she would have done in the wild.

The dish/nest continued to be filled with dry material for several days. It did not matter what type of dry material was supplied. Precious methodically emptied the dish/nest. Then the dish/nest was left empty. Precious hopped into the dish and sat there contentedly for 15 to 20 minutes at a time. She continued to spend time in the dish, with her head facing the front of the aviary, for several days. Her buddy, Mr. Green Birdie, came to visit her. He sat on the dish's perch while she sat in the dish/nest. They were a bonded pair, but not a breeding pair. Years later Precious continues to spend time in the dish/nest a couple times a week. She does not sleep there, she just sits there contentedly. It is likely that this activity fulfills a maternal psychological need. (See Photograph 7.10.)

Yellow Birdie, also a female, is the only other bird Precious allows to use the dish/nest. Yellow Birdie prefers to have the dish/nest dangling from one hook instead of two, so that it is tilted a bit. As soon as the dish is straightened, Yellow Birdie unlatches one of the hooks again, usually the one facing the back of the aviary. Precious has not expressed any objections to Yellow Birdie's customization of the dish/nest. When Yellow Birdie sits in the dish/nest her buddy, Lilac II, comes to visit her and sits on the dish's perch.

None of the other female budgerigars engaged in this behavior, even though empty oblong dishes were installed in all 3 budgerigar aviaries. (Daisy occasionally lays an infertile egg, but she does not use the dish/nest; instead she lets the eggs drop onto the newspaper.) Precious and Yellow Birdie are the only two. Yellow Birdie probably learned this behavior by watching Precious. On occasion Slate II will dump food on the floor. In his case, this is a rapid messy behavior during meal time. He is looking for a particular item, like an oat groat or canary grass seed, and simply pushes everything else out of the way and onto the floor.

Precious gets along well with all the other budgerigars because of her even temperament. She doesn't try to boss anyone around and no one tries to boss her around. Precious was not part of a bonded pair the first year she was here. It is unclear whether or not Precious was part of a bonded pair before she was adopted. Shortly after Mr. Green Birdie (a normal green budgerigar) arrived in late 2005, he and Precious became a bonded pair. At the time Mr. Green Birdie was about 1 year old

Photograph 7.10 In the Dish/Nest. Precious pops her head up to check out the commotion outside her dish/nest. Precious repeatedly "carved" out a nest in her oblong dry food dish. Once it was learned that Precious was "carving" out a nest, and not simply dumping her food on the floor, one dish was purposely left empty. Precious sits in the dish for 15 to 20 minutes a few times a week. Her head faces the front of the aviary and her tail points to the back. © D. Herrmann

and Precious was about 6 years old. That was a significant difference in age considering the normal life span of budgerigars, but neither of them seemed to mind. Precious and Mr. Green Birdie did all the things bonded pairs of budgerigars normally do: allo-preening, allo-feeding, playing with the same toy, sitting and napping on the same perch, and just hanging out together. If another budgerigar crowded Precious when she was on the loft resting her foot, Mr. Green Birdie shooed it away. (See Photograph 7.11.)

Precious and Mr. Green Birdie remained a bonded pair for several years even though other budgerigars moved in and out of their aviary. Mr. Green Birdie had the same cheerful, easy going temperament as Precious. He was not as inquisitive as Precious, but she could usually coax him into trying new things. Overnight Mr. Green Birdie liked to sleep on a swing with an acrylic arch. As he settled in to sleep, Mr. Green Birdie tilted his head upward and arched his back a little. None of the other budgerigars ever slept that way. Precious wanted to be near Mr. Green Birdie when it was time to settle in for the night. There wasn't room on the swing's perch. So she flew to the top of the swing and lay across the arch while holding onto the small metal loop with her foot. This position didn't look very comfortable, but that wasn't the point. Precious could not hold onto the metal loop all night since she only has one foot. An hour or two after the lights in the aviary were turned off Precious could be heard flying to a perch or the loft. No 2-legged budgerigar has ever tried to sleep on top of an arched acrylic swing.

In the fall of 2009 Mr. Green Birdie developed a small fluid sac on the exterior of his upper breast. The fluid sac appeared suddenly, but stayed small in size. Mr. Green Birdie's activity level, energy level, and appetite remained normal. Three months later the fluid sac expanded rapidly. Mr. Green Birdie had trouble flying from the bottom of the aviary to the top, due to the extra weight. He had to break the journey into two pieces. First, he would fly from the bottom to a perch. Then he would fly from the perch to the top of the aviary. The veterinarian drained the fluid sac. Lab tests indicated that Mr. Green Birdie had a rare blood cancer for which there was no treatment. During the hour long journey home from the veterinarian Mr. Green Birdie sat still or stumbled around like a person who has lost a lot of blood. Within four hours the fluid sac was full again.

Photograph 7.11 A New Pair. Precious and Mr. Green Birdie rest on the loft shortly after they became a bonded pair in 2006. Bonded pairs go through their daily activities together. © D. Herrmann

A separate aviary was quickly set up for Mr. Green Birdie and Precious. A grill was placed across the horizontal ledges to cut the distance from top to bottom in half. Newspaper and bubble-wrap were placed over the grill and all their favorite toys, perches, and swings were moved into the new aviary. Then Precious and Mr. Green Birdie moved in. Mr. Green Birdie mostly slept for the next 24 hours. After that he tried to continue doing everything as before including singing, but his flying days were over. A few ladders were added at key locations and Mr. Green Birdie quickly became adept at climbing wherever he wanted to go, even to his swing. The food and water dishes were moved to the grill/floor, so Mr. Green Birdie didn't have to worry about maintaining his balance. Mr. Green Birdie came running whenever breakfast was served in the morning. His walk made a distinct sound. He would take six quick steps; the seventh step was always louder and slower. Precious adjusted to the new environment quickly and continued to hang out with Mr. Green Birdie just like before.

After 8 weeks Mr. Green Birdie's condition declined rapidly. He could no longer climb, even with the help of a ladder. Instead he spent the day on the grill/floor. Mr. Green Birdie ate smaller meals, but more of them. He was given a fresh millet spray daily to make sure he was getting enough calories. When it was time to sleep, Mr. Green Birdie snuggled up in one of the back corners. It was difficult to change his paper since he was always on it. However, Mr. Green Birdie quickly learned when to go to the right side or left side of the aviary to make the paper changing process as quick and easy as possible. Precious began taking care of Mr. Green Birdie. She cleaned his feathers daily and fed him every now and then.

The last few days Mr. Green Birdie got weaker and weaker. He slept more and fell over once in a while when he was walking, but he never gave up. Precious continued to take good care of him. On the morning of January 30, 2010 Mr. Green Birdie was found laying on his side next to his millet spray. He had died during the night. Precious was there sitting next to him.

Mr. Green Birdie and Precious were a bonded pair for over 4 years, essentially Mr. Green Birdie's entire adult life. The loss of a mate is a traumatic event for any bird, especially those like budgerigars that mate for life. Precious exhibited all the behaviors characteristic of the two-week mourning period for birds: reduced activity level, reduced appetite, and a downcast spirit, which was very unusual for her. She wouldn't even come out of the aviary for flying time. Some birds will search for their lost mate by calling out to them and visiting all their favorite places over and over again. Since Precious was with Mr. Green Birdie when he died, she did not do that.

After the 2-week mourning period Precious' aviary was reconfigured into a full size budgerigar aviary and some new toys were added. Precious began to perk up and be her old self again. During this time Precious received extra attention and was hand-fed treats. A month later two budgerigars moved into her aviary. She seemed to enjoy their company but did not bond with either of them.

Instead, Precious took a fancy to Yellowhead who was in the aviary next to hers. Yellowhead looks like a normal green budgerigar, except that his entire head is yellow, not just the top. It was obvious that Yellowhead was attracted to Precious as well. Whenever it was his turn to fly he would fly to Precious' aviary. Then he would climb around and around on the bars trying to figure out how to get in. Precious followed him around from the inside. Even when they were both in their own aviaries they sat facing each other. This pattern of behavior continued for a couple weeks.

Before moving them in together an experiment was conducted. The budgerigars in Yellowhead's aviary and Precious' aviary were let out to fly at the same time. This provided an opportunity to see how well Precious and Yellowhead got along, and if they wanted to room together.

It didn't take long to get an answer. As soon as Precious' door was open, she went zooming out the door after Yellowhead. She flew straight from the middle perch through the open door without even stopping on the landing perch. Precious and Yellowhead played together and flew around together. When it was time to go home Yellowhead moved himself into Precious' aviary. The other budgerigars returned to their normal homes. Precious was happy. Yellowhead was happy and he got along well with Precious' two roommates.

All was well with this young romance for about a month. Then Yellowhead's enthusiasm for Precious got the best of him and he became too rambunctious. He followed Precious around everywhere and wanted to play nonstop. While Precious enjoyed his company, she needed some rest; in particular she needed to rest her one leg frequently. Precious was 10.5 years old. Yellowhead was 1 year old; he just didn't understand.

Precious began to show signs of exhaustion, so unfortunately they had to be separated. Yellowhead was put back in his old aviary. It was sad for both of them. Precious was sad, but relieved. After a week or so she resumed her normal cheerful routine. Yellowhead was beside himself for a couple weeks. In effect he had lost a mate, but he could still see and hear her. Yellowhead continued to try to figure out how to get in Precious' aviary whenever he was out flying. After a while Precious did not reciprocate his visits. Even though several years have passed Yellowhead has not formed or attempted to form another pair bond.

Dandelion, a normal green budgerigar, moved in Precious' aviary a few weeks after Yellowhead moved out. He was 1 year old at the time and is very gentle. His temperament is similar to Mr. Green Birdie's. It didn't take long for Dandelion and Precious to become a bonded pair.

November 15, 2010 Precious celebrated the sixth anniversary of her homecoming. Given that she was about 5 years old when she arrived, that meant she was at least 11 years old. That is not bad considering she only has one leg and the average life expectancy of indoor budgerigars is 6 to 8 years. Precious has never been ill and her only visit to a veterinarian was her initial check-up and nail trim. Somehow she has managed to keep her nails trim since then, even though her nails do not come in contact with perches. Precious is still going strong and doing all of her favorite things: flying, swimming, hanging upside down from the loft, scooting along the ledge, rattling the whiffle balls, chewing kabobs, sitting in the dish/nest, and resting on the toy with 2 pegs. Precious takes more naps now than she did six years ago, but she had earned them, just like she has earned her name. (See Photograph 7.12.)

Photograph 7.12 Takeoffs and Landings. Precious is an energetic precision flyer. Unlike other the other budgerigars, Precious has to find a place to land where she can hold on and balance with one leg. But that doesn't slow Precious down at all, as shown here. © D. Herrmann

7.5 Slate, Lilac, Sunflower and Sky

This quartet of budgerigars arrived May 1, 2005 in the midst of the avian flu scare. There was so much misinformation flying around many people became concerned that their pet birds might be carriers. The few avian flu outbreaks that did occur around the world were due to raising poultry in over-crowded and unsanitary conditions. The projections of mass human epidemics due to the annual migration of waterfowl never materialized. Had government epidemiologists consulted with avian veterinarians, ornithologists, and bird enthusiasts, they would have realized the fallacy of their projections. Similarly, the probability of a budgerigar or other pet bird contracting or transmitting avian flu, when the bird has spent its entire life indoors is nil. Unfortunately, that fact was lost in the noise and many loyal, decent pet birds were needlessly put up for adoption.

Parrot rescue/adoption organizations are frequently inundated with small birds in the spring. Many parents have not learned that pet birds are not animated toys that do not need batteries. Rather pet birds are live sentient beings with intellectual, social, physical and emotional needs. The birthday and Christmas presents that parents insisted on buying, even though the child the pet was being purchased for never expressed any interest, become "throw-away pets", when no one in the family wants to take the time to care for the little bird anymore. Or, the parents never expected the bird to live more than a year or two, it did (somehow), and now they want to get rid of it. Since budgerigars only cost $22, they are dumped en masse on rescue/adoption organizations. That is, the lucky ones. Those less fortunate are simply released into the wild, where they might survive 24 hours before starving to death or being eaten by a predator.

In April 2005 an email was received from the local parrot rescue/adoption organization. Could we please take in some more budgerigars, even on a temporary foster care basis? The organization was swamped with birds; many others were still waiting to be picked up. We said sure and asked for a week to get a new aviary set up.

This quartet of budgerigars arrived mid-afternoon totally frazzled. The birds had gone from their original homes to an intake center to our aviary in a matter of days. The birds had no idea what was going on or why they were uprooted and moved twice in quick succession. The four birds were from different homes, where most likely they were the only pet bird, living alone in a cage that was too small. The four budgerigars did not know or appear to like each other. Being thrown together with three strange birds in a large aviary in a strange room was just too much. The budgerigars were nervous, tired, and disoriented. As a result there was a lot of fussing and fighting over a particular perch, swing, and food dish, even though there were many of each item in the aviary. These budgerigars had never lived with other birds or in a large aviary where there were multiple items to choose from. It took the budgerigars several days to sort all this out and get used to their new environment.

The budgerigars did not even have names when they arrived. They were just four random budgerigars of many the parrot rescue/adoption organization placed that year. We gave them names based on their appearance and guessed their age. Sometime later the parrot rescue/adoption organization called to see if we could continue providing foster care. We replied that we wanted to adopt the four budgerigars. They had slowly settled in and we could not envision or condone putting the birds through the trauma of being rehomed again.

Slate

Slate was a young dark or slate blue female budgerigar, probably two years old. Slate was totally traumatized by being rehomed. The multi-bird environment seemed to make her think she was back in the pet store. A few weeks later Slate was still disoriented, seemed depressed, in spite of extra attention, and was not eating enough. So Slate was taken to the veterinarian. The veterinarian performed an examination and did not find anything wrong. He recommended taking an x-ray. He neglected to mention that involved clipping a bird's wings. Before I could say no, Slate's flight feathers were trimmed. The technician posted the x-ray when it was ready. I looked at it. As soon as I said that I did not see anything wrong, the veterinarian snatched the x-ray. He said that Slate needed an MRI instead and called to make an appointment at another veterinarian clinic, again without asking me. In the meantime he recommended a regimen of antibiotics to perk Slate up and boost her appetite.

I left with an uneasy feeling. Was the veterinarian faking it because he did not know what was wrong, or was he being abrupt because he was over-booked, or both? Slate was not sneezing or wheezing, nor did she have diarrhea or ruffled feathers—the typical signs of a bacterial infection in a pet bird. Antibiotics are often (too often?) prescribed to boost human health and energy, so I reluctantly decided to follow his recommendation. Since antibiotics are dispensed in a pet bird's water, Slate had to be moved into a small cage by herself.

When Slate was taken out of the travel cage she took off flying, still nervous from the journey to and from the veterinarian's office. Or at least she tried to fly. Because the veterinarian had clipped her wings, Slate could only fly about 3 feet (1 m) high before coasting to the floor. Slate kept trying, three, four, five, six times before giving up. The fact that she could no longer fly totally demoralized her. Slate stood still when she was picked up and put her in her new private cage. Slate sat on the same perch and did not move, eat, or sing all evening, even though the other birds were calling to her.

In a few days it was obvious that Slate preferred being alone in the small cage. She became excited when her cage was set on the floor to clean it or on the pass-through so she could have extra attention. At those times Slate would clown around and put on a little show, like hanging in a corner of her cage and turning inside out several times quickly, while chattering away. Slate had obviously been a child's pet. (See Annex D.)

When Slate suddenly began doing all these tricks once she was in a small private cage, it was easy to understand the environment she had come from. It was sad that Slate and the child had been separated. It was no wonder Slate reacted to being rehomed the way she did. Solitary birds bond to their primary caretaker. When they lose their primary caretaker, it is like losing their best friend, but they do not understand why this loss has occurred. Many pet birds go through a ~ 2-week mourning period after being rehomed.

Slate began eating more, slowly. Some mornings she would eat as many as four baby spinach leaves, if I would hold them for her. Unfortunately antibiotics are not an appropriate prescription or remedy for sadness and loss. Instead of curing Slate, the antibiotics killed the good bacteria in her digestive system. Although she was consuming the right amount of food and water, she could not digest or absorb any of it. Several proposed therapies were tried to restore the good bacteria; none were successful. In the end Slate died, essentially from abandonment, starvation, having her wings clipped, and too many changes in such a short life. This is one case where I never should have taken a bird to the veterinarian. Likewise, the veterinarian should have admitted that he did not know what was wrong and at least have considered the possibility that the problem was emotional and not physical. Slate paid the price for both our errors. (See Annex E.)

It is an established fact in human medicine that one's emotional state is an important factor in one's overall health. People that are happy, content, confident, and relaxed tend to be healthier and live longer. In contrast, people who are nervous, stressed out, fearful, or depressed tend to be more susceptible to disease and take longer to recover. A person's mental state can suppress or weaken their natural immunity.

Consequently, it is logical for veterinarians to inquire about a pet bird's mental state when evaluating its overall health. This is particularly important when treating pet birds that have recently experienced the trauma of losing a mate or being rehomed. Outwardly a pet bird may exhibit signs of an illness or other medical condition, when in reality it is working through a difficult emotional period.

Lilac

Lilac was a beautiful translucent white budgerigar. In the correct light, strains of light blue or lavender were visible, especially on her front and back. Hence she was given the name Lilac. She was a very young female budgerigar, not quite a year old; probably an abandoned Christmas present. Lilac was not tame, nor was she afraid of people. Lilac simply tolerated people. Lilac was reasonably active playing by herself, but very skittish around other birds.

Not long after the four budgerigars arrived, small puddles of clear water 1.25 inch (3.17 cm) in diameter began to appear on the paper. This is not normal for budgerigars. During the day budgerigars generally have one dropping per hour, a small round semi-solid dropping that is about 0.10 inch (0.25 cm) in diameter, black or brown on the outside and white in the middle. They do not pass clear water, certainly not 1.25 inch (3.17 cm) in diameter. On occasion if a budgerigar has drank a lot of water or eaten a watery fruit, the dropping might be surrounded by a little water.

Since there were four budgerigars in the aviary, it took a while to determine which bird was making the puddles. It was Lilac. She expelled water every hour or more frequently, but droppings only every two or three hours. Worse, whenever she expelled water her vent became noticeably prolapsed, then returned to normal. However, she did not appear to be straining or in pain. She was losing a lot of water, but not enough droppings. Yet her food and water intake was normal. Why hadn't anyone noticed this before me, especially since she had been kept as a solitary bird. Didn't anyone notice this unusual behavior? Or, didn't anyone care? Pets that have medical conditions are often dumped on rescue organizations by people who do not have or who are unwilling to commit the time, money, and emotional resources needed to help their pet get well.

That night I did a series of Internet searches and consulted all the bird medical and first aid books I had. All the sources agreed. The excessive loss of water, temporary prolapsed vent, and insufficient droppings pointed to the same thing: inadequate elimination of wastes. For whatever reason, Lilac's digestive system was not expelling all the accumulated wastes. Immediate treatment was recommended. If the waste matter accumulated, over time it would lead to toxemia and death. Plans were made to take Lilac to the veterinarian first thing in the morning.

But, it was too late. In the morning Lilac was laying on the paper on her side. She thrashed around for 10–15 seconds in pain, then stopped exhausted. Lilac thrashed once more, let out a loud painful cry, and was gone. Lilac was not with us very long. I hated to see her suffer so. I felt as it I and all the humans who had touched her short life had let her down. If I had known earlier or been told when Lilac arrived that she needed medical treatment, I would have taken her to the veterinarian immediately. Lilac did not deserve to die such a painful death and at such a tender age.

Both Slate's and Lilac's deaths were unnecessary, occurred at a very young age, and were due to human mistakes. Neither Slate nor Lilac lived long enough to pick out a favorite toy or become buddies with another bird. All of this bothered me a lot. They both should have had a full happy fun life that every pet bird deserves. To honor these two little guys who got a raw deal at human hands, I picked out two new budgerigars that looked like Slate and Lilac. They were named Slate II and Lilac II, so the original birds could have a second chance and live on, at least in name. Lilac II turned out to be a male. He is feisty and fearless and just celebrated his ninth homecoming anniversary. Yellow Birdie is Lilac II's buddy. Slate II, who just celebrated his sixth homecoming anniversary, is a bit shy. Slate II is a strong flyer and likes to sit on the top of the larger birds' houses, as long as their door is shut. Slate II currently holds the record for being the fastest to demolish an entire cuttlebone. (See Photograph 7.13.)

Sunflower

Sunflower was a middle-aged female budgerigar when she arrived, probably five years old. Sunflower had bright lemon yellow wings and a yellow head. Her front and back were yellow with a slight greenish tinge. Sunflower had the opposite reaction to being rehomed than Slate. In contrast, Sunflower was bossy and aggressive. Sunflower was a very determined bird; she knew what she wanted and was determined to have it, whether it was a perch, toy, swing, or food dish. Sunflower did not have a problem with sharing. She never tried to hoard all the toys or dishes. But, if Sunflower wanted a particular toy, swing, perch, or food dish, the other birds were expected to get out of her way.

Sunflower even tried to steal Blue Boy from Uzziel, but was unsuccessful. Uzziel was the only bird Sunflower could not boss around. Because Sunflower was so aggressive the other birds would not play with her. Certainly none of them wanted to be her buddy, though she did not seem to understand

Photograph 7.13 Cuttlebone Chewing Champion. Slate II is always the first to check out a new cuttlebone and the fastest to demolish one. Cuttlebones are used to supplement an indoor budgerigar's diet with minerals and trace elements that wild budgerigars obtain from the soil. Indoor budgerigars enjoy chewing cuttlebones and it helps keep their beaks trimmed. Wild budgerigars do not engage in this activity. © D. Herrmann

why. Sunflower was a fast flyer, but not very precise. Sunflower was not afraid of anything, including people. She was not interested in pats. Being self-sufficient, Sunflower did not want or need a lot of attention. Having been abandoned by people twice before, Sunflower had steeled herself against bonding with them.

Sunflower was a champion chewer. She chewed through a 0.5 inch (1.27 cm) wooden dowel in less than three days. She destroyed a wooden budgerigar-sized abacus toy in a few days. It was replaced with the parrot-sized version of the same toy. It was essentially shredded in little over a week. Sunflower was the only budgerigar to chew through either a perch or an abacus toy. Sunflower loved a swing that had a dangling 0.5 inch (1.27 cm) cotton rope tassel. She attacked the rope with enthusiasm, chewing and pulling out individual threads. Several female budgerigars have enjoyed doing this also. The males seem to prefer dangling upside down from the rope. Sunflower never shredded her paper (too easy), but did give Slate II some serious competition in the race to see who could demolish a cuttlebone the fastest. Sunflower would walk back and forth sideways across her perch very fast whenever she was excited. (See Photograph 7.14.)

In late 2007 Sunflower ended up in a flight cage by herself for a short time, a time-out for being too bossy. She was clearly unhappy being by herself—it is no fun being bossy, if there is no one to boss around!

A new group of homeless birds arrived about then. We monitored everyone's temperament while they passed their quarantine period. Then we established three budgerigar aviaries: one for gentle

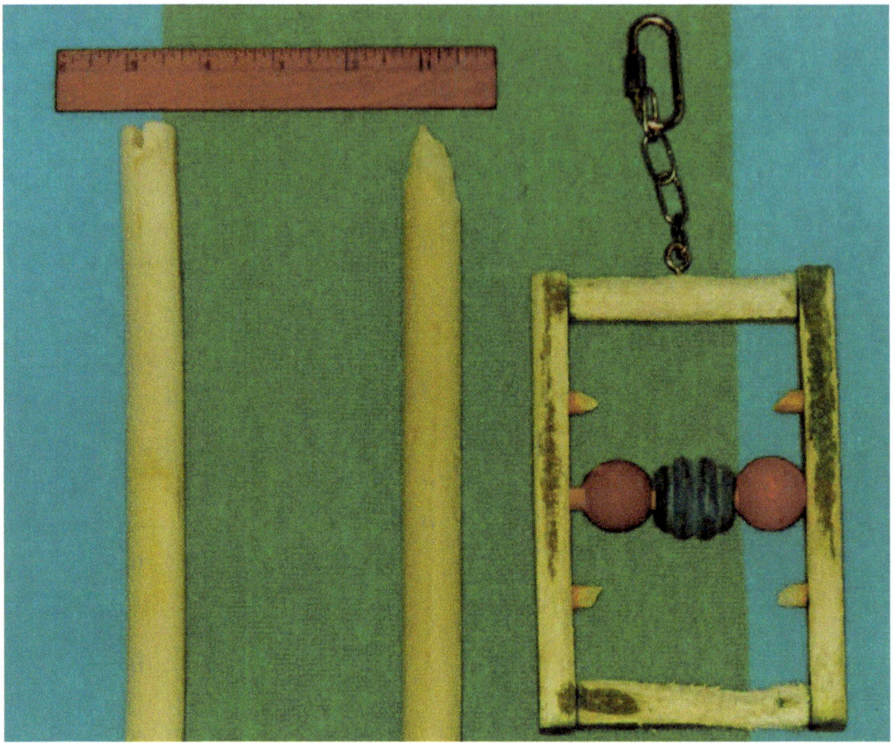

Photograph 7.14 Chewed Wooden Toys. Sunflower loved chewing anything made of wood. This behavior is consistent with wild budgerigars carving out tree holes for nests. Sunflower's chewing spurts were sporadic. But once she started a chewing project Sunflower didn't quit until she was finished. It took less than 3 days for her to chew through the 0.5 inch (1.27 cm) wooden dowel shown in the center, which made the perch unusable. Sunflower shredded the parrot-sized abacus toy shown on the right in less than a week. That was after she demolished the smaller budgerigar-sized version of the same toy. Sunflower was the only budgerigar to chew wooden toys this thoroughly. © D. Herrmann

birds, one for rowdy birds (usually the younger birds who are overly energetic, but not aggressive), and one for aggressive bossy birds, like Sunflower. Everything went fine with the first two aviaries. Everyone got acquainted quickly and began playing together almost immediately. But the bossy cage was another story. Each of these four birds had been the "boss" of their previous cage. They expected, assumed, and demanded that their new cagemates defer to them. But it did not work out that way. It was hilarious to watch as each bird in turn tried to assert over and over again that they were the one and only boss. In the end they all learned to get along and in fact seemed to enjoy having someone around that was as bossy as they were.

Sunflower always held her head high, even in the end. In the fall of 2009 she was still holding her own when playing games with the other bossy birds. But by early spring 2010 it was obvious that Sunflower was slowing down a bit. The first sign was when she did not come out to fly in the evenings. (See Table 7.7.)

The first two times Sunflower did not want to come out to fly I did not think much about it; she was a very independent bird. The third time Sunflower did not come out, it was obvious she was slowing down. About the same time Sunflower began spending the majority of her day on one of two perches. Sunflower still flew down for food or water. She ate and drank a normal amount, her droppings were normal, and she was alert. It was best just to let her be; Sunflower was in her house with her buddies.

Table 7.7 Let the Birds Decide.

I do not believe in forcing a pet bird to do something or forcing myself on a pet bird. Some sources insist that birds should be forced to learn the step-up command. Other sources insist that birds should be forced to become tame immediately after arriving home, like breaking a horse. Others recommend withholding food, so a bird will be so hungry she will do anything in order to obtain the treat.
Instead, I let each bird decide how much interaction and what kind of relationship it wants to have with its human companions. Whether it is a lot of attention, a little attention, or somewhere in between, it is the bird's decision. Whether they want to be a snuggle bug, will tolerate occasional head scratches, or something in between, I respect the bird's decision. The ability to make choices is an important part of a bird developing emotional and intellectual maturity.
I also recognize a bird's right to change their mind over time. This flexibility is particularly important when dealing with rehomed birds. A rehomed bird may want to be a snuggle bug, but it may take a year or more before she trusts her new caretakers enough to enter into that type of relationship. Or, a pet bird that was a snuggle bug may suddenly be more interested in playing with another bird than spending time with a human. Birds, like humans, go through different phases during their life. In this instance, a pet bird is just going through a new growth and learning experience. She has not forgotten her human companions.
And so it is with play time. If a bird does not want to come out to play or fly, I do not force it to. That would not be consistent with the definition of play time.

After a few days Sunflower was no longer flying down to the food and water dishes. She was climbing down to them. A couple times the other birds were too rowdy with Sunflower and she fussed at them to leave her alone. It was sad to see this proud domineering bird have to behave so. I thought about moving Sunflower to Uzziel's condo. In the end it was decided to leave her in the aviary, as long as she was getting enough food and water and the other birds were not bothering her too much. There was a concern that it would be too demoralizing for Sunflower to have to move to a small cage.

At 1 a.m. a few days later I heard a commotion in the aviary and went to see what was happening. Sunflower had fallen off her high perch and onto the cage bottom. She did not make a sound. She looked at me for a while. Then with a very determined look she took off flying. She tried to fly to her high perch, but could not reach it. Instead she fell back onto the cage bottom. Sunflower tried to fly back up two more times, but was unsuccessful. Then she climbed up to a low perch and went to sleep. At 2 a.m. Sunflower fell down again. I stayed with her until she climbed back up and went to sleep.

Sunflower was still on her perch in the morning. I put her breakfast and a clean paper (with bubble-wrap) in Uzziel's condo. I went over to Sunflower's aviary and watched her for a while. This move was not going to be easy emotionally for Sunflower. Before I opened her aviary door, Sunflower fell off her perch again. Only this time she did not try to climb back up. Sunflower walked around slowly, then stumbled. Sunflower walked some more and stumbled again. She lay there for a while. Sunflower never thrashed around or cried out in pain, even though I am sure she was very uncomfortable. I patted her and talked to her, which she seemed to appreciate. Sunflower did not want to go. With a determined look in her eyes, Sunflower rallied three times; then she was gone.

Sky

Sky was a middle-aged female budgerigar approximately six years old when she arrived. Sky was sky blue and had a white patch on her front that resembled a cirrus cloud. Sky was a sweet happy bird. She liked to clown around, but was a bit clumsy. Sky bumped into the top of her cage every now and then when flying and may have had a slight vision problem. Sky was delighted with and would play happily with any toy. She was not fussy about anything. Sky played with all the other birds but was never part of a bonded pair.

The process of being rehomed left Sky unsure of herself. She was grateful for the slightest bit of attention. It was almost as if Sky could not believe that someone was actually paying attention to her. Never bossy or aggressive, Sky resided in the gentle aviary. Sky did not shred her paper. Instead, she would wait until another bird had finished shredding the paper. Then she would play with the shredded pieces, arranging and rearranging them, like building sand castles on the beach.

Sky figured out that water is coldest right after it is served and spinach wilts quickly. As a result, Sky would come running whenever I changed the water and take one or two big drinks with glee. She would also eat a baby organic spinach leaf or two, if I held it for her. Sky always ate the stem first, then the leafy part. Wild birds often eat the stems of plants and leaves, rather than the leaf itself.[12, 186] What appears to be random crop damage is actually a rather smart activity on the birds' part. The concentration of nutrients is higher in the stem than the leaf itself, since the stem serves as an aqueduct for transporting nutrients.[12, 186] Humans, on the other hand, often throw stems away because they are tough.

Early in 2007 Sky's beak began to grow abnormally long. At first it was just a bit too long, then it was noticeably too long. Sky was just as active as normal and had a good appetite. She was also a regular at cuttlebone chewing parties. For all appearances Sky was her happy healthy self. To make sure, Sky was taken to the veterinarian for an examination and beak trim. Sky, who was reasonably tame, had a panic attack upon being placed in a travel cage. Nothing would calm her. Sky thrashed around and around trying to fly, until she was exhausted and broke a blood feather. There seemed to be only one explanation for her unusual behavior. Having been taken out of her home, the gentle aviary, and away from her friends, Sky was terrified of being rehomed again. The fear and bad memories were still vivid, even though Sky had been with us for two years.

Sky arrived at the veterinarian in bad shape. The veterinarian treated the broken blood feather, then let Sky rest for a while. He came back later to do the examination and beak trim. Everything checked out normal. The veterinarian explained that excessive beak growth at that stage of life was a sign of aging, or to use the cold clinical term, the degenerative process. Excessive beak growth is an end of life issue for birds outside the parrot family as well.[128] Sky's liver function was starting to slow down and as a result was not processing keratin, a substance which the beak is composed of, correctly. No remedies were available. I was advised to make sure Sky continued to eat a healthy diet and had a normal appetite. Also, I was told to plan on bringing Sky in regularly for a beak trim, or sooner if her beak became so long that it interfered with her ability to eat.

Exhausted, Sky sat still on the trip home. Fortunately she did not bleed anymore. When we walked into the aviary Sky perked up. When Sky was placed back in the gentle aviary with her buddies Sky was ecstatic. She looked at me and began singing and flying from perch to perch. Sky was home. She was really home. No, Sky had not been abandoned again.

We nicknamed Sky, Sky Long Beak, and she did not mind. Her new name sounded more like a parrot belonging to a pirate, but she loved the extra attention. Because of Sky's traumatic experience with the travel cage, going back to the veterinarian every couple months for a beak trim was not an option. We were lucky that this time she had only broken a blood feather. I was not about to chance her injuring herself again. So I learned how to do beak trims for budgerigars. Whenever Sky needed a beak trim, I took her into a small room. We played together for a while. Then I wrapped Sky in a small white washcloth, with her little head sticking out. I gave Sky a small piece of wood, the size of a Q-tip™ to chew on; this kept her busy and calm. It also helped to separate her upper and lower mandibles (or maxilla and mandible if you prefer), so I could quickly and easily trim her beak. After I was finished, we played some more. Then it was back to the gentle aviary. Instead of a traumatic hour long journey to and from the veterinarian's office in a small travel cage, Sky and I spent some quality time together.

Sky continued to be her happy self throughout the summer and fall. Our beak trimming sessions became part of her normal routine. There wasn't any change in her activity level or appetite and she continued to fly around as usual. The evening of December 23, 2007 there was more than the usual tussling in the gentle aviary about who got to sleep on what perch for the night. In the morning Sky was gone. Sky had not had a prolonged debilitating illness; nor did she have to be moved to a small cage her last days—that would not have worked out at all. Sky, who cherished the slightest bit of attention, had died alone.

The story of this little quartet of budgerigars is not as happy as it should have been. I have often wondered how their four lives might have been different if the people who originally bought them at the pet store had really understood the intellectual, social, physical, and emotional needs of pet birds, known how long pet birds live, and had made a life time commitment to their feathered friend. What might have been different if the public had not been inundated with a sea of misinformation about the avian flu non-epidemic?

How might their four lives have been different? Slate might have had a normal long and happy life, instead of dying of a broken heart from being abandoned at an early age. If someone had cared enough, Lilac might have received medical treatment early enough to prevent her painful premature demise. In spite of being rehomed, Sunflower lived a long life. Sunflower was her own bird. You have to wonder though, had she not been rehomed maybe Sunflower would have been a bit mellower and welcomed at least one human into her world. Sky was blessed with a long life too. Sky would have been Sky, but a confident, assured version of herself. (See Table 7.8.)

Table 7.8 Relinquishing and Adopting Pet Birds: Too Easy or Too Hard?

Is it too easy to relinquish pet birds? The first time a bird acts up or someone is too tired to change their paper, the poor bird can be dumped on a parrot rescue/adoption organization as easy as it is to dump trash down a chute? The Montgomery County Humane Society on average takes in 5 to 25 new animals every day.[4] In a good month they adopt out 250 to 300 animals.[4] Due to space limitations, if an animal is not adopted or placed in foster care within two weeks, its chances for a normal life span are slim. Multiply this situation across the country and it is apparent why some five to six million homeless pets are euthanized in the United States each year.
Most county and city humane societies do not have the knowledge, facilities, or skills to accommodate homeless pet birds. Instead, homeless pet birds are turned over to parrot rescue/adoption organizations. Miraculously, these organizations operate almost exclusively through kind hearted energetic volunteers and generous donors. And this is in spite of the fact that there are approximately 5,000 pet birds in foster care and/or awaiting adoption in the United States alone.
There are valid reasons that cause a pet bird to be relinquished. For example, the primary care giver becomes too old or contracts a prolonged illness that prevents him or her from properly caring for a pet bird. In this situation having to relinquish a life-long feathered companion is a painful event for both parties.
On the other hand, we do not want to make it so difficult to relinquish a pet bird, that people simply release them into the wild. While a fortunate few may survive, like those documented in "Wild Parrots of Telegraph Hill" by Mark Bittner,[21, 187] the vast majority will not, dying from starvation or being killed by a predator. Pet birds that have spent their entire lives in a cage, with someone bringing them food and water, will not adapt overnight to life in the wild. Pet birds have not developed immunity to the bacteria, viruses, and diseases that wild birds have. Like Blue Boy, they will get infections. Nor are pet birds used to the dramatic changes in the weather outdoors from day to night and season to season.
It is too easy to buy pet birds. Anyone can walk into a pet store, buy a bird, and walk out in 5 minutes without providing any evidence that they know anything about pet birds or how to care for them. One chain pet store in our area used to have a policy about buying pet birds (other than budgerigars or finches). If someone was interested in buying a particular bird, they put down a $50 deposit. Then they were required to meet with the manager of the bird department and visit the bird at least three more times. At the end of this process, if the bird department manager agreed, they could take the bird home. Undoubtedly this process prevented many spur of the moment purchases. Unfortunately because this process cut into sales and profits, the policy was dropped after a year.
Whenever I am approached about where to get a pet bird, I always refer people to the local parrot rescue/adoption organization. Usually this works out fine. On occasion people have complained about having to take a 2-hour class before they can adopt a bird and gave up on the idea. That is fine too. Because if someone does not have time to take a 2-hour class they certainly do not have time to care for a bird properly or meet its intellectual, social, physical and emotional needs.

7.6 Other Interesting, Surprising, and Amusing Observations

You see a lot of different things when you have the opportunity to observe 28 budgerigars over a 30-year period. In particular you gain an appreciation of which behaviors are common to (indoor) budgerigars and which are unique to individual birds. You also spot behaviors that cross parrot families. Here are some of the more interesting, surprising, and amusing behaviors that were observed. These behaviors highlight individual thought patterns and preferences. Sometimes only one budgerigar was involved; other times two or three budgerigars participated. The observations are presented in chronological order. An indication is given of the frequency with which each behavior was observed and changes in the behavior over time, if any.

In 1986 I made a 5-day business trip to Switzerland. Beforehand I made reservations to board Greenie I and Oliver at a family owned pet store that specialized in birds. On my way to the airport I stopped at the pet store to drop off Greenie I and Oliver. It was the weekend and the regular store manager wasn't there. The substitute manager couldn't find the reservation, wouldn't call the owner at home, and refused to take Greenie I and Oliver. I called everyone I could think of; no answer. This was before voice mail and cell phones. In the end, I had no choice but to take Greenie I and Oliver back to the aviary. I put triple the usual amount of food and water in their aviary and rushed to the airport. When I returned 5 days later, their aviary was a bit messy, but they were both fine. However, instead of their usual cheerful greeting, Greenie I and Oliver turned their backs to me and did not look or speak to me for 2 hours. I freshened up their food, water, and paper first thing, but that didn't make any difference. Greenie I was used to me making one or two day business trips, but this was the first long trip. This was the first and only time Greenie I and Oliver exhibited this behavior. None of the other budgerigars have ever done this.

Blue II was an agile flyer. Given a choice he preferred to fly laps within his aviary to flying in an open room among the aviaries. Blue II seemed to relish the increased challenge of navigating around tight quarters at full speed to the relative ease of open flying. Blue II came out of his aviary to fly, but rarely stayed out more than 30 to 45 minutes. Blue II exhibited this behavior throughout his life. None of the other budgerigars have ever indicated a similar preference. Blue II's behavior is similar to Squeaky, a Meyer's parrot testing her flying skill by flying full speed directly toward cactus, only to veer off at the last minute. (See Section 9.3.)

The budgerigars were served crushed gravel in dishes. The gravel dishes were white or colorful plastic caps from fruit juice bottles. Asher loved to drag the gravel dish from one side of his aviary to the other and back again like it was a hockey puck. This was a distance of 37.5 inches (95.2 cm). Asher did this several times a day; his favorite time was right after a clean paper was put in his aviary. Asher preferred to play with the gravel dish, even though he had a small whiffle ball. Every now and then he would hit a bump and spill the dish. When that happened Asher sat the dish upright and kept on going. Asher engaged in this activity throughout his life. He is the only budgerigar to have done so.

Birds that eat seeds and insects collect food in their crop. Later the food in their crop is mashed before traveling to their stomach to be digested. The technical term for this process is maceration. This mashing process makes a crunching sound. The sound is particularly noticeable late in the evening when all else is quiet in the aviary. The mashing process and the associated noise appear to be a social activity, since several birds join in at the same time, even different species of birds. The combined crunching sound resembles a percussion team practicing and at times is almost rhythmical. Greenie II liked to crunch along with me whenever I ate something dry like crackers or popcorn. Greenie II is the only budgerigar to time his crop crunching routine to coincide with mine. Greenie II engaged in this activity every evening he could throughout his life.

Budgerigars are native to the Australian grasslands. Growing greens, a product that has been available for decades, contains a variety of grass seed mixed with vermiculite inside a plastic pot. Once water is added and the grass has grown about 4 inches (10.1 cm) tall, the pot is attached to the side of an aviary. Growing greens were even available during Perky's time and like all budgerigars, he loved to nibble on them. I mist growing greens before putting them in the aviary. Uzziel loved to brush her feathers against the moist greens before eating them. This behavior approximated wild budgerigars bathing in the morning dew.[99] Uzziel was the only one to do this.

One time the grass seeds were grown in potting soil and 2 inch (5 cm) round nursery pots, the kind gardeners use to start plants indoors. When the grass was mature the pot was hung in Uzziel's and Asher's aviary. Uzziel and Asher devoured the greens as usual. They also dug up all the dirt and threw it all over their paper, from one end of their aviary to the other. There wasn't any way to tell if this behavior was a one-time response to a new item having been introduced into the aviary. So the process was repeated. Again, as soon as the greens were devoured the soil was scattered everywhere. Both Uzziel and Asher dug through the soil rapidly with their beaks. This behavior was puzzling, since neither of them was in the habit of throwing food or gravel around. Budgerigars in the wild have regular contact with soil. Were Uzziel and Asher simply having fun? Or, did their behavior indicate that they were seeking some trace elements only found in soil, even though their diet included standard avian vitamin and mineral supplements?

Budgerigars are very social birds. In the wild budgerigars travel in flocks up to 1,000 birds. Hence, it is not surprising that they like to socialize when kept together in an aviary. Three interesting variations of this social behavior have been observed over the years.

First, whenever the budgerigars in one aviary are let out to fly, the first thing they do is to fly to the other two budgerigar aviaries and visit the budgerigars there. They walk up and down the sides of the aviary (front, back, and top). They try to play with the toys inside, all while chattering away. The birds inside reciprocate and follow them around. If a budgerigar on the outside taps a kabob toy, the budgerigars on the inside come over and start playing with the kabob. All of the birds out flying visit with all the budgerigars inside the other two aviaries before they start flying laps or go anywhere else. This behavior occurs even though all the budgerigars in the three different aviaries can see and hear each other throughout the day. Hence, more than curiosity is at work. There must be a need to socialize in close proximity.

Second, not all of the large birds enjoy having budgerigars land on top of their aviary. Three of the four Indian ringneck parakeets lunge at the budgerigars and shoo them away. One Indian ringneck parakeet will tolerate a budgerigar sitting on his landing perch. In contrast, Yellow Birdie struck up a social bond with a Meyer's parrot. Whenever Yellow Birdie sits on the Meyer's parrot's landing perch, she comes over to visit. There is a little chatter and head bobbing back and forth. Once in a while they attempt allo-feeding. Given the different beak sizes, that usually doesn't work very well. These visits usually last about 10 minutes. The Meyer's parrot is indifferent to other budgerigars that stop by unless they land on top of her aviary. In that case she shoos them away. What causes an inter-species social bond to form? It would be fascinating to know.

Third, the mitred conure also welcomes visits by budgerigars. They may sit on his aviary and perches at will. Budgerigars can even fly into the mitred conure's aviary. The mitred conure becomes excited whenever he has visitors. He is the only conure in the aviary and does not have another bird like himself to socialize with. In return the mitred conure visits the budgerigars whenever he is out. He sits and walks on top of their aviary and spends a lot of time on their landing perch. During the budgerigars' morning and evening chorus, the mitred conure adds a few notes of his own. While the mitred conure interacts with the budgerigars, he does not interact with any of the other parrots. Again, it would be fascinating to know why this inter-species social interaction takes place. (See Photograph 7.15.)

As mentioned earlier Slate I put on a show by turning inside out while hanging in the corner of his small private cage. Slate I grasped the cage bars with his feet while facing outward. Then, without moving his feet, Slate I tucked his head and body downward until he was facing the inside of the cage—in effect turning inside out. Slate I did this 2 or 3 times in a row rapidly. Five years later Sunflower II started doing this. There was no contact between Slate I and Sunflower II, yet they both engaged in this behavior. They are the only two budgerigars to do so.

Daisy joined the flock in the spring of 2010. She is a small late middle-aged female yellow green budgerigar. Daisy appeared in the pet shop one day in the midst of the young budgerigars. She

Photograph 7.15 Inter-species Social Play Bonds. A few budgerigars formed social play bonds with other parrots, in spite of significant differences in size. These social play bonds continued for several months. (Top) A mitred conure sits on a perch outside a budgerigar aviary. His buddy Yellowhead comes closer for a chat. (Bottom) A mitred conure sits on a perch outside the conure aviary. Dandelion is visiting inside the conure aviary. Mitred conures are native to South America. Budgerigars are native to Australia. These two species would never meet in the wild. Yet in an indoor setting they are able to form a social play bond. © D. Herrmann

may have been breeding stock that was sold because of the recession. Daisy looked miserable in the small overcrowded cage. Despite her tiny size Daisy spent her first month fighting with all the other birds. Obviously the breeding situation Daisy had been kept in was over-crowded and she was used to having to fight for everything. Daisy was moved from aviary to aviary. Finally she noticed the other birds weren't fighting and that there were enough food, toys, perches, and swings for everyone, and she calmed down. Daisy took a fancy to a wooden ladder in the back of each aviary and chewed one wood peg loose, so that it dangles from the ladder. Daisy only chewed one peg loose from each ladder—not all of the pegs. The other budgerigars had been exposed to the ladders a year before Daisy. But, none of them chewed any pegs loose. Daisy was the only one to do this. Her behavior is similar to Sunflower I chewing through the wooden perch and abacus toy. Was Daisy having a chewing frenzy, or was she customizing her environment? She has not chewed any wooden toys since. (See Photograph 7.16.)

Photograph 7.16 Chewed Wooden Ladder. Lilac II and Slate II play on a wooden ladder. Daisy chewed one peg loose on two separate ladders shortly after her arrival. On one ladder it was the third peg. On this ladder it was the fifth peg. The other budgerigars played with the ladders for a year before Daisy's arrival, yet none of them chewed any pegs loose. Daisy hasn't chewed any pegs loose since then. © D. Herrmann

Budgerigars are very energetic and can be very rowdy. However, they rarely have knockdown, drag out fights with each other. In fact it is doubtful that budgerigars in the wild fight. Most likely fighting is a result of being confined in a cage or aviary. Only two instances of serious fighting were observed. One instance was between two females. The other instance was between two males.

In October 2009 Yellow Birdie and Sunflower I suddenly started fighting. They both had been living in the aggressive aviary together for over 3 years at that point. While they were both aggressive and rowdy, they had never fought before. Yellow Birdie was part of a bonded pair with Lilac II. Sunflower I was not part of a bonded pair. It was unclear why they suddenly started fighting. They fussed at each other, but no biting was involved. Instead they flew at full speed directly toward each other. At the last minute they turned upward so that their undersides collided. Yellow Birdie and Sunflower I did this over and over again, often dropping to the floor after they collided. They ignored attempts to calm them down. To prevent the birds from injuring themselves, Yellow Birdie and Lilac II were moved to the rowdy aviary. A few days later Yellow Birdie and Sunflower I seemed to miss each other. They continued to visit each other whenever one of them was out flying.

In October 2010 Greenie III and ShortTail, both young normal green males, suddenly began fighting. This time it was clear what the fight was over. Both Greenie III and ShortTail had taken an interest in Sunflower II and were trying to entice her to become part of a bonded pair. Sunflower II is pretty and has a pleasant disposition. However, she didn't care for all the extra attention and wasn't interested in being part of a bonded pair at that time. The two green males had grown up together as juveniles. Now they were fighting over Sunflower II. Greenie III and ShortTail flew collision course after collision course, just like Yellow Birdie and Sunflower I had done. They continued tussling after they crashed on the paper. The only time they stopped was when they were both exhausted. The second day things took a turn for the worse. ShortTail bit Greenie III on the head. It may or may not have been an accident, since they were moving so fast. The left side of Greenie III's head was covered with blood, but they kept on fighting. Sunflower II was immediately taken out of that aviary and placed in the gentle aviary, where she is doing fine. ShortTail and Greenie III quit fighting and are buddies again.

The tray is removed and a grill is inserted when it is time to change the paper in an aviary. Young budgerigars find this transition fascinating. They frequently hop onto the grill as soon as it is inserted, walk back and forth quickly, and stare at the floor below. Sometimes 3 to 4 budgerigars do this at the same time. They immediately recognize the fact that their "floor" has changed and there is yet another floor underneath. Interestingly this sudden change in their habitat piques curiosity and not fear. Around middle age budgerigars lose interest in this phenomenon.

Many, but not all, budgerigars learn to enjoy sitting on their caretaker's hand and being hand fed. Usually this is a gradual process as the budgerigar learns to trust their caretaker. Some budgerigars prefer to avoid hands altogether; probably due to lingering memories of hands: (a) taking them away from their parents when they were very young, (b) stuffing them in a small crowded box for shipment to the pet store, and (c) scooping them up at the pet store and putting them in another box to travel home with their new caretaker. Also, they would have seen all their buddies at the pet store disappear into hands one by one. Hence, by the time they are only a few months old, budgerigars have learned to be wary of hands. Two budgerigars did not exhibit this inhibition. While Yellow Birdie never sat on my hand, she ate millet sprays out of my hand from her first day here and throughout her life. Dandelion hops on my hand as soon as I put it in his aviary, whether or not I have food. He has done this since the first day he arrived. Dandelion sits on my finger, walks around on my hand, and chews on my sleeve. The other budgerigars gradually learned to develop trust. Dandelion's willingness to interact was used as a way to teach the other birds in his aviary how to eat from my hand.

One final observation has quite interesting implications. Daisy quickly learned to let herself out through the gap between the tray and the bottom of the aviary. This is the gap the grill slides in and out of when it's time to change the paper. Daisy tried this several times. If Daisy saw me watching her, she stopped. Before long Daisy mastered the technique and if I wasn't watching out she came. Daisy never flew very far, just over to the next budgerigar aviary. I let her stay out for a while then put her back in her aviary. This pattern of behavior continued for over a week. Then one day Daisy checked herself out of her old aviary and into a new aviary—she had figured out how to get into an aviary through the gap as well. In effect Daisy moved herself out of one aviary and into another aviary with a new set of roommates. Since then Daisy has stayed put. The two aviaries are identical, even the toys, perches, swings, dishes, and so forth. What explains Daisy's desire to move from one aviary to the other? It must have something to do with the roommates. Through this behavior Daisy indicated ingenuity and her preference of companions.

Not long afterward Sunflower II began letting herself out through the gap on the bottom of her aviary. She was not housed in either of the two aviaries Daisy was. But Sunflower II probably saw Daisy slipping out. Several budgerigars have peeked out through the gap when they are walking on the paper. Even Precious has done that. But Sunflower II is the only other bird to let herself out. She just seems to be exploring and doesn't have any particular destination in mind. So far Sunflower II has not learned how to let herself back in through the gap. She lets herself out a couple times a week. Daisy and Sunflower II were the only two birds to let themselves out through the bottom gap. (See Photograph 7.17.)

Photograph 7.17 Minding the Gap #1. Yellow Birdie peeks out through the 0.75 inch (1.9 cm) gap between the cage bottom and the tray. Budgerigars often peek through the gap when they are walking around on the paper. Daisy and Sunflower II are the only two who learned how to let themselves out through this gap. Daisy is the only one who learned how to let herself back in through this gap. Daisy did this when she decided to move into a new aviary and change cagemates. © D. Herrmann

A variation of this behavior was observed during an early experiment. Two budgerigars, Greenie I and Blue II, were individually housed in separate 24 x 18 x 24 inch (61 x 46 x 61 cm) cages. The furnishings in the two cages were identical. The birds lived in the two separate cages side by side for a few weeks. Then a small tunnel was installed connecting the two cages through the side doors. The birds could walk through the tunnel to and from each other's cage. The birds learned how to do this quickly. Soon they were going back and forth visiting each other and eating out of each other's dishes. They spent most of the day together. But at night, when it was time to go to sleep, they both went back to their original cage.

7.7 Discussion

Now let's see what these observations tell us about the intelligence, behavior, and individuality of wild and captive budgerigars.

Budgerigar Learning and Intelligence

Budgerigars have an average life span that is about one tenth that of humans. Accordingly young budgerigars acquire knowledge and become self-sufficient rapidly. A 6-month old budgerigar is a young adult and self-sufficient. That is equivalent to a 5-year old human child being self-sufficient!

Young budgerigars acquire and assimilate a broad range of knowledge and survival skills quickly. Some skills they teach themselves. Some information they learn by observing their parents and other budgerigars. And some knowledge is passed on genetically. The "subjects" young wild budgerigars study (and must pass) include:

- Flying
- Feeding themselves
- What to eat and not eat
- The need to consume small bits of gravel and trace minerals
- How and where to find food and water
- Predator detection and avoidance
- Social norms and acceptable behavior
- How to identify their nuclear and extended flocks
- How to identify themselves to their nuclear and extended flocks
- Vocal communication skills

Young budgerigars teach themselves how to fly. Often a few days before they fledge young budgerigars start to stretch their wings. Up to this point their world has been the nest. Now they begin to be curious about what is beyond the nest and start to isolate sensory feedback from their wings. As they get older young budgerigars see their parents entering and leaving the nest. One day they decide to follow their parents. The first flight of fledgling budgerigars is not very precise or strong. However, they soon develop precision navigation skills, endurance, and strength.

The process is the same for wild and indoor budgerigars—both must teach themselves how to fly. The difference is the time frame and the level of expertise required. Young wild budgerigars must master the art of flying quickly to survive. Predators are nearby and going back to the nest is not an option. Hence their learning interval is measured in minutes and hours.

In contrast, young indoor budgerigars have the option of mastering the art of flying at a leisurely pace. The process of learning to fly may be delayed due to the practice of breeders and retailers clipping the wings of pet birds. The first flight of indoor budgerigars is as wobbly as that of their wild cousins; but the consequences are nil. They simply bump into something or drop to the floor. Indoor budgerigars develop the same or better precision than their wild cousins as they navigate around manmade objects within a confined space inside and outside their aviary. Indoor budgerigars do not develop the same level of endurance as their wild cousins since they fly around 30-foot (9 m) living rooms, and their food and water dishes are inches (centimeters), not miles (kilometers), apart. Nor do they learn how to fly rapidly in close formation like wild budgerigars. Four to eight indoor budgerigars flying at the same time learn how to avoid crashing into each other, but not how to fly in unison. That feat requires enhanced levels of flying and communication skills.

An experiment was conducted in regard to budgerigar flight. A group of young juvenile budgerigars were kept in an aviary together and all of them had clipped wings. They could not see any other birds. In this environment they learned how to run and climb quickly. The young budgerigars stretched their wings now and then, but they made no attempt to fly. Then one of the young juvenile budgerigars with clipped wings was placed in an aviary with budgerigars of the same age who did not have clipped wings. All of these budgerigars were adept flyers. Upon seeing this, the young budgerigar with clipped wings immediately began trying to fly. It continued to fly and crash (on bubble wrap) for several days before it reverted back to climbing, until its flight feathers grew in. Each time it crashed it tried flapping its wings faster and faster to compensate.

Fledglings must learn how to perch the same time they are learning how to fly. This involves learning how to land on and grasp a variety of objects that are different sizes, shapes, and textures. This is similar to the hand/eye coordination humans must develop, since parrots' zygodactyl feet function as both hands and feet. Budgerigar nestlings are flat footed and gravity holds them in place. Fledgling budgerigars must develop strength and coordination in their legs and feet to land on a particular branch or stem and hold onto it securely, so they don't slide off or crash. This learning process takes some practice, usually a couple days. In the interim fledglings may appear to be a bit clumsy. Wild budgerigars must teach themselves to land and perch quickly to survive.

Young indoor budgerigars are frequently kept in large flat cages without perches. They spend their first few weeks walking around flat footed, rather than flying, climbing, or perching. Some of the budgerigars that were acquired from pet stores had never climbed up and down cage bars or sat on a perch, even though they were several months old. These concepts, including perching on a food or water dish, puzzled them at first. It took at least 3 days for these birds to develop the strength and coordination necessary to master these skills. The older the birds were when they came home from the pet store, the longer it took for them to learn these skills; perhaps because these skills are supposed to be learned early in life.

Nestling budgerigars are fed a predigested mash by their parents. Fledgling budgerigars follow their parents around for about 10 days and continue to beg for food. Gradually the parents start pointing out food sources, rather than actually feeding the fledglings. Fledglings must then learn how to crack open small grass seeds since they are granivores. These seeds are tiny. A certain degree of dexterity is required between beak and tongue to ensure that the seed contents fall into their mouth and are swallowed whole, while the hull falls outside onto the ground, and not the reverse. Most likely a practice period is required before this skill is mastered. That might explain why fledglings continue to beg food from their parents for a few days.

This learning process is the same for wild and indoor budgerigars. The consequences of how quickly and how well wild versus indoor fledglings master this skill varies. Wild budgerigar fledglings must master this skill quickly or go hungry. The food they drop on the ground will be eaten by other birds. Indoor budgerigar fledglings don't have the same time pressures. If they drop a seed, it falls into a dish and they can try again. There may be other birds around, but a conscientious breeder will make sure there is plenty of food for everyone.

Fledgling and juvenile budgerigars must learn what to eat and equally important what not to eat—items that are poisonous or otherwise hazardous to their health. Initially this information is acquired by observing their parents and other members of their flock. Fledglings and juveniles watch what other birds eat and imitate their behavior. As they mature juveniles also experiment to learn what foods should and should not be eaten. They sample a variety of items to see what is and isn't acceptable. Some items they cannot crack open; other items don't taste right and are spit out before they are swallowed.

Parrots have between 300 and 400 taste buds; this is among the highest number of taste buds among birds.[18] Parrots' taste buds are located on the roof of their mouth, not on their tongue.[113] This is why it is common to see members of the parrot family roll a new food item around in their mouth before they eat or reject it. Like other birds a budgerigar's taste buds are designed to help them discern which foods are safe and healthy, not to evaluate epicurean delights. This process of learning through observation and experimentation is the same for wild and indoor budgerigars. However, each must learn to avoid the food hazards that are unique to their environment.

At the same time wild juvenile budgerigars discover that it is beneficial to consume minute amounts of gravel, soil, and possibly tree bark. Gravel aids the maceration of hard seeds in their crop. Small amounts of soil and possibly tree bark provide a ready source of essential minerals and trace elements, like iodine, that are not present in seeds.[102] Juvenile budgerigars learn this activity by observing their parents and other members of the flock; there may be some genetic intelligence at work as well.

This is an area where wild and indoor budgerigars part company. Breeders take young budgerigars away from their parents at a very young age, usually as soon as they are weaned. Hence, young indoor budgerigars have little opportunity to learn the necessity of or how to consume gravel, soil, or tree bark by observing their parents. Nor do they have access to soil or tree bark. Indoor budgerigars must obtain essential minerals and trace elements elsewhere.

The traditional approach has been to provide cuttlebones and mineral blocks for indoor budgerigars to chew. The thinking is that budgerigars will obtain the minerals and trace elements they need by chewing cuttlebones and mineral blocks and at the same time keep their beaks trim. On the surface this approach sounds reasonable. But, when examined in detail some questions are raised.

An examination of three popular mineral blocks indicated that the primary ingredient is calcium (20–26%). Other minerals are listed as being .01% or less. No standardized tests are conducted to detect the presence of harmful impurities in mineral blocks, like the melamine that popped up in dog and cat food with fatal consequences. Furthermore, how useful can mineral blocks be when the exact essential mineral and trace element requirements of the parrot family have not been determined?[6, 53, 113] Instead, the well-known mineral requirements of chickens have been applied to psittacine nutritional needs without any scientific verification.[6, 53, 113] Budgerigars are not chickens. They don't live in the same climate, they don't eat the same food, they don't... One would think that after 170 years some person or organization could have taken the time to determine the exact mineral and trace element requirements for budgerigars (and other parrots).

Even if the mineral blocks were formulated to correspond to budgerigars' exact needs, there is the question of dosage. Is a budgerigar supposed to devour a mineral block in 1 week, 1 month, or 1 year? How do you determine if they are consuming too much or too little? A wild budgerigar knows when it is time to consume a beakful of soil. In contrast an indoor budgerigar simply chews on a mineral block for fun whenever they feel like it, or maybe never. Some budgerigars discussed in this chapter never touched a mineral block their entire lives. Other budgerigars scraped a mineral block into oblivion in a day or two. It is unclear how much, if any, of the minerals they actually swallowed, since most of the powder ended up on the floor. Other budgerigars took a single bite every 2 or 3 weeks. This hardly makes for consistent mineral consumption, even though they all have the same mineral requirements.

The same pattern was observed in regard to cuttlebones. Cuttlebones, a supposedly tried and true tradition of aviculture, also raise some questions. Cuttlebones are the dried internal calcified shell of cuttlefish. Cuttlefish are squid-like mollusks that live in the oceans, primarily in the Indian and Pacific oceans from the Bay of Bengal across southern Asia and down to northern Australia.[150] Most of the cuttlebones sold for pet birds come from parts of the ocean that are very polluted. Questions have been raised about how many of these pollutants are being passed on to pet birds.[118]

Neither mineral blocks nor cuttlebones are laying about in the Outback. Wild budgerigars obtain essential minerals and trace elements from the soil and possibly tree bark.[102] The practice of wild birds eating soil and clay is referred to as geophagy. Wild budgerigars keep their beaks trim by chewing on wood. Over the decades indoor budgerigars have learned to chew on mineral blocks and cuttlebones. These are not normal behaviors that occur in the wild. It is questionable whether they are beneficial.

Budgerigars are not migratory birds in the true sense of the term. Rather, budgerigars are nomadic and travel locally from place to place for food and water. Nest building activities are synchronized to coincide with the availability of food and water. Hence, there is usually abundant food and water when juvenile budgerigars are being weaned. At first finding food and water simply means following Mom and Dad around.

As they mature young budgerigars join temporary flocks of various sizes as they seek food and water. Through this experience young budgerigars learn:

- Where food is located
- When each food source is and is not ripe

- Where water sources are located
- When each water source is full or empty
- Which water source is nearest each food source

Budgerigars are not scavengers or hunters. Budgerigars learn and remember where their food and water are located and, just as important, what time of year each is available. For example, it would be costly and perhaps fatal to fly miles (kilometers) to a food source before the seeds were ripe or to a water source during the dry season. Over the years budgerigars also learn about alternate sources of food and water for use during environmental challenges like droughts. That information is stored away for emergencies. At some point budgerigars learn that, because of their salt gland, they can drink brackish water on a temporary basis if fresh water cannot be found. It is not clear when budgerigars become aware of this option. Perhaps a combination of biological intelligence and genetic intelligence come into play whenever this dire situation is encountered.

Knowing how and where to find food and water is another area where the learning experiences of wild and indoor budgerigars diverge. As just discussed, the learning process for wild budgerigars is complex and involves acquiring, remembering, and perhaps updating a large volume of geographical, botanical, and meteorological information throughout their lives. Indoor budgerigars have a totally different and much simpler learning experience.

A young indoor budgerigar must get used to its new environment when it comes home from the pet store or breeder. Part of this orientation process involves finding out where their water, dry food, and fruit/vegetable dishes are now located. Usually the size, shape, and color of the new dishes are different. A newcomer generally finds their dishes within 2 to 4 hours in a small baby/quarantine cage. It may take 24 hours for a newcomer to find their dishes in a large aviary, if they are by themselves. The newcomer learns quickly if other birds are present.

At this point a young indoor budgerigar only needs to remember that the water dish is on the left and the food dishes are on the right. Young indoor budgerigars do not have to learn and remember multiple locations for food and water, when each is and is not available, or how to deal with droughts and other environmental challenges. An astute indoor budgerigar will learn when the food and water are the freshest and that on occasion a millet spray or other treat appears in a special location.

As a result the potential intellectual resources of indoor budgerigars are barely tapped when it comes to learning how and where to find food and water. To compensate for this less challenging environment, water, dry food, and fruit/vegetable dishes and treats were moved around. In addition the contents of the dishes were varied. This approach added an element of surprise, which many budgerigars seemed to enjoy. Each morning when the dishes were replenished, the budgerigars would go from dish to dish to see what they contained. Swapping the food and water dishes and moving dishes higher or lower generated the most activity and interest. The process of finding and sampling food in another bird's aviary was equally as interesting to many budgerigars. They particularly enjoyed eating someone else's millet spray from outside the aviary. Perhaps because it was more challenging or considered a special find.

In one experiment a food dish was filled with gravel in order to observe the budgerigars' reaction. The majority of the budgerigars hopped on the dish, saw that it contained gravel, and left for another dish. Two budgerigars in separate aviaries decided that there must be some food under the gravel—after all, this was a food dish. They proceeded to dump the gravel on the floor one beakful at a time until the dish was empty. They even tapped the bottom of the empty dish to make sure there wasn't any food before they flew away.

The ability to detect and avoid predators is an essential survival skill, one that young budgerigars must learn immediately. This skill is learned primarily by observing their parents and other members of the flock. Predator detection and avoidance also involves biological intelligence—the extended hearing range of budgerigars—and genetic intelligence that is passed from generation to generation in the form of shared historical memories of encounters with predators. For example, one location is known to have more predators than others. Hence, that location is visited less frequently, only in emergencies, or not at all.

Predator detection and avoidance is a survival skill that is operational 24 x 7 when budgerigars are flying, feeding, drinking water, bathing, and roosting. Since budgerigars are small and prey they cannot afford to take their predator detection and avoidance algorithms offline. Because of this situation predator detection and avoidance is both an individual activity and a cooperative flock activity.

Young budgerigars quickly learn predator detection and avoidance skills and their role in each cooperative flock activity. In the air a flock adopts a swirling flight pattern to confuse predators. On the ground several flocks gather in a large assembly when feeding or drinking water. The spectacle of a large number of budgerigars landing and taking off at the same time deters predators and provides multiple simultaneous lookouts. At midday or overnight budgerigars roost at a safe height; one that is high enough to safely foil ground-based predators, yet at the same time shielded from raptors. Budgerigars have a standardized alarm call which all individuals and flocks understand. It is likely that there are subtle variations in the alarm call which indicate the type of imminent danger, variations that humans have not yet detected.

Predator detection and avoidance skills come in a different flavor for indoor budgerigars than for their wild cousins. Indoor budgerigars do not need to learn how to avoid raptors and tree snakes, even though some historical memories probably still linger. Instead indoor budgerigars must learn to avoid other hazards in their environment, like house cats, dogs, glass windows and mirrors, ceiling fans, white walls, corners, electrical cords, cacti, and so forth.

Until they are about 6 months old budgerigars are not fearful. Before this age their curiosity and happy go lucky disposition outweigh any concept of fear. After this age budgerigars are more cautious around hands and people, especially new people. That is why it is important to build a trust relationship with a new budgerigar as soon as possible. Even indoor budgerigars understand that they are prey. They have to learn whether or not the humans and other mammals who share their environment are predators. Adopted budgerigars have to recalculate this equation since their whole world has changed. Consequently, adopted budgerigars should be given adequate time and space to reassess the predator/prey relationship in their new environment.

Indoor budgerigars also have an alarm call. All budgerigars, young and old, understand this call, even though it is used infrequently. Other types of birds in the aviary also take note whenever a budgerigar makes an alarm call. It is unclear whether or not this alarm call is the same as that of a wild budgerigar.

Throughout their lives, from nestling to fledgling to juvenile to young adult, wild budgerigars live in a group situation. A young budgerigar shares a nest with its clutch mates. It interacts with them and its mother. Shortly before fledging a young budgerigar meets its father. A fledgling follows its parents around and gradually becomes aware of other budgerigars in the flock. As a young adult a budgerigar participates in flock activities, like traveling to food and water sources, contributing to predator detection and avoidance, and roosting midday and overnight. In other words, a budgerigar is never alone; it is always part of a community.

As a result a young wild budgerigar learns the social norms of the group and what is and is not considered acceptable behavior quickly. This is particularly important given the dynamic non-hierarchical non-territorial nature of budgerigar flocks. A flock may consist of 10 birds one day, 100 birds the next day, or more than 1,000 birds later on. As a result budgerigars are constantly meeting and interacting with new birds. In this environment refined social skills, like communication and cooperation, are essential for the survival and well-being of the individual and the flock. A high level of intelligence is required for an individual to function in large complex societies like this.[189, 190]

Indoor budgerigars go through a different socialization process. They may hatch in a colony breeding situation where many other birds are present or in a nuclear family situation if a specific color pattern or physical trait is being sought. In either case the opportunities for socialization and learning acceptable behaviors are limited. The vast majority of birds are removed from their parents as soon as they are 6 weeks old and weaned, so that they can be sold. It is questionable whether or not the parents had an opportunity to learn a complete set of social skills either prior to being put in a breeding situation. Before they have learned to fly or been given an opportunity to acquire social skills young budgerigars are shipped to pet stores. There they are placed in cages with several other young budgerigars. Their new cagemates may or may not be from the same family or breeder. In essence, a group of children are raising themselves without any adults present.

This is not an optimal situation. The long-term effects of this common practice depend on:

- Whether the cage is clean and spacious or overcrowded and dirty
- Whether there is enough food and clean water for everyone
- The length of time a young bird spends in this environment

Young birds that spent a short amount of time in a pet store with clean spacious cages and adequate food and water, adjusted quickly and easily when brought into the aviary. In contrast, young birds that spent a long time in a crowded less than hygienic environment, were initially either timid or aggressive. The longer a young bird spent in a negative environment, the longer it took to adjust to life in the aviary. Daisy's lengthy adjustment period is a classic example. It was also noted that the adjustment period was faster when an aviary contained a mixture of birds of different ages.

The development of social skills is further stunted when a young budgerigar is taken home to live as a solitary bird. The budgerigar may be pampered and considered part of a human family, but it has no opportunity to learn how to interact or socialize with other budgerigars. A solitary budgerigar may become territorial—which does not happen in the wild—if it lives alone for a long period of time. The lengthy adjustment period when Slate, Lilac, Sunflower, and Sky arrived is a classic example of this scenario. None of them had any experience or understanding of the need to share or take turns when it came to perches, swings, toys, and dishes. Again, the older the bird was when it arrived in the aviary, the longer the adjustment period needed.

An experiment was conducted to determine the optimum number of budgerigars per aviary. All three budgerigar aviaries are identical. The number of budgerigars per aviary was varied from two to six over a period of several months. No negative behaviors were observed when two or three budgerigars were in an aviary. However, the budgerigars were less active; particularly if only two budgerigars were present. This observation was somewhat surprising since there is more room to fly and play when there are less birds and less competition for the toys.

The activity level and vocalizations nearly doubled when four to six budgerigars were in an aviary together. Six turned out to be the optimal number of budgerigars and provided enough varied companions and interactions to keep everyone interested and active. There was one exception. If the six contained a newly bonded pair and outsiders, there were tussles among the males on occasion. If the six contained pair bonds that had been established for some time, outsiders were welcome.

For a period of several months six budgerigars resided in each aviary and everyone was happy with that arrangement. Then a new experiment was conducted. The budgerigars from all the aviaries were let out to fly at the same time. At first each bird returned to its original aviary. After a couple weeks individual budgerigars decided to trade places and moved to a new aviary. Sometimes two budgerigars would be in one aviary and eight in another. After a few more weeks all the budgerigars (12) moved into one aviary—leaving the others empty. The first couple of nights there was a little tussling about who got to sleep on the swings and so forth, but no fussing over toys, food, or water dishes. Since then, individual budgerigars change aviaries and cagemates if and when they want to. In summary, once the budgerigars had a chance to become acquainted and learn social skills, they demonstrated a preference for living in a large group setting. This is consistent with the minimum wild flock size of 10 budgerigars. (See Photograph 7.18.)

Photograph 7.18 Swing Time. Yellowhead and Sky II share a swing. Budgerigars are very social birds. This swing is only 4 inches (10.16 cm) wide. On the horizontal axis there is 180° freedom of movement. On the vertical axis there is 360° freedom of movement. As a result it takes a lot of coordination and communication between the two birds to fly and land on the swing together. This is evidence of a strong social bond between these two birds. There are four identical swings in this aviary. Yet both birds chose to sit together on this swing. Yellowhead and Sky II were both 5 years old when this photograph was taken. They were seen sitting on a swing together frequently from this age through their senior years. Budgerigars less than 6 months old usually experiment with two birds sitting on a swing together, sometimes with the birds facing opposite directions. By the time they are 9 months old budgerigars generally prefer private swings and stop doing this. © D. Herrmann

As a whole budgerigars are better off than other young indoor parrots in this regard. Young budgerigars are always kept in a group situation, even if it is less than optimal. As such, they develop at least rudimentary social skills. Other species of young parrots go from being weaned to being put in a cage by themselves. Some are even put in a cage by themselves before they are weaned—when they are still being hand fed. These birds have no opportunity to learn social skills from their parents or the indoor flock. This practice is the source of many of the behavioral problems seen later in pet parrots.[113]

Wild budgerigars must learn to identify their flock and recognize flocks that they want to join, even on a temporary basis. This ability is part and parcel of the daily life of a wild budgerigar. Given the nomadic nature of budgerigar flocks and the fact that several small flocks may come together to form a large flock for a limited time, it is important to stay in touch. A solitary budgerigar that is left behind won't survive very long.

A wild budgerigar must also learn how to identify itself to members of its flock. The structure and size of flocks may change, but bonded pairs always stay together. Hence, bonded pairs have a means of identifying themselves to each other amidst all the chatter and wing flapping. It was noted that bonded pairs had a unique call they made to each other. This call was made most often when the pair was out flying and became separated. Somehow whenever this call was made only the one bird for whom it was intended answered. Budgerigars also made this call during the 2-week mourning period as they searched for their lost mate.

This raises two questions:

- How does a budgerigar identify itself?
- How does a budgerigar recognize the identity of others?

In other words, how does one normal green budgerigar identify itself or recognize the identity of another when they are both in the midst of 100 or even 1000 nearly identical normal green budgerigars?

I was able to tell a group of four to six normal green budgerigars apart. Sometimes their black dots were larger or smaller or one bird had molted a dot. Sometimes their purple cheek patches were larger or smaller or brighter. Sometimes one bird's bright yellow head feathers stopped in a neat line where the barring started, while another bird's yellow head feathers were fluffy and extended over the bars on the back of its head. Sometimes one bird was very fastidious about preening, while another's tail feathers were always unkempt. There were also behavioral clues. To notice all this I had to observe the birds up close for several days. That would not be possible with an energetic flock of 40, 100, or 1000 budgerigars who are always on the move.

How do budgerigars identify themselves and recognize the identity of others in their dynamic, nomadic, and noisy environment? It is an established fact that birds use a combination of vocal and visual communication techniques.[74, 137, 154] Some of these techniques are effective over short distances. Other techniques are effective over long distances. Most likely the fluorescence of certain feathers plays a part, as do other visual communication techniques that have yet to be identified. For example, do budgerigars have a greeting ceremony like Canada geese pairs? Or, does beak touching serve that purpose? Certainly, vocal communication skills play a major role. Although, beyond the contact call just mentioned, the specifics have not yet been identified. It would be fascinating to know what types of names or labels budgerigars (and all birds for that matter) use to identify themselves and others. I wonder what name or label they have assigned to humans.

Wild budgerigars only come in one color pattern: green with a yellow head. That is referred to as a normal green budgerigar. Through 170 years of captive breeding and genetic manipulation there are now some 70 budgerigar color mutations. That poses another question. How and why do indoor budgerigars recognize each other as budgerigars given all these color mutations? Wild budgerigars are not "programmed" to recognize or accept blue, white, or yellow budgerigars. Would they do so?

Nine functional categories of vocalizations have been identified for wild budgerigars:[59, 130]

- Normal chatter
- Prior to the group taking flight
- Alarm/warning
- Conflict
- Roosting chatter
- Courtship warbling
- Distance call
- Nestling/fledgling begging for food
- Nesting female begging for food

Interestingly, this list does not mention the contact call between indoor bonded pairs that was observed.

A hatchling begins to vocalize almost immediately when it is hungry. This call is not loud since the mother rarely leaves the nest. Hence, this call does not attract predators. This call is passed on through genetic intelligence; it is not learned. It is the same for wild and indoor budgerigars. The first hatchling has no one to learn this vocalization from, nor is there time to learn it. Subsequent hatchlings hear the other nestlings begging for food. But their auditory skills are not yet acute enough to isolate and repeat the sound. All hatchlings vocalize when they are hungry. There may be slight variations or imperfections in their calls. It doesn't matter, perfection is not needed. Their mother gets the message, just like a human mother understands the less than precise speech of a 2-year old child. Several of the long-lived budgerigars made this vocalization as their farewell song an hour or two before their death, even though they had not made or heard this sound for over a decade.

In budgerigar families the mother feeds the nestlings and the father feeds the mother. Just as the nestlings let the mother know when they are hungry, the mother lets the father know when she is hungry. A special vocalization is used to convey this message. It is unclear whether this vocalization is learned or passed on genetically. If it is learned, who does a mother bird learn it from and when is it learned? It is plausible that this vocalization is somewhat unique to each bonded pair. This is a vocalization the father understands and responds to, but does not make himself.

A young budgerigar begins to expand its vocabulary after fledging. From then on it will be exposed to a variety of different vocalizations that are used in a variety of different settings. Like the other information a young budgerigar must learn quickly to survive, this knowledge is acquired and perfected before it is a year old. At first a young budgerigar just hears a variety of vocalizations. Gradually it learns what these vocalizations mean and when they are used. Eventually a young budgerigar learns how to make these vocalizations itself, although a practice interval is required. A young male also develops his own unique variation of the courtship warbling song.

Lastly, a special vocabulary is developed that is reserved for use within a bonded pair. Studies have shown that vocal imitation in budgerigars is sex-biased; males in newly formed pair bonds learn vocalizations from their partner, but not the reverse.[90] This behavior, referred to as call convergence, helps strengthen a pair bond. This behavior explains why male pet budgerigars are considered better "talkers"—they mimic what they hear as part of the call convergence process while bonding with their caretaker. A female pet budgerigar probably expects a caretaker to converge their vocalizations to match hers!

The different socialization process of indoor budgerigars, compared to their wild cousins, has an impact on their acquisition of a vocabulary. To illustrate, if indoor budgerigars are taken away from their parents when they are six weeks old and placed in a cage with other very young budgerigars, who do they learn native budgerigar vocalizations from? As an analogy, assume a group of infants were isolated from adults and never heard human speech. Regardless of the continent or ethnic group, these children would end up speaking some form of human gibberish that they all understood, but no one else. So, just as young indoor budgerigars are raising themselves, they are also teaching themselves vocalizations. As a result their vocabularies are more limited and improvised than their wild cousins.

An interesting observation was made in this area. If eight or less budgerigars were housed in the same room, their vocalizations were limited to the juvenile chatter heard in a pet store, with an occasional pair bond contact call or infrequent alarm call. Once the population of budgerigars in the same room was expanded to more than eight, their vocabularies increased. A significantly greater variety of vocalizations were heard when there were 10 to 12 or more budgerigars in the same room, especially if their ages were mixed. This behavior is related to the fact that the smallest flock size seen in the wild is 10 budgerigars. In particular, the males spent more time perfecting and customizing their courtship warbling calls. Beforehand, when the population was smaller, the males never even made this call. This observation points back to the principle that greater intelligence is required to live in larger and more complex societies, part of which is enhanced communication skills. Budgerigar vocalizations can be heard on Disk 2 Track 4.

A further note is warranted in regard to the male budgerigar courtship warbling call. Several publications consulted implied that the *undulatus* half of a budgerigar's scientific name refers to the wavy barring of the feathers on their back. It is too late to ask Mr. Gould what his intentions were. It is the author's position that *undulatus* refers to the male's courtship warbling call. This call does indeed rise and fall, consistent with Webster's definition of undulate.

Budgerigar Behavior

Avian behavior is more complex than simply responding to various internal and external stimuli, like hunger and predators. To understand the behavior of an individual bird or group of birds it is critical to explore what these birds do *and* why they do what they do. Behavior consists of an intricate and diverse combination of voluntary and involuntary actions that are ongoing throughout one's life. Behavior entails a steady stream of choices, preferences, and decisions, even in situations where there don't appear to be any choices or choices are limited. Isolated behaviors must be scrutinized within the context of a lifetime of behavior to discern their true meaning. Simply put, it is not possible to interpret an isolated behavior correctly in a vacuum. To compare the behavior of wild and indoor budgerigars it is essential to examine the totality of how both populations interact with the different environments in which they live. Specifically, it is necessary to compare a spectrum of different scenarios such as their:

- Diet
- Lifestyle
- Emotional states
- Daily routine
- Common activities
- Individual behaviors

Diet is a good place to begin a comparison of the behavior of wild and indoor budgerigars. Diet involves choices, preferences, and memory, within anatomical and environmental constraints. Budgerigars are granivores. Wild budgerigars primarily eat ripe grass seeds found on or near the ground.[137] Twenty-one different types of grass seeds have been found the crops of wild budgerigars.[59] Favorites include spinifex, Mitchell grass, porcupine grass, saltbush, and occasionally ripening wheat.[41, 94]

The fact that twenty-one different types of seeds have been found indicate that wild budgerigars adjust their diets to the types of grass seed that are available at different times of year and in various locations.[183] These actions involve choice and memory. Several sources indicated that spinifex is the mainstay of a wild budgerigar's diet. The prevalence of spinifex in a wild budgerigar's diet may be due to preference, the abundance of spinifex in the Outback, or a combination of the two. Perhaps wild budgerigars prefer spinifex and only consume small amounts of the other twenty types of seeds to add variety to their diet. Or, perhaps wild budgerigars eat more spinifex because it is available in more locations throughout the year. Either way the budgerigars decide what to eat when and how much to eat, based on what types of seeds are available.

An examination of multiple commercial budgerigar diets indicates that they are composed of three main types of seeds: millet, canary grass seed, and oat groats. Usually a mixture of white proso millet, red millet, and golden German millet is provided. Minute amounts of specialty seeds like flax, thistle, buckwheat, hemp, poppy, sesame, amaranth, and rapeseed may be included. The protein content listed ranges from 12% to 14%. The fat content listed ranges from 4% to 6%. And the crude fiber content listed ranges from 8% to 14%.

Here is the problem. None of these seeds are native to the Outback. None of these seeds are among the twenty-one types of seeds wild budgerigars are known to eat. On the contrary, during the past 170 years the tradition of feeding budgerigars millet, canary grass seed, and oat groats arose because these seeds were readily available in countries where budgerigars were kept as pets—not because wild budgerigars eat these seeds. In fact, some publications from the 1800's suggested feeding budgerigars bread soaked in milk and cooked vegetables.[81, 168]

Early on the first budgerigars in captivity learned to eat millet, canary grass seed, and oat groats to survive. Over time subsequent generations of indoor budgerigars learned to recognize millet, canary grass seed, and oat groats as acceptable or normal food. Even so individual preferences still come into play. Some of the budgerigars discussed in this chapter preferred millet to canary grass seed, or vice versa. Some of the budgerigars preferred white or golden millet to red millet, or vice versa. Other budgerigars preferred oat groats to either millet or canary grass seed. All of the budgerigars agreed that millet sprays were preferable to a dish of seeds; probably because fresh seeds, still on a branch, are closer to what would be eaten in the wild.

An urban legend has developed and been perpetuated to the effect that seed diets are bad for indoor budgerigars and cause diseases like fatty liver. This is an example of the faulty logic that is often employed in marketing hype. How can a seed-based diet be bad for granivores? Some sources have claimed that these seeds are high in fat content. However, if we consult the Nutrition Almanac the following nutritional content is listed for 1 dry uncooked cup of millet and oatmeal.[52] (Oatmeal is flattened oat groats.) Similar percentages are stated in other and earlier publications dating back to 1961.[53, 108, 178]

	Millet	Oatmeal
weight	228 grams	80 grams
calories	746	300
protein	22.6 grams	10 grams
carbohydrate	166 grams	54 grams
fat	6.8 grams	6 grams
fiber	7.3 grams	8 grams
calcium	45.6 mg	26 mg
magnesium	369 mg	72 mg
phosphorous	709 mg	228 mg
potassium	980 mg	169 mg

Hence, the fat content of both millet and oat groats is low. The lipids in these grains are 67% unsaturated and 33% saturated. Neither grain contains any cholesterol. In fact, in human nutrition oatmeal consumption is recommended to lower cholesterol and other lipids in the bloodstream.

It is not logical to assert that a seed diet is bad for granivores like budgerigars. Fatty liver and other diet-related diseases are not due to a seed-based diet, but rather the fact that indoor budgerigars do not have access to top soil or tree bark to supplement their diet with adequate minerals and trace elements, like iodine. Seeds found in commercial budgerigar seed mixtures are not grown in the same type of soil that is found in the Outback; hence different types and levels of minerals are present in the seeds. Commercial budgerigar seed mixtures do not contain the twenty-one types of seeds budgerigars eat in the wild; neither does any of the commercial budgerigar pellets examined.

Budgerigars have a high metabolic rate; this creates a critical need for iodine and other trace elements. A high metabolic rate causes dietary deficiencies to occur faster.[99] A lack of iodine and other minerals causes fatty liver and other avian diet-related diseases.[115] A lack of iodine causes hypothyroidism in birds and humans.[99] The thyroid, which controls metabolism, cannot function properly when there isn't enough iodine in the diet.[6, 52, 99, 167] Iodine is essential for the production of the hormone thyroxine.[6] Iodine deficiency can contribute to the inability to synthesize cholesterol and hardening of the arteries.[52, 167, 185] Iodine is found in concentrated form in kelp and other seaweeds or as an impurity in sodium nitrate deposits—not seeds, nor fruits and vegetables. That is why iodine is added to table salt in human diets.

In summary, eating seeds is not the problem—the lack of access to iodine and other minerals found in soil is the culprit.[6, 99] Greenie I and her generation ate a seed-based diet. They lived to be 10 years old or older. They did, however, have a mineral supplement that contained iodine added to their water daily.

Most likely indoor budgerigars, through genetic intelligence, know that they need to consume a small amount of soil on occasion to obtain essential minerals. Recall the incident where Uzziel and Asher dove into the potting soil, not once but twice? However, most indoor budgerigars do not have access to soil. The need for iodine coupled with the lack of access to soil may explain why some budgerigars explore mineral blocks and cuttlebones, even though this behavior does not occur in the wild. In cases of severe mineral deficiency indoor budgerigars will consume a small quantity of dried droppings. This behavior is an attempt to regain essential minerals lost during the digestive process. This behavior mimics wild budgerigars consuming small amounts of soil.

Dietary deficiencies and the associated diseases may be congenital. There is no way to know whether or not the parents of a budgerigar were fed a nutritionally complete diet. Hence, the origin of a mineral deficiency may precede a bird's hatch day.

Several sources indicate that indoor budgerigars should be given fresh fruits and vegetables, especially green leafy vegetables.[31, 48, 70, 99, 108, 123, 124, 139, 167, 178] The intent is to supply additional nutrients that may be missing from the diet of indoor budgerigars, even though wild budgerigars do not eat these foods.

Indoor budgerigars were given a variety of fresh fruits and vegetables and the opportunity to decide over a period of several months what they did and did not like. This is another area where choice is important. Budgerigars were less inhibited and more likely to try new foods when they were in a group situation versus housed separately. If one budgerigar in an aviary tried something new, the other birds soon followed. Some, but not all, budgerigars were more likely to try new foods when they were younger, or when they just arrived in the aviary. In the latter case, since everything around a new bird had changed, they were more open to trying new foods.

The menu included apples, bananas, blueberries, cantaloupe, oranges, peaches, raisins, strawberries, arugula, baby spinach, broccoli, celery leaves, cherry tomatoes, corn, dandelion leaves, dried greens, green peas, kale, red and green leaf lettuce, mache, parsley, shredded carrots, and watercress. Everything was served in budgerigar sized portions. Whenever possible, organic produce was supplied. Unlike larger parrots, budgerigars did not attempt to wash fruits and vegetables in their water dish. Like larger parrots, budgerigars did not eat the peel or skin of fruits.

Individual fruit and vegetable preferences were noticeable, as summarized in Table 7.9. Other reports have also noted individual food preferences in budgerigars.[53] For example, Greenie I, Blue, and Oliver were the only ones to eat celery leaves, blueberries, and strawberries. Everyone liked raw green peas. Frequently the birds played with the green peas, like they were soccer balls, before eating them. Shredded carrots and green leaf lettuce were also popular. For some reason no one would eat red leaf lettuce.

Individual dietary preferences also surfaced in regard to eating a pellet diet formulated for budgerigars. About half of the 23 budgerigars that were given an opportunity to eat pellets would eat some natural pellets as a side dish. None of the budgerigars would eat natural pellets as an entree or only course, even though they were exposed to pellets for more than a year. Likewise, none of the budgerigars would eat the colorful pellets. Obviously other budgerigars eat the colorful pellets, since they have been marketed for years.

Baby spinach was recommended by several sources because it supplies Vitamin A, iron, and other micro-nutrients. Baby spinach quickly became a favorite of all the budgerigars. In 2010 spinach

Table 7.9 Individual Budgerigar Fruit and Vegetable Preferences.

	No birds ate	**1 to a few birds ate**	**Half or more birds ate**	**All birds ate**
I. Fresh Fruits				
apple				X
banana			X	
blueberries		X		
cantaloupe		X		
grapes	X			
orange	X			
peaches			X	
raisins	X			
strawberries		X		
II. Fresh Vegetables				
arugula		X	X	
baby spinach				X
broccoli			X	
celery leaves		X		
cherry tomato	X			
corn			X	
dandelion leaves	X			
dried greens			X	
green peas				X
kale	X			
leaf lettuce				
• green				X
• red	X			
mache			X	
parsley		X		
shredded carrots				X
watercress		X		

was banned from the aviary. Recent reports indicate that spinach, like cabbage and beet greens, is high in oxalic acid.[167, 178, 203] Oxalic acid causes the production of calcium oxalate by binding calcium with other minerals and trace elements.[203] The consequences of budgerigars consuming oxalic acid ranges from calcium deficiency to kidney and/or liver failure and death, depending on the quantity consumed and the time interval.[203] The effects of some plant toxins are cumulative.[6]

Oxalic acid is a common component in houseplants that are considered toxic to budgerigars, like philodendron and pothos.[19, 69] Rhubarb leaves are poisonous to humans because of their high oxalic acid content. In the wild budgerigars know which plants are poisonous and avoid them. Wild budgerigars have had 5 million years to acquire this knowledge and pass it on from generation to generation. Indoor budgerigars do not have an opportunity to learn this information from their parents and flock. Indoor budgerigars rely on their human caretakers to keep poisonous plants away from them. As a case in point all of the budgerigars liked baby spinach even though it contains oxalic acid which is harmful to them.

Wild budgerigars lead an active nomadic life style.[36] Their daily movement from place to place is tied to the availability of food and water. Given the dry conditions in the Outback, food and water sources may be miles (kilometers) apart.[59] Neither pairs nor flocks of budgerigars are territorial. They frequent the same food and water sources and roosting areas repeatedly throughout the year, but do not exhibit any territorial behaviors, like trying to drive other birds or flocks away.

Wild budgerigars are very social and choose to participate in dynamic non-hierarchical flocks. Other than the fact that bonded pairs stay together, flock membership is not fixed and fluctuates over time. There is no hierarchy or structure to budgerigar flocks. Flock size varies considerably. Flocks of 10 to 100 budgerigars are common; however, several small flocks may come together to make a flock of 1000 or more birds.[59, 60] Because of their nomadic lifestyle and the dynamic nature of flocks, budgerigars are constantly meeting and interacting with new birds. As a result they develop sophisticated social and communication skills early in life.

Indoor budgerigars lead a totally different life style. If they are lucky they may have an opportunity to journey around a 30-foot (9 m) living room. For the most part, however, indoor budgerigars live in a confined space—the opposite of their wild cousins. Indoor budgerigars fly by choice, when given the opportunity. Indoor budgerigars fly wherever they want to for fun. They are not searching for food or water or avoiding predators. Indoor budgerigars simply fly somewhere to visit another bird or stretch their wings. The flight of indoor budgerigars is ad hoc since they have no particular destination or purpose in mind. The flight of indoor budgerigars is that of individuals, not a flock. There are no concerns about staying with the flock or fears of being left behind. The flight of indoor budgerigars is characterized by sudden flights over short distances, not the sustained flight of their wild cousins who travel miles (kilometers) at a time.

The social interactions of indoor budgerigars are as restricted as their flight. Most indoor budgerigars live their entire lives as solitary bids, or with a single companion they didn't choose. In comparison the minimum flock size of budgerigars seen in the wild is 10 birds. Since they have little opportunity to meet and interact with new birds, the social skills of most indoor budgerigars are rather limited. That situation led to the establishment of designated gentle, rowdy, and aggressive budgerigar aviaries. Later, when given the opportunity all the budgerigars moved in together in one aviary.

The budgerigars had distinct temperaments, which ranged from gentle to aggressive. However, not all temperaments were static. The temperament of several individuals changed over time. In particular, some of the more aggressive birds mellowed with age. Also, the group dynamics changed whenever a new bird was added to an aviary, or one was removed.

Each of the birds exhibited social preferences in terms of which bird(s) they wanted to be buddies with versus those they merely tolerated. The most extreme examples of social preferences were illustrated by:

- Uzziel being excited about Blue Boy, versus ignoring Asher
- Precious flying out the door after Yellowhead
- Daisy letting herself out of one aviary and into another, so she could room with a different set of birds

It is difficult if not impossible to study the emotional states of wild budgerigars. That is not the case in an up close observation over a period of three decades. Twenty-one of the twenty-eight budgerigars were observed their entire lives. As a result, changes in an individual bird's emotional state were discernable. As Dr. T. J. Lafeber stated in 1977:[99]

> "... people working with birds see an assortment of personalities ranging from a nice average to both extremes of friendship and hostility. This reconfirms our knowledge that birds are emotional. Without feelings, there could be no fear, hostility or the opposite—friendly, outgoing, attention seeking attitudes."

Some budgerigars are more demonstrative than others. However all budgerigars exhibited different emotional states at different times during their lives, even within the same day. The activity level, posture, and eyes of a budgerigar change to reflect to their emotional state. For example, pet budgerigars are known to express a relaxed state by letting their eyelids droop a bit, which makes their round eyes appear almond-shaped. This behavior is referred to as "almond eyes". Vocalizations may be louder or quieter, frequent or infrequent, or strangely absent. The breathing rate may be faster or slower. Feather position is also an indicator of emotional state. A budgerigar that is very nervous pulls its feathers flat against its body and appears very skinny. Proximity may also give a clue about a budgerigar's emotional state. If a budgerigar is interested in interacting with a person, object, or other bird it will come close to them. If a budgerigar is fearful of a person, object, or other bird it will move in the opposite direction. Or, if a budgerigar is not interested in interaction, it will simply stay away.

At times emotional states were an individual experience. Other times emotional states were a group experience. One bird may be sad and lonely while it is going through the 2-week mourning period. At the same time the other birds are happily playing. One new bird overcomes its fear and decides to interact with a human right away. At the same time the other new birds retain their fear and avoid human contact for a while. Or, a group of birds may be fearful because of a noisy thunderstorm during the night, especially if there are flashes of lightning nearby. A group of birds becomes excited upon seeing their swimming pool and they all rush over to it.

Table 7.10 summarizes the spectrum and prevalence of emotional states observed in the budgerigars. Positive emotions are the standard set of emotions that are experienced with the normal ups and downs of ever day life. Negative emotions, if left unchecked, can lead to destructive behavior, like feather plucking. With one exception, none of the negative emotions commonly found in humans or large parrots were observed.

Young males that were courting a particular female or part of a newly formed pair demonstrated possessiveness, by actively chasing away any males that came too close to their partner. This pattern of behavior was not seen in established pair bonds and may have indicated insecurity on the part of the young male about the permanence of the new pair bond. None of the female budgerigars exhibited possessiveness. When Sunflower I tried to steal Blue Boy away from Uzziel, Uzziel did not react. Nor did Precious react when Blue Boy parted company and bonded with Uzziel.

Emotions are related to or attributions of cognitive and physiological states. Hence, it is not unreasonable to assert that wild budgerigars experience many of the same emotional states as indoor budgerigars.

Table 7.2 summarizes the typical daily routine of wild budgerigars. Indoor budgerigars adhere to essentially the same routine, if they experience the natural day/night cycle. Just before sunrise indoor budgerigars wake-up and start stretching and preening. At first there is limited quiet chatter. As more birds wake up and the sun rises the chatter becomes louder and more energetic. Budgerigars from all three aviaries participate in one combined morning chorus.

The first meal of the day is usually an hour after sunrise. Indoor budgerigars do not have to travel miles (kilometers) to a food source. But they fly laps around the aviary before breakfast anyway. Indoor budgerigars learn to wait until their food and water dishes are freshened up. Afterward they eagerly hop onto the dish of their choice and munch away. Like their wild cousins, indoor budgerigars take a small drink of water after their crop is full.

From mid-morning to mid-day indoor budgerigars play, socialize, fly, hop, and climb from perch to toy to swing or run around on the paper. They take short naps from time to time, but mostly they are active. And, as a group they are constantly vocalizing. The rapidity and volume of the chatter varies, depending on how many birds are napping. Even while napping, indoor budgerigars engage in quiet chatter. Budgerigar vocalizations can be heard on Disk 2 Track 4.

Late afternoon there is a sudden burst of energy and everyone is up and about. The last meal of the day coincides with sunset. Once in a while indoor budgerigars have a second drink of water at this time. As it begins to get dark there is a lot of fluttering as everyone decides which perch or swing they want to spend the night on. There is some final group chatter. Then it becomes quiet as the budgerigars fall asleep. Occasionally, the budgerigars can be heard changing perches after dark.

Table 7.10 Spectrum and Prevalence of Emotional States Observed in Indoor Budgerigars.

	All budgies	**Most budgies**	**A few budgies**	**No budgies**
I. Positive Emotions				
excited	x			
happy	x			
sad[1]	x			
confident	x			
timid/unsure[2]			x	
lonely[1]			x	
curious	x			
puzzled		x		
playful/eager	x			
anxious/fearful[2]		x		
aggressive/impatient[3]			x	
gentle/patient		x		
II. Negative Emotions				
anger				x
hatred				x
jealousy				x
grumpy				x
overwhelmed				x
stressed out				x
possessiveness			x	
self pity				x
depression				x
boredom[4]				x

Notes:

1—Most obvious during 2-week mourning period. Solitary birds may try to compensate for loneliness by playing with mirrors.
2—Most obvious after being rehomed, upon arrival in the aviary, during a major weather event like a thunderstorm, or when presented with a new situation like being put in a travel cage.
3—Actual physical aggression was limited to the two brief incidents mentioned in Section 7.6.
4—Boredom was not a problem because there were many toys and other birds to interact with.

Indoor budgerigars eat more during the colder months and less during the warmer months. During the colder months they frequently have a small mid-day meal. Regardless of the season or time of day, indoor budgerigars enthusiastically sample a millet spray or other treat as soon as it is placed in the aviary.

During the winter months when sun lamps provide an additional 4 hours of sunshine there is a slight variation in this routine. The extent of the variation depends on when the sun lamps are turned on. If the sun lamps are turned on before sundown and there is no break in the daylight cycle, the indoor budgerigars automatically adjust their evening routine to coincide with the artificial sunset. If the sun lamps are turned on after sundown, the indoor budgerigars treat the extra 4 hours of sunshine like a mini-day. There is a sudden burst of energy and increased activity and chatter. Also, the budgerigars have a small meal before retiring.

A comparison of activities common to wild and indoor budgerigars is also warranted. Eating is a community activity for wild budgerigars. One or more flocks of budgerigars find and travel to food sources together. Each individual bird is responsible for finding and removing the hulls from the seeds it eats. But the flock eats together at the same time and in the same place. Indoor budgerigars eat as a community too, except for the infrequent small mid-day meals. Snacks like millet spray are devoured as a group.

Many flocks of wild budgerigars gather together as a large flock at water sources to drink and bathe. Wild budgerigars stay close together at this time. When a swimming pool is installed in an aviary all the indoor budgerigars rush over to it at the same time. This behavior approximates that of wild budgerigars. In contrast, indoor budgerigars do not rush toward or congregate at their water dish. This behavior is different than that of wild budgerigars.

Table 7.5 identifies the common activities observed in bonded pairs of indoor budgerigars. These activities include allo-preening, allo-feeding, head bobbing, strutting, and beak touching. These five activities are common to both wild and indoor budgerigars. Wild budgerigars carve out a nest in tree holes or stumps and line it with wood dust. Indoor bonded pairs of budgerigars approximate that behavior by chewing on wooden toys and paper. Indoor female budgerigars also pull threads from cotton rope toys. An indoor budgerigar pair may shred paper and wooden toys without actually building a nest or laying any eggs. This activity seems to satisfy a psychological need. Both indoor and wild bonded pairs of budgerigars sit together and go through their daily activities together.

One activity related to indoor budgerigar pairs may or may not occur in the wild. When a young male is strutting back and forth and bobbing his head up and down to impress a female, he will chase away other males that get too close to his partner. This activity occurs most often with newly formed pair bonds and during courtship. This activity may be the result of being confined and having a fixed number of budgerigars around. If so, it would not be expected to occur in the wild.

All adult indoor budgerigars formed lasting pair bonds, even though they did not raise a family. These pair bonds were for life. Hence, these pair bonds had social and emotional significance. Why else would a bird go through the ~ 2-week mourning period after losing a mate? It is unclear whether or not wild budgerigars form lasting pair bonds without raising a family.

Table 7.4 lists the common play activities young indoor budgerigars use to learn about themselves and their environment. These activities include hanging upside down, tail stepping, laying on their backs, punching toys/toys punching back, and learning the difference between food, fingers, and toys. Play is a natural behavior in young birds.[99] Young indoor budgerigars have the luxury of playing in a predator-free environment. Young indoor budgerigars also have more time to play. As a result they are free to explore every nook and cranny of their home and try experiments like laying on their back.

One play behavior has been documented as far back as 1977.[99] That is the common practice of indoor budgerigars to wear a toy bell like a hat.[99, 112] Most of the budgerigars engaged in this play activity, particularly Uzziel. It is unclear what if any wild behavior this activity approximates. Maybe it is just for fun, because the bird's eyes are covered. Or, perhaps it is similar to using leaves to hide from predators or the noon day sun.

Play and experimentation are the process by which young birds learn about themselves and their world. Undoubtedly young wild budgerigars play too, like hanging upside down. They may or may not engage in tail stepping or laying on their backs. Wild budgerigars learn the laws of physics by landing on branches that move and touching a stem of a plant, rather than by punching toys. Wild budgerigars learn the difference between a branch, a pod, a seed, and a hull, instead of exploring food, fingers, and toys. Most likely young wild budgerigars have other unique play activities that have yet to be documented.

Since play is an essential part of learning, both indoor and wild budgerigars engage in play activities, especially when they are young. Some play activities are common to both indoor and wild budgerigars, like hanging upside down. Other play activities are different between indoor and wild budgerigars because they correspond to the specific environment each inhabits.

Budgerigar Individuality

Lastly let's examine the individual behaviors observed in indoor budgerigars. These behaviors are particularly informative because they highlight creativity, individuality, and problem-solving skills. The majority of these behaviors do not occur or have a corollary in the wild. Only one or two budgerigars exhibited each of these behaviors.

A total of 42 unique behaviors were observed; these are in addition to social and dietary preferences. Twenty-four of the unique behaviors were random or not specifically related to a life stage. The remaining eighteen behaviors reflected a reaction to a specific life stage or event, like aging, an illness, or a disability. The twenty-four unique random behaviors fell into three categories:

- Interacting with humans (6)
- Interacting with other birds (7)
- Playing, exploring, and learning (11)

Budgerigars, like other indoor parrots, view their human caretakers as part of their flock. Budgerigars attempt to communicate and interact with their human caretakers. The difference between budgerigar and human social norms and communication styles leads to some interesting interactions. However, budgerigars usually get their point across. This was certainly the case when:

- Greenie I and Oliver ignored me after I returned from a 5-day business trip
- Greenie II placed his beak on my finger (without biting) to express his displeasure about something
- Perky let me know I should pay more attention to him by continually landing and taking off from my pen/pencil while I was writing
- Blue Boy would go into his speech routine whenever he heard a political speech he liked
- Uzziel let me know she valued me as a member of the flock by having lunch "with me" and singing a song afterward
- Greenie II made sure his evening crop crunching routine coincided with my dinner

Budgerigars are very sociable by nature. Given adequate opportunities they enjoy interacting with other birds, whether or not they are budgerigars. This was demonstrated by Uzziel's excitement upon seeing and hearing the rose-colored house finches and singing to them year after year. Furthermore Uzziel sang along with the backyard birds CD during the winter months year after year, even when she was part of a bonded pair. Yellow Birdie demonstrated her interest in interacting with other birds by visiting the Meyer's parrot whenever she was out flying. Precious went to great lengths to demonstrate her fondness for and loyalty to Mr. Green Birdie by sitting on top of his swing at night. This was difficult for Precious to do since she only has one leg. No two-legged budgerigar has ever tried to do this.

Daisy let herself out of one aviary and into another just so she could interact with another group of birds. This was not a onetime event. With practice Daisy taught herself how to exit an aviary through the gap between the tray and the cage bottom. After she mastered this skill Daisy taught herself how to enter an aviary through the same gap. Both of these exercises took determination and a practice period of several days. Later Sunflower II also learned how to exit an aviary through the gap so she could visit other birds. To date Daisy is the only one who has learned how to enter an aviary through the gap.

Indoor budgerigars have more time for play activities than wild budgerigars. As a result they are continually engaged in exploring and learning throughout their lives. Hence the constant need for new and different toys. Blue II preferred the challenge of flying within the confined space of an aviary to flying in an open room. Uzziel preferred the challenge of trying to balance on a thin wire to the ease of balancing on a wooden perch, after turning her swing upside down. Slate I and Sunflower II

(5 years later) turned inside out for fun and to attract attention while hanging in a corner of their aviary. Asher dragged his gravel dish from one side of the aviary to another like a hockey puck.

Sky I arranged and rearranged shredded pieces of paper, like building sand castles on the beach. Sky I also ate the stem before the leaf when dining on baby spinach. Sunflower I chewed through a 0.5 inch (1.27 cm) wooden dowel and a wooden parrot-sized toy. Daisy chewed one peg loose on two separate ladders. Precious "carved" a nest out of the dry food dish. Uzziel and Asher dove into the potting soil after dispensing with the growing greens, not once but twice. Yellow Birdie liked to unlatch the dishes so that they dangled by only one hook.

Like humans budgerigars have to make some adjustments in their senior years. For example, Greenie I learned to take bubble-wrap naps to stay warm. There is no parallel to this behavior in the wild. Both Uzziel and Sunflower I had to switch from flying to climbing late in life. (Other budgerigars flew until the end.) Uzziel learned how to preen while hanging perpendicular from the side of her cage to compensate for her reduced ability to balance on one foot on a perch. Uzziel learned how to sleep with her right foot/nails looped around the mirror's wires and her left foot on the perch. Furthermore Uzziel learned that she could take naps on the rotating mirror or on the paper with her head propped up on the tray. These adjustments were made through creativity and determination due to the realization by each of these birds that their physical condition had changed.

Indoor budgerigars also recognize an illness, their own and that of others, especially a prolonged or terminal illness. Mr. Green Birdie learned to deal with his changed circumstances once he developed a fluid sac. At first he could no longer fly due to the additional weight of the fluid sac. To compensate Mr. Green Birdie climbed everywhere. As the fluid sac expanded Mr. Green Birdie could no longer climb and had trouble balancing on a perch. At this point Mr. Green Birdie had to adjust to spending his entire day on the paper, albeit the paper was 4 feet (1.21 m) above the floor. This meant Mr. Green Birdie spent his whole day walking around flat footed, with little opportunity to rest either leg or foot. It also meant Mr. Green Birdie had to get used to eating and drinking out of new dishes that were on the paper. Instead of sleeping on his favorite swing, Mr. Green Birdie learned to snuggle up in a corner so he could at least rest his head and neck. Throughout this ordeal Mr. Green Birdie never gave up. On the contrary, Mr. Green Birdie learned and adjusted.

Mr. Green Birdie and Precious were a bonded pair at this time. It is equally instructive to note that throughout Mr. Green Birdie's illness Precious groomed Mr. Green Birdie (he could not do this himself), fed him, and stayed right next to him. On the morning of Mr. Green Birdie's death Precious was found still sitting right next to him on the paper.

Precious has faced several challenges of her own throughout her life, since she only has one leg. Precious taught herself how to compensate for this disability, since there weren't any role models or genetic intelligence to consult. Precious not only taught herself how to compensate for her disability, she learned how to do a few things normal 2-legged budgerigars cannot. Precious learned some of this behavior before she was adopted. For example, before she arrived Precious had learned to clench all four toes together and curl her leg under the center of her body to balance or hop along a perch. Precious learned to hold her leg folded forward toward her beak, and not straight back like other birds, when she flies. This way her center of gravity is co-located with her beak, making it easier to balance her flight. Otherwise, there would be more weight on her right side.

Precious learned several new behaviors after she arrived because she was in a new environment. Precious learned to lay on her side on the loft to preen, stretch her wings, and rest her leg/foot. Precious learned how to climb using her solitary leg and beak. Precious learned to swing from side to side while hanging by her beak, when more momentum was needed to cross gaps of an inch (2.54 cm) or more. Precious taught herself to scoot along the ledge by laying on her left side and pulling herself forward with her right foot. Precious converted a palm leaf globe into a rest platform. In addition, Precious figured out how to lay across 2 pegs of a toy, pushing the third peg out of the way. Precious invented all of these behaviors herself and is the only one to engage in them. The process of devising and learning each of these behaviors required creativity, problem solving skills, and persistence.

Table 7.11 tabulates the distribution of unique random behaviors and unique life stage behaviors by budgerigar. Some budgerigars did not or have not yet demonstrated any unique behaviors. It is interesting to note that the female budgerigars account for the majority of the unique behaviors, even though they represented only 39% of the total budgerigar population.

Table 7.11 Distribution of Unique Behaviors by Individual Budgerigar.

Budgerigar	**Unique Random Behaviors**	**Unique Life Stage Behaviors**	**Total**
Asher (m)	2		2
Blue II (m)	1		1
Blue Boy (m)	1		1
Daisy (f)	2		2
Greenie I (f)	1	1	2
Greenie II (m)	2		2
Mr. Green Birdie (m)		4	4
Oliver (m)	1		1
Perky (m)	1		1
Precious[1] (f)	2	8	10
Sky I (f)	2		2
Slate I (f)	1		1
Sunflower I (f)	2	1	3
Sunflower II (f)	2		2
Yellow Birdie (f)	2		2
Uzziel[1] (f)	5	5	10
Total[2]	27/59%	19/41%	46/100%
Female unique behaviors	19/70%	15/79%	34/74%
Male unique behaviors	8/30%	4/21%	12/26%

Notes:

1—Highest number of unique behaviors.
2—The totals reflect the fact that some behaviors were exhibited by two birds at different points in time.

Summary

A 30-year study of 28 budgerigars, 21 of whom were observed their entire lives, provides a statistically significant sample. An objective interpretation of these observations yields the following conclusions:

- Budgerigars are individuals. They have individual preferences and temperaments and they exhibit individual emotional states.
- Budgerigar behavior is the result of a combination of biological, genetic, and applied intelligence. Some behavior is common, some behavior is learned, and other behavior is unique to individuals.
- Budgerigars are intelligent. Throughout their lives budgerigars have the ability to learn new things, remember past events, places, inanimate objects, and living subjects, compare alternatives, make decisions, deal with novel situations, and express choices.
- After 170 years of separation wild and indoor budgerigars are not all that dissimilar in terms of behavior or intellect.

Budgerigars are most definitely individuals. Nearly every source consulted referred to budgerigars as being "as individual as people".[7, 70, 99, 112, 115, 123, 124, 139, 203] The observations made in this chapter confirm those assertions. For example, there was a marked difference in the reaction to being rehomed by the seven budgerigars that were adopted. Slate, Lilac, Sunflower, and Sky exhibited varying degrees of being unnerved or demoralized by the process of being rehomed. In contrast, Precious, Blue Boy and Yellow Birdie enjoyed their new home environment from day one.

Budgerigars consistently demonstrated distinct individual preferences in several areas, such as:

- Social preferences
- Pair bond preferences
- Play preferences
- Toy preferences
- Dietary preferences
- Sleeping perch preferences
- Human interaction preferences

The process of developing and exhibiting distinct individual preferences involved memory and making decisions and choices. Preferences were not ingrained or inherent. Rather individual preferences only surfaced after a bird was exposed to a certain item over an extended period of time and had accumulated some experience with it. In some cases preferences, once established, remained static throughout a bird's life. In other cases, a bird changed a preference later in life. Individual choices, preferences, and behaviors indicate intelligence and self-awareness.

Evidence of individuality was apparent in others areas as well. Each bird had a unique temperament and displayed a spectrum of emotional states throughout its life. Flight was also an individual experience. Even though a group of budgerigars were given the opportunity to fly at the same time, each bird pursued an individual flight plan. Each budgerigar decided whether or not it wanted to come out to fly, where it went when it was flying, how long it stayed at each location and what it did while there, and when it was time to return home. Pair bonds demonstrated a degree of individuality too. Over time specific habits and a specific vocabulary were acquired that were restricted to an individual pair bond.

It is apparent that budgerigar behavior is the result of a combination of biological, genetic, and applied intelligence, within anatomical and environmental constraints. Some behaviors were common across the population of budgerigars. Some behaviors were learned. Other behaviors were unique to one or two budgerigars. Some behaviors were common to wild and indoor budgerigars. Other behaviors were unique to one population or the other.

All indoor and wild budgerigars enthusiastically jump at the opportunity to bathe in a pool of water. This is one example of genetic intelligence influencing behavior due to the necessity of taking excellent care of one's feathers. No indoor budgerigars were taught to bathe. Yet they all rushed at the first and every chance to do so.

Another example of genetic intelligence influencing behavior is limited to indoor budgerigars. All of the indoor budgerigars ate millet sprays before loose millet seeds in a dish. This was an immediate choice, not an acquired taste. Indoor budgerigars seem to understand that fresh seeds still on a branch are closer to what they would eat in the wild.

Indoor budgerigars do not eat the peel or skin of fruits. This practice is consistent with what was observed in other indoor parrots. However, there is no corollary since wild budgerigars do not eat fruit. Through applied intelligence indoor budgerigars seem to understand that the peel or skin of a fruit might be dirty or contain harmful chemicals.

Wild budgerigars know which plants in their native environment are poisonous or harmful and avoid them. Indoor budgerigars do not live in their native environment. Nor have they had time to learn or pass on knowledge about potentially harmful plants. In contrast, they are dependent on their caretakers to remove hazards from their environment.

Several behavioral characteristics were found to be common to both wild and indoor budgerigars, albeit with slight variations due to the difference in habitat. Wild and indoor budgerigars are very social birds. Neither exhibit territorial behavior. Both wild and indoor budgerigars follow essentially the same daily routine. Bonded pairs of wild and indoor budgerigars engage in the same activities, like allo-preening, allo-feeding, head bobbing, strutting, and beak touching. Young wild and indoor budgerigars engage in play activities to learn about the world around them.

Forty-two unique behaviors were observed. These behaviors demonstrated individuality, creativity, resourcefulness, ingenuity, and choice. In particular, the 18 unique behaviors that were related to a specific life stage or event evinced self-reliance, determination, the ability to improvise, and flexibility. Greenie I's and Uzziel's unique adjustments in their senior years, Mr. Green Birdie's accommodation to his illness, and Precious' self-taught disability-related behaviors are classic examples of this.

Lastly, it should be noted that budgerigar behavior was not static or hard-wired. As an example, when Daisy arrived she exhibited very aggressive behavior toward all the birds. Over a period of a few months she replaced this negative behavior and learned how to share, cooperate, and interact with the other birds in a positive manner. Uzziel remembered the wooden swing from her baby/quarantine cage four years later, even though she and the swing were in a different cage.

Budgerigars have a high degree of biological intelligence, genetic intelligence, and applied intelligence. The anatomy and physiology of wild budgerigars is fine-tuned to correspond with the dry conditions of their native habitat. Their salt gland, efficient use of water, and ability to fly miles (kilometers) between food and water sources demonstrate this.[137] Like other birds, budgerigars have the ability to pass on essential survival skills and historical memories genetically. This genetic intelligence is complemented with an impressive ability to learn new things throughout their lives. When confronted with a novel situation budgerigars draw upon memories of past events and places as inputs to their decision making and problem solving process. The rapidity with which budgerigars acquire knowledge should not be overlooked either. Not only can budgerigars learn new behaviors, they can also learn new vocabularies throughout their lives. Recall that the vocabulary of the budgerigars increased dramatically when 10 or more birds were housed in the aviary at the same time. Furthermore, budgerigars exhibit two traits that are acknowledged as a sign of intelligence in mammals:

- The ability to function in large complex societies, like flocks of 100 to 1000 or more birds
- Their insatiable curiosity and aptitude for engaging in play activities

After 170 years of separation wild and indoor budgerigars are not all that dissimilar in terms of behavior or intellect. Some physical differences, like color, have developed over the years by accident or on purpose. However, a normal green indoor budgerigar would be recognizable by his wild cousins. Wild and indoor budgerigars share many common behaviors although they live in different habitats. Undoubtedly each population has developed vocabularies that are unique to their environment. For example, wild budgerigars have vocalizations to indicate the presence of spinifex and tree snakes. In contrast, indoor budgerigars have vocalizations that refer to swings and vacuum cleaners. Wild and indoor budgerigars may have different accents or dialects, but most likely they can still understand each other. Wild and indoor budgerigars share a capacity to learn new things throughout their lives. Wild and indoor budgerigars learn different things because they live in different environments.

8

Zebra Finch

Poephila guttata castanotis

8.1 Background

It is amusing that such a tiny bird should have such a long name. But, if these three Latin words are deciphered the logic of this name becomes clear. Poephila literally means a friend of grass.[174] This term makes sense since zebra finches are considered to be grass finches.[117] Guttata means spotted.[174] There is a small rust patch with white spots on the flanks of a normal gray adult male zebra finch; although juveniles, females, and many hybrids do not have these spots. Castanotis means chestnut colored ears.[174] A zebra finch's ears are hidden by feathers, like other birds. The ears of a normal gray adult male are covered by a chestnut cheek patch. In fact, the zebra finch is known as the Chestnut-eared finch in Australia.[36, 117] Again, juveniles, females, and many hybrids do not have rust cheek patches. If all three words are put in context, the Latin name makes sense, at least for male normal grey zebra finches.

Originally zebra finches were referred to as *Taeniopygia guttata*, as part of the group of weaver finches.[174] As late as 1973 publications still referred to zebra finches as weaver finches.[85] Later it was decided that zebra finches do not actually do any weaving during nest building activities and had other unique identifying traits.[174] Now zebra finches are grouped with grass finches and parrot finches.[174] As we will see in Sections 8.3 and 8.5, the assertion that zebra finches do not engage in weaving during nest building activities may not be true in all cases. Earlier, in the late 1800's zebra finches were often referred to as *Amadina castanotis* and considered to be part of the sparrow family.[73, 79]

There is an ongoing debate about the origin of the name "zebra" finch. There are three dominant theories as to how the name zebra finch came about. The first theory refers to the black and white barring on an adult zebra finch's tail. These black and white bars are quite noticeable—they can be up to 0.25 inch (0.635 cm) wide. Yet, these black and white bars do not line up; instead the bars are offset by up to 0.125 inches (0.317 cm). The second theory points to the tiny black and white stripes on the breast of an adult male zebra finch. These are indeed black and white stripes, but they are horizontal—not vertical like a zebra's stripes. The third theory zeroes in a zebra finch's eyes. All zebra finches, male, female, juvenile, and hybrid, have a tiny black vertical line under each eye. The line is the same width as the eye and about 0.33 inches (0.83 cm) long. In front of this black line is a small white patch which runs from the eye to the beak. Hence, the eyes of a zebra finch look like those of a zebra, especially when viewed from the front. The author subscribes to the third theory.

Zebra finches are believed to be the third most popular pet bird, behind budgerigars *Melopsittacus undulatus* and canaries *Serinus canaria*.[117] Zebra finches are considered to be the most popular of all domesticated finches.[16] Publications from the late 1800's describe zebra finches as being hardy, active, bold, fearless, gregarious, and the most popular of tiny finches.[33, 73]

Zebra finches have been bred in captivity for many years.[41] European ornithologists have been aware of zebra finches for about 200 years.[174] Zebra finches arrived in Amsterdam in 1859.[174] Viellot, a French ornithologist, is believed to be the first European to keep and breed zebra finches.[174] Several reports of successful captive breeding appeared in the late 1800's and early 1900's, such as:

- 1872—German publication Die Gefiederte Welt[174]
- 1879—Foreign Cage Birds, volume 2, by C. W. Gedney[73]
- 1879—Prachtfinken by Dr. Karl Russ[174]
- 1891—Favorite Foreign Cage Birds by W. T. Greene[79]
- 1908—Foreign Birds for Cage and Aviary, volume 1, by A. G. Butler[33]

Since then many color mutations have been developed, including the first white zebra finch which was bred in 1921 in Sydney, Australia.[41, 174] In spite of the many color mutations captive zebra finches still recognize and accept each other.[85] It is unknown whether or not wild zebra finches would do this.

Fortunately, zebra finches continue to live in the wild. Exact population figures are unknown. They are not currently considered threatened, in spite of agricultural expansion and human encroachment. The main threat zebra finches face today is drought. A key reason zebra finches continue to live successfully in the wild is the fact that the Australian government had the foresight to ban the export of zebra finches in 1960.[97, 117]

8.2 Characteristics

Zebra finches share several physical and behavioral characteristics. Let's examine these characteristics, such as:

- Habitat and range
- Diet
- Physical characteristics
- Vocalizations
- Daily activities and socialization
- Family life

Habitat and Range

Zebra finches live in grasslands, open plains that are dotted with trees or bushes, land that has been partially cleared for cultivation, dry savannahs, pastures, and occasionally forests—all of which border water sources.[36, 97, 203] Zebra finches are nomadic since droughts are common in their native habitat.[174]

Zebra finches are found throughput most of the Australian continent, except the coastal areas.[36, 41, 85, 97, 174, 203] Zebra finches are not found in Tasmania.[41, 174] A closely related subspecies, *Poephila guttata guttata* lives in Indonesia and Timor-Leste.[203] Zebra finches have been introduced in Portugal, Puerto Rico, and parts of the continental U.S.[203]

Diet

Wild zebra finches are considered to be ground dwelling birds.[33, 203] As zebra finches go about their normal day to day activities, like flying, roosting, and foraging, they usually do not venture too far above ground. As a result it is no surprise that zebra finches are ground feeders.[33, 36, 97, 174] Zebra finches consume primarily the seeds of various grasses that are native to Australia, like spinifex (*Triodia*).[36, 41, 97, 174] On occasion wild zebra finches add a few insects, like termites, small moths, and caterpillar larvae, to their diet for protein.[174] Like budgerigars, wild zebra finches can consume a limited amount of brackish water without negative consequences.[85, 174] This feature comes in handy given the prevalence of droughts in their native habitat.

Pet zebra finches are fed a different diet, usually a seed mixture based on red, white, and yellow millet, canary seed, poppy seed, flax, and thistle, with some fresh greens on the side. Most publications from the late 1800's agreed that millet and canary seed were the mainstay of a captive zebra finch's diet.[33, 73, 79] Too much thistle, however, can cause liver problems.[174] It is doubtful whether many pet zebra finches receive insects as part of their diet, at least not live ones. Because of their high metabolic rate, zebra finches cannot go more than an hour without food.[174]

Physical Characteristics

The gray zebra finches found in the wild are considered to be "normal" zebra finches. Hence, this variety will be used to describe the physical characteristics of zebra finches. Zebra finches are from 3.9 to 4.25 inches (9.9 to 10.7 cm) long.[16, 85, 97, 117, 203] The straight square tail accounts for 1.25 inches (3.17 cm) of a zebra finch's length.[16] Wild zebra finches are slightly smaller than those that are raised domestically.[174] Male and female zebra finches are the same size.[203] Zebra finches weigh between 12 and 13 grams or less than a half of an ounce.[18] Sixty-three percent of their weight is water.[18] Zebra finches have a wingspan of 4.25 inches (10.7 cm).[174] Three toes face forward and one backward in an anisodactyl arrangement; the middle toe is longer than the other two.[174] Zebra finches have dark eyes. The placement of the eyes, and the flexibility of the neck, gives them almost a 360° view of their world. Zebra finches tend to tilt their head in all sorts of directions when they want to focus on something in particular, since their eye movement is limited.[174]

Zebra finches have a pointed symmetrical beak similar to cardinals (Chapter 5).[174] Zebra finches are sexually dimorphic. Males have a bright red beak, while females have a medium orange beak. As mentioned earlier, male zebra finches have a chestnut colored cheek patch; females do not. Both have orange legs and feet that are about 0.8 inch (2 cm) long.[16] Males have thin black and white horizontal stripes on their breast. A single thick black stripe is underneath this. Females do not have these stripes. Zebra finches have offset black and white bars on their tails. However, the barring is more pronounced in males. The crown, back, and wings are a light to medium gray. The underside is white. To complement their chestnut cheek patch, male zebra finches have a white spotted chestnut patch on their flanks.

Since zebra finches come from a warm climate, they do not tolerate cold temperatures very well. The minimum tolerable temperature is listed to be somewhere between 55°F and 65°F.[112, 174] Below these temperatures zebra finches cannot eat and digest food fast enough to produce sufficient energy to stay warm.

The maximum life span for zebra finches living in the wild is believed to be 10 years.[174, 203] The average life span for zebra finches living in captivity is closer to 5 years.[174, 203] The difference in diet, lack of access to adequate sunshine, and confinement are major contributing factors to the shorter life span.

Vocalizations

Zebra finches are known for making a variety of beeping, buzzing, and humming sounds, none of which are very loud. Male zebra finches also sing more melodic songs to impress the females.[112, 137, 174, 203] Individual male zebra finches develop their own customized songs.[203] It is believed that male zebra finches pass these songs on from generation to generation, as the sons imitate and enhance their father's songs.[203] This behavior is not unlike indigenous peoples passing oral histories from generation to generation. Young zebra finches learn new vocalizations best when they are 15 to 50 days old.[137] Young zebra finches can be heard practicing new sounds and songs at this time.[137, 203] Young zebra finches break new sounds into small pieces (syllables?) as they learn them. As a result, their vocalizations sound a bit disjoint at first.[137, 174]

Daily Activities and Socialization

Zebra finches, with their high activity level, appear to be the avian equivalent of a perpetual motion machine. With a hop here and a short flight there, they are always on the go. Even when standing on a perch, zebra finches move up and down, backward and forward, left and right in a never ending series of postures. Even the flight of zebra finches is rapid and direct.[174] This non-stop movement can be attributed to a high energy level and a high degree of curiosity. Whenever anything new was introduced into the aviary the tiny zebra finches—not the larger birds—were the first to check it out, as discussed in Section 8.3. Several sources consulted indicated that captive zebra finches do not need toys to play with. As shown in Sections 8.3 and 8.4 that is not necessarily true.

Because of their tiny size many pet owners mistakenly place zebra finches in small cages. This is the exact opposite of what zebra finches need. Experts recommend that zebra finches should be housed in cages that are 2 x 4 x 2 feet (60 x 122 x 60 cm); at a minimum a single pair needs a cage that is 18 x 30 x 18 inches (45.7 x 76.2 x 45.7 cm).[112, 155, 174] Zebra finches need lots of horizontal flying space because of their high activity level.[112, 174] Sadly, many zebra finches die each year from being confined to cages that are way too small.[112]

Wild zebra finches travel in pairs, small groups, or flocks year round.[36, 97] They go through their daily routine as a group.[174] The groups are smaller during nesting time, and larger afterward.[174] The normal non-nesting size of a flock is 40 to 100 birds.[174] However, during a severe drought as many as 1,000 zebra finches may congregate, as several smaller flocks join forces in the search for food and water.[174]

Zebra finches are social birds. However, the males are known to bully and pick on smaller weaker birds on occasion.[97, 112] This behavior is observed more often in captivity than in the wild. This behavior may be related to the stress of confinement and/or having a limited number of other birds with which to interact.

Zebra finches form same sex social bonds when they are not nesting.[112] Zebra finches frequently sit close to each other on a perch, so that their bodies are touching. This behavior is known as contact sitting.[97] Bonded pairs and unaffiliated birds do this when they are roosting. Often three or more birds will crowd together in a line on a single perch. This behavior occurs year round, not just during the colder months. This behavior fills a combined social and security need. If the birds are sleeping and a predator or other danger approaches, the movement of one bird will instantly wake up everyone. Contact sitting is similar to clumping, but it is undertaken for different reasons.

As Gedney noted in 1879, zebra finches "have no inclination to spend their nights on a perch".[73] Instead zebra finches like to sleep or rest in what is referred to as sleeping nests.[97, 112, 203] An individual bird or a bonded pair might take a short nap in a sleeping nest. This unusual behavior serves two purposes. First, it allows a zebra finch to temporarily "power down" its high energy system. This practice is similar to switching a laptop computer to hibernate mode. The sleeping nest provides a place for a zebra finch to rest. It also provides a degree of safety by concealing the zebra finch's presence. (See Photograph 8.1.)

Family Life

Wild zebra finches are believed to mate for life.[174] Bonded pairs engage in allo-preening, go through their daily activities, and stay together year round.[112, 174] Zebra finches nest in the rainy season, to ensure the availability of food and water for their new family.[85, 97, 203]

Photograph 8.1 Zebra Finch Sleeping Nests. Sleeping nests provide an opportunity for zebra finches to rest, since they don't have to balance or hold onto a perch. Wild zebra finches construct sleeping nests of natural materials like twigs and leaves. Indoor zebra finches use empty dishes. Some zebra finches decorated their sleeping nests with pieces of string or bits of shredded paper. Once a zebra finch or bonded pair has selected a sleeping nest, they are not likely to let other finches use it. © D. Herrmann

Wild zebra finches build rather elaborate domed nests from dried grass and other small vegetation.[36] Both parents participate in nest construction.[174] The nests have a side entrance and are lined with down.[36] Zebra finches place their nests in low thick bushes and shrubs.[36] In certain areas of southwest Australia zebra finches nest almost exclusively in thorny bushes which helps keep predators away.[174]

Zebra finches lay 3 to 6 thin white eggs.[36, 97, 117, 174] The eggs are slightly under 0.5 inch (1.27 cm) long and the shell is paper thin. The eggs are laid at least 24 hours apart, sometimes longer. (See Photograph 8.2.)

Photograph 8.2 Zebra Finch, Budgerigar, and Indian Ringneck Parakeet Eggs. Zebra finch eggs are tiny, narrow, and less than 0.5 inch (1.27 cm) long. The white shell is paper thin. Notice the difference in size and shape between a zebra finch egg and a budgerigar egg (top) and a zebra finch egg and an Indian ringneck parakeet egg (bottom). © D. Herrmann

The mother incubates the eggs while the father stands guard nearby. The mother rarely leaves the nest and then only for short periods of time. Incubation lasts between 12 and 15 days.[16, 97, 117, 174, 203] The eggs hatch in the order they were laid.

Zebra finches are very attentive parents.[117] Both parents feed the hatchlings.[97] They take care to feed the youngest hatchling first, so soon there is no difference in size within a clutch.[174] Zebra finch parents will vigorously protect and defend their hatchlings against any danger in spite of their small size.[33, 203] In fact, sources from the late 1800's report male zebra finches in a mixed bird aviary chasing birds 5 or 6 times their size away from the nest.[33, 73] Zebra finches are known to foster orphaned or lost hatchlings.[117]

Upon hatching zebra finches are extremely tiny, being less than 0.5 inch (1.27 cm) long. Hatchlings are fed dried grass seeds, green (immature) grass seeds, and insects for protein.[97, 174] Unlike many songbird hatchlings, zebra finch hatchlings do not make a begging call; instead they just open their tiny beaks quietly when they are hungry.[174] The sources consulted indicated that zebra finch hatchlings have dull black beaks. As we will see in Section 8.4, that is not entirely true.

Zebra finch hatchlings grow very quickly and double in size every few days. Zebra finch hatchlings are ready to fledge when they are about 3 weeks old.[16, 174, 203] Zebra finch fledglings still depend on their parents for at least some food for a while. By the time they are about 5 weeks old, young zebra finches are able to fend for themselves.[174] Young zebra finches tend to stay in the same flock with their parents. Young zebra finches acquire their adult plumage after their first full molt.[117]

8.3 Adult Zebra Finch Behaviors

Now let's look at the following traits of captive zebra finches and compare them to those of wild zebra finches:

- Temperament
- Vocalizations
- Daily activities
- Play activities and toy preferences
- Diet preferences
- Socialization patterns
- Reaction to novel situations
- Life stages and emotional states

Behaviors are presented as they occurred. The analysis, comparison, and interpretation of these behaviors are reserved for Section 8.5 Discussion. These observations took place at the locations and during the timeframes described in Annex C.

Before jumping into a discussion of temperament, let's take a moment to identify who is who among the 16 captive zebra finches. This will put the discussion in context and help readers keep references to individual birds straight. Table 8.1 lists the 16 zebra finches by name. The coloring, gender, temperament, and source of each bird are specified.

Temperament

A range of temperaments, from aggressive to passive to somewhere in the middle (neither aggressive nor passive), were observed among the zebra finches. There were no clear patterns as to what caused aggression or passiveness. Some of the sources consulted indicated that only males engaged in bullying activities.[97, 112] That was not the case for the 16 birds discussed in this chapter. Males only accounted for 57% of the aggressive birds. Neither the color nor the source of the birds was a key factor in determining aggression either. Seventy-one percent of the aggressive birds were grey; the rest were not. Forty-two percent of the aggressive birds were adopted; the rest were not.

Aggressive behavior took two main forms: (1) chasing another bird at top flight speed away from a perch, swing, food dish, or water dish, or just around and around the aviary until one of the birds dropped from exhaustion, and (2) pecking another bird's wings, often until they bled. Both males and females were attacked at various times by an aggressive bird. And, contrary to what was expected, increased cage size did not alleviate aggressive behavior. On more than one occasion moving to a larger

Table 8.1 Zebra Finches: Who's Who.

Name	Color	Gender	Temperament	Source
Ami	fawn	female	passive	pet store
Tzvi[1]	grey	male	aggressive	pet store
Puff	white	female	passive	adopted
Buzz	white	female	aggressive	adopted
Cinnamon	grey	male	aggressive	adopted
Ginger	grey	male	aggressive	adopted
Tiny Guy (father)	grey	male	midway	adopted
Little Bald Guy (uncle)	grey	male	midway	adopted
Ami II[2]	fawn	female	passive	adopted
Lady Finch	grey	female	passive	pet store
Ami III	fawn	female	passive	pet store
Lady Finch's Sister (mother)	grey	female	passive	pet store
Little Boy	grey pied	male	aggressive	hatched 7/23/2010
Little Girl	fawn pied	female	midway	hatched 7/28/2010
Naomi	fawn	female	aggressive	hatched 7/28/2010
Lady Silver	silver grey	female	aggressive	pet store

Notes:

1—Tzvi was close to being a true normal grey zebra finch. He had a red beak and a thick black bar across his chest. He did not have chestnut cheek patches or thin black and white barring across his chest.

2—Ami II had abnormal droppings upon arrival. The veterinarian determined that she had a hernia, probably due to a tumor. Ami II died 3 weeks after being adopted.

cage made things worse—a larger cage allowed the pursuing bird to chase another bird faster. If the situation got out of hand, the aggressive bird was removed from the aviary for a temporary time-out.

A few special observations related to aggression deserve to be mentioned.

Tzvi was a very aggressive bird from the day he arrived from the pet store. He flew like a dare devil at top speed around and around the aviary, looping around perches and through swings. When he tired of that, Tzvi began chasing Ami around unmercifully. Ami would move off of whatever toy, perch, or dish she was sitting on. But that wasn't enough. Her passive nature seemed to encourage Tzvi's aggression. Until one day Tzvi was flying at break neck speed, clipped a perch, and broke his left wing. His wing heeled. Tzvi was able to fly from perch to perch or perch to cage bottom, but that was all. His dare devil days were over. Overnight Tzvi's aggressive nature disappeared. He also stopped pecking Ami. Ami endured Tzvi's aggressive behavior for over a year. Once the tables were turned she could have retaliated, but she didn't. Instead, Ami remained her gentle-natured self. More than that Ami accommodated Tzvi's new needs as he learned to live with his disability. (See Photograph 8.3.)

Puff, Buzz, Cinnamon, and Ginger arrived together as juveniles. Puff was repeatedly chased and pecked by the other three birds. Sometimes one bird would harass her; sometimes all three would. It was not uncommon to find blood spots on her wings. Puff was a little bit smaller and a little bit weaker than the others. When they matured the four birds formed social pairs: Buzz and Cinnamon, Puff and Ginger. There was less bullying, but it did not disappear entirely. The bullying was now directed at the pair, not just Puff. On several occasions Ginger pushed a millet spray into a corner and he and Puff hid behind it like it was a barricade, until the "attack" was over. As the birds aged, they all mellowed and the aggressive behavior dissipated.

Photograph 8.3 Ami and Tzvi. Ami (right), a fawn female zebra finch, was gentle throughout her life. Her mate, Tzvi (left) was close to being a normal grey zebra finch. He had a red beak and a thick black bar across his chest. But he lacked chestnut cheek patches, spotted flanks, and thin black and white stripes across his chest. Tzvi was an aggressive dare devil until he broke his left wing, as shown here. Afterward Tzvi's temperament mellowed. © D. Herrmann

For a period of three months the zebra finches were separated into same sex aviaries to see how that would affect aggressive behavior. There was less chasing and less pecking but the separation did not eliminate aggressive behavior altogether. There still seemed to be a need on the part of the aggressive birds to establish their dominance.

Four of the six birds who were passive were strong and healthy. They could have fought back if they wanted to. For some reason they chose not to. Likewise, all three of the midway birds were strong and healthy. They selectively chose when to fight back and when not to.

Vocalizations

Male and female zebra finches made different vocalizations. All of the zebra finches seemed to understand each other's vocalizations, whether or not they were in the same aviary. (See Table 8.2.)

The females made a variety of quiet buzzing and humming sounds. About six or seven variations of these sounds were heard. Although the sounds were always quiet, the volume changed depending on the circumstance. The loudest sound female zebra finches made was a single syllable drawn out "anh". Sometimes it was repeated after a long pause. This vocalization was used as a contact call. It was heard most frequently when one aviary was moved out of sight for cleaning or an aggressive bird had been temporarily separated from the other birds.

Male zebra finches made three categories of vocalizations. These vocalizations were consistent (at least to human ears) across the male population. The first vocalization, "t-da-da t-da-da", seemed to serve two purposes. This vocalization was frequently used as a contact call, back and forth between aviaries. One male would initiate the call. One or more males would respond. This vocalization was also used by a single male when he was trying to impress a particular female. A male would make this vocalization over and over, sometimes increasing the volume, as he moved closer to a particular female. Tiny Guy and Little Bald Guy repeated this vocalization over and over. Early publications often referred to this vocalization as trumpeting.[73]

Table 8.2 Zebra Finch Vocalizations.

Sound	Who Makes It	Meaning and Context
humming and buzzing	females	• a variety of quiet sounds that correspond to different situations
"anh"	females	• contact call
"t-da-da t-da-da"	males	• contact call, especially between aviaries • courtship call
"du-du-du-du-du-du-du"	males	• general warning that the flock should be on alert
"tuh-tuh-tuh-tuh-tuh-tuh-tuh"	males	• urgent warning of a more immediate danger
beep beep	females and males	• answering a sound heard in their vocal frequency range, like a microwave oven beeping • number of beeps is one more than heard

Note: Zebra finch vocalizations can be heard on Disk 1 Track 4.

Male zebra finches made an alarm call: "du-du-du-du-du-du-du". This vocalization was loud and always repeated seven times. This vocalization seemed to be a warning signal that something in the immediate environment had changed and the flock should be on alert. Only one male made this vocalization at a time.

A third vocalization, "tuh-tuh-tuh-tuh-tuh-tuh-tuh", was used to indicate a more immediate danger. The vocalization was louder and faster than the alarm call. One or more male zebra finches might make this vocalization at the same time. This vocalization usually caused a lot of scurrying in the aviary.

A fourth category of vocalizations were context specific. Whenever any of the zebra finches heard a sound that was in their frequency range—like a microwave oven beeping or a truck beeping while backing up—they would "answer". This vocalization sounded more like a loud "beep beep". If the sound the zebra finches heard had X beeps, they frequently responded with X + 1 beeps. If the sound the zebra finches heard was repeated, they responded with the same number of iterations.

If there was only one male zebra finch in an aviary he was not very vocal and/or he limited his vocalizations to sounds that were similar to the females. The number and variety of male specific vocalizations increased dramatically once there were two or more males in an aviary. In contrast, the number and variety of female zebra finch vocalizations changed in proportion to the total number of birds in an aviary—not how many females there were. Zebra finch vocalizations can be heard on Disk 1 Track 4.

Daily Activities

Zebra finches engage in a variety of activities on a daily basis. Zebra finches stretch, preen, bathe, eat, nap, fly, explore, vocalize, and socialize daily. However, there isn't a fixed pattern for these activities. Zebra finches are not on a sunrise/sunset schedule, primarily because of their high metabolic rate.

Zebra finches stretch one wing at a time, while holding onto a perch and leaning forward. Zebra finches stretch several times throughout the day. They stretch first thing after getting up in the morning, after a nap, and before they intend to do some serious flying. Like everything else zebra finches do their stretching routine is quick.

Zebra finches preen their breast, wing, and tail feathers in the morning after they awake. This is a basic preening session; it does not consume much time. Zebra finches seem to enjoy preening others more than themselves. Allo-preening sessions last longer and are more involved as one bird carefully preens the head, face, and around the eyes of another bird. There is not a fixed time of day for allo-preening sessions—they happen when they happen. Zebra finches also groom their legs and feet a couple times a week and again whenever they bathe. Zebra finches can be seen rubbing their beak back and forth on their perches several times a day. This action helps to clean, polish, and trim their beak. (See Photograph 8.4.)

Zebra finches are enthusiastic bathers and will bathe daily given the opportunity. The zebra finches took turns splashing around until all the water was gone. Often a bird would splash around until it was soaked. Next it would fly up to the top perch, flap its wings to shed the excess water, and preen. Then, the same bird would fly back down to the water and repeat the process two or three times. All the zebra finches were equally comfortable bathing in an oblong plastic dish that was placed on their paper, as they were in a 5-ounce (150 ml) water dish on the side of the cage. The zebra finches were visible when they were bathing in the oblong dish. In contrast, the 5-ounce (150 ml) stainless dish concealed their presence. All that was visible were the water drops flying overhead.

Photograph 8.4 Allo-preening. Zebra finches seem to enjoy preening others more than themselves. Allo-preening sessions last longer and are more involved as one zebra finch carefully preens the head, face, and around the eyes of another. Here Little Boy preens Naomi when they are about 15 months old. Note the different male and female beak and leg colors. © D. Herrmann

Zebra finches eat several meals throughout the day, usually less than an hour apart. A large meal for a zebra finch may be seven or eight seeds. It takes two beak maneuvers to open most seeds. Zebra finches are ready to eat the first thing in the morning when the lights come on, regardless of whether it is 5 a.m. or 8:30 a.m. Likewise, zebra finches will hop down for a quick meal if the lights come on at 10 p.m. or 11:30 p.m. Zebra finches take several drinks of water throughout the day. On average one zebra finch consumes more water per day than one budgerigar.

Zebra finches consume gravel on a daily basis. Each zebra finch consumed 3 or 4 granules of gravel a day to assist the digestion of their food. The fact that all the zebra finches repeatedly consumed gravel throughout their lives implies that their digestive juices are able to dissolve the gravel gradually. This makes sense for several reasons. First, the gravel is a source of minerals in a zebra finch's diet. Second, if the gravel was not being dissolved, there would be no need to consume it on a daily basis. Third, if the gravel was not being dissolved, its accumulation would represent a significant increase in weight. This additional weight would complicate a zebra finch's flight.

Zebra finches consume cuttlebone on a regular basis. On average one zebra finch consumed twice as much cuttlebone as one budgerigar. There was no difference in the volume of cuttlebone consumed between males and females. Instead of biting a cuttlebone like a budgerigar, zebra finches engage in a chiseling action because of their different beak structure. Zebra finches use their tongue to catch the fine powder. For some reason zebra finches drop less cuttlebone powder than budgerigars. Since dried cuttlebones are not present in Australia, wild zebra finches must obtain these minerals by other means. On the other hand, the zebra finches consistently ignored mineral blocks. (See Photograph 8.5.)

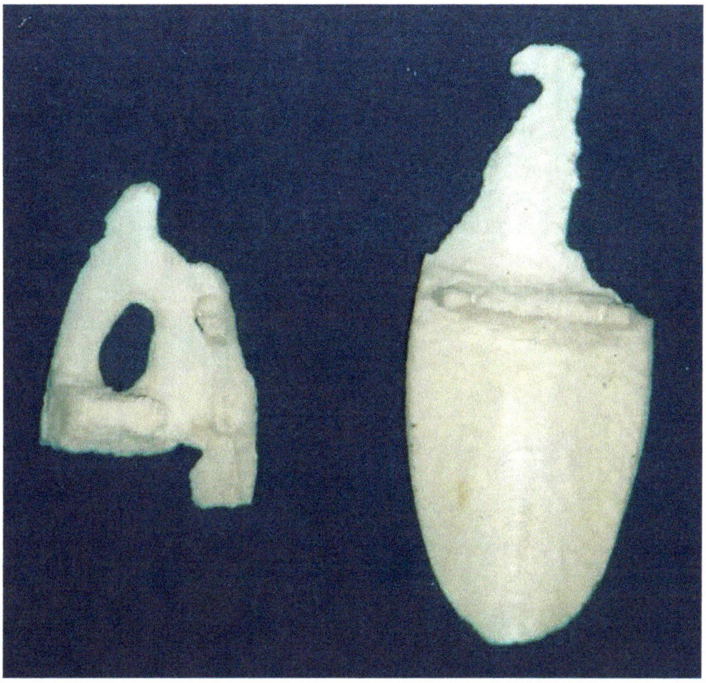

Photograph 8.5 Cuttlebone Remnants. A zebra finch consumes about twice as much cuttlebone per day as a budgerigar. Budgerigars bite and scrape cuttlebones, often spilling a lot of powder in the process. In contrast, zebra finches engage in a chiseling action, because of their different beak structure, and catch the powder with their tongue. As a result the remnants of their respective cuttlebones are different. On the right is a cuttlebone which was bitten and scraped down to the hard outer layer by budgerigars. On the left is a cuttlebone which was chiseled to a small fragment by zebra finches. © D. Herrmann

The zebra finches are comfortable eating from an oblong dish placed on their paper. This is not surprising since wild zebra finches are ground feeders. Two to four birds eat at a time like this without any problems. The zebra finches learned to eat from tube feeders and 5-ounce (150 ml) stainless dishes that were hung half-way up the side of their aviary. A single zebra finch or pair of zebra finches eats from one stainless dish at a time. The tube feeder was viewed more as a toy at first. Ami and Tzvi in particular would peck away until all the food was spilled on the floor. Then they would eat the food off the floor. Ami and Tzvi could empty an 8-ounce (240 ml) tube feeder in minutes. In the beginning Little Girl decided that one of the stainless dishes would also make a good sleeping nest. She would eat, then plop down in the dish for a nap. If Little Girl was hungry when she woke up, she didn't have far to go.

Zebra finches take several naps throughout the day. Some naps are very brief; others last 15 to 20 minutes or more. Zebra finches tend to roost by themselves for short naps. If a longer nap is contemplated, a zebra finch will snuggle up to another bird on the same perch. All the finches involved push toward each other until they are touching. Some of the finches sleep with their head forward; others turn their head around and tuck it under a wing. All of them perch very low, so that their feet are covered. Aggressive and passive birds nap together.

Flight is a major daily activity of zebra finches. Wild zebra finches spend a lot of time flying and cover considerable distances as they look for food and water, avoid predators, and find a place to roost during the day and sleep overnight. Captive zebra finches are also constantly on the move (if they have sufficient room), flying from one side of the aviary to another, perch to perch, cage bottom to swing, etc. Zebra finches can hang onto the side of an aviary, but they do not climb up and around on the bars like members of the parrot family. Zebra finches fly or hop whenever they want to go somewhere. They do not walk or climb.

Tiny Guy frequently let himself out to go joyriding about the room in which his aviary was located. Tiny Guy was small, even for a zebra finch; he fit through any 0.5 inch (1.27 cm) square opening. Tiny Guy flew around and around the room and in and out of other (larger) birds' aviaries—he fit through the bars. Sometimes Tiny Guy even wandered into adjacent rooms. Tiny Guy landed on the plants near the windows, but he never attempted to fly through the windows. Two of his favorite places to land were on the African violets and on the milk cactus. Tiny Guy was the only zebra finch to go joyriding, even though the others had the opportunity to do so. (See Photograph 8.6.)

Photograph 8.6 Tiny Guy Joyriding. Tiny Guy sits on a milk cactus near some African violets; these were two of his favorite plants. Tiny Guy was so small he fit through any 0.5 inch (1.27 cm) square opening. Tiny Guy frequently let himself out to go flying around the room, including in and out of larger birds' aviaries—he fit through the bars! Unlike many indoor and outdoor birds Tiny Guy never attempted to fly through the windows. © D. Herrmann

Play Activities and Toy Preferences

The zebra finches seemed to enjoy playing games. They played a friendly nonaggressive form of tag, especially in their younger years. Two or more finches would chase one bird until they caught and "tagged" it. Then the birds would regroup and they all chased another bird until it was tagged. The group would carry on like this for 10 to 15 minutes until everyone was tired.

Another game involved flying vertical loops. One zebra finch would jump straight up from its perch, fly a vertical loop about 8 inches (20 cm) high, then land in the exact same place it had taken off from. Afterward a second finch or pair of finches repeated this performance. The original finch repeated its performance two or three times. The other finches followed suit. Next the original finch changed locations and the game started over. These sessions usually lasted 10 to 15 minutes. Puff, Buzz, Cinnamon, and Ginger were the most active participants in this game.

A third game was reserved for the male zebra finches. It seemed to be some sort of challenge. Tiny Guy or Little Bald Guy would grasp a millet spray in his beak and fly with it to the top perch. Often the millet spray weighed as much or more than the zebra finch and more than one take-off attempt was necessary before becoming airborne. Once Tiny Guy or Little Bald Guy had the millet spray on the top perch, the other male zebra finches tried to steal it from him. There was a little tug-of-war, the millet spray fell to the bottom of the aviary, and the game started over again. The game seemed to be more about possessing or holding the millet spray than eating it.

Little Bald Guy devised a variation of this game. He would carefully tear a piece of paper about 0.5 inch (1.27 cm) wide by 4.0 inches (10 cm) long from the newspaper on the cage bottom. Little Bald Guy grasped the strip of paper in his beak and flew with it up to the water dish. Little Bald Guy dipped the strip of paper in the water briefly. Then, he wove it into the aviary bars near the water dish, methodically inserting the paper strip over or under each bar, until the entire strip was held in place. Sometimes Little Bald Guy wove two paper strips together like this. Little Bald Guy was the only zebra finch to engage in this behavior. This behavior may have been an invitation to engage in nesting activities.

Several sources consulted indicated that zebra finches do not need and/or do not play with toys. All 16 zebra finches proved these assertions to be false. All 16 zebra finches engaged in play activities on a daily basis and exhibited a high degree of curiosity by exploring new items as soon as they were introduced into their environment.

The challenge wasn't to find a toy that would interest the zebra finches. Rather, the challenge was to find toys that were appropriately sized for zebra finches. On occasion a toy that was designed for a large bird became a multi-finch toy. For example, one parrot toy consisted of 3 horizontal acrylic hoops with increasing diameters. The hoops were strung about an inch above each other. Instead of a parrot toy that was too large, it became a multi-tiered multi-finch round swing. One finch perched on each level. Of course, they each had to learn where to land so that the swing remained balanced.

As shown in Table 8.3 the zebra finches played with a variety of different toys. Their favorites included string toys, toys made with natural fibers like sisal and dried palm leaves, and mirrors. All of the zebra finches chewed, preened, and spread out cotton toys. Cotton ropes were preened into individual threads. Natural fiber toys received almost as much attention. Zebra finches sat and looked at themselves in mirrors and pecked the mirrors much like budgerigars do. Several of the finches played with a toy that consisted of thin flexible plastic "strings" and beads of various shapes. The finches preened the plastic strings like a cotton toy and moved the beads up and down. The zebra finches did not show a preference for or aversion to any toy color. None of the zebra finches played with bells, rattles, whiffle balls, or foot toys. (See Photograph 8.7.)

Tiny Guy, Little Bald Guy, and Lady Finch's Sister took the string toy preference one step further. On two separate occasions a loose string about 12 inches (30.48 cm) long was placed on the bottom of the finch aviary. The same behavior was observed each time. At first each of the three finches took turns bunching the string up into a wad and then flying around with it. One bird would drop the string. Another bird would pick up the string and start the game over.

A few days after the string was introduced Little Bald Guy carried it up to an empty hoop for 5-ounce (150 ml) stainless dish. The zebra finches like to perch on empty hoops because they are narrower than the other perches. Instead of chewing and dropping the string, Little Bald Guy began weaving the string around the hoop, much like a person crocheting. He used up about half of the

Table 8.3 Zebra Finch Toy Preferences.

Toy	No finches played with	1 to a few finches played with	Half or more finches played with	All finches played with
bells	x			
cotton ball puffs				x
foot toys	x			
ladders			x	
mirrors				x
natural fiber toys				x
push pull toys			x	
puzzle toys		x		
rattles	x			
shredding toys				x
small whiffle balls	x			
string toys				x
swings				x

*Toy color did not affect toy preferences.

Photograph 8.7 Zebra Finch Toys: Before and After. Contrary to what most publications state, all 16 zebra finches played with a variety of toys. Their favorites included cotton puffs, mirrors, natural fiber toys, shredding toys, string toys, swings, and push pull toys. Sometimes one zebra finch played with a toy; other times multiple zebra finches played with a toy at the same time. Here we see the before and after shots of some natural fiber toys. The zebra finches transformed these toys over a period of 2 to 3 months. © D. Herrmann

string, then moved onto another activity. Over several days Little Bald Guy, Tiny Guy, and Lady Finch's Sister continued to weave the string, rearrange, and improve it until the entire string was woven around the hoop. Even after that, the three finches continued to poke the string over several weeks, like they were adding some finishing touches. More weaving behavior took place months later. Little Bald Guy and Naomi wove a string around the bars of a platform loft. Little Boy and Naomi wove a string around a (depleted) millet spray. There were no other birds in the aviary from which the zebra finches could have learned how to weave. This behavior and Little Bald Guy's weaving activities with strips of newspaper disprove the assertion that zebra finches do not weave. In April 2014 Little Girl even wove a string around her left leg about 0.5 (1.27 cm) long. It resembled an open-toed sock. (See Photographs 8.8 and 8.9.)

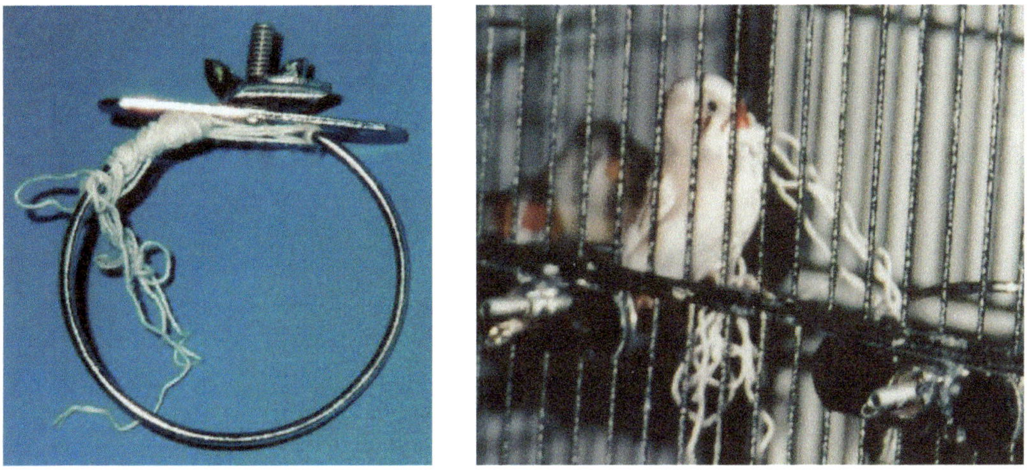

Photograph 8.8 Zebra Finches Weave #1. (Left) Supposedly zebra finches do not weave. However, Little Bald Guy, Lady Finch's Sister, and Tiny Guy wove a loose string around a metal hoop. All three contributed to this effort. The stitching resembles knitting or crocheting. Except the only implements the zebra finches had to work with were their beaks. (Right) More than a year later Little Girl attempts to add to the weaving on the hoop, while Lady Finch's Sister and Little Boy watch in the background. In April 2014 Little Girl even wove a string around her left leg about 0.5 inch (1.27 cm) long. It resembled an open-toed sock. © D. Herrmann

Diet Preferences

The diet of wild zebra finches is quite different than that of captive zebra finches. Even so dietary preferences were noticeable among the 16 zebra finches. One unexpected observation was made in this regard. For over a year a spider plant hung over the zebra finch aviary. The zebra finches regularly pulled small leaves off the spider plant and ate them. The zebra finches did this even if they had just eaten shredded carrots, leaf lettuce, or spinach. The challenge of trying to pull a spider plant leaf that was just about out of reach through the aviary bars piqued their interest. All of the zebra finches, male and female, young and old, engaged in this behavior. Since spider plants are considered a pet bird safe plant this behavior was not discouraged. (See Photograph 8.10.)

As shown in Table 8.4, the zebra finches did not show much interest in fruit. Ami and Tzvi were the only finches who sampled fruit. All of the finches liked vegetables. Green leafy vegetables were welcomed, as long as they were not too tough or too bitter. Spinach was eliminated from the zebra finches diet due to concerns about oxalic acid, as discussed in Chapter 7. On average one zebra finch ate twice as much leafy greens per day as one budgerigar. Like the budgerigars zebra finches would not eat red leaf lettuce. All of the zebra finches readily ate shredded carrots. Only Ami and Tzvi tasted sliced cherry tomatoes. No one tried fresh green peas.

Photograph 8.9 Zebra Finches Weave #2. In a more complicated endeavor, Little Bald Guy (top) and Naomi (bottom) attempt to weave a string around the bars of the platform loft. © D. Herrmann

Zebra finches bite a leaf, then shake their head to get rid of the remainder of the leaf. After all, a piece of leaf lettuce is about the same size as a zebra finch. That action sends the leaf flying toward another finch, who repeats the process. Soon the individual leaf, or what is left of it, is flying around the aviary and all the zebra finches are chasing after it.

Buzz devised a different strategy for eating green leafy vegetables. She would bite a leaf, then step on it to rip off the excess. That way the leaf didn't go flying around and she was first in line for a second and third bite. Buzz was the only one of the 16 finches to figure this out. This behavior is not something zebra finches in the wild would learn, since the fresh leaves they eat are still attached to roots. Buzz's behavior is similar to how Lady Finch's Sister and Tiny Guy resized the shredded paper and cotton puffs to fit their nest. (See Section 8.4 Week 1 7/23/2010). Though, shredded paper and cotton puffs do not occur in the wild either.

Photograph 8.10 Spider Plant Salad. All of the zebra finches, young and old, male and female, enjoyed nibbling on spider plants. Perhaps because the spider plant was fresh or they picked it themselves it tasted good. Here Naomi helps herself to a spider plant salad. © D. Herrmann

Socialization Patterns

Zebra finches are social birds in spite of temporary bouts of aggression. Wild zebra finches associate in flocks of 40, 60, to 100 or more birds.[174] Captive zebra finches have a tendency to sit close together on hoops, perches, and swings. Captive zebra finches take turns bathing without any conflict. Two or more random zebra finches eat out of the same food dish without any problems. Each zebra finch has its preferred place to sleep overnight and the other birds acknowledge that. Zebra finches experience life as a member of a group from the time they hatch. As a result they learn the norms of behavior and how to function in a social setting from an early age. Hence, behavioral problems like those common to large parrots are essentially non-existent among zebra finches.

The zebra finches were housed in groups of 2 to 8 birds. Eight was the maximum number of birds the aviary would support without overcrowding. The finches were less active if there were only 2 birds in an aviary. However, the number of birds in an aviary had no effect on aggressive behavior. Sometimes, but not always, there were problems if there were an odd number of birds in an aviary and all but one bird were part of a bonded pair. Sometimes there were problems if there was a big difference in age. This was a result of different energy levels, not aggression.

For example, when Puff, Buzz, Cinnamon, and Ginger arrived they were only a few months old. Ami and Tzvi were already in their senior years by then. Puff, Buzz, Cinnamon, and Ginger were at the age where they had more energy than they knew what to do with. They played and flew around nonstop all day long. Ami and Tzvi played like that when they were that age. But at this point in time they just couldn't keep up. Puff, Buzz, Cinnamon and Ginger did not understand that. They had no reference point or experience with the limitations of old age. Eventually the two sets of finches had to be separated so that Ami and Tzvi could get some rest.

Table 8.4 Captive Zebra Finch Fruit and Vegetable Preferences.

	No finches ate	1 to a few finches ate	Half or more finches ate	All finches ate
I. Fruits				
apple		x		
banana		x		
blueberries		x		
cantaloupe	x			
grapes	x			
orange		x		
peaches	x			
raisins	x			
strawberries		x		
II. Vegetables				
arugula			x	
baby spinach				x
celery leaves		x		
cherry tomato		x		
dandelion leaves	x			
dried greens				x
green peas	x			
kale	x			
lettuces—red	x			
lettuces—green				x
mache			x	
parsley		x		
shredded carrots				x
watercress		x		

Tiny Guy and Little Bald Guy arrived at the same time. Both were mellow middle-aged normal grey zebra finches. They were also buddies. Prior to arriving they had established a social bond. Tiny Guy and Little Bald Guy frequently sat on the same swing together or napped together on the same perch. Tiny Guy and Little Bald Guy shared an aviary with Lady Finch's Sister. For about a year just the three of them were in the aviary. Tiny Guy and Lady Finch's Sister became a bonded pair. In many cases that would have created a problem. But, since Tiny Guy and Little Bald Guy had a longstanding social bond, the three of them got along fine, even after Tiny Guy and Lady Finch's Sister became parents. (See Section 8.4.)

On occasion an aggressive bird was separated from the others for a temporary timeout. The solitary bird repeatedly called out to the finches back in its original aviary. The solitary bird also spent a fair amount of time trying to figure out how to escape and get back to them, as an expression of separation anxiety. This behavior, of not wanting to be separated from the flock, also indicates the social nature of zebra finches.

Some, but not all, bonded pairs of zebra finches have an unusual method of showing affection. The male will gradually over preen the back of the female's head until there is a small bald spot. The bald spot is noticeable when a finch turns its head around to nap. Once in a while the female will over preen the male, if she is the dominant member of the bonded pair; hence, Little Bald Guy's name. If two unattached males have formed a social bond, sometimes one of them will over preen the other. This over preening behavior seems to be related to solidifying a pair bond and showing commitment. None of the pairs who engaged in this behavior split up. It is unknown whether or not wild zebra finches engage in this behavior.

Reaction to Novel Situations

The zebra finches reaction to novel situations was quite interesting, especially considering their petite size. Whenever anything new was introduced into the aviary the zebra finches immediate reaction was immense curiosity. As soon as the new item was installed, the zebra finches flew to the item and began examining it with their beak. After a poke here and a prod there the zebra finches would decide: (a) what the new item was (a toy, perch, food, etc.), (b) how they could interact with it, and (c) how it fit into their environment. In contrast to the larger birds in the aviary, there was rarely any fear, hesitation, or a "let's think about this for a while first" attitude.

The zebra finches were not alarmed by events that often caused a fear response in the larger birds in the aviary. This counter intuitive observation may be due to the fact that zebra finches are smaller than most everything in their environment. For example, the parrots have a mild panic attack whenever a large item, like a box or chair, is carried through the room. The zebra finches do not panic in this situation. Apparently the zebra finches have come to terms with the fact that most things are larger than they are. Likewise, the zebra finches were not as alarmed by loud noises as the parrots, especially the second or third time they heard the same loud noise. In this situation the zebra finches would become quiet for a few seconds, but that was about all.

The zebra finches did not like it when another bird (a non-zebra finch) landed on the top or side of their aviary. That action almost always caused a high state of anxiety. The sight of a larger bird flying toward them and landing on their cage was too similar to a predator attacking. Even though zebra finches have been bred in captivity for decades, these memories persist. A close-up view of a vegetarian parrot's beak provokes the same reaction as a close-up view of a carnivorous predator's beak. The size of the visiting bird did not make a difference. The zebra finches did not appreciate tiny budgerigars landing on their aviary any more than they liked larger parrots landing there. The zebra finches reacted to all the visiting parrots the same way—with one exception.

Rudy, a mitred conure *Aratinga mitrata mitrata*, visited the zebra finch aviary frequently. (See Chapter 11.) Rudy spent a lot of time flying and the top of the zebra finch aviary was his "home away from home" whenever he was out. Rudy stretched, preened, napped, and ate his snacks on top of the zebra finch aviary. At 15-inches (38.1 cm) in length Rudy was considerably larger than the zebra finches. To be more precise, Rudy's beak was about the same size as a zebra finch. Yet, for some reason, the zebra finches accepted Rudy and did not view him as a threat. Whenever Rudy landed, the zebra finches would look up, then continue about their business. On the other hand, Rudy didn't have to worry about the zebra finches nipping his feet while he was napping. In essence, the zebra finches and Rudy formed an inter-species social bond. (See Photograph 8.11.)

The zebra finch aviaries were configured identically. When a zebra finch moved from one aviary to another there was no hesitation or disorientation. The zebra finch acclimated to its new surroundings and new neighbors immediately. The zebra finches also acclimated quickly when their aviary was moved into a new room. The acclimation process was a bit slower if a single zebra finch was in an aviary by itself when it was moved to a new room. The ability of captive zebra finches to acclimate to new situations quickly is consistent with the nomadic life style of wild zebra finches. In contrast, some captive parrots are change phobic and need time to adjust to a new aviary and other changes in their immediate environment.

Photograph 8.11 A Trusted Friend. Rudy visits the zebra finches. The zebra finches go on high alert whenever any birds land on the top or side of their aviary. This behavior is based on historical memories of predators in the wild swooping down from above. The one exception was Rudy, a mitred conure. Whenever Rudy visited the zebra finches continued about their business. After 140-plus years of captivity zebra finches can still distinguish between a friend they trust and others they do not. This ability is an essential survival skill. Zebra finches and mitred conures would never meet in the wild because they live on different continents. © D. Herrmann

As a rule zebra finches do not like to be held. However, on more than one occasion an individual zebra finch was able to overcome this tendency and realize that in certain circumstances there are advantages to being held in a human hand.

Ami liked to play with a string toy. Sometimes she would sit on the top part of the string toy while she preened it. One day the back nail of her left foot became tangled in the threads. She tried to fly away several times, but ended up hanging upside down dangling by her foot. This experience was similar to being caught in a trap and must have been terrifying.

Ami was too panicked to be freed while she was in the aviary. So, the toy was unhooked and Ami (and the toy) were carried into a small room. Ami did not try to flee out of my hand; she laid still. The thread that was stuck on Ami's foot was cut and she was carried back to the aviary. Ami pulled out the tiny piece of string that remained on her foot. Then she took off flying as usual. Ami had realized she was in an unusual situation and needed help. As a result, Ami was willing to trust a human. This was the only time Ami consented to being held.

Life Stages and Emotional States

Few if any publications acknowledge the life stages of zebra finches, other than hatchling, fledgling, and adult. Perhaps this is because a zebra finch's appearance does not change as it ages; only its behavior changes. The truth is zebra finches experience different stages of life like all living entities as a normal part of the aging and maturation processes. Zebra finches also have to deal with an illness every now and then. As shown in Table 8.5 zebra finches experience five stages of life beyond hatchling and fledgling:

- Juvenile
- Young Adult
- Middle-aged
- Senior
- Impending Death

Table 8.5 Life Stages of Captive Zebra Finches.

Characteristics of Life Stages
Juvenile (5 weeks–3 months) • Curious, trusting, somewhat fearless • Likes being with a group of finches • Explore their world with enthusiasm • Open to new food and toys
Young Adult (3 months–1 ½ years) • Confident, but more cautious around new people • Have limitless energy and curiosity • Very vocal • Aggressive birds try to assert dominance • Open to new food and toys
Middle-aged (1 ½–4 ½ years) • Energy level decreases a little • Curiosity level remains high • Less aggression is displayed • Usually part of a bonded or social pair, but interacts freely with other zebra finches
Senior (4 ½–6+years) • Energy level decreases noticeably; while they still fly and hop around they do so less often and are content to stay in one place longer • Take more and longer naps throughout the day • Eat smaller meals and slower • Don't chatter as much as before • Flying strength wanes a little, especially when trying to fly to high places • Grip is not as secure • More sensitive to cold temperatures
Acknowledge Impending Death (last 24 to 36 hours) • Stay on low perches or on the paper • Start to eat, then stop • Snuggle up closer than usual to its mate • Some finches want to be held—the hand is seen as a warm secure nest • Snuggle up in the corner of their cage for security, like they are back in the nest • Mate often feeds their partner, grooms them and sits next to them to keep them warm, even on a low perch or cage bottom
Mourning Period (~ 10 to 14 days) • Exhibits a profound sense of sadness and loss, withdrawn at times • Reduced appetite • Bird calls out to and searches frantically for lost mate or buddy

Zebra finches are on their own and able to take care of themselves by the time they are five weeks old. Juvenile zebra finches are very curious and explore the world around them with energy and enthusiasm. They are very open to checking out all sorts of new items like toys, perches, and food. Young zebra finches are very trusting, since they haven't yet developed the notion of fear or that there are things they should be afraid of. Juvenile zebra finches have a strong sense of belonging to a family unit and tend to stay near their parents and clutch mates. A fair amount of time is spent perfecting flying skills, like landing, taking off, and turning.

By the time zebra finches are three months old they are young adults. Young adult zebra finches have the same limitless energy and curiosity of a juvenile. By now they have mastered the art of precision flying and begin playing games with each other, like a high speed game of tag. Young adult zebra finches shed any notion of dependency and instead display a high degree of confidence. Aggressive birds begin to assert their dominance.

Young adults are more vocal than juveniles and they begin to exercise caution around people. The young adult stage lasts until a zebra finch is about 1.5 years old.

Zebra finches spend half or more of their life in middle age. Middle-aged zebra finches are just as curious as before, but a bit less active. Aggressive birds tend to mellow out. Most zebra finches are part of a bonded pair and/or a social pair bond by this time.

As zebra finches enter their senior years, around 4.5 years of age, there is a noticeable decrease in their energy level. Senior zebra finches still fly and hop around, but they do so less often and are content to stay in one location longer. As they age senior zebra finches lose some of their flying strength. Beforehand they could fly directly from the cage bottom to the top perch. Now that journey has to be broken into two stages. Likewise, the grip of a senior zebra finch is not as firm. Senior zebra finches can be seen adjusting their grip more often; occasionally they slip off a perch. This condition may be similar to arthritis in humans. Senior zebra finches take more and longer naps throughout the day. They eat smaller meals and more slowly. Senior zebra finches do not vocalize as often as before. They are much more sensitive to cold and sudden changes in temperature, up or down. Senior zebra finches adjust quickly to their changed condition, just like they have adjusted to all the changes they encountered previously throughout their life. There is no doom or gloom, self-pity or remorse. Rather, there is an acceptance of this is the way things are now.

Zebra finches show signs that they are aware they are about to die about a day beforehand. The first sign is that a zebra finch sits on low perches and quits trying to fly up to the high perches and swing. It is important for a caretaker to spot this behavior quickly. Because a failing bird may not be able to reach the food and water dishes if they are up high on the side of a cage. Another common sign is for a zebra finch to snuggle closer than usual to its mate. This behavior probably indicates that the zebra finch's grip is weak and/or it is having trouble staying warm. Closer to the end a zebra finch stays on the paper. Frequently a zebra finch will try to snuggle up in the corner of a cage for security, like when it was a hatchling in the nest. Often their mate understands what is going on. The mate will sit by the failing zebra finch—even on the cage bottom—and try to feed and groom them. Some zebra finches welcome being held in an open hand the last couple hours. A warm hand and a few pats comfort them and provide a sense of security.

Zebra finches go through a mourning period when they lose their mate or buddy from a social bond. The mourning period usually lasts 10 to 14 days. During this time a zebra finch eats much less and exhibits a profound sense of sadness and loss. It alternates from being withdrawn to frantically searching for and calling out to the lost bird. Afterward the zebra finch gradually rejoins the activities of the flock. It may or may not take another mate or become buddies with another bird.

Tzvi went from flying around top speed like a dare devil to learning to adjust to a broken left wing. Tzvi also faced significant end of life challenges. How Tzvi and Ami—and the other zebra finches—dealt with these issues is rather instructive.

About eight weeks before his death Tzvi's strength waned. At first he could no longer fly to the top perch or swing. Before long he couldn't fly to the lower perches either. All the perches and food and water dishes were moved as low as possible, about an inch (2.54 cm) above the paper. That was as much height as Tzvi could manage. Ami would fly down to eat or drink some water, then fly back up to the high perches. After a minute or two Ami noticed that Tzvi was still on the bottom of the aviary and she flew down to join him.

Even more telling was what happened when it was time to go to sleep for the night. Ami would fly up to the high perch and sleeping nest where they normally slept. Ami waited for Tzvi. When he did not come Ami flew down to Tzvi and they spent the night together on the paper. This behavior says a lot about the strength of their pair bond. A bird's natural behavior is to sleep up high for safety in order to avoid predators.

About four weeks before his death Tzvi developed seizures. Since none of the other birds developed seizures, this was not due to environmental conditions, like lead poisoning. The veterinarian was unable to determine the cause or propose a therapy. Tzvi had an attack about every hour or two. He would flap around wildly on his paper until he was upside down; then he laid there exhausted. Afterward Ami always came down and sat next to him. She frequently groomed him and tried to feed him. In the last few days the attacks came more frequently. If I held Tzvi the attacks were less violent and of shorter duration.

Ami and Tzvi had never been separated. They arrived together as young adults. Ami went through a prolonged mourning period after Tzvi's death. She died six weeks after Tzvi.

Lady Finch was a normal grey zebra finch. She arrived as a juvenile with Ami III and Lady Finch's Sister. Twenty months after Lady Finch arrived she developed a sudden weakness and could not keep up with the other finches. Lady Finch was moved to a small private cage to recuperate. All the perches and dishes were placed low in the cage. After a couple weeks Lady Finch appeared to have regained her strength. She was returned to the aviary. Within two days Lady Finch was back on the bottom of the aviary. Again she was placed in a small private cage. Lady Finch quickly lost the ability to fly and was limited to hopping around on the paper. Given her condition a small cage was big enough. Lady Finch became blind in her left eye at the same time due to a tumor.

From April 10th until September 10th Lady Finch soldiered on, never giving up or feeling sorry for herself. She could eat and drink ok, but her head was too heavy to hold up for long periods of time. Lady Finch would lean over and rest her head on the paper. Before long she rolled over and ended up sleeping on her back. More than once we thought Lady Finch was gone—only to happily find out she was just taking a nap upside down.

Lady Finch's paper was changed twice a day since she was living on it. While Lady Finch's paper was being changed, she was placed in a small cardboard box that was lined with a fluffy white towel. Lady Finch liked the warm soft towel and immediately curled up for a nap, like she was in a sleeping nest. We let Lady Finch stay in the box as long as she wanted. Her naps were usually 15 to 20 minutes long. When Lady Finch awoke and was ready to go back to her cage/paper, she would tap the side of the box with her beak: peck-peck, peck-peck, peck-peck-peck. Lady Finch always repeated the same tapping pattern.

A cuttlebone was placed on the paper so she could reach it. Lady Finch stood on the cuttlebone and pecked it every day or two. Lady Finch enjoyed being misted twice a week, since her precarious balance made it difficult to preen. A small amount of charcoal was added to Lady Finch's gravel to help remove the toxins from her system. That perked her up a bit. Throughout the last week Lady Finch began to sing again, even though by then she was blind in both eyes.

When Ginger was elderly he showed few signs of being a senior zebra finch. He still flew and hopped everywhere, although not as fast as before. There was only one indication that Ginger was elderly and facing an end of life issue. His bottom beak or mandible grew longer than the top. While this condition is common in elderly parrots, especially budgerigars, as their liver loses the ability to process keratin, it is somewhat rare in zebra finches. Ginger's bottom beak was trimmed since it interfered with his ability to eat. The long beak did not grow back and Ginger continued his normal routine. One morning about a month later Ginger was hopping and flying around as usual. Fifteen minutes later Ginger was gone. Ginger hid his illness to the end to avoid attracting predators, like a wild zebra finch would do.

Lady Finch's Sister and Tiny Guy were both visibly tired by the time their three young finches had fledged. They were eating a healthy diet and receiving vitamin and mineral supplements. But, feeding and caring for three rapidly growing hatchlings did not leave any time to rest. Not to mention, Lady Finch's Sister and Tiny Guy were nearing their senior years when they had their family.

Lady Finch's Sister hopped and flew and did everything she did before she raised her family. She just wasn't as energetic. Lady Finch's Sister died about a year after she laid the first eggs of what became her family. Other than having less energy there were no other warning signs. That last morning Lady Finch's Sister was noticeably weak and couldn't fly up to the high perches. Several times she started to eat something, then stopped. I picked Lady Finch's Sister up and held her in my open hand. Lady Finch's Sister sat down and snuggled up like she was in a warm nest. She stayed there until the end.

Four years later Lady Silver exhibited the same behavior. Lady Silver had always been an active somewhat aggressive bird. She had never been interested in being held or interaction with people. On her last day, a few hours before her death, Lady Silver began making a loud version of the zebra finch hatchling begging call. She repeated the call over and over. When I picked her up and held her in my hand, she snuggled up, became silent, and went to sleep as if she were back in the nest. Memories of being a hatchling, warm and secure in a nest, must have comforted her. She stayed there until the end. (It is important to remember that zebra finch hatchlings do not make a begging call unless: (a) they know their parents are nearby, and (b) more than one zebra finch hatchling is present.)

Tiny Guy went through the normal mourning period. The three offspring of Lady Finch's Sister and Tiny Guy hatched, fledged, and lived in the same aviary their entire lives up to that point. They had never been separated from their parents or each other. This was the first loss the three young finches had experienced. Their loss and grief was very pronounced, especially that of Little Boy. For several days in a row he frantically searched all over the aviary for his mother, called out to her, and scanned the room in case she was out there somewhere. It took all three of the young finches a long time to understand that their mother was gone and she was not coming back. The father and the uncle were there and they were all in the same aviary; but the mother was not. The young finches were only 11 months old at the time. They did not understand why their mother was gone.

Tiny Guy aged rapidly after he raised his family. For several months he took long naps in the food dish. Little Boy tried to play tag and other games with his father, but Tiny Guy could not keep up. Instead Tiny Guy hung out with his buddy Little Bald Guy. In the last couple weeks Tiny Guy had trouble perching. To compensate he leaned into the aviary bars. He may have had the avian equivalent of arthritis.

Little Bald Guy remained strong and active for a senior zebra finch. He did not show any major signs of decline until the last 24 hours. When he could no longer fly up to the high perches, Little Bald Guy snuggled up in the corners of the aviary on the paper. He even stretched and spread a cotton puff over him like a blanket. Neither Tiny Guy nor Little Bald Guy wanted to be held at this time.

Within a 4 month period the three young zebra finches lost their mother, father, and uncle, from whom they had never been separated. Nothing else in their world had changed from the day they hatched. It took all three young zebra finches several months to understand this loss and return to their normal routine.

Table 8.6 summarizes the spectrum and prevalence of emotional states observed in the 16 zebra finches throughout their lives. No negative emotions were observed, only positive ones. Changes in posture, behavior, activity levels, breathing rate, and daily routine, as well as an increase or decrease

Table 8.6 Spectrum and Prevalence of Emotional States Observed in Zebra Finches.

	All finches	**Half or more finches**	**1 to a few finches**	**No finches**
I. Positive Emotions				
excited				x
happy	x			
sad[1]	x			
confident	x			
timid/unsure[2]			x	
lonely[1]			x	
curious	x			
puzzled				x
playful/eager	x			
anxious/fearful[2]			x	
aggressive/impatient		x		
gentle/patient		x		
II. Negative Emotions				
anger				x
hatred				x
jealousy				x
grumpy				x
overwhelmed				x
stressed out				x
possessiveness				x
self-pity				x
depression				x
boredom[3]				x

Notes:

1—Most obvious during the ~ 2-week mourning period after the death of another zebra finch.
2—Occurred briefly after a sudden new loud noise, a strange person entered the room, or an aggressive bird was isolated from the rest of the zebra finches.
3—Boredom was not a problem; there were many toys and many other birds to interact with.

in vocalizations and/or interactions with other finches were the primary indicators of different emotional states. One trait of zebra finches that pet owners find endearing is the way they express contentment. To do so a zebra finch adjusts its grip on a perch and settles in, opens its beak just a little, and swallows twice. This behavior is the zebra finch equivalent of a dog licking his chops. This behavior often occurs after a caretaker has lavished verbal praise on a zebra finch.

8.4 Development and Behavior of Zebra Finch Nestlings (0–3 months)

Tiny Guy, Little Bald Guy, and Lady Finch's Sister shared an aviary by themselves for over a year. Tiny Guy and Little Bald Guy were buddies. Tiny Guy and Lady Finch's Sister were a bonded pair. All three finches got along together well. Lady Finch's Sister laid an egg or two in her sleeping nest every now and then. After a few days she pushed the eggs out of the nest. Usually this is a sign that the eggs are infertile.

In early July 2010 Lady Finch's Sister laid two eggs in her 5-ounce (150 ml) stainless steel sleeping nest. This time Lady Finch's Sister did not push the eggs out; instead she laid three more eggs and began to sit on the eggs. The notes below chronicle the development and behavior of these zebra finch hatchlings during their first few months of life. Several events, like the first flight, occurred earlier than the average dates stated in the literature. Events that are not supposed to happen, like returning to the nest, are also recorded. This chronology illustrates the hatchlings learning new and more complex tasks through trial and error learning or experimentation much like a human child. Likewise, the parents on occasion encouraged this learning and experimentation. Table 8.7 summarizes the key observations.

Table 8.7 Development of Young Zebra Finches (0–3 months).

Stage	Time frame	Events
Egg	Days 1 to ~ 14	• Eggs are incubated by both parents
Nestling	Day 1	• Altricial babies hatch, are fed by both parents • Hatchlings open their beaks when hungry, but do not make a sound until parents start feeding one of the hatchlings • Only have quills on head
	Day 5	• Have quills on most of body, but not wing stubs
	Day 8	• Begin pushing themselves along bottom of nest with legs and feet
	Day 9	• Begin flapping featherless wing stubs
	Day 10	• Ears visible, quills start to appear where tail feathers will be
	Day 11	• Eyes open, begin imitating preening behaviors, like preening featherless wings
	Day 12	• Begin recognizing the sight of large items, start trying to stand up
	Day 14	• Recognize the direction sounds come from, begin to tuck head under wing during naps
Fledgling	Day 17	• Take first flight from nest
	Day 20	• Begin to scratch head and beak on perch, begin to explore toys
	Day 21	• First attempt to imitate adult vocalizations, begin to explore swings, first attempt to eat seeds
	Day 22	• First attempt to eat gravel and leafy vegetables
	Day 23	• Spent first night outside nest
	Days 27–31	• Begin feeding themselves, first eat cuttlebone • Beaks begin changing to adult color
Juvenile	Day 35	• Cease taking food from adults
	Day 54	• Male fledglings develop white spotted chestnut patches on flanks
	Day 60	• Male fledglings develop orange cheek patches and stripes across their chest • Beak assumes adult color
Young Adult	Day 90	• Attain full size and adult coloring • Are now expert flyers

Week 1

7/23/2010 4:30 p.m.

Lady Finch's Sister was still sitting on the eggs about two weeks later. A decision was made to move the eggs to a larger "nest"—a 10-ounce (300 ml) stainless steel dish. If all five eggs hatched there wouldn't be enough room in the 5-ounce (150 ml) dish. The nest was retrieved when Lady Finch's Sister left to eat some food. There were only four eggs in the nest. The eggs had a slight bluish tint and felt moist like a quartz rock. The four eggs were carefully moved to the new larger nest.

There were no droppings in the nest, only some small pieces of shredded paper. When picking up the shredded paper I felt something soft and squishy. At first I thought it was some shredded carrots that had been carried up to the nest. But no—it was a minuscule baby finch that had hatched that day!

The hatchling was not as long as the tip of my little finger and half as wide. The hatchling wiggled when touched. Its tiny feet, head, wing stubs, eyes (not open yet) and beak were all discernable. Its head was smaller than a fresh green pea and had a few tiny dark gray quills on top. Its eyes were the size of a pencil dot. Its beak, no more than 0.125 inch (0.317 cm) long, was pointed like a finch's beak and a dull yellow color. The rest of the body was pink because no down had grown in yet. I carefully put the hatchling in the new nest, then put the new nest in the same place as the old nest.

The parents came over to the new nest immediately. They did not approve of the way the new nest was furnished. The parents removed the cotton puff and shredded paper from the nest so energetically there was a concern that one of the eggs or the hatchling might fall out. Fortunately they did not. The zebra finch parents proceeded to shred the pieces of paper and cotton puff into smaller and smaller sizes. Then the parents carried the pieces of paper and cotton back up to the nest one at a time. Shredding paper or cotton puffs is easy with two hands and ten fingers. It is more difficult to do with a beak. The zebra finches stretch the cotton puff (or shredded paper) with their beak, then step on it with their feet to trim it to the desired length. After about an hour the zebra finch parents had the new nest arranged the way they wanted it.

7/24/2010–7/27/2010

The zebra finch parents continued to sit on the nest studiously. They kept the remaining eggs and the hatchling snuggled up together. One parent stayed on the nest for 45 to 60 minutes at a time; then they traded places. During their time off the nest the zebra finch would fly, preen, drink water, and eat. While they were nesting the zebra finch parents consumed twice as much leafy greens, twice as much cuttlebone, and the same amount of seeds. Every now and then one of the parents could be seen swishing food around in its throat. Then the parent would tuck its head under its body to feed the hatchling. The hatchling never made a begging call. It just opened its beak quietly.

When it was Tiny Guy's time-off he hung out with Little Bald Guy. Tiny Guy let Little Bald Guy sit on the edge of the nest so he could see what was going on. Because of this Little Bald Guy was nicknamed the uncle. When Tiny Guy was on nest duty, Little Bald Guy sat in his sleeping nest and imitated Tiny Guy's behavior. Lady Finch's Sister did not let Little Bald Guy sit on the edge of the nest. However, she laid the eggs in the nest closest to Little Bald Guy's sleeping nest, even though there were other nests to choose from.

7/28/2010 11 a.m.

A faint sound was heard coming from the zebra finch nest. It sounded like one bird and did not last very long. The nest was inspected when the parents stepped out. The hatchling has doubled in size. It has tiny dark grey quills but no feathers over most of its body, except for the wing stubs. There were also two new tiny hatchlings! Both were only a few hours old. There were no egg shell fragments in the nest. The parents must have removed them. Two eggs remain.

7/28/2010 5:30 p.m.

A faint sound was heard briefly coming from the zebra finch nest again.

Week 2

7/31/2010 12 p.m.

The zebra finch parents have started to leave the nest unattended for a few minutes every now and then. All three hatchlings are the same size now, about an inch (2.54 cm) long. They move their heads and feet frequently and have a rapid breathing cycle. The oldest hatchling can be identified since it has more tiny grey feathers.

Both adults continue to share parenting duties. The three hatchlings do not make a begging call when they are hungry. Instead they simply open their tiny beaks. However, if a parent starts to feed one of the hatchlings all three make a faint buzzing sound. The parents keep trying to line the hatchlings up so they all face the same direction, probably because it is easier to feed them that way. Regardless, the hatchlings keep wiggling around and end up facing every which way. The parents are doing a good job of keeping the nest clean.

One unhatched egg was discarded out of the nest by the parents. This egg disappeared by the next morning. A few days later the fifth egg was also pushed out of the nest. The egg contained a partially developed chick. For some reason the growth and development of this zebra finch embryo stopped about half way through the process. This behavior indicates that the zebra finches know the time interval in which the eggs are supposed to hatch.

8/1/2010

The hatchlings are pushing themselves forward along the bottom of the nest with their energetic feet. Their legs appear to be rather strong even though they cannot yet stand or perch. The fact that the leg muscles and movement develop so early may explain why several young chicks fall from their nest each spring. The same situation has been observed in connection with house sparrow (*Passer domesticus*) nestlings. Fortunately for these zebra finches, their stainless steel nest has high sides which prevent them from falling out. The oldest hatchling has begun flapping its featherless wing stubs energetically twice an hour.

8/2/2010

The ears of the oldest hatchling are becoming visible. The three hatchlings tend to stay together in a clump, even though there is non-stop movement. If the hatchlings are spread out in the nest, they wiggle around until they are back together again. The three hatchlings are now too big for a parent to sit on, so they spend less time in the nest. Tiny Guy spends more time socializing with Little Bald Guy. While in the nest the parents attempt to straddle the hatchlings, standing over them and putting their wings around them. Both parents participate in feeding duties, which occur more frequently. Twice as much shredded carrots and thistle are consumed now. The hatchlings are warm to touch, even though they have limited down feathers. The hatchlings are getting wider, but not much longer. Some tiny sprouts have appeared where their tails will be.

8/3/2010

The oldest hatchling opened its eyes. This hatchling has started to move its head sideways, as if it were preening its wings, several times a day. But so far it only has white quills—no feathers—on its wing stubs. The oldest hatchling's beak is a translucent light pink and two-thirds the size of an adult. All three hatchlings are beginning to look like zebra finches.

8/4/2010

Today the father woke up a hatchling that slept through the first feeding. Usually the hatchlings wake up when the sun comes up or the lights are turned on, whichever happens first. This is probably in response to increased light and noise. The hatchlings' crops are visible. Their crops are full when they contain four to six seeds.

The oldest hatchling looks up and seems to notice when a person walks or stands near the nest. This is the first indication that it is beginning to recognize individual items. The oldest hatchling began trying to stand up. The second hatchling began flapping its wing stubs and imitating preening behaviors. The father occasionally pulls out a baby or natal down feather.

8/5/2010

The oldest hatchling is very alert. It stands up to stretch its wings, scratch its head, and imitate other preening behaviors even though it doesn't have a full set of feathers. The oldest hatchling has mostly white wings. Its beak is a light translucent pink. The younger two hatchlings have dull black beaks. The eyes of the youngest two hatchlings are discernable, but not yet open. All three hatchlings have strong legs. They push and try to stand up frequently. If the rim of the nest/dish were not as high as it is, they would probably have fallen out by now.

The parents continue to pull out uneven body down. This seems to annoy the hatchlings. Little Bald Guy visits the nest frequently. He sits on the rim and even sleeps in the nest with the hatchlings when the parents are away.

8/6/2010

All three hatchlings are alert. The youngest two hatchlings' eyes are open now. The hatchlings recognize the direction a sound is coming from when they are spoken to. For the first time the oldest hatchling tucked its head under its wing to take a nap. The three hatchlings still like to stay snuggled together in a clump. One of the younger two hatchlings is also developing white feathers on its wings.

Week 3

8/7/2010–8/8/2010

The oldest hatchling is becoming aware of activities outside the nest, like the parents and uncle flying around, people walking by, new sounds, etc. It is almost fully feathered. The youngest two hatchlings can see, but they remain focused on the world inside the nest. The hatchlings are half the size of an adult and have most of their wing feathers, but none on their back. All three hatchlings will sit still in my hand while I pat or talk to them. The uncle visits more often now that the parents don't stay in the nest. The adults' consumption of carrots and leafy greens is back to normal, but seed consumption has doubled. The hatchlings generally go to sleep for the night between 7:30 and 8:30 p.m. regardless of whether the lights are on.

8/9/2010 12:45 p.m.

The oldest hatchling hopped up onto the rim of the nest for the first time and looked around. One adult hopped onto the rim, then back on the nearest perch two or three times, as if to encourage the hatchling. The other two adults flew to that perch also. The hatchling hopped around the rim of the nest for a while. Finally it attempted to fly to the perch where the parents were waiting. The hatchling undershot the perch and ended up hanging on the side of the aviary. From there it flew to the other side of the aviary, finally landing in a front corner on the bottom of the aviary. The hatchling sat there for a while and let me pat it.

One of the adults flew down to check on the hatchling. For some reason that startled the hatchling and it flew out the small opening that holds the grill. The hatchling continued flying until it had covered a distance of 12 feet (3.65 m) and settled on the landing perch on the front of an Indian ringneck parakeet's *Psittacula krameri manillensis* aviary. After the hatchling had some time to rest, it was picked up and returned to the nest. The hatchling—now fledgling—made no attempt to resist or escape and welcomed pats. The hatchling's flight was strong, but not very precise. The leg muscles were strong, but its grip was weak. So far it had only sat flat-footed in a nest. The oldest hatchling made no further attempts to leave the nest today. Just before going to sleep for the night it briefly sat on the rim of the nest. (See Photograph 8.12.)

The flat-footedness of zebra finch and budgerigar hatchlings (Chapter 7) contrasts with the firm grip American robin (*Turdus migratorius*) and Common grackle (*Quiscala quiscala*) hatchlings demonstrate from the moment they fledge. If handled an American robin or Common grackle hatchling will grip a finger tightly as if it were a perch immediately after fledging. It is unknown if the flat-footedness observed in captive zebra finch and budgerigar hatchlings occurs in the wild.

Photograph 8.12 First Flight. The oldest zebra finch hatchling (Little Boy) made his first flight from the nest on 8/9/2010 at 17 days of age. He flew from the rim of the nest to one side of the aviary, back to the other side of the aviary, to the bottom of the aviary, out a small opening, and 12-feet (3.65 m) across the room to an Indian ringneck parakeet's landing perch! Little Boy's flight was strong but not very precise. The notions of stopping, turning, and landing had not yet registered. Little Boy sits low on the perch because his toes and feet have not yet developed a firm grip. Prior to this he had been sitting flat footed in a nest. Note that Little Boy has a translucent light pink beak—not the dull black beaks attributed to zebra finch hatchlings. At this young age Little Boy is missing his chest stripes, spotted flanks, and chestnut cheek patches. © D. Herrmann

The parents continued to look for the hatchling on the bottom of the aviary after it was returned to the nest. The parents feed the hatchlings less frequently now, but a larger quantity than before. It takes both parents to complete one feeding cycle; before they alternated.

None of the three hatchlings look like the parents (or the uncle). The parents and the uncle are all normal grey zebra finches. The oldest hatchling is a grey pied with white wings. One of the younger hatchlings is fawn. The other is a fawn pied with white wings.

8/10/2010 10 a.m.

The oldest hatchling flew from the nest to a nearby perch, then other perches, following the adults. Eventually the hatchling coasted down to the paper to rest, where it let me pat it. The adults have begun vocalizing again. They suspended their vocalizations once their nesting duties began. Perhaps this is a practice wild zebra finches follow in order to not give away the location of the nest. The oldest hatchling made its first attempt to imitate the adults' vocalizations.

8/10/2010 11 a.m.

The oldest hatchling returned to the nest on its own. The youngest two hatchlings are very active in the nest and flap or stretch their wings frequently. They are beginning to be curious about the world outside the nest. The youngest two hatchlings stick their heads on the rim of the nest so they can see out; but, they haven't made any attempt to leave. All three are still fed by the parents.

8/11/2010 10 a.m.

The fawn pied hatchling sat on the rim of the nest briefly, then flew for the first time to the nearest perch. Ten minutes later the hatchling returned to the nest.

8/11/2010 11:35 a.m.

The fawn and the fawn pied hatchlings both flew out of the nest at the same time, without sitting on the rim first. Both missed the perch and ended up landing on the paper on the bottom of the aviary. They sat there for a while, taking in all the new sights. All three hatchlings still ask the parents for food. The oldest hatchling is developing the foot strength and coordination needed to perch securely. The youngest two hatchlings, who are used to sitting and walking in a flat bottomed nest, do not yet have this ability.

8/11/2010 1 p.m.

The fawn and the fawn pied hatchlings returned to the nest on their own.

8/11/2010 9:05 p.m.

The fawn pied hatchling left the nest again and went to sit with the parents on a perch. It made several unsuccessful attempts to return to the nest, and ended up spending the night on a wide perch. The other two hatchlings slept in the nest.

8/12/2010

The fawn pied hatchling returned to the nest at daylight. All three hatchlings spent most of the day outside the nest. The grey pied and the fawn pied can now perch reasonably well. They have begun to scratch their heads and beaks on a perch, after seeing the adults do this. The fawn hatchling does not have a very good grip yet. The fawn pied explored a sisal toy for the first time and sat on the hoop next to the water tube, after seeing an adult drink from it. None of the three hatchlings are eating or drinking on their own yet.

8/13/2010

All three hatchlings spent the night in the nest together. They spent most of the day outside the nest. All three are developing a better grip, especially on sand and cement perches. Wood perches are still a bit slippery. The grey pied and the fawn pied landed on the swing for the first time. They also explored a second sisal/bead toy and a cotton/bead toy. The three hatchlings tend to stay together. The mother still feeds the hatchlings, wherever they happen to be. The grey pied attempted to pick up loose seeds that were on the paper, but it had trouble trying to open them.

Week 4

8/14/2010

All three hatchlings spent the night together in the nest; although they did not return to it during the day. All three hatchlings attempted to eat loose seeds off the paper. The grey pied experimented until it learned how to eat leaf lettuce. The parents still provide most of the hatchlings' food. The three hatchlings spend most of the day together and sit on the same perch together. Sometimes the parents join them.

8/15/2010

The three hatchlings spent the night huddled together on a perch. The outer two faced one direction and the middle one faced the opposite direction, similar to how they slept in the nest. The father stayed with the hatchlings for the first hour, then spent the night with the mother. All three hatchlings are proficient landing on any perch now, including swings. All three hatchlings actively explore sisal, cotton, and plastic toys. They try to eat off the paper, but are still fed by the parents. Lady Finch's Sister spends a lot of time flying, like she is making up for the time she spent sitting on the nest.

8/16/2010

The flight of all three hatchlings is becoming more precise. They continue to spend their time together. Tiny Guy is spending more time with Little Bald Guy.

8/17/2010

The hatchlings have learned how to sleep on a swing.

8/18/2010

The three hatchlings do everything the parents do now, except eating by themselves. The three hatchlings sleep and roost together, but they are starting to play and explore the aviary independently. The three hatchlings' energy level is increasing. They sit on the food dish, but don't eat from it. The mother is still very attentive; the father less so. He spends more time with the uncle.

Week 5

8/21/2010–8/23/2010

The three hatchlings have not increased in size. All three can now eat from the food dish, eat leaf lettuce, and drink from the water tube on their own. However they are still fed by the mother two or three times a day. She is still very attentive. The father does not feed the hatchlings anymore. He seems worn out by the nesting duties and often falls asleep standing on the food dish. The hatchlings saw themselves in a mirror toy today for the first time. They will no longer let me hold or pat them.

8/24/2010–8/26/2010

The grey pied is aggressive toward the father on occasion and engages in beak boxing with him. All three fledglings visited the cuttlebone for the first time today. The beaks of all three fledglings are starting to change color. The oldest, who has a translucent light pink beak, has developed a small dark spot on its beak. The youngest two, who have dull black beaks, now have a small light spot on their beaks. The spots are on the upper mandible near the nares. The fledglings have learned to hover briefly. The fledglings now fly quickly and strongly. Then they hover briefly while they decide where they want to land. Often at the last minute they decide to land on an occupied perch and end up on top of another bird, until the group reshuffles. All three fledglings are very energetic and spend their time exploring new things.

8/27/2010

All three fledglings have stopped taking food from the mother. They still snuggle up together to sleep.

Week 6

8/28/2010

The three fledglings took their first baths today. After watching the mother splash around, the oldest fledgling jumped in and started splashing. It didn't take long to master the technique. Afterward the other two fledglings jumped in too.

9/3/2010

The beaks of the three fledglings are half the color of a hatchling and half the color of an adult.

Week 7

9/6/2010–9/8/2010

The grey pied tried to imitate the father's vocalization. Instead of a fluid song of various tones and pitches, it was a series of individual notes and a bit choppy.

Week 8

9/13/2010

The grey pied has grown a small speckled patch on its right side near the wings, like an adult male zebra finch. There is not a speckled patch on the left side yet.

9/14/2010

The three fledglings learned how to pull spider plant leaves through the aviary bars so they could eat them.

9/16/2010

The grey pied has white speckled chestnut patches on both flanks now. The barring on the tail feathers of all three fledglings have become more pronounced.

Week 9

9/22/2010

The grey pied is starting to develop tiny black and white stripes across its chest. It is also just developing an orange patch over its left ear. The grey pied's beak is now red. The fawn and fawn pied fledglings have light orange beaks. All three are very energetic and curious and explore new things.

Week 18

11/19/2010

The perches were moved and spread out to allow more flying and hopping room horizontally and vertically, now that the three young zebra finches are full size and expert flyers. Two new toys were added at the same time. For the first 20 to 30 minutes the adults and young finches continued to fly to where the perches had been and hovered over that location, even though they could see where the perches had been moved. It was as if they were flying on auto pilot based on a stored memory of the location, rather than using current visual input. After about an hour all the birds began flying directly to the perches in the new location. The young zebra finches flew straight up and down to the new perches or wide arcs from one new perch to another. They appeared to be repeating the motions in order to store the new perch locations in their memory. All six zebra finches also examined the new toys immediately. (See Photographs 8.13 and 8.14.)

8.5 Discussion

Now let's see what these behaviors tell us about the intelligence, behavior, and individuality of wild and captive zebra finches.

Photograph 8.13 Finch Family #1. Here we see the zebra finch family when the young finches were 5 months old. The top photograph features (L-R) Lady Finch's Sister (mother), Little Boy, Naomi, and Little Bald Guy (uncle). The bottom photograph features (L-R) Little Boy, Tiny Guy (father), Naomi, and Little Girl. Note that Little Boy does not have a right cheek patch. © D. Herrmann

Photograph 8.14 Finch Family #2. Here we see the zebra finch family when the young finches were 9 to 12 months old. The top left photograph features Lady Finch's Sister and Naomi. The top right photograph features Little Boy. The bottom left photograph features Little Girl. The bottom right photograph features Little Girl, Little Boy, and Naomi. © D. Herrmann

Genetic Intelligence

The zebra finches normal daily activities provided several examples of genetic intelligence. The most obvious examples were when a zebra finch engaged in an activity—or knew it was supposed to engage in an activity—even though it had not been exposed to a given situation before nor taught to do so. To illustrate, none of the six zebra finches that were purchased and none of the seven zebra finches that were adopted had been exposed to swimming pools or cuttlebones. Yet all 13 zebra finches knew they were supposed to bathe and consume cuttlebone and did so at the first opportunity. Bathing is an activity that occurs in the wild; consumption of cuttlebone does not. Hence the immediate interaction with cuttlebone is all the more interesting.

The six zebra finches that were purchased had no previous experience with gravel. Yet they began consuming gravel immediately. It is not known whether or not the seven adopted zebra finches had access to gravel before they arrived. They too began consuming gravel immediately. It appears that genetic intelligence is driving this behavior.

Genetic intelligence is apparent throughout the development of nestlings. The first hatchling did not make a begging call; it simply opened its tiny beak. After the second and third hatchlings hatched, the hatchlings made a quiet begging call—but only after a parent had returned to the nest with food. In other words, the hatchlings do not give their location away unless a parent is there to defend them. A days old hatchling would not be able to reason that out. Hence, this behavior must be attributed to genetic intelligence.

Equally interesting is the fact that both parents feed the youngest hatchling first, regardless of whether there are two, three, or more hatchlings. This behavior indicates that the parents can distinguish individual hatchlings and the order in which they hatched. This behavior also indicates that the parents at some level understand that the youngest hatchling has some catching up to do. The origin of this behavior must be genetic intelligence, since all zebra finch parents do this.

Nestlings stay clustered together in a clump. If the nestlings are spread out in the nest, they will wiggle around until they are all back together again, before and after their eyes are open. The nestlings do not care who is facing which direction—much to their parents' dismay—but they do understand the need to be close together. The nestlings also seemed to have a basic understanding of how many other nestlings belonged in their nest. If one nestling was temporarily removed, the remaining two nestlings kept wiggling around looking for it. This behavior is a precursor to contact sitting, which adult zebra finches engage in for a combination of safety and social reasons.

When they were eight days old nestlings began pushing themselves along the bottom of the nest as a precursor to hopping. On the ninth day zebra finch nestlings began to flap their featherless wings. On the eleventh day zebra finch hatchlings began to imitate adult preening behaviors even though they: (1) did not have any feathers to preen, and (2) could not see their parents preening. Again these activities must be driven by genetic intelligence. There hasn't been enough time for the nestlings to reason out the necessity or purpose of performing these behaviors. Instead the nestlings just start doing them.

Applied Intelligence

All of the zebra finches exhibited a high degree of curiosity about new items, sights, and sounds in their environment. The level of curiosity remained constant throughout their lives. Curiosity is a sign of an inquiring mind and intelligence. Curiosity indicates the ability to distinguish between sights, sounds, and items that are normally encountered and those that are new. The act of exploring a new item and subsequently classifying it as a new toy, perch, plant, or food, etc. indicates a keen sense of situational awareness.

The zebra finches were aware of the changing seasons and changing weather patterns since they could see out the windows. The zebra finches adjusted their daily activities to correspond to longer or shorter and warmer or colder days. For example, the first meal of the day was whenever they woke up and the lights came on, regardless of whether that was 5:30 a.m. or 8:30 a.m. Likewise, the last meal of the day occurred as the lights went out. The zebra finches adjusted their dining schedule to correspond to the availability of light. Also, the quantity of food consumed increased on colder days and decreased on warmer days, in proportion to their varying energy needs.

Sudden changes like thunderstorms, high winds, and snow caught their attention, as did unusual sounds like sirens and dogs barking. If the unusual sound was in their vocal/frequency range, the zebra finches attempted to communicate. For example, when a truck backed up and beeped, the zebra finches beeped back. The same thing happened whenever a CD of wild bird song was played.

This behavior raises a question: why would a tiny zebra finch attempt to communicate with an unknown unseen entity? Did the zebra finches think the "other" was another type of finch because the vocalization was similar to their vocal/frequency range? Were they attempting to communicate, socialize, or exchange information? Regardless of motive, the zebra finches indicated that they were curious about and not afraid of other finch-like birds. The zebra finches also indicated that they understood that predators cannot make these sounds. Otherwise, the zebra finches would not have given their location away.

The zebra finches were curious about new items that were brought into their room, like a new plant, a new (non-zebra finch) bird, or a new inanimate object. They watched intently as the item appeared. Afterward there was a little chatter back and forth.

The zebra finches explored anything new that was introduced into their aviary immediately, without hesitation. The exploration seemed to be focused on determining what the item was, how it fit into their environment, and how they were supposed to interact with it. Usually several zebra finches explored new items at the same time. New toys and new foods were greeted with extra enthusiasm.

The zebra finches learned how to use tube feeders quickly. The zebra finches learned how to play with a variety of toys. As juveniles and young adults the zebra finches invented and participated in games. None of these behaviors are common to wild zebra finches. Rather, the zebra finches learned and developed these behaviors in response to the unique indoor environment in which they lived.

The process of learning how to fly is the result of a combination of genetic intelligence and applied intelligence. Before their eyes are open zebra finch nestlings start to flap their featherless wings. This behavior can be attributed to genetic intelligence. On the other hand, curiosity prompts zebra finch nestlings to take their first flight. At some point zebra finch nestlings—like all altricial nestlings—become curious about their parents comings and goings and the world beyond the nest. At first nestlings just peer out over the top of the nest. Then one day, with the parents encouragement, each nestling takes that first wobbly flight. Each fledgling fine tunes its coordination, flying, navigation, and landing skills through lots of practice.

What was unusual is that all three zebra finch fledglings returned to the nest after their first flight. This behavior probably does not happen in the wild. For a few days all three would fly for a while, then return to the nest to rest or sleep overnight. Also, all three zebra finch fledglings abandoned the nest on the same day even though they fledged on different days.

The zebra finch fledglings learned some things by watching and imitating their parents. These activities, like bathing and cracking open seeds, took some practice. Eventually the fledglings mastered the necessary skills. The zebra finch fledglings taught themselves other skills like how to eat lettuce leaves or pull spider plant leaves through the aviary bars.

When the fledglings were 3.5 months old the perches in the aviary were spread out and some new toys were added. Surprisingly, for the first 20 to 30 minutes the fledglings and adults continued to fly to where the perches had been and hovered over that location even though they could see where the perches had been moved. It was as if they were flying on auto pilot based on a stored memory of the location, rather than using current visual input. After about an hour all the birds began flying directly to the perches in the new location. This behavior would indicate that zebra finches rely on memory more than current visual input when flying to familiar locations. That would free up the brain to focus on other tasks like predator avoidance.

The weaving activities of Little Bald Guy, Lady Finch's Sister, and Tiny Guy indicate applied intelligence. Little Bald Guy shredded pieces of paper, then wove them in and out the vertical bars of the aviary. This was after Little Bald Guy dipped the paper strips in water to make them more pliable. Little Bald Guy, Lady Finch's Sister, and Tiny Guy all three participated in weaving a string tightly around a metal hoop.

Supposedly zebra finches don't weave. That is why zebra finches are (now) grouped with grass and parrot finches. Yet, Little Bald Guy, Lady Finch's Sister, and Tiny Guy all engaged in weaving activities. They wove around structures that are not found in the wild: vertical aviary bars and a horizontal metal hoop. They wove with materials that are not found in the wild: newspaper and string. Little Bald Guy had to manipulate the newspaper to make it weave-able, by shredding the paper to the right size and dipping it in water. The behavior of Little Bald Guy, Lady Finch's Sister, and Tiny Guy indicate that not only did they know how to weave, but that they were able to expand their weaving skills to incorporate new structures and new materials. There were no other birds in the aviary from which these three zebra finches could have learned how to weave.

One behavior of wild zebra finches in particular indicates intelligence. The normal non-nesting zebra finch flock size is 40 to 100 birds.[174] However, during a severe drought as many as 1,000 zebra finches have been known to come together, as several smaller flocks join forces in the search for food and water.[174] In other words, when facing a dire situation multiple flocks of zebra finches come together, collaborate, and cooperate to increase everyone's chance of survival. That is the exact opposite of human behavior. When faced with a catastrophe our highly intelligent human society dissolves and descends into lawless chaos and panic as everyone tries to fend for themselves.

In contrast, multiple zebra finch flocks come together. Undoubtedly there is more going on than commiserating or thinking that it is better to have 1,000 pairs of eyes looking for food and water than 40. Most likely the various flocks are sharing information about where they each have been and not found food or water, as well as historical memories of where food and water were found during previous droughts. By sharing information each flock saves a lot of precious wasted time and energy searching barren areas, and the chance of survival for all the flocks is increased.

Zebra finches have been raised in captivity for at least 140 years.[33, 41, 73, 79, 174] Captive zebra finches retain some memories of life in the wild. For example, captive zebra finches do not like other birds to land on the top or side of their aviary. This action evokes shared historical memories of predators swooping down from above, even though that hasn't happened for over 140 years. At the same time the zebra finches were able to distinguish between birds they trusted and those they did not ("friend or foe"). The zebra finches went on high alert when any bird landed on their aviary—except Rudy. Ironically Rudy, a mitred conure, was the largest bird to land on the zebra finch aviary. Mitred conures and zebra finches would not meet in the wild. They live on different continents.

Other mixed species experiments yielded similar results. For three months two male zebra finches, two budgerigars, and a pair of young Canary-winged parakeets *Brotogeris versicolorus versicolorus* shared an aviary. The birds shared a common water dish, but had different food dishes because of their different diets. None of the birds exhibited any hostility or fear, only *curiosity*. At first they just ignored each other. Then slowly they began exploring each other's toys and food dishes. The zebra finches tried to eat orange and apple slices for the first time and the budgerigars tried to eat squash for the first time, after seeing the Canary-winged parakeets do so. Moreover, the Canary-winged parakeets learned to sit on a swing from the zebra finches—they had never done that before. By the end of the first month the different species would sit on the same perch. Indeed, the white male zebra finch sang his courtship call repeatedly to the female budgerigar who was mostly white (Prism). The green budgerigar and the white male zebra finch sat on the water dish together. The white male zebra finch even let the green budgerigar (Topaz) preen his head and around his eyes. This behavior is significant; because of the difference in size and beak structure it implied trust.

Individuality

The sixteen zebra finches displayed a high degree of individuality in many aspects of their lives. The differences in temperament were the most obvious. Forty-four percent of the zebra finches were aggressive. Thirty-seven percent were passive. Nineteen percent were neither aggressive nor passive. Temperament could not be traced to the gender, color, or source of a zebra finch. Temperament was simply a characteristic of an individual zebra finch. Not all temperaments were static. Many aggressive birds mellowed by middle age.

The sixteen zebra finches displayed individuality in terms of preferences, such as toy preferences, diet preferences, social preferences, and even the preferred place to sleep overnight. All the zebra finches played with cotton puffs, mirrors, natural fiber toys, shredding toys, string toys, and swings.

Several zebra finches played with ladders and push pull toys. Only a few zebra finches played with puzzle toys. All the zebra finches ate dried greens, green lettuces, and shredded carrots. Several zebra finches ate arugula and mache. A few zebra finches ate celery leaves, parsley, and watercress. Only Ami and Tzvi sampled fruit and cherry tomatoes. All of the zebra finches had distinct social preferences as well. Ami's loyalty to Tzvi after he broke his wing and during his illness was impressive. The social bond between Tiny Guy and Little Bald Guy was strong, as was the pair bond between Tiny Guy and Lady Finch's Sister. All three adults got along well and stuck together before, during, and after Tiny Guy and Lady Finch's Sister raised their family.

Emotional States

The sixteen zebra finches exhibited a variety of emotional states throughout their lives. Changes in posture, behavior, activity level, breathing rate, and daily routine, as well as an increase or decrease in vocalizations and/or interactions with other finches were the primary indicators of different emotional states. The most noticeable emotion was a sense of loss and grief after the death of a bird. The three young zebra finches went through a prolonged mourning period when they lost their mother, father, and "uncle", from whom they had never been separated, within a four month period. For a long time the young finches alternated between being withdrawn to frantically searching for and calling out to their lost relatives.

Perhaps the best indicator of the zebra finches emotional health was how they dealt with changes and challenges throughout their life. At no time did the zebra finches display any of the negative emotions, like anger, self-pity, or depression that humans often do when confronted with similar crises.

Tzvi mellowed overnight when he broke his wing. His aggressive dare devil ways simply disappeared. In the last six weeks Tzvi lost the ability to fly and suffered through recurring seizures. Yet Tzvi did not give up or sit in the corner, hold his head low, or feel sorry for himself. Instead Tzvi adjusted to his changed circumstances—for which genetic intelligence could not have prepared him—and tried to maintain as normal a daily routine as possible.

Lady Finch soldiered on for six months as she lost the ability to fly, became blind in one then both eyes, and was too weak to hold up her head. Yet throughout it all Lady Finch expressed pleasure and comfort at little things, like being held or patted, being able to take a nap on the warm fluffy towel in the box while her paper was being changed, and being misted. Lady Finch even began to sing again the week before she died.

In summary, zebra finches are intelligent social individuals. Captive zebra finches share many traits with their wild cousins. At the same time captive zebra finches have developed several new behaviors that correspond to the unique environment in which they live and demonstrate their ability to deal with novel situations, like playing with toys, learning to eat from tube feeders, weaving around manmade structures with items that do not occur in the wild, and being comforted by being patted and held in a warm hand. Captive zebra finches recognize as one of their own several hybrid color mutations that do not occur in the wild. It is not known whether or not wild zebra finches would do this.

9

Meyer's Parrot and Red-bellied Parrot

Poicephalus meyeri and *Poicephalus rufiventris*

9.1 Background

The Poicephalus family is one of three families of short-tailed African parrots.[8] The Poicephalus family is not well known in the wild or in captivity.[8, 25] Very little has been published about members of the Poicephalus family. This chapter is an attempt to remedy that situation.

The Poicephalus family is composed of the nine species of small stocky African parrots listed below. The Yellow-fronted parrot and Niam-Niam parrot are rare in the wild and unknown in captivity.[8, 94]

- Meyer's parrot (*Poicephalus meyeri*)
- Red-bellied parrot (*Poicephalus rufiventris*)
- Brown-headed parrot (*Poicephalus cryptoxanthus*)
- Jardine's or Red-fronted parrot (*Poicephalus gulielmi*)
- Cape or Levaillant's parrot (*Poicephalus robustus*)
- Rueppell's parrot (*Poicephalus rueppellii*)
- Yellow-fronted parrot (*Poicephalus flavifrons*)
- Niam-Niam parrot (*Poicephalus crassus*)
- Senegal parrot (*Poicephalus senegalus*)

Meyer's parrots were named in 1827 after the German ornithologist Bernhard Meyer.[203] There are six subspecies of Meyer's parrots, each with slight variations in coloring and geographical ranges.[60, 94, 203] A theory was put forth in 1990 and confirmed through a DNA study in 1998 that seven members of the Poicephalus family are descended from Meyer's parrots.[170] A second research project is being conducted to independently confirm these results.[25] Three observations gave rise to this theory:[25]

- The range distribution of Meyer's parrots covers that of all other members of the Poicephalus family
- Meyer's parrots are "food item generalists", while other members of the Poicephalus family have a more specialized or limited diet
- Meyer's parrots have been known to hybridize with other members of the Poicephalus family in the wild and in captivity

Meyer's parrots inhabit central and eastern sub-Sahara Africa.[60, 94, 202, 203] Meyer's parrots appear on stamps from Botswana.[199] Red-bellied parrots have a smaller range in eastern Africa, primarily Tanzania, Kenya, Somalia, and Ethiopia, but not in coastal areas.[60, 94] Both are normally local birds, but they may become nomadic during times of drought.[60] Meyer's parrots and Red-bellied parrots live in plateau woodlands, savanna, wooded grasslands, forests bordering water sources, closed canopy forests, secondary growth, and Acacia scrub land.[60, 94, 202, 203] Meyer's parrots and Red-bellied parrots generally stay at altitudes under 2,200 feet (670.5 m),[60, 147, 203] and seldom venture far from water.[60, 94] Sparrow hawks (*Accipiter nisus*) and inland Osprey are the major predators of concern.[25]

Towering baobab trees (*Adansonia digitata*) are a favorite habitat of Meyer's parrots and Red-bellied parrots.[8, 34, 60, 94] Baobab trees are believed to be some of the oldest trees on the planet, with some estimated to be 1,000 years old or more.[131] The fruit and seeds of baobab trees are edible.[131] The damp spongy bark of baobab trees can be consumed by Meyer's parrots and Red-bellied parrots during a drought to replenish fluids.[131] These features have fostered a long-term symbiotic relationship between baobab trees, Meyer's parrots and Red-bellied parrots.

The London Zoo acquired its first Meyer's parrot in 1855.[34] Be that as it may, a book published in 1909 declared Meyer's parrots to be a "rarely imported parrot" and rare in private collections.[34] While a few captive Meyer's parrots and Red-bellied parrots have escaped or been released in metropolitan areas, no feral flocks have been established.[147]

The population of Meyer's parrots and Red-bellied parrots in the wild has been decreasing rapidly since 1990 due to: (1) illegal capture of wild birds (smuggling), (2) habitat destruction as a result of deforestation, desertification, and agricultural expansion, and (3) persecution as a crop pest.[8, 25, 94, 181, 203] The Poicephalus family is one of the most heavily trapped wild parrots, with approximately 735,775 birds entering international trade since 1981.[202] Importation to the United States has been banned since 1992.[8] Meyer's parrots and Red-bellied parrots are relatively new to the world of aviculture, only appearing since the late-1980's.

9.2 Characteristics

Now let's look at the characteristics of wild Meyer's parrots and Red-bellied parrots, such as their:

- Physical characteristics
- Diet
- Family life and daily activities
- Vocalizations

Physical Characteristics

Published descriptions of Meyer's parrots and Red-bellied parrots do not match in all details the birds discussed in this chapter. This highlights the problem identified by Kaufman of ornithologists relying on written descriptions by authors who relied on earlier written descriptions without any visual observations of their own to confirm the material.[95]

For example, the majority of the publications consulted indicated that Meyer's parrots are 8.25 inches (21 cm) in length. Red-bellied parrots were listed as being 9 inches (23 cm) in length. In contrast to these published numbers, the two birds living in the aviary are both 7.5 inches (19 cm) long. Likewise several publications described Meyer's parrots and Red-bellied parrots as having brown or ashen brown feathers. Neither the Meyer's parrot nor the Red-bellied parrot discussed in this chapter have any feathers that appear to be even remotely brown in natural sunlight, incandescent, fluorescent, or full spectrum lighting. In addition, several sources indicated that the eyes of Meyer's parrots and Red-bellied parrots are reddish orange as adults and brown as juveniles. The reverse was observed in the two birds discussed in this chapter. As juveniles both parrots had reddish orange eyes. Both parrots acquired orangish brown eyes as they matured.

Consequently, when published descriptions of the characteristics of Meyer's parrots and Red-bellied parrots agree with the visual observations reported in this chapter, citations are provided. When published descriptions of the characteristics of Meyer's parrots and Red-bellied parrots are inconsistent with visual observations, information from direct visual observation is reported.

As stated earlier Meyer's parrots are 7.5 inches (19 cm) from head to tail and about 2 inches (5 cm) wide across the breast. The wingspan is 12 inches (30 cm). Meyer's parrots have a short square tail about 2 inches (5 cm) long and 2 inches (5 cm) wide. The beak is 0.75 inch (1.9 cm) long and the lower mandible is moderately pronounced. Like other parrots Meyer's parrots have zygodactyl feet with the outer toes being 0.875 inch (2.22 cm) long. The inner toes, front and back, are 0.125 inch (0.31 cm) shorter. Their legs are short. The average weight of a Meyer's parrot is 4.3 to 4.5 ounces (120 to 126 gr).

The head, neck, upper breast, upper back, wings, and tail of a Meyer's parrot are a rich medium gray. The head feathers are particularly soft to the touch. The lower breast is a variegated pattern of light green with subtle gray highlights. The lower back has a metallic turquoise sheen, particularly in full spectrum lighting. Adult birds have a 0.5 inch (1.27 cm) wide yellow patch on their head that runs from eye to eye just above the cere. Juveniles have large yellow patches on their wing coverts. These patches shrink considerably as the bird matures. The cere and beak are the same gray color as the head feathers, as are the legs and feet. The eyes are a dark orangish brown. There is no difference in appearance between males and females.

From the back a Meyer's parrot and a Red-bellied parrot are almost indistinguishable, except for the metallic turquoise lower back which isn't visible unless the wings are spread.

From the front it is easy to tell a Red-bellied parrot and a Meyer's parrot apart. A Red-bellied parrot does not have any yellow patches on its head or wings. In a female the lower breast is a variegated pattern of light green with subtle orange highlights. A male Red-bellied parrot has a bright orange lower breast. Hence, it is curious that these birds are named Red-bellied and not Orange-bellied parrots. The orangish brown eyes of an adult Red-bellied parrot are slightly brighter than a Meyer's parrot when seen in full spectrum lighting. Other than these features, Red-bellied parrots resemble Meyer's parrots. Some sources acknowledge two subspecies of Red-bellied parrots.[60]

Diet

Wild Meyer's parrots and Red-bellied parrots eat fruits, seeds, nuts, berries, buds, and occasionally cultivated grain crops.[60, 94, 202, 203] Some favorites include pecans, cashews, peanuts, millet, acacia seeds, maize, figs, mango, ziziphus berries, and oranges.[8, 94, 203] Meyer's parrots and Red-bellied parrots often eat the seeds and juice of fruits rather than the actual pulp.[25, 94] During a study conducted in 2004/2005 Meyer's parrots were observed consuming 38 different food items.[25] Meyer's parrots associated freely with other birds that were feeding at the same time and in the same location; however, they ate unripe fruit at times to avoid competing with baboons and monkeys.[25]

Family Life and Daily Activities

The life span of wild Meyer's parrots and Red-bellied parrots is 20 to 25 years.[202] Meyer's parrots and Red-bellied parrots travel in pairs or as an extended family of 4 to 5 birds.[8, 60, 94, 202] Meyer's parrots and Red-bellied parrots may form flocks of 10 to 50 birds outside breeding season and when food is plentiful.[34, 60, 94, 202, 203] The flight of wild Meyer's parrots and Red-bellied parrots has been described as "swift and direct", "straight and powerful", and "bullet-like".[34, 60, 94] In their native environment Meyer's parrots and Red-bellied parrots are considered shy, wary, and inconspicuous. This is another reason why they are not well known.

Meyer's parrots and Red-bellied parrots roost and nest in tree holes.[34, 60, 94, 202, 203] Meyer's parrots and Red-bellied parrots frequently recycle woodpecker nests.[25, 60] They prefer tree holes that are at least 10 to 39 feet (3 to 12 m) above ground.[25, 94] The primary breeding season is April through June, with courtship and nesting activities beginning in March.[25, 199] Bonded pairs have been observed dueting near their chosen tree hole.[199] Meyer's parrots and Red-bellied parrots are believed to establish life time monogamous pair bonds.[8] Meyer's parrots and Red-bellied parrots first form pair bonds when they are about 5 years old.[8]

Meyer's parrots have 3 to 4 eggs per clutch, while Red-bellied parrots have 2 to 3 eggs per clutch.[60, 94, 202, 203] The eggs are round, white, and approximately 1.1 by 1 inch (2.7 by 2.5 cm).[60, 94, 202, 203] The female incubates the eggs for 27 to 28 days.[60, 94, 202, 203] Nestlings fledge 60 to 63 days after hatching.[60, 94, 202, 203] Juveniles may grow up together in crèches that are attended to by alternating adults.[8]

Vocalizations

Both Meyer's parrots and Red-bellied parrots are considered quiet—for parrots that is.

Wild Meyer's parrots have been described as making a high pitched "che-che-che" sound.[60] The Meyer's parrot discussed in this chapter did not make that sound. Instead she made a "d-hee" sound that was repeated 4 to 6 times and a higher pitched "ps-tee" sound that was repeated 3 to 5 times. The "d-hee" sound is often heard after receipt of a favorite food or treat.

Wild Red-bellied parrots have been described as making a high pitched "screeeak" sound.[94] The Red-bellied parrot discussed in this chapter does not make that sound.

In 1998 a study was made of the distress calls of 40 members of the Poicephalus family. The Poicephalus distress call has been described as sounding like a low pitched growl.[170] In contrast, the alarm call is high pitched. The acoustical structure of distress calls within the Poicephalus family was found to be very similar.[170] The distress calls were found to be composed of a series of short (3 to 6 msec) harmonic pulses which followed one another very rapidly.[170] Interestingly, there was a difference in the duration of the distress calls by gender. In a sample of 10 Meyer's parrots the average duration of distress calls was:[170]

- Females: 349.2 ± 11.3 msec
- Males: 303.5 ± 6.6 msec

In a sample of 6 Red-bellied parrots the average duration of distress calls was:[170]

- Females: 459.3 ± 29.7 msec
- Males: 218.4 ± 6.2 msec

For obvious reasons neither parrot discussed in this chapter ever made a distress call.

One issue in regard to vocalizations warrants further discussion. The majority of non-predatory birds are known to make a distress call. Distress calls differ from alarm calls. An alarm call is a warning of impending danger, like an early warning system. In contrast, a distress call is made when a bird is actively under attack or confined by a predator; it is a combination of a real-time warning and a plea for help.[170] Hence, the fact that Meyer's parrots and other birds make distress calls indicates that they have an awareness of self. If birds did not have an awareness of self they would not make a distress call or fear death. Likewise, why would a bird make an alarm call if it did not have an awareness of others. Usually only one bird makes an alarm call, yet all the birds within hearing range understand and respond accordingly. An alarm call does not benefit the bird who makes it. On the contrary, an alarm call may give away that bird's location.

Now let's look at the characteristics of indoor Meyer's parrots and Red-bellied parrots in comparison to their wild cousins, such as their behavior, daily activities, and stages of life. Sections 9.3 and 9.4 discuss the behavior of individual birds, for example play activities, socialization, vocalization, and reaction to changes and challenges throughout their lives. The behaviors are presented as they occurred. The analysis, comparison, and interpretation of these behaviors are reserved for Section 9.5 Discussion. These observations took place at the locations and during the timeframes described in Annex C.

9.3 Squeaky

Squeaky joined the aviary August 30, 2001. Squeaky is a female Meyer's parrot. Meyer's parrots are considered to be hanging parrots and love to spend part of their day hanging upside down by one or both feet, especially when playing with toys. Squeaky came with a health certificate that listed her hatch day as February 25, 2001, recorded her leg band number as BRO 2143, and stated that she had the necessary tests and vaccinations. (See Figure 9.1.)

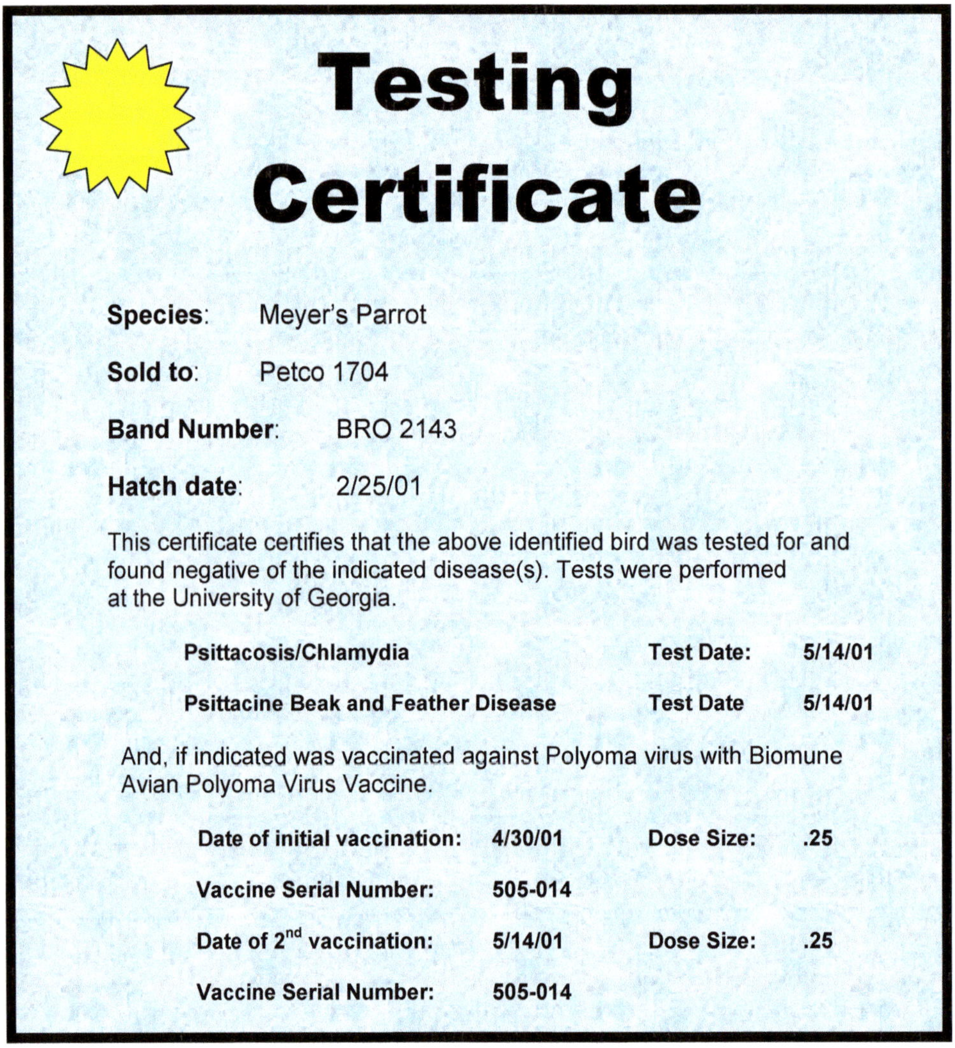

Testing Certificate

Species: Meyer's Parrot

Sold to: Petco 1704

Band Number: BRO 2143

Hatch date: 2/25/01

This certificate certifies that the above identified bird was tested for and found negative of the indicated disease(s). Tests were performed at the University of Georgia.

Psittacosis/Chlamydia	**Test Date:**	**5/14/01**
Psittacine Beak and Feather Disease	**Test Date**	**5/14/01**

And, if indicated was vaccinated against Polyoma virus with Biomune Avian Polyoma Virus Vaccine.

Date of initial vaccination:	**4/30/01**	**Dose Size:**	**.25**
Vaccine Serial Number:	**505-014**		
Date of 2nd vaccination:	**5/14/01**	**Dose Size:**	**.25**
Vaccine Serial Number:	**505-014**		

Figure 9.1. Meyer's Parrot Health Certificate. This is Squeaky's health certificate. Most responsible breeders provide a health certificate for all the birds they sell; some breeders only provide a health certificate for larger birds. A health certificate indicates the hatch date, band number, and the medical tests and vaccinations that were performed before the breeder offered the bird for sale. This information should be shared with an avian veterinarian.

Upon arrival Squeaky was placed in her new baby/quarantine cage. Squeaky was not tired, nervous, or afraid. Instead, Squeaky was very calm, which became her trademark character. Squeaky seemed to be taking everything in with a mixture of curiosity and contentment.

The next morning Squeaky was sitting on her dry food dish methodically emptying the entire dish on the floor, one beakful at a time without eating anything or making a sound. I picked the food up and put in back into her dish. Squeaky dumped the food on the floor again. I picked it up again, two or three more times. The night before I selected an all-purpose parrot seed mixture with pellets, dried fruit, and dried vegetables. The sales person told me that Squeaky would never eat that. I should buy a different product—the same food Squeaky was fed in the store. Not wanting to subject Squeaky to too much change at once, I agreed and bought the other food. Only to be confronted the next morning by a very resolute Squeaky. By 10 a.m. I was off to the pet store (again) to buy the food I originally intended to buy the night before. Squeaky devoured it. The birds outside were given the other bag of food. They did not seem too crazy about it either...

Squeaky was raring to go that first morning so she was let out to play. Immediately she began to sing a song and do a little dance, walking around in a circle and bowing her head. Her song did not sound like any bird I had heard before—a medium pitched "do-do, do-do." Then without any prompting, Squeaky decided to show me that she knew the "step-up" command. She stepped up on one hand, then the next over and over again, like an impromptu tread mill. It even took a while to get her to stop and stand still. Next Squeaky lowered her head, an indication that she wanted a head scratch.[8, 18] She was offered head scratches and pats which she loved. Then Squeaky was taken on a walk, like Uzziel (Section 7.3) had been, from room to room and window to window. Squeaky took in all the new sights and sounds with curiosity, never making any attempt to fly away. (See Photographs 9.1 and 9.2.)

One day when I was taking Squeaky for a walk not long after she arrived I stopped by a mirror. Squeaky stared at the mirror and tilted her head to the right. She noticed that the "other" bird did the same thing at the same time. Next Squeaky tilted her head to the left. Again she noticed that the "other" bird did too. Squeaky turned around and looked at me, then at the other "person" in the mirror. As she turned around to look at me again, I tilted my head right and left. Squeaky watched me and the other "person" intently. Squeaky tapped the mirror twice with her beak, then turned around. She had quickly figured out that this was a mirror and she was seeing our reflections. It was apparent that Squeaky was one smart bird. Many birds never figure out glass or mirrors, often with fatal consequences.

Squeaky loved to play and hang-out together. Whenever I let her out to fly or walked into the room when she was out, Squeaky would immediately fly right to me without any cues. Squeaky loved to sit on my shoulder. She would stay there an hour or more when I was writing, reading, or working on the computer.

Squeaky liked to play catch, at least the parrot version of it. Her favorite ball had a colorful lattice design with a rattle inside. Squeaky would throw the ball and I would catch it. Squeaky had a mean curve ball until one day she crunched the ball with her powerful beak. I panicked, afraid she might swallow a sharp piece of broken plastic and cut herself internally. Somehow I managed to convince Squeaky to spit out all the broken plastic pieces.

Like most birds Squeaky had her favorite sleeping perch.[115] It was a small braided cotton rope perch in the back right-hand corner of her cage. Squeaky groomed the perch every night, then turned around in a circle three times before settling in. Dogs often do this before they lay down for a nap. But, I had never seen a bird circle around like that before. Nor have I seen any birds other than Squeaky do that since. How and why she picked up this habit remains a mystery. As Squeaky got older, she performed this ritual less frequently.

Early on Squeaky made her fresh food preferences known. Most books recommend offering parrots green leafy vegetables because of their high vitamin content.[31, 48, 70, 99, 108, 123, 124, 139, 167, 178]

Photograph 9.1 Squeaky as a Baby. Squeaky, alert and inquisitive, explores the top of her baby/quarantine cage shortly after her arrival on August 30, 2001. She was 6 months old at the time. Note the large yellow patches on her wings and the absence of any yellow feathers on her head. Also, her original wooden swing is still intact! © D. Herrmann

Photograph 9.2 Squeaky as a Juvenile. Squeaky is out exploring 9 months later. Note that the yellow wing patches are gone, while a few yellow head feathers have begun to appear. © D. Herrmann

Squeaky, like most parrots in the aviary, throws green leafy vegetables on the floor—they are too wimpy. She and the other big birds prefer something that has more substance and is a challenge to bite. Instead, apples, oranges, bananas, peaches, carrots, and dried papaya are favorites. Once in a while Squeaky will actually eat some broccoli. Most of the time, however, she opts for treating broccoli as a shredding toy. Squeaky's real weakness is almonds. She has one raw almond delivered to her perch daily, two on special occasions. If I need Squeaky to do something she does not want to do, like get in her travel cage, I know I can always entice her with an almond. In contrast, Squeaky has little interest in peanuts.

Squeaky routinely threw unopened sunflower seeds on the floor—our floor, not her cage bottom. I picked up the sunflower seeds and put them back in her dish. Squeaky threw the seeds on the floor again. I picked them up again and again. In actuality Squeaky was pressure testing the sunflower seeds, before wasting the time and energy to open each one. She would squeeze each seed slightly, applying just enough pressure to determine if there was a seed inside or whether the shell was empty. The empty ones she threw on the floor without opening them. Humans on the other hand open every seed or nut. I do not know how many times I have cracked open an empty almond or filbert and been reminded of Squeaky's superior nut cracking method.

Every morning Squeaky methodically took a couple pieces of dry corn out of her seed dish and deposited them in her water dish. One weekend I was able to observe what she was really doing. Squeaky would pick up a dry kernel of corn, puncture it slightly with the sharp tip of her beak, then place the corn in her water dish. About ten minutes later Squeaky would retrieve the kernel of corn, which was soft by now and much more edible because the puncture hole allowed water inside the kernel. This behavior was rather ingenious since: (1) Squeaky was not even a year old at the time, and (2) the pet store had fed her a pellet only diet. Other birds dip their food in water to moisten it before eating it, especially pellets which are dry and hard. But puncturing corn and letting it sit in the water for a while to soften is a higher skill level.

Squeaky's third food insight was equally amazing. The first time I popped popcorn after Squeaky arrived, she went nuts. She danced, rocked back and forth from foot to foot, and beeped a "please, please, please" type of beep. I could not imagine what was causing her behavior. Sure Squeaky could smell the corn popping, but where in the world would she have encountered popped popcorn before? Certainly not at the pet store. Certainly not at the breeder.

I gave Squeaky a dishful of popcorn anyway (with no salt which is bad for birds), and she loved it. At first Squeaky wanted me to hand her each piece of popcorn, one at a time. Before long, Squeaky learned to eat popcorn by herself. Other birds that came after Squeaky learned to like popcorn; but, none of them let me know they wanted popcorn beforehand. If it were not for Squeaky, popped popcorn would never have been considered an appropriate treat for indoor parrots. Later I read in a non-parrot book that it is traditional in rural parts of central and eastern Africa to throw popped popcorn at weddings, similar to the custom of throwing rice at weddings in the West. But how did Squeaky know that? There must have been some genetic intelligence at work, similar to the way Squeaky knows what other foods she is supposed to eat. Even now, 14 years later, if Squeaky is out playing she will fly over immediately for some fresh popped popcorn.

One of Squeaky's favorite toys was a cylindrical blue metal bell, about 2 inches (5 cm) long that hung on a long chain, also blue. This bell made a very melodic sound and was one of the few bells from which Squeaky could not remove the clapper. Five minutes or less after I installed other bells Squeaky proudly deposited the clapper on the bottom of her cage. Squeaky has a penchant for disassembling toys, especially complex puzzle toys, rather than simply playing with them. When presented with a new toy Squeaky generally stares at the toy from all different angles for a while, sometimes for a couple days. Then, all at once, she lunges forward and takes the toy apart, piece by piece. Afterwards she struts to the top perch.

This particular blue bell presented a challenge; so Squeaky found a novel way to play with it. Toys are designed with the intent that pet birds will play with them in a certain way. People purchase the toys assuming their pet birds will play with the toys in a certain way. But no one consults the birds. Often I have been amazed and amused to find a bird playing with a toy in a fashion never intended or envisioned by its designer. So it was with Squeaky and the cylindrical blue metal bell. Simply tapping the bell with her beak to make it ring was too easy for Squeaky. Instead she climbed onto the long chain, hung upside down, and flapped her wings. The resulting air turbulence was sufficient to ring the bell. In sequence Squeaky climbs on the bell while it is still, starts the momentum by turning upside down, then flaps away at full power. (See Photograph 9.3.)

Another toy was called a ring toss. It consisted of a bright orange 1.5 inch (3.8 cm) plastic ring hanging from a 3-inch (7.6 cm) metal chain and two 1-inch (2.54 cm) posts. Every morning I hung the ring on one of the posts, while Squeaky watched curiously. Every night I found the plastic ring hanging on the other post. On the weekends I left the ring dangling on the chain. Squeaky continued this routine for a couple of weeks. Then one day I found that Squeaky had tied the metal chain in knots, making it about a third of its original length. I took the toy out and looked at it. I could not figure out how Squeaky had managed to intertwine and tangle up the chain links without any hands or tools. I had to remove the plastic ring in order to untangle the chain links. Squeaky also learned how to push the ring through the aviary bars. (See later Photograph 9.12b.)

Most books, magazines, and catalogs recommend buying the biggest cage you can afford for an indoor bird. The "big is better" motto is a step forward. Ten or fifteen years ago the common guidance was to buy a cage big enough so that a bird could fully stretch out its wings. I am not sure who originated this bit of wisdom. But that is equivalent to saying that because a human can stand up and turn around, it can spend its whole life in a center coach class seat on an airplane. Now there are minimum size recommendations for cage length, width, and height by type of bird, based on their size, wingspan, activity level, and so forth. For example, the minimum recommended cage size for a tiny zebra finch (*Poephila guttata castanotis*) is 30 inches (76 cm) long by 18 inches (45 cm) wide by 18 inches (45 cm) high.[174] Unfortunately, many pet stores still sell cages that are way too small. Why? Because large cages take up too much floor space. The online pet suppliers, in contrast, offer several large cages.

When it came time for Squeaky to move from her baby/quarantine cage to an aviary I took the "big is better" advice to heart. A 48-inch (122 cm) long by 32-inch (81 cm) wide by 72-inch (183 cm) high stainless cage was purchased. The bar spacing was listed as 1.25 inches (3.1 cm) and the bar diameter was listed as 0.1875 inch (0.47 cm). Bar spacing is supposed to represent the interval at which horizontal and vertical bars are installed. The distance between the bars is then calculated as the bar spacing measurement minus the bar diameter. The recommended bar spacing and bar diameter increase in proportion to a bird's size and beak strength. As it turned out the distance between bars, not the bar spacing, was 1.25 inches (3.1 cm). This meant Squeaky could stick her head through the bars (but not her body), and pull it back inside the cage safely; as long as she did not turn her head sideways, in which case her beak would get stuck. Fortunately, this only happened once.

Sudden changes in their environment can be frightening to indoor birds.[8, 99, 115] To avoid this Squeaky's baby/quarantine cage was placed next to her new cage. Every day Squeaky was shown a new perch or toy, then it was installed in the new cage. Next I picked Squeaky up, walked in front of the new cage, and touched the new cage while saying, "This is Squeaky's new house." This pattern was continued for a week. Slowly Squeaky's favorite toys from her baby/quarantine cage were moved into her new home. Then it was time to move Squeaky. I let Squeaky out to fly. While she was flying—but watching me—I moved the last of her things from her baby/quarantine cage to her new aviary. The baby/quarantine cage was placed in another room out of sight. Then I opened the door of her new aviary. Squeaky understood what was going on right away. She flew over and landed on top of

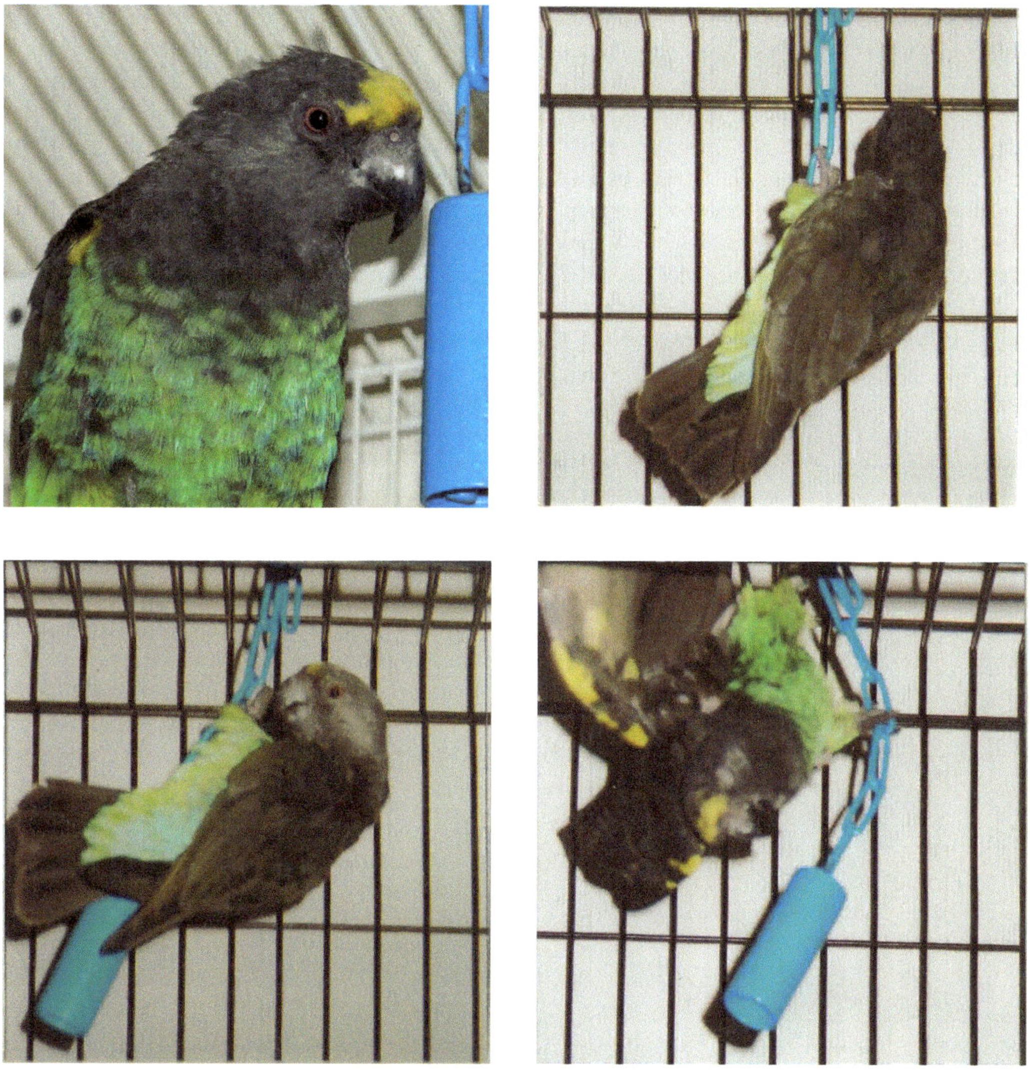

Photograph 9.3 Parrot Aerobics. Squeaky takes the cylindrical blue metal bell out for a ride. First, Squeaky climbs onto the bell while she is upright. To start the momentum Squeaky hangs upside down. Then it's time to rev the engines and flap away at full power. The resulting air turbulence is sufficient to ring the bell. This is an example of birds playing with toys in creative ways that the toy designer never intended. Squeaky engaged in this behavior as a juvenile; she continues to do so 14 years later. It is unlikely that this behavior occurs in the wild. © D. Herrmann

her new aviary. Squeaky climbed along the top of the aviary, peeking through the bars. Before long, Squeaky began exploring the inside of the aviary. This transition was successful because Squeaky was allowed to: (1) get used to the new environment gradually, and (2) decide when to move in.

There are limits to the "big is better" motto. Sometimes a cage can be too big, such that it is intimidating. I thought a 48-inch (122 cm) cage would be good for Squeaky. It was a short flight from one side to the other, giving her ample room to play. But something about that cage intimidated Squeaky. She was comfortable sitting on top of it when out flying, but she was not comfortable inside it. Squeaky never went to any of the perches or toys in the bottom half of the cage. Squeaky had regularly played with them in her baby/quarantine cage and had gone down to retrieve dropped pieces of food. In her new cage Squeaky climbed from side to side rather than flying. She avoided her swing until I moved it to one side of the cage. Squeaky would not enter the cage through the front door after flying; instead I had to open one of the small side doors. Squeaky even quit having "flying attacks." About once a day indoor birds hold onto a perch or the side of their cage and flap their wings vigorously, as if flying full speed, for exercise. This activity boosts a pet bird's morale and is referred to as "flying attacks." Members of the Poicephalus family in particular are known for becoming lethargic if they are cage bound and not given enough time to fly and play outside their cage.[112]

Something about that cage bothered Squeaky and dampened her spirit. Was it too big and thus intimidated her? Was the bar diameter threatening? Was the interior too dark because the cage was stainless? Did the stainless bars reflect UV light or shadows in a manner that was scary? Or, all of the above? I did not know what the problem was. I just knew Squeaky was not happy. The situation was analogous to a plant being over-potted.

A long time passed and there was no improvement in Squeaky's behavior. A new white powder coated steel cage, 37.5 inches (95 cm) long by 27.5 inches (70 cm) wide by 63 inches (160 cm) high, with thin bars and 0.5 inch (1.27 cm) bar spacing was ordered. Squeaky's things were quickly moved into the new cage. Squeaky moved herself in the first chance she had. Now she is back to being Squeaky: playing on the swing, flying mini-laps around the cage, and entering/exiting through the front door during flying time. Yes, pet birds are very sensitive to their environment. (See Photograph 9.4.)

The stainless cage was well made, new, and in perfect condition. It was donated to a parrot rescue organization. Hopefully it will make a good temporary home for a cockatoo or macaw.

Squeaky inherited the 10-ounce (300 ml) stainless dishes that were too big for Uzziel and the budgerigars (*Melopsittacus undulatus*) in Chapter 7. They fit perfectly and are a size she is comfortable with. A 30-ounce (900 ml) dish makes a perfect swimming pool for Squeaky. The 10-ounce (300 ml) dishes have become the standard for the Poicephalus parrots and Indian ringneck parakeets (*Psittacula krameri manillensis*) in Chapter 10. The Mitred conures (*Aratinga mitrata mitrata*) in Chapter 11, being larger, prefer the 20-ounce (600 ml) size.

Photograph 9.4 Home Sweet Home. Squeaky sits on the open door of her new white aviary. Squeaky is more comfortable in an aviary this color and size. Squeaky found a larger stainless steel cage to be intimidating. Note the solid yellow patch on Squeaky's forehead, indicating that she is a young adult. © D. Herrmann

Earlier I noticed that budgerigars are sensitive to color, in particular the color of clothing that I wear.[118] In fact, some books recommend wearing the same outfit or color of clothing when trying to tame a new bird so it will recognize you.[174] The same was true with Squeaky. Yellows, greens, blues, and earth tones are agreeable colors. Red and black cause anxiety, especially all red or all black outfits. If I wear a black coat or suit and walk through the aviary, that usually triggers a minor panic attack, until the birds realize it is me. If, while wearing an all black outfit, I walk directly in front of a cage quickly, that will trigger an all-out panic attack. Black or dark navy seem to be associated with predators. A dramatic change in hairstyle or suddenly wearing a hat or glasses causes a similar reaction.[8, 128] Hats and glasses seem to be associated with an aggressive display. The one exception was a pair of cockatiels (*Nymphicus hollandicus*). They, of course, have a crest that can be raised or lowered at will. The cockatiels were not intimidated by hats, headphones or glasses.

Red clothing, on the other hand, seems to convey a warning about lurking hazards. In the aviary, red is viewed as a warning sign. Some sources postulate that the reason woodpeckers have red heads is to warn predators to stay clear. Woodpeckers are clinging birds and they are very vulnerable when they are on the ground. Scarlet macaws and other brightly colored birds of the tropics live in a different environment. In their case the brightly colored feathers help them to blend into the scenery.

Birds recognize each other through their vocalizations at a distance and/or their coloring up close. Unlike humans birds do not just look at the head to identify another bird; they look at the whole bird: beak, wings, tail, back, and so forth. When a new bird joins the aviary they are often confused when they see me in different outfits, especially during the same day. At first they think each outfit is a different person. After a while the birds realize that I have just "molted". They are intrigued when, standing in front of them, I take off a jacket, sweater, or shoes and set the item down so that it is clearly no longer part of me.

Squeaky has a calm yet inquisitive temperament. She has never been hyper, fussy, nervous, aggressive, or complaining. On occasion she seems lost in thought, like an absent minded professor solving a scientific problem. More than once I have seen Squeaky staring at the track lights that house the full spectrum sun lamps. She seems to be trying to figure out why there is one sun outside the window during the day and six suns on the ceiling in the evening. Due to her calm contemplative nature, Squeaky is rarely frightened by thunderstorms or things that scare other birds in the aviary.

Because of that I was surprised the first time Squeaky got frightened. We were rearranging things and I carried a chair into the aviary. It was not near Squeaky's cage, but she panicked and flew from side to side. I stopped and sat the chair down. I talked to Squeaky while touching the chair. It took a while to calm Squeaky. I couldn't imagine why Squeaky was afraid of the chair being moved. None of the other birds, including much smaller zebra finches and budgerigars, have ever been alarmed when furniture or boxes were moved.

Why was Squeaky scared? There was one possible explanation. In the wild people do not pick up large trees and move them. Squeaky does not mind if I move her cage or other bird cages. But, if I need to move a piece of furniture or carry a large box through the aviary that causes a lot of anxiety The same thing happens if a repairman comes through the door with a ladder. Squeaky has the same response after all these years; hence, it must be a pre-programmed self-defense mechanism. Meyer's parrots are not far removed from the wild. Unlike budgerigars, Meyer's parrots and Red-bellied parrots have not been bred in captivity long enough for them to be considered domesticated. On the contrary, their genetic intelligence is still very much intact.

To prevent Squeaky from panicking and possibly hurting herself, we have worked out a simple routine. If I need to carry something large into or through the aviary, I bring it to the entrance and set it down. I touch the item several times and say "it's ok, it's ok." Then I pick the item up and say "coming through, coming through" while carrying it across the aviary. This routine almost always works. On the other hand, if I forget to do this and get in a hurry Squeaky panics. Squeaky is a big bird and one of the original residents of the aviary. If she panics the other smaller birds reconsider whether they should panic too.

Like the budgerigars Squeaky loved to sit on and hang upside down from the hoops that held her stainless steel food and water dishes; so much so that a couple extra empty hoops were installed just for her to play with. The fact that the hoops are round, hollow, and have narrow rims, which are good for exercising foot muscles and building coordination, makes them an intriguing toy.

Squeaky had three favorite toys in her early years: a wooden swing, an octopus perch, and another bell toy. Squeaky's first swing was a colorful 6 by 8 inch (15.24 by 20.32 cm) wooden swing with small rope tassels on the bottom. The perch part was a single piece of wood with a varied diameter. The arch consisted of a number of round and square wooden beads of different sizes and colors. The clasp at the top allowed the swing to rotate 360 degrees. Squeaky took to that swing immediately when she arrived in 2001. She rocked back and forth by stepping sideways from foot to foot. The faster she stepped, the faster the swing rocked. Every morning Squeaky rocked on the swing for ten to fifteen minutes before and after breakfast, and again before going to sleep in the evening. This activity is still part of her daily routine. Sometimes Squeaky even hung upside down from the swing and flapped her wings.

Eventually Squeaky chewed up all the wooden blocks on the arch. I searched in vain for a replacement wooden swing for Squeaky. The only ones I could find were 10 by 12 inch (25.4 by 30.5 cm) or 12 by 14 inch (30.5 by 35.5 cm), much too large for a Meyer's parrot. It wasn't until December 2009 that I found Squeaky a genuine replacement wooden swing. Like Uzziel, Squeaky remembered the swing she had as a juvenile and was delighted with the new wooden swing. In the interim she made due with an acrylic swing.

Squeaky had an octopus perch in her early years. This bright yellow perch was a combined perch and toy. At the end of the 6 inch (15.24 cm) long wooden perch there was a 2-inch (5 cm) round wooden ball. The ball turned around in a circle and had four colored cotton ropes and four colored sisal ropes dangling from it, each with a small wooden bead at the end. Squeaky loved to sit on top of the wooden ball and chew on the ropes, which she held in her foot. First she untied the knot at the end of each rope and removed the small wooden bead. Then she systematically chewed each rope until it disappeared. Long after the ropes and beads were gone, Squeaky still sat on top of the wooden ball. I searched everywhere to find Squeaky a replacement octopus perch, but had no luck.

The marketing strategy behind parrot toys is puzzling. The toys the birds in the aviary play with the most are usually only marketed for a year or two. While toys they are ho-hum about are sold for years. There is a difference between toys that humans purchase and toys parrots actually play with. Toys that humans like are only purchased once. In contrast, toys parrots actually play with are purchased over and over again—if they are available. I am not sure marketers have recognized that yet. On the other hand, over the years I have learned not to give up on a particular toy too soon. Sometimes two or three birds will totally ignore a toy, while a fourth bird goes crazy over it. Or, the same bird will suddenly take an interest in a toy it ignored four or five years earlier. It just goes to show that birds are individuals and like humans their tastes can change over time.

Squeaky had another toy that consisted of a 6-inch (15.24 cm) wire filled with colorful wooden and plastic balls and blocks, with a silver bell at the bottom. This toy was hung next to Squeaky's sleeping perch when she was a juvenile. Squeaky immediately removed the clapper. Regardless, that bell became her best friend. Squeaky talks to that bell every morning and every evening and wears it on her head on occasion. Squeaky rarely lets any other birds near that bell. That bell moved with Squeaky from her baby/quarantine cage to the stainless cage to her white cage. Neither the beads nor the bell are as shiny as they once were, but Squeaky still loves it just the same. (See Photograph 9.5.)

Pet birds chew their toys for a variety of reasons. Since birds do not have hands, they explore the world with their beak. Their beak and tongue provide tactile feedback about the objects they touch. Chewing, particularly chewing wood and mineral blocks, helps birds keep their beaks trim and tone the muscles associated with their upper and lower mandibles (or mandible and maxilla). This is essential for birds to be able to eat, preen, and climb. Chewing is also a natural behavior. In

Photograph 9.5 An Old Friend. This bell has been Squeaky's favorite toy since she was a baby, regardless of which aviary she was living in and whether she was living by herself or with Sydney. Some bird toys come and go or are only interesting for a short period of time. However, most pet birds have one favorite toy that remains a best friend for life. Squeaky removed the clapper before she was a year old. Buster, a mitred conure, exhibits the same behavior in Photograph 11.10. © D. Herrmann

their native environment parrots chew on branches to obtain food and hollow out tree trunks to make nests. In addition to chewing for entertainment or out of necessity, birds in captivity also chew on things to customize their environment. This behavior is not unlike humans. Twenty people who have the same office cubicle layout or apartment floor plan will customize their environment differently. Because of that realization I have learned not to replace items that have been chewed too quickly. The chewing may be part of a bird's decorating scheme.

Squeaky's serious chewing spurts are sporadic. When woven palm leaf piñata toys first appeared, I bought Squeaky a 9 inch (22.86 cm) one. The sales person assured me that his cockatiel loved the piñatas and would demolish one in a day or two. A piñata was hung in Squeaky's cage where she would be sure to see it. The piñata hung there for four months. Squeaky totally ignored it. Finally I decided to take the piñata out and replace it with another toy. That evening I found that Squeaky had demolished the entire piñata, which now lay on her cage bottom in pieces smaller than an inch (2.54 cm). What accounted for the delay? Why did Squeaky decide to shred the entire piñata the very day I was going to remove it? Now days when I hang a piñata in Squeaky's cage it is completely shredded in a day or two.

A similar behavior was observed in regard to a calcium perch. Squeaky's new white aviary had a calcium perch with a tapered diameter. The perch was meant to be chewed and thus provide calcium and other essential minerals. In the meantime the tapered diameter would exercise her foot muscles and trim her nails. Squeaky loved to sit on that perch, it was her favorite. Squeaky rubbed her beak back and forth on the perch, but she never chewed it. Over a year went by.

Then one day Squeaky chewed up the entire 6-inch (15.24 cm) perch. Nothing was left but the center nail. The powder on her cage bottom was insufficient to reconstruct the perch; Squeaky had obviously ingested the majority of the minerals. What prompted Squeaky to devour the entire calcium perch that particular day? Did it suddenly taste good? Did the perch need to age a bit before it was edible? Why not just nibble a little bit every day? Squeaky has never been one to chew on mineral blocks or cuttlebones. Maybe this calcium fix was meant to last a long time. The calcium perch was replaced that evening. Squeaky has not chewed it at all although several years have passed.

Hand feeding or eating a small portion while a bird is watching is a good way to get a pet bird to eat or at least try something new. Squeaky had not been exposed to fruit or vegetables at the pet store. Hand feeding seemed like a good way to start and it was. At first Squeaky wanted me to hold the fruit while she nibbled on it. After all, fruit can be very slippery. For some unknown reason Squeaky would take a few bites, move to a new location, take a few more bites, and move again. Squeaky expected me to move the fruit every time she did and I complied. Most likely this behavior is due to genetic intelligence; such as remembering that it is not a good idea to stay on the ground too long eating in any one place, in case predators are nearby. Even now when Squeaky eats fruit out of a dish, she eats some of it at the dish, then carries the rest somewhere else to eat. (See Photograph 9.6.)

Squeaky readily tries new food items as soon as they are given to her. That doesn't automatically mean she likes them or dislikes them. Squeaky will often roll an item around in her mouth, then drop it. That may mean she's not interested in the item at all. Or, it may mean she's not interested in the item now and will eat it later. Squeaky rolls a new item around in her mouth because a parrot's taste buds are on the roof of their mouth, not their tongue.[18] A parrot's tongue is dry.

Meyer's parrots are known for their calm gentle dispositions. They also have two interesting and unique traits. The first approximates the human "terrible two's"; the second is the ability to be potty trained.[8, 49] True to her family reputation, Squeaky exhibited both of these traits.

Squeaky became totally uninterested in pats and head scratches when she was about 18 months old. At the same time she lost interest in going on walks from room to room. Squeaky still flew to me, but she did not stay on my shoulder or arm long. It was obvious Squeaky was entering her terrible

Photograph 9.6 Snack Time. Squeaky samples a piece of egg bread. A good method to encourage a pet bird to try something new is to hand feed them, or eat a small portion yourself while the bird is watching. Squeaky readily tries new food items when they are presented to her in this manner. Squeaky often rolls a new item around in her mouth first. She does that because a parrot's taste buds are on the roof of their mouth, not their tongue.[18] © D. Herrmann

two's. In humans the terrible two's is the time when a child begins to differentiate itself from the parent. Young Meyer's parrots go through a similar experience as they start to see themselves as an individual member of the flock and less dependent on their parents. Often the shift is quite dramatic, almost overnight. What was once a sweet cuddly baby bird becomes an aggressive assertive independent spirit who thinks they no longer want or need human attention. The Meyer's parrot equivalent of the terrible two's lasts six to eight months. Squeaky navigated through her terrible two's and emerged a more mature version of herself.

The terrible two's is only reported in connection with members of the Poicephalus parrot family. One explanation for this behavior is that because members of the Poicephalus parrot family are known for being calm and gentle, their behavior during the terrible two's is more pronounced and observable. In contrast, parrots that are known for being noisy, assertive, and aggressive, like many of the parrots from South America, would seem to be their normal selves during the terrible two's. Hence, the changes and behaviors would be much less pronounced or noticeable. A second explanation for this behavior relates to the fact that wild juvenile Meyer's parrots generally grow up together in a crèche attended to by alternating adults.[8] In this environment juveniles would go through a tussling phase as they begin to assert their individuality and at the same time learn how to live in a flock environment. This experience mirrors that of young children learning how to interact with others in a day care center. (See Table 9.1.)

Some readers may be surprised to learn that some parrots can be potty trained. Meyer's parrots like Squeaky are known to have this trait.[8, 49] Shortly after Squeaky arrived I noticed that she started to get a bit uneasy if she were out more than thirty minutes at a time, especially when I was holding her. Squeaky's actions and looks indicated that she wanted to go home. I would set her inside her cage, leaving the door open. Squeaky would go potty, then come back out to play. The first couple of times Squeaky did this I was amazed. I had not made any attempt to potty train her. Later I found a couple books that indicated Meyer's parrots were one of the easier parrots to potty (or paper)

Table 9.1 Developmental Stages of Young Meyer's Parrots and Red-bellied Parrots.

Stage	Time frame	Events
Egg	Days 1–28	• Eggs are incubated by mother
Nestling	Day 1	• Altricial babies hatch, are fed by mother • Babies start to sing when hungry
	Day 3	• Down starts to appear
	Day 20	• Eyes open, are redish orange for both
	Day 30	• Feathers begin to grow. Male/female coloring is evident in Red-bellied parrots
	Days 56–59	• Start to stretch wings and move about
Fledgling	Day 60–63	• Take first flight from nest
	Days 60–90	• Continue to follow and be fed by parents
Juvenile	3–9 Months	• Young Meyer's parrots have solid gray head, large yellow patches on wing coverts • Both are very quiet • Both are curious and explore their environment; continue to practice and perfect flight
	10 Months	• Meyer's parrots lose yellow wing patches
	12 Months	• A few isolated yellow feathers appear on head of Meyer's parrot
	18–24 Months	• Enter terrible two's period during which they become independent and assertive; this period generally lasts 6 to 8 months
Young Adult	2 to 5 years	• Eyes of both turn orangish brown • Both expand repertoire of vocalizations • Yellow patch on head of Meyer's parrot expands slowly until it is solid from eye to eye • Bird is now a young adult

train.[8, 49] Now days Squeaky has progressed to the point where, if she is out and about in the aviary, she either looks for the nearest newspaper or flies back home to take care of business.

Parrots are attuned to their primary care giver's emotional state.[99] They perceive when a person is happy, sad, tired, angry, nervous, and so forth. Often a parrot will commiserate with their primary care giver by emulating the care giver's mood through their vocalizations and actions. I have frequently noticed the birds in the aviary echoing my sentiment, whether it is grousing about a miserable commute or laughing at something funny. It is always enlightening and humbling to see that they understand me so well. In the wild this capability helps a parrot determine whether or not the actions of a potential predator or intruder are hostile.

Many publications talk about the concept of height dominance. According to this concept parrots and other birds in the wild establish a social hierarchy according to which bird is on the highest perch. Proponents of this theory discourage people from allowing their pet birds to perch on their shoulder or head, since they believe this would signal parity between the pet bird and their caretaker. This concept has been repeated so often it has almost become sacrosanct.

In the 30-plus years I have been observing birds in their natural habitat in the wild and in indoor settings I have never seen any behavior that conforms to this theory. Let's look at wild birds for a moment. Four patterns of behavior are observable on a repeated basis.

First, when it is time to settle in for the evening birds generally seek the highest location above the ground. Not fifty stories up on the Empire State Building, but a location that is "high enough" to avoid predators during the night, like snakes, cats, and other mammals. This behavior is related to safety, not one bird dominating the other.

Second, the high enough criteria is tempered with the fact that the branch a bird perches on must be able to support its weight. Branches on the top of a tree are new growth and the most delicate. Small birds can perch there; large birds cannot. That does not mean small birds dominate large birds. Rather, it simply means the laws of physics are at work, laws to which all birds are attuned.

Third, when a group of birds settle in for the night, there is a lot of chattering and movement from branch to branch as they make sure everyone is there and catch up on the day's gossip. As sunset turns to darkness, things gradually become quiet. None of this has anything to do with dominance. Instead, this behavior relates to individual preferences: which branch each bird decides it wants to spend the night on (it may be a different branch the next night) and which birds it wants to roost next to.

Fourth, when a group of birds settle in for the night or a rest during the day, they do not pack themselves in like sardines on the subway. In contrast, they leave a polite distance between birds. This even spacing is particularly noticeable when a large flock settles on telephone lines. If one bird lands too close, the group reshuffles itself. This behavior also has nothing to do with height dominance. The flock and each individual member are simply being practical. Should the flock or individual members need or decide to takeoff in a hurry, the spacing between birds accommodates their wingspans, like the width of a runway corresponds to the wingspan of the largest aircraft that is allowed to land and takeoff there.

Likewise pet birds living indoors exhibit similar behaviors, whether they are living singly or in a group. Most pet birds choose a favorite sleeping perch and return to this perch each night.[115] When a bird's cage is moved to the other side of the room or near a new object, they may choose a new sleeping perch, if it makes them feel more secure. Sleeping perches are usually the highest perch, or in some cases swing, in the aviary. Birds in a multi-bird aviary may have designated sleeping perches that they all agree on. Or, the birds may alternate sleeping perches on different nights. Usually this process, which may take a few iterations, works out to everyone's satisfaction. When it does not, there is some brief tussling before the lights go out. Zebra finches, which prefer to sleep in sleeping nests, are an exception. Since sleeping nests conceal their occupants they do not have to be the highest item in a zebra finch cage. If zebra finches do not have sleeping nests, they will seek out the highest perch for a sleeping perch.

This safety-related behavior is part of a bird's genetic intelligence and it appears early in life. Very young parrots migrate immediately toward the highest perch at night once they are able to climb up cage bars. About a fourth of the young budgerigars discussed in this book took the height = safety maxim a bit too literally at first. They climbed to the highest perch and started to go to sleep. Then, all of a sudden they woke up and climbed up to the top of their cage and hung upside down parallel to the cage bottom and settled in for the evening. Once in a while a variation of this behavior is seen when a young bird hangs overnight on the highest point on the side of their cage, often in a corner. Either position requires the bird to use both feet and their beak to hold on, which is not very restful. (Normally a bird alternates feet during the night while sleeping.) Either the bird falls off during the night (most likely), or the bird is very tired the next day. Usually a young pet bird will only do this for three or four nights before opting to sleep on a perch.

Squeaky selected her sleeping perch without any guidance or coaching. She understood the height = safety maxim. Likewise, when Squeaky comes out to play she prefers to land on my arm, then walk up and sit on my shoulder. This behavior has nothing to do with dominance. On the contrary, three other precepts are at work. Birds understand these precepts. Humans, on the other hand, have been slow to grasp these precepts due to the tendency to impose human psychological motifs on bird behavior.

First, what Squeaky and other parrots are doing is trying to facilitate communication by maintaining eye contact; that is, having a consistent human to parrot eye level. A toddler's eye level approximates their parent's knees. The same is true for a pet dog. Birds do not have that limitation. They can fly next to whomever they wish to communicate with, human or bird.

Second, by coming in close proximity to their human counterpart, Squeaky and other parrots are expressing friendship and trust. They want to be near someone, human or bird, they like.

Third, it is important to remember that not all predators that attack birds come from the ground up. Some predators, like birds of prey and humans, swoop down from above. Towering over pet birds and insisting that they remain below the shoulder does not establish height dominance. Quite the contrary. This posture evokes a predator-prey relationship and birds know they are prey. Fear, however submerged, is not a good basis for establishing a trust relationship with a pet bird.

Whenever I talk to birds in the aviary, especially when they are in their cage, I drop down to their level. The birds instantly understand my attempt to communicate with them. I avoid talking to the birds from above or putting my hands on the top of their cage, particularly with the smaller birds. Because no matter how comfortable they are with me, towering from above evokes instant predator-prey unease. I hold my hands behind my back, as if my "wings" are retracted. This approach has a calming effect. Humans are not the only ones who respond to body language.

Occasionally Squeaky will land on my head; after all it is a larger runway than my arm or shoulder. Once there, Squeaky likes to preen my hair. This behavior, known as allo-preening, is a sign of affection so I do not discourage her. A couple times Squeaky accidently slid down my face. Not wanting Squeaky's nails and my eyes to have a close encounter, I now wear a baseball cap whenever Squeaky is out flying to give her firmer material to land on. Squeaky lands on my cap and attempts to preen it. She particularly likes to play with the grommets, which are just the right size to be interesting. Every now and then Squeaky walks up front and sticks her head under the bill, just to make sure I am still there. One day Squeaky made an attempt to land on the bill. She quickly learned that the bill is in restricted air space and for good reason—the cap went flying…

The intelligence of birds, particularly parrots, has not been recognized or acknowledged until recently. Bird enthusiasts have long noted the intelligent behaviors, problem solving, and reasoning abilities of birds. But the scientific community tended to dismiss these observations as "anecdotal". Historically the scientific community applied characteristics of human intelligence to birds: Humans communicate via a spoken and written language that is represented by letters and sounds. Humans read and write books and spend decades studying in schools. Humans use advanced technology to perform advanced mathematical calculations, like those needed for flight. Humans have all sorts of degrees and titles after their names, which prove they are intelligent. Birds do not have any of these traits; therefore birds are inferior intellectually. Or so the theory went.

This theory failed to recognize that just as there are different forms of life, there are different forms of intelligence.[72] After all it would not make much sense for a bird, which is designed to be light weight and airborne, to try and fly around with a complete set of the *Encyclopedia Britannica* on its back. What would make sense, however, would be for a bird to store all the knowledge, reasoning, and problem solving skills it needs, not just for survival but for enjoying life too, in a miniature biological nano-circuit in its very portable brain. Or, to put it another way, I have never seen a bird pecking away for days on end on a super computer to calculate the aerodynamics of wind foil, wind shear, wake turbulence, or lift parameters, yet their flight is fine and much less error prone than that of humans. Nor did it take 6,000 years for birds to learn how to fly.

Several episodes mentioned earlier hinted at Squeaky's intelligence. A few more deserve to be mentioned as well.

Most days Squeaky announces the impending sunset. Squeaky has announced sunset since 2002 when she was one year old, no matter:

- What aviary she was living in
- Where her aviary was located in the room in relation to windows
- Which direction her aviary was facing (north, south, etc.)
- What building her aviary was located in
- What season it was

At sundown, but before sunset, Squeaky does a special "beep-beep, beep-beep" two or three times. This is the only time Squeaky makes this particular sound. It is as if Squeaky, like the town crier, is letting the other birds in the aviary know sunset is approaching. Why is sunset important? The time interval between sundown and sunset is when birds in the wild reconnect with their flock and find a safe place to spend the night using the height = safety principle.

Squeaky does not make this sound mid-morning or mid-afternoon when it suddenly becomes dark because the sun is obstructed by clouds. Somehow she knows the difference. Nor does Squeaky make this sound if the sun lamps in the aviary are turned on before sunset. Squeaky is the only bird in the aviary that announces sunset. Squeaky was the first big bird in the aviary. Most likely there is only one "town crier" designated per flock. If Squeaky were not here, probably another big bird would assume this role.

In her younger years Squeaky demonstrated her intelligence and verbal communication skills another way. Whenever I stood by another bird's cage Squeaky would make the vocalizations that bird normally made. Squeaky was the only Meyer's parrot; the other birds were different species and had different vocalizations. Not only were they different birds, but they came from different continents, meaning that Squeaky would not have met these types of birds in the wild. If I stood in front of the budgerigars' cage, Squeaky would emulate the song of a budgerigar. If I stood in front of an Indian ringneck parakeet's cage, Squeaky would emulate the song of an Indian ringneck parakeet, and so forth quite accurately. Zebra finches were the only birds Squeaky had trouble emulating. For some reason the high pitched buzzing sound zebra finches make was hard to emulate.

Squeaky did all of this on her own initiative. I made no attempt to teach her to sing like the other birds. Nor did I solicit her response when I stood in front of the other birds' cages. By this act Squeaky communicated three ideas, each of which represented a different thought process:

- Squeaky demonstrated that she knew what sound each bird made
- Squeaky demonstrated that she knew how to make that sound, even though it was not part of a Meyer's parrot's vocabulary
- Squeaky demonstrated that she wanted me to know that she knew what sound each bird made and could repeat that sound accurately

Squeaky demonstrated an understanding of communication protocols whenever we had a conversation and as she was learning various beeps and whistles. I thought this was just a random event the first time it happened. But Squeaky continued to demonstrate this behavior consistently over a span of several years. Here is what happens. I start a conversation by saying something: a phrase, a series of beeps, or a whistle. Pause. Squeaky "says" something in response. Pause. We repeat this exchange three or four times. The phrases, beeps, or whistles change. If by chance one of the pauses is too long, and we both start to say something at the same time, Squeaky immediately stops and lets me finish. In essence, Squeaky is demonstrating that she knows:

- The difference between transmitting and receiving information
- That it is an impediment to communication for both parties to talk at the same time

These are rather sophisticated concepts. Unfortunately I know a lot of humans who have never figured this out.

For a while one of the tiny zebra finches liked to slip out and go flying about the aviary. Since zebra finches can fit anywhere that is 0.5 inch (1.27 cm) square, it is easy for them to escape. Zebra finches are so tiny they fit or get stuck in a lot of unusual places, like under the refrigerator, behind bookcases, or under a planter with casters. This little guy enjoyed flying from here to there, up and down, and room to room. He could land on tiny branches and the branches would not even move because he was so light. And he was fast! I would look at him, blink, and he was 25 feet (7.62 m) away before I even moved a muscle. His name was Tiny Guy.

One day Tiny Guy flew into Squeaky's cage. Squeaky was still in the stainless cage at this time and the bar spacing was wide enough for a zebra finch to fly through. The first time Tiny Guy flew into Squeaky's cage I thought he was history. Squeaky opened her beak like she was going to swallow a flying insect. At the last minute Tiny Guy tweeted. Squeaky closed her beak, but had a surprised look. Tiny Guy landed on Squeaky's perch, swing, and dry food dish. Squeaky took it all in stride, like a little cousin visiting.

Another day Tiny Guy seemed to think he was beyond reach because he was in Squeaky's cage. After a few unsuccessful attempts to retrieve Tiny Guy, Squeaky figured out what was going on. I tried once more to catch Tiny Guy. He flew toward Squeaky for safety, only to have Squeaky lunge toward him. Tiny Guy made a quick retreat from Squeaky's beak right into the net. It appeared that Squeaky had deliberately shooed Tiny Guy in my direction so I could catch him.

Squeaky has always been a master swimmer; there is not much she enjoys more. Some birds stand on the edge of their swimming pool and throw water up on their backs and wings. They may dip their head and feet into the water, but not too deep. Budgerigars will walk around and around in shallow water, but do not put their heads underwater—except for Precious. Squeaky has no such inhibitions. In fact she splashes so vigorously I have to put out her "beach towel" beforehand to catch the overflow. As far as Squeaky is concerned, the deeper the water the better. Instead of a delicate or careful approach to swimming Squeaky goes entirely under water. Afterward she flaps her wings and shakes, like a dog shakes after its bath to expel the excess water.

Squeaky's favorite swimming pool is a 30-ounce (900 ml) stainless dish, which she can empty in a matter of minutes. Most birds rush down to go swimming as soon as I put their swimming pools in their aviary. The birds swim for a few minutes and are finished. In contrast Squeaky goes down to swim, back up to eat or rest, and back down to swim two or three times in one swimming session. I learned to leave her swimming pool in place for a couple hours and refill it as needed. Is Squeaky demonstrating an individual preference, or is her behavior related to the lack of frequency with which abundant water is available for serious bathing in east central Africa?

Squeaky has always loved flying. Squeaky, like all Meyer's parrots, has a wide wing span. Having figured out the principles of aerodynamics a few millennia ago, birds either have a wide wing span and a short tail, or a short wing span and a long tail, but not a wide wing span and a long tail. Squeaky's wide wingspan makes it more challenging for her to navigate around the aviary than a budgerigar, but she relishes the challenge. Squeaky likes to fly laps around the aviary. Sometimes she goes clockwise. Sometimes she goes counter-clockwise. Other times she does aerial figure eights. Squeaky likes to fly about 18 inches (46 cm) from the ceiling and as close as she can get to the objects she is flying past without touching them; almost as if it is a contest.

Before the aviary expanded there was a 3-tier plant stand with sun lamps across the room from Squeaky's cage. It was filled with cacti, many with sharp spines. Squeaky had an innate understanding that flying birds and cacti need to respect each other's distance. At the same time Squeaky could not resist a challenge. At least five or six times a night Squeaky would fly full speed directly toward the largest cactus with the biggest and sharpest spines. Then, at the very last second, she would veer at a right angle away from the cacti. This was not simply an exercise in flying from one point to another. Instead, Squeaky was purposely testing and perfecting her precision flying. She loved the mental and physical challenge. (See Photograph 9.7.)

Photograph 9.7 Nature Walk. Squeaky frequently visits plants when she is out flying. Squeaky seems to understand that plants are real and part of nature. While she visits plants Squeaky has never attempted to eat them. Other parrots in the aviary have nibbled on spider plants, African violets, jade trees, and sansevieria. Wild birds know which plants are poisonous and avoid them. Indoor birds haven't had time to acquire this knowledge or the opportunity to pass it on; hence, the necessity of having only bird safe plants in an aviary. © D. Herrmann

Squeaky is a master swimmer, a precision flyer, and an energetic climber. But she does not like floors. Some birds enjoy walking around on their paper and playing with balls and rattles while there, particularly budgerigars. Some birds will hop down on their paper to retrieve a piece of dropped food. Other birds enjoy shredding their paper or walking underneath it. A few birds will even walk around on the carpeting in the aviary when they are out to play. Not Squeaky. For some reason she avoids floors, inside her cage and in the aviary. Squeaky will climb down the side of her cage and grab a piece of food off the bottom tray with her beak. But neither her beak nor her feet touch the paper. On a couple of occasions Squeaky has climbed along the outside of her cage to within a few inches (centimeters) of the carpeting. She leans down to touch it with her beak, then pulls back at the last minute. Possibly there are some lingering memories about ground-based predators at work. Squeaky was visibly nervous the first few times I vacuumed the aviary after her arrival. It was not the size or noise of the vacuum that bothered her, rather the long black cord that was slinking around behind me.

One day I was hanging a large toy from the top of Squeaky's cage. Squeaky likes to "help" whenever I am working on her cage, so she was right beneath me. All of a sudden the toy slipped out of my hands. I was afraid the toy would hit Squeaky on her head or back. But before I could catch the toy Squeaky did something amazing. Most birds would have panicked and flown away at full speed. In contrast Squeaky looked up and saw the toy falling in her direction. Instead of trying to fly away (there was not enough time), Squeaky simply loosened her grip on her perch (a metal hoop). Working with the laws of gravity, Squeaky went from being upright at a 45 degree angle to being upside down like a bat with her straight back parallel to the falling object. After the toy hit the bottom of the cage, Squeaky swung upright. Squeaky expressed no fear or panic. Quite the contrary. Squeaky thought this was a game and she was ready for another exciting round. Squeaky has exhibited this behavior three or four times since then under similar circumstances.

Through her pioneering professional publications and seminars Dr. Irene Pepperberg reported the mathematical abilities of her African grey parrot, Alex. As related, Alex understood basic numerical concepts such as quantities (up to 6), more than, and less than. Alex did not just understand these concepts; he could communicate his understanding of these concepts in English when asked specific questions.[134, 135] Alex also understood shapes and colors.[134, 135]

While Squeaky has never said the words one, two, or three (I never taught her), she has demonstrated her understanding of basic mathematical concepts. For example, Squeaky knows she is supposed to have one almond before breakfast. I say, "One for Squeaky." She comes and takes the almond from me and eats it. Squeaky does not make a fuss or come back for a second or third almond. She knows she is only supposed to have one almond. The only times Squeaky has two almonds before breakfast is on special occasions. On those days Squeaky makes it clear that she knows two almonds are better than one. When Squeaky was younger I gave her three fortified treats every evening, one at a time. If I stopped at two treats, Squeaky got fussy. If I indulged Squeaky and gave her four treats, Squeaky bobbed her head up and down rapidly. And, if I offer her two pieces of the same thing, she will inevitably choose the larger piece, unless it has a defect.

Squeaky also acknowledges quantities in her beeps and whistles. Squeaky has a series of beeps and whistles that she taught herself. You can hear her practicing them randomly during the day. Squeaky can also imitate a microwave oven beeping, a telephone ringing, the fire alarm, and the beeping sound trucks make when they back up. If the microwave oven beeps three times, Squeaky beeps three times. If the fire alarm goes off four times, Squeaky repeats the fire alarm sound four times. Squeaky matches the number of times a particular sound is repeated. In this manner she has "counted" up to seven.

At the same time we need to step back a minute and remember that the complex calculations Squeaky and all birds perform in order to fly, navigate, land, and takeoff are performed in real-time without the aid of onboard computers calculating lift, thrust, fuel mixture, wind speed, wind direction, when to raise and lower flaps, and so forth. Nor did Squeaky or any other bird spend years at a university studying sophisticated theorems or formulas in order to learn how to do these calculations. This is another example of genetic intelligence at work, in this case applied mathematics.

Squeaky has demonstrated a practical knowledge of applied spatial geometry, again without using a calculator. On occasion I hand Squeaky, through the cage bars, a treat, something to chew, or an item she has dropped. Squeaky reaches out with her beak to pull the item through the bars. If the item is too wide to pull straight through the bars, Squeaky immediately adjusts her algorithm and pulls the item through at an angle. This behavior exhibits astute mathematical reasoning on Squeaky's part. In contrast, I don't know how many times I have seen people tug and tug at something that will not fit through a given space, until the item becomes stuck or breaks, never even considering that they need to adjust the angle.

One example of biological intelligence is common to Meyer's parrots. Meyer's parrots are one of the few birds that exhibit a symmetrical molt.[8] Whenever Squeaky molts a feather from her left wing, she molts the corresponding feather from her right wing (or vice versa) within twenty-four hours. The first time this happened I thought it was a random event. But, every time Squeaky has molted wing feathers in the succeeding thirteen years the same thing happens—a symmetrical molt. On the surface this may not seem like much. In fact, it is a very ingenious concept which ensures flight stability. If you are a bird you do not want a brand new strong flight feather on one wing and a worn out or absent feather on the other wing; this would cause an imbalance during flight.

The human corollary to a symmetrical molt are all the regulations and standards which govern when flaps and other mechanical structures on airplanes that are subject to stress and the physics of failure need to be replaced. These regulations and standards were only written after decades of human flight, a few crashes, and extensive modeling and simulation of reliability and wear out phenomenon. And yes, if the left widget is replaced, the right widget is replaced at the same time. Through biological intelligence Meyer's parrots have incorporated this important concept into their feather biology. And they did not need laws, regulations, international consensus standards, or regulatory bodies to mandate that they do it.

Squeaky was the first big bird in the aviary. She has been here as homeless birds continued to join the aviary in one's and two's. Like Uzziel, Squeaky is intrigued by the arrival of each new bird. She is curious about their coloring, size, and song. Squeaky views each new bird as a potential new playmate. Squeaky was particularly interested in a yellow female budgerigar that arrived June 1, 2006, named Yellow Birdie. Whenever Squeaky was out flying she would go over to visit Yellow Birdie. Squeaky would hang on the side of Yellow Birdie's aviary or sit on her landing perch. Yellow Birdie would come right over to play and talk, in spite of the difference in their sizes. Squeaky would stay ten to fifteen minutes; she did not pay any attention to the other budgerigars in this aviary. If Yellow Birdie were out flying, she would reciprocate by visiting Squeaky's aviary. Recently Squeaky began playing like this with a male normal green budgerigar named Daffodil. (Yellow Birdie became part of a bonded pair with Lilac II.) There are other normal green budgerigars in the aviary, but Squeaky does not play with them.

What prompts two birds of a different species, from different continents, and of such different sizes to form these play/social relationships? This is not a courtship, nesting, or parenting relationship; nor is it a precursor to forming a bonded pair. Instead this is strictly a social relationship built around some mutual interest. But what? In contrast, Squeaky may or may not let other birds land on top of her aviary or on her landing perch when they are out flying.

As an original member of the aviary Squeaky has been through a couple moves. Squeaky takes the confinement in a travel cage and all the noise and commotion associated with a move in stride. One interesting behavior was noted that occurred with each move. At first Squeaky's cage was on a wall that ran north/south. She looked out on windows that faced west and north. Later Squeaky's cage was on a wall that ran north/south. However she looked out on windows that faced east and south. The first couple of mornings when Squeaky woke up in the new aviary, she stood waiting for the sun to come up in the wrong direction. Squeaky was looking for the sun to come up on the same side of her cage as it always had. Squeaky was puzzled for a day or two, then she re-calibrated her internal compass. Similar behavior has been observed in budgerigars a couple of times after moves. They too realize the sun has "moved" after a day or two.

Once when Squeaky's cage was moved from one side of the aviary to the other, another interesting behavior was observed. Squeaky was let out to fly in the evening as usual. She roamed around and went in and out of her cage for a drink of water or a snack. But, when it was time to go home for the evening and get settled in so the lights could be turned off, Squeaky landed on top of the cage that was now located where her cage had been. This cage was the identical design, size, and color; however, it had different occupants. I tried unsuccessfully to get Squeaky to go home. She knew her food and water were in the new location, but for some reason "home" still registered at the old location. In the end I had to pick her up and take her home. This pattern continued for three weeks. Squeaky was not acting up, playing games, or being ornery. She thought the old location, where she had resided for two years, was still her home. During the fourth week Squeaky's memory cells updated "home" to point to the new location. Squeaky's behavior was similar to that reported for zebra finches in Section 8.4—Week 18. In this case the zebra finches continued to fly to the location where their perches had been for about an hour even though they could see that the perches had been moved.

In contrast, none of the budgerigars have trouble returning home after their cage has been moved or after they are moved to a new cage, regardless of age. Perhaps that is because budgerigars are nomadic in the wild and "home" is always the current location.

Another interesting observation was made in regard to moves. During two moves all the birds in the aviary were placed in travel cages and flew in the cabin of an airplane for 4.5 hours. The birds could not see out the windows. When the plane flew straight the birds were quiet. If the plane changed direction, for example west to south, the birds perked up and chattered among themselves. A change in altitude did not cause any chatter, only a change in direction. And, the birds were less stressed out by a 4.5 hours plane ride than they are by a half hour car ride.

If I need to enter the aviary when it is dark, after the birds have been asleep for a few hours, I do not immediately flip on the lights. After I close the door, I repeat a series of whistles and say, "It's me." After the second or third repetition Squeaky usually (but not always) repeats the whistle. The longer she has been a sleep, the longer she takes to respond and/or is less likely to respond. Then, I repeat the whistle and say, "Going to turn the lights on." Again, Squeaky usually responds. Then I slowly turn on only the lights that are needed.

Turning lights on abruptly, especially when the birds have been asleep for a couple hours, can cause panic attacks. This approach allows the birds to wake-up gradually. Since Squeaky was the first big bird in the aviary, her assurance that it is me entering the aviary at this late hour, and not a stranger, calms the other birds.

Squeaky migrated to a 16-ounce (480 ml) water bottle after she had been here for two years. The use of water bottles with pet birds has advantages and disadvantages; they are not right for every bird. Drinking water from a metal tube is not a normal bird behavior and not all birds will learn how to drink from one. Also, water bottles drip and they can drip dry; especially if a bird decides to stick something in the tube so that the flow of water is unimpeded. On the other hand, the use of water bottles ensures that a bird is drinking clean water. Water bottles also come in handy if the primary caretaker has to be away a day or two. The trick is to make sure a bird knows how to use a water bottle and is comfortable drinking all the water she needs from it before removing her water dish.

For Squeaky the move to a water bottle was a good one. Squeaky took a grand total of five minutes to master the art of drinking from a water bottle. She even carries items she wishes to wash or moisten before eating over to the water bottle. She presses the item against the tube ball until the item is "watered" enough to her liking. Then she consumes it. Squeaky also presses her head against the tip of a water bottle on occasion in order to take a mini-shower.

In comparison, most other big birds in the aviary take three or four days to learn and become comfortable using a water bottle. A couple big birds have refused to learn; after all, it is unnatural. A hybrid approach was followed for two birds. They each have both a water bottle and a water dish. These birds drink clean water from the water bottle; they seem to know that this water is clean. Then, these birds use the water dish to rinse off certain food items before they eat them or for an occasional short swim when they can't wait for the swimming pool.

Luckily Squeaky has never been ill, not at all. Squeaky has never been to a veterinarian's office either. On two separate occasions Squeaky had an appointment to go to the veterinarian for a routine physical; she never made either appointment. Squeaky watched as I rounded up the other birds that were scheduled to go that day and put them in their travel cages. But when it was Squeaky's turn she made it clear that she had no intention of getting in her travel cage. These are the only two times in fourteen years that Squeaky has refused to do something. She was perfectly healthy, so I decided not to force the issue and break our trust relationship. If Squeaky had been ill or had experienced some trauma, that would be a different story.

Squeaky celebrated her fourteenth hatch day February 25, 2015.

9.4 Sydney

Sydney joined the aviary September 12, 2007. Sydney came from the same chain pet store as Squeaky. That store, like many, has a policy of rotating unsold birds around from store to store every six months. The intentions of this policy are good: to maximize the opportunity for each bird to be sold and end up in a good home situation while it is still young. However, this policy must be traumatic or at least unsettling to some of the birds involved. Sydney was one such bird.

I spotted Sydney a few days after she had been transferred from another store in the area. Unlike Squeaky who was visibly happy, dancing around, beeping, and putting on a show the first time I saw her, Sydney was terrified when anyone got near her cage and cowered in a back corner. The contrast was striking. Why would a member of the Poicephalus family, known for being calm and gentle, behave like this? It didn't take long to find out.

A sales person saw me looking at Sydney and came over. She said I didn't want Sydney and tried to talk me into a tame sun conure. I said thanks, but no. At that point the sales person grabbed a perch. With her head and one arm leaning over the top of Sydney's cage, she began chasing Sydney all over the cage with the perch, while yelling "step-up, step-up, step-up ..." Terrified, but without making a sound, Sydney ran and climbed as fast as she could in every direction to get away from that menacing perch.

I told the sales person that was fine. If Sydney didn't want to be held just then that was ok. I could tell she was very energetic by how fast she ran ... I decided to "rescue" Sydney from this horrible situation. I put down a deposit and told the sales person I'd be back to pick Sydney up over the weekend. The sales person was dumbfounded.

Sydney is a female Red-bellied parrot. Her health certificate listed her hatch day as December 7, 2006, recorded her leg band number as FX 3148, and stated that she had the necessary tests and vaccinations. Sydney came from a different breeder than Squeaky.

Sydney's baby/quarantine cage was set up to mirror her cage at the pet store, knowing how traumatized she was already. Dishes and perches were in the same place. Even identical toys were obtained.

I convinced the sales person to let me put Sydney in the pink travel cage. I did not cover the travel cage out of concern it would provoke a panic attack. I felt it was better to let Sydney see what was going on around her. Sydney sat still, breathing rapidly, eyes open wide, and silent during the entire trip home.

Upon seeing and hearing the other birds in the aviary Sydney tilted her head, but remained silent. I did not want to handle Sydney just then since she was already in a state of high anxiety. I put the travel cage inside her baby/quarantine cage and opened the door. That way Sydney could climb out on her own when she was ready. After about ten minutes she did and the travel cage was removed.

Sydney sat very still the first day. She drank a little water but didn't eat anything. Multiple dishes with different kinds of fresh and dry food were placed in Sydney's cage. These were all Squeaky's favorites, but Sydney didn't touch anything. By the second morning Sydney still wasn't eating. I was concerned and decided to try spoon feeding her. This had worked when Blue II (Chapter 7) was new and didn't eat.

I took some natural pellets and soaked them in hot water to make a mash. Then I held a spoonful a few inches (centimeters) away from Sydney. I didn't want her to be threatened by the spoon. I also wanted her to make the decision to come over and try the warm mash. After a few minutes Sydney walked slowly over to the spoon and took a small bite. Before long she devoured the entire spoonful and a second.

I continued to spoon feed Sydney for several days. At the same time I put a dish of natural pellets underneath where I spoon fed her. Gradually I put some of the mash on top of the pellets, so Sydney would learn to associate the two foods. By the end of the week Sydney was eating on her own, but only pellets. For several months Sydney only ate natural pellets. Sydney slowly began to try new items, including fresh fruits and vegetables. Then she discovered that pellets roll because they are round. Thereafter Sydney considered pellets to be a toy and she threw them on the floor so she could play with them.

Parrots have one of the longest life spans of any birds.[137] For their size and body weight parrots are much longer lived than comparable mammals.[137] Parrots, wild and indoor, are able to learn new things throughout their lives. New locations, new vocalizations, and new behaviors can be learned at any time. Negative behaviors can be ameliorated. However, parrots have long memories and will remember the actions and events or person that triggered a negative experience.[99]

As a result I decided not to actively interact with Sydney right away, because of her negative experience at the pet store. Instead of beginning immediately like I did with Squeaky, I decided to take it slow and let Sydney decide when she wanted to interact with me and to what degree she wanted to interact. Letting pet birds make choices about their daily activities is very important to their emotional and intellectual wellbeing; it also minimizes boredom and frustration.

At first I just left her cage door open for a few hours each day. Sydney learned to climb around the sides and sit on the top. Sydney decided when she came out and when she went back in. (See Photograph 9.8.)

Next I wanted Sydney to overcome her fear of hand held perches. When Sydney was sitting on top of her cage (and not in a confined space) I held a small perch steady a few inches (centimeters)

Photograph 9.8 Sydney as a Baby. Sydney, curious but cautious, explores the top of her baby/quarantine cage shortly after her arrival on September 12, 2007. She was 10 months old at the time. Red-bellied parrots are sexually dimorphic and male/female coloring patterns are discernable by the time a young bird fledges. © D. Herrmann

away from her. If and when Sydney wanted to interact she could hop on the perch. If not Sydney could avoid the perch by walking away from it or going inside her cage. After a few days Sydney was comfortable with the perch and I was able to take her for walks around the aviary, like I did with Uzziel and Squeaky.

By being in the right environment Sydney made a lot of progress, compared to her days at the pet store when she was terrified of people. Before long Sydney became a real snuggle bug and loved being held close and patted. She even climbed inside the sleeve of a sweatshirt or sweater on occasion.

At night when it was time to go to sleep Sydney climbed underneath a cotton rope toy. Her tail and wing tips pointed up, while her head pointed down. Eight years later Sydney still prefers to sleep in that position overnight.

Gradually Sydney became very inquisitive and explored her new environment with enthusiasm. Toys, dishes, papers, plants, new food items, and such were subjected to close scrutiny. Unlike Squeaky, Sydney does not study something new before checking it out. Instead Sydney walks right up to a new item. One of the first things Sydney explored was the gap between the cage and the cage bottom when the tray is removed. Sydney was intrigued that this opening appeared and disappeared whenever the paper was changed. At first Sydney just stuck her head through the opening for a look around. Later Sydney learned to climb out through this opening. (See Photograph 9.9.)

Photograph 9.9 Minding the Gap #2. Sydney, still a baby, explores the gap at the bottom of her baby/quarantine cage that is created when the tray is removed to change the paper. Letting Sydney decide when and whether or not to come out to play was a tremendous boost to her confidence and morale. The opportunity to make choices for herself helped Sydney overcome the traumatic experiences she had at the pet store. © D. Herrmann

Sydney quickly taught herself how to swim, without having watched any other birds. Sydney is as enthusiastic about swimming as Squeaky. Unlike other birds Sydney rarely went swimming in her water dish. She always went swimming in the designated swimming pool, no matter what color or shape it was or what location the pool was is. In contrast, some indoor birds get used to a particular style of swimming pool or having the swimming pool in a particular location and take a while to adjust to any changes.

Like Squeaky, Sydney loves to hang upside down, by one or both feet. Sydney will hang upside down from anything: a swing, a perch, a toy, a hoop, a trellis, and open door, and so forth. Sydney often hangs or tries to hang upside down with one foot attached to one item and the other foot attached to something else. This of course is more challenging than having both feet grasping the same item. Sometimes this works. Other times it doesn't and Sydney ends up doing the splits. (See Photograph 9.10.)

Photograph 9.10 Parrot Splits. Red-bellied parrots, like Meyer's parrots, are considered to be hanging parrots. They frequently climb and play while hanging upside down or sideways, often by holding onto one item with the left foot and another item with the right foot. Here Sydney explores a treat stick while hanging upside down. © D. Herrmann

Sydney entered her terrible two's abruptly around 18 months of age. Almost overnight Sydney went from a gentle snuggle bug to an aggressive dare devil. Unlike Squeaky, Sydney did not exit the terrible two's at the end of 6 months. Instead she had a prolonged terrible two's. At times it seemed like Sydney would never exit her terrible two's and progress to a normal adult behavior pattern. Possibly as a reaction to being mistreated at the pet store, Sydney belatedly overcompensated for the fear she felt then by lunging at and attacking everything and everyone. Sydney was 9 months old when she came home from the pet store. She spent 50% more time there than Squeaky and she was bounced from one store to another.

Sydney was intelligent and curious, but her actions began to take on an ornery streak. When she finished eating Sydney picked up her dishes and hurled them on the floor. Sydney played soccer with her grapes before eating them. The grapes were then cut in half so they wouldn't roll. In particular Sydney was a dare devil when she was out flying. She flew lap after lap at full speed, changing course and direction midair, almost crashing into something or another bird. And for some reason it took Sydney a long time to learn how to go back home after flying. In the beginning Sydney would hop on her training perch and let me take her back. Later she became too independent for that.

Sydney seemed to be oblivious or at least unaware that her actions sometimes annoyed other birds. For example Sydney frequently landed on top of Maku's aviary. (See Section 10.3.) Maku absolutely does not like it when another bird lands on top of her aviary; she doesn't even like budgerigars to land there. Whenever another bird lands on her aviary Maku lunges at them with her beak open in a threatening manner. Other birds get the message right away and don't come back; or, if they do they stay on the landing perch out of Maku's reach. Not Sydney. She watched Maku lunging at her, nipping at her feet, and just sat there. After a while Sydney left, but she came back two or three more times.

When a young bird is taken away from its parents and clutch-mates too early, then placed in a cage by itself at a pet store and mistreated, it is unrealistic to expect the bird to instantly develop good socialization skills. Most publications speak of socialization in terms of a solitary bird leaning how to interact correctly with one or more humans. From my observations over the past 30 years it is more important for an indoor bird to learn how to interact appropriately with other birds. Once a bird has mastered the social skills needed to interact appropriately with its peers, learning how to interact with humans is straight forward.

Sydney finally exited her terrible two's in March 2011. She is not a cuddly snuggle bug anymore, nor is she aggressive.

Sydney and Squeaky went through a slow gradual introduction process, in part because of Sydney's prolonged terrible two's. Initially their cages were on opposite sides of the aviary. Sydney and Squeaky could see and hear each other from a distance. Sydney's cage and furnishings were identical to Squeaky's. Several months later their cages were moved side by side. They could see and hear each other up close. Squeaky seemed curious about Sydney. But Sydney wasn't curious about Squeaky, at least not at first.

Squeaky and Sydney were let out to fly and play one at a time. Squeaky would immediately go over to Sydney's cage to visit before flying. Sydney only visited Squeaky's cage once in a while when she was out flying. After several months Squeaky and Sydney were let out to fly at the same time. The first few times did not go well. Both birds became aggressive toward each other if one of them tried to enter or land on the other's cage.

Until then Squeaky and Sydney had lived by themselves their entire lives. Neither of them understood the concepts of sharing or taking turns. Swings, perches, toys, and dishes were viewed as their private property, unlike wild Poicephalus parrots which are not territorial.

We went back to letting Squeaky and Sydney out to fly one at a time for several months. Everything was fine. Eventually we let them out to fly together again. Squeaky and Sydney cautiously went in and out of each other's cage. But when it was time to go home, they each went back to their cage. One day Sydney moved herself into Squeaky's cage and stayed put. Squeaky did not object. The door to Sydney's cage was left open so she could return if she wanted to. A month passed and Sydney did not return to her cage, so it was removed. Squeaky and Sydney have been roommates since.

Overall this move has been beneficial for both Squeaky and Sydney. Although, there have been some tiffs along the way. Like the first time Sydney decided to play with Squeaky's swing and her favorite bell toy, the one Squeaky has had since she arrived. Sydney has learned things from Squeaky and vice versa. Unlike Squeaky, Sydney has no inhibition about walking on or under her paper or on the carpeting. Squeaky watched this activity with interest. Then for the first time Squeaky took 3 steps on the paper in August 2010. However, Squeaky did not start visiting the paper regularly until December 2012. May 2015 Squeaky began to shred her paper for the first time. Squeaky never ate millet sprays until 2010 after watching Sydney do so; now they are a favorite treat.

Squeaky helped Sydney navigate her terrible two's and grow up by providing a model of acceptable behavior. Sydney had a lot to learn when she first moved in. For example, if Squeaky was eating an almond and Sydney wanted one, she snatched it out of Squeaky's beak rather than find her own. The first few times Sydney preened Squeaky's head, she over preened leaving a bald spot. At first Squeaky tolerated these youthful indiscretions; later on Squeaky popped Sydney whenever she acted up.

Squeaky and Sydney have formed a social play bond. They sleep side by side when it is nap time, but not overnight. When it is flying time Squeaky and Sydney fly around and play together. Squeaky and Sydney explore new things together and end up in strange places, like clinging to an air conditioning vent or smoke alarm, sitting on top of the thermostat, or holding onto a calendar. The smaller an item the more interesting it is on which to try to perch. Sydney and Squeaky explored a new rope ladder the first day it was hung in the aviary. Instead of sitting on the wooden perches like the toy designer intended, they both held onto the cotton ropes on the sides. Perhaps that is more challenging. (See Photographs 9.11 and 9.12.)

Photograph 9.11 Visiting the Rope Ladder. Sydney checks out the rope ladder shortly after it was hung in the aviary. Squeaky and Sydney investigated the rope ladder that first afternoon, often at the same time. This indicates a keen awareness of changes to their environment, as well as curiosity. Sydney and Squeaky hold onto the cotton ropes on the side, rather than sitting on the wooden perches as the toy designer intended. Perhaps that technique seems more natural or is more challenging. © D. Herrmann

Sydney has learned to share since she moved in, albeit with limitations. One water bottle is sufficient for both birds. But two separate dry dishes are needed. If a favorite fruit, vegetable, or treat is being served (like popcorn), two dishes are needed. If a food item is being served that they are both ho hum about, one dish is sufficient. Squeaky and Sydney do not have designated dry food, fruit/vegetable, or treat dishes. They both eat out of all four dishes. Squeaky and Sydney start by eating from different dishes. Then they walk around sampling all the other dishes. By the end of the meal they have traded places. This behavior may be due to curiosity about what the other dish contains. Or, this behavior may be a variation of Squeaky's safety practice of moving from place to place when she eats fruit.

Squeaky makes a variety of vocalizations and has taught her herself a series of beeps and whistles. In contrast Sydney is silent for the most part—a rare trait among parrots. Before they moved in together Sydney began mumbling quietly to herself. She only does this once in a while and only for a minute or less at a time. Instead of a parrot-like sound Sydney mumbles in English. Her first

(a) Cotton Rope Toy

Here we see a cotton rope toy before and after Squeaky and Sydney chewed it. Squeaky and Sydney prefer thick cotton rope toys that they can chew over a number of days. The cotton fibers are similar to the soft fibers of a baobab tree which they would chew in the wild. Sometimes Squeaky and Sydney chew a toy at the same time; other times only one bird participates. It took about three weeks to transform this toy. © D. Herrmann

(b) Ring Toss Toy

Here we see a ring toss toy before and after Sydney played with it in 2014. The toy designer intended that the orange ring would be placed on one of the blue pegs on either side of the toy. Instead, Sydney knotted up the chain and placed the orange ring on the middle of the toy—a more difficult task. Another time Sydney pushed the ring through the bars and outside the cage. Squeaky exhibited the same behaviors years before Sydney arrived. © D. Herrmann

Photograph 9.12 contd....

Photograph 9.12 contd.

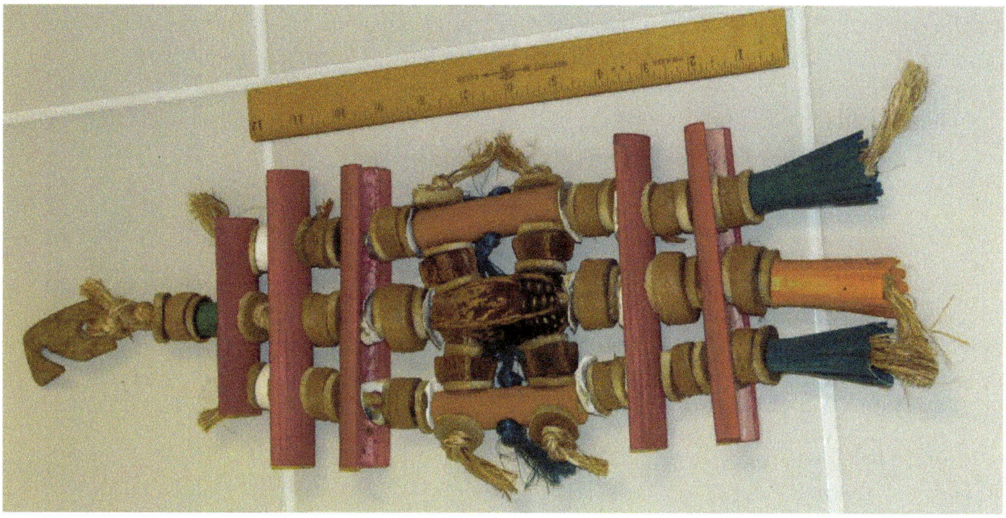

(c) Shredding Toy

Here we see a shredding toy before and after Squeaky and Sydney chewed it in 2014. Whether one or both birds were chewing on the toy, they followed a methodical process. First they chewed the soft coconut fiber disks. Next they chewed the hard center wood piece. Then they chewed the smaller hard wood pieces. Lastly, they chewed the cuttlebone discs and colorful fibers. In essence, Squeaky and Sydney sorted the pieces by type prior to chewing them. This process took 10 days. The fact that the toy was twice their size did not bother Squeaky or Sydney. They began to chew the toy immediately after it was placed in their aviary. © D. Herrmann

Photograph 9.12 Meyer's and Red-Bellied Parrot Toys: Before and After.

vocalization was "it's ok, it's ok"—the standard phrase used in the aviary to calm the birds after something scary. No attempt was made to teach her this phrase; she picked it up on her own. Other vocalizations include a random assortment of words and syllables. They almost sound like she is practicing. Recently Sydney began imitating some of Squeaky's vocalizations. Perhaps Red-bellied parrots only learn vocalizations from their parents and flock.

Sydney is still learning and growing emotionally, socially, and intellectually. Fortunately Sydney has a good teacher in Squeaky. Sydney celebrated her eighth hatch day December 7, 2014.

9.5 Discussion

Now let's see what these observations tell us about the intelligence, behavior, and individuality of wild and captive Meyer's parrots and Red-bellied parrots. It is important to remember that neither Meyer's parrots nor Red-bellied parrots can be considered domesticated at this time. Both parrots are not far removed from the wild. Neither has been bred in captivity in any numbers for any length of time. In particular let's examine:

- Captive behaviors that approximate wild behaviors
- Memory, reasoning, and problem-solving behaviors
- Common behaviors
- Unique behaviors
- Temperament
- Dietary preferences

Captive Behaviors that Approximate Wild Behaviors

To begin let's examine behaviors that were observed that approximate behaviors of wild Poicephalus parrots.

Meyer's parrots and Red-bellied parrots roost in tree holes.[60, 94, 202, 203] Shortly after her arrival Sydney began sleeping under a cotton rope toy. Sydney chewed the individual cotton ropes to lengthen and soften them. At night when it is time to sleep, Sydney arranges the ropes over her back. Then she goes to sleep, with her tail and wingtips pointing up and her head pointing down. Sydney does this year round; she is not sleeping under the cotton rope toy for warmth. Sydney has slept under four different cotton rope toys over the years; even two different toys within the same time frame. Thus, this behavior is not the result of Sydney bonding with a favorite toy. Sydney does not nap under the cotton rope toy during the day. She only sleeps there at night. Sydney's behavior approximates wild Red-bellied parrots roosting in tree holes for safety. Squeaky does not do this. In addition Sydney climbs inside a hollowed out kabob toy. Squeaky participates in chewing the kabob toy. But, she does not climb inside the toy. Only Sydney does that. (See Photographs 9.13 and 9.14.)

Wild Poicephalus parrots do not stay on the ground very long in any one place when eating. They tend to move around quickly to avoid attracting predators like Sparrow hawks and inland Osprey.[25] Small items like seeds can be consumed quickly in one bite. Large items like nuts or chunks of fruit have to be carried to a place of safety before they can be consumed. From the beginning whenever Squeaky was fed fruit she moved from place to place. It didn't matter if she was being hand fed or eating the fruit on her own. The type of fruit (apple, orange, banana, etc.) has no impact on her behavior. Squeaky also moves around when eating almonds. Both Squeaky and Sydney move from dish to dish and dish to perch when eating anything larger than a single beakful. They carry these items from place to place. Squeaky did this years before Sydney moved in; hence this behavior is not related to possessiveness. Squeaky and Sydney's practice of moving around when eating large items, mirrors wild Poicephalus parrots not staying on the ground too long in any one place when eating.

Photograph 9.13 A Safe Night's Sleep. Wild red-bellied parrots roost in tree holes for safety. Sydney mirrors that behavior by sleeping under a cotton rope toy. Sydney chews the individual cotton ropes to soften and lengthen them. Each night Sydney arranges the ropes across her back. Then she goes to sleep with her tail and wingtips pointing up and her head pointing down. Sydney has done this consistently since her arrival in 2007. © D. Herrmann

Like other birds, wild Poicephalus parrots are attuned to their environment and develop sophisticated predator detection and avoidance skills. Squeaky's fear of large moving objects is an extension of this behavior. A large box or piece of furniture that is carried into or through the aviary is viewed as a potential threat. Squeaky knows that it is a foreign object that doesn't belong in the aviary. The size also makes it intimidating. Predator avoidance skills tell Squeaky to flee the threatening object. She can't because she is confined in an aviary. So she flies rapidly from one side of her aviary to the other while making an alarm call. Sydney began exhibiting this behavior too, once she moved in with Squeaky.

However, if a certain procedure is followed neither Squeaky nor Sydney panic. First the item is placed at the entrance of the aviary. I touch the item several times and say "it's ok, it's ok". Then I pick up the item and say "coming through, coming through". This routine almost always works, indicating that hardwired behaviors can be overruled at will by reasoning.

Similarly Squeaky and Sydney are very sensitive to colors, particularly the color of outfits humans wear. Dark colors like black and navy seem to be associated with predators. Red seems to

Photograph 9.14 Exploring a Kabob Toy. Squeaky and Sydney enjoy chewing kabob toys. Sometimes they chew on a kabob toy at the same time. Sydney is the only one that climbs inside after the kabob toy has been hollowed out. This behavior mimics carving and climbing inside tree holes in the wild. Note the similarity between this play behavior and that of the budgerigars in Photograph 7.7. © D. Herrmann

be associated with a warning; perhaps it infers that an animal is injured and a predator is nearby. This behavior is also a reflection of wild Poicephalus parrots being attuned to their environment.

Wild Meyer's parrots and Red-bellied parrots are believed to establish lifetime monogamous pair bonds.[8] They generally travel as pairs or as an extended family of 4 to 5 birds.[8, 60, 94, 202] Squeaky was 6 months old when she arrived in the aviary. Sydney was 9 months old when she arrived in the aviary. Both birds exhibited an identical behavior shortly after they arrived, even though they arrived years apart. The first time each bird was let out to play it danced in a circle (the diameter of the circle was about 12 inches (30 cm)), while bowing its head. Each bird went around the circle twice going clockwise, then twice going counter-clockwise. Squeaky did this for several minutes the first morning she was here. Squeaky repeated this behavior on the second morning. Sydney repeated this identical behavior years later the second morning she was here. This was not a courtship behavior, since both birds were much too young to engage in a pair bond at the time. Instead, this behavior must have something to do with family and identifying oneself upon being reunited with the flock. Neither bird has engaged in this behavior since.

Wild juvenile Meyer's parrots and Red-bellied parrots are believed to grow up in crèches that are attended to by alternating adults.[8] In such an environment the juveniles would go through a tussling phase as they begin to assert their independence and individuality, while at the same time learn how to live in a flock environment. Both Squeaky and Sydney went through a similar phase during their terrible two's. Squeaky was living by herself when she went through her terrible two's. There weren't any other juvenile parrots around, so she directed her assertiveness and independence toward her human counterparts. When Sydney first entered her terrible two's she was living alone. Sydney also directed her assertiveness and independence toward her human counterparts. Sydney was still in her terrible two's when she moved in with Squeaky. Squeaky was long past her own terrible two's by then. Squeaky provided a mature role model of acceptable behavior, which helped Sydney to grow up.

Table 9.2 summarizes the behaviors that were observed that approximate those of wild Meyer's parrots and Red-bellied parrots.

Table 9.2 Behaviors that Approximate Those of Wild Poicephalus Parrots.

Wild Behavior	Equivalent Indoor Behavior
• Roosting in tree holes	• Sleeping under cotton rope toys
• Not staying on the ground too long in any one place when eating	• Moving around when eating fruit and nuts • Moving from dish to dish and dish to perch during a meal
• Predator detection and avoidance	• Fear of large moving objects • Sensitivity to colors
• Lifetime monogamous pair bonds, traveling in pairs or as a nuclear family	• Doing a circular dance and bowing upon arriving in the aviary
• Juveniles growing up in crèches	• Going through the terrible two's

Memory, Reasoning, and Problem Solving Behaviors

Throughout their lives Squeaky and Sydney have engaged in behaviors that indicate memory, reasoning, and problem solving skills. The majority of these behaviors do not correlate to behaviors that could be expected to occur in the wild. As a result they cannot be attributed to genetic intelligence. Some of these behaviors have been exhibited by one bird; others have been exhibited by both birds. Squeaky has engaged in more of these behaviors than Sydney. This is to be expected since Squeaky is older than Sydney. A detailed discussion of nine of these behaviors follows, including the significance of each behavior and the frequency with which it was observed.

Squeaky and Sydney pressure test sunflower seeds before they open them. They squeeze a seed with their beak, applying just enough pressure to determine whether or not there is a seed inside. If not, they throw the seed on the floor unopened. This way Squeaky and Sydney do not waste the time or energy to open empty seeds. This practice indicates reasoning and memory. First, Squeaky and Sydney indicate that they are aware that some sunflower seeds do not contain seeds and are empty.

Second, Squeaky and Sydney created a method to determine if there is a seed inside without opening each one. Third, Squeaky and Sydney have demonstrated that they understand that it is more efficient to test a seed than to open it. Squeaky engaged in this behavior early in life, shortly after her arrival. There is no one she could have learned it from. Sydney engaged in this behavior from a young age also, while she was still living by herself. It is possible, but unlikely, that Sydney learned this behavior from Squeaky. None of the other parrots in the aviary engage in this behavior.

Squeaky punctures a dry kernel of corn, then places it in a water dish. Squeaky retrieves the kernel of corn from the water dish about 10 minutes later. This practice allows the interior of the kernel to become soft and thus more edible. Squeaky began doing this shortly after her arrival when she was quite young. There is no one she could have learned this behavior from. Sydney does not do this. This behavior indicates reasoning and an understanding of the concept of time. First Squeaky indicates that she knows the difference between the hard yellow exterior and the white powdery interior of a dry kernel of corn. (Parrots suck the white milky fluid from a kernel of fresh sweet corn; they do not consume the outer yellow skin.) Second, Squeaky shows an understanding of time. Simply dunking a punctured kernel won't soften the interior. An interval of several minutes is needed to do that. Squeaky comes back later when the corn is soft.

Both Squeaky and Sydney dodge falling objects by hanging upside down. According to the "fight or flight" theory bird behavior is hardwired and Squeaky and Sydney should fly away to avoid falling objects since they can't fight them. However, neither Squeaky nor Sydney do this—indicating again that hardwired behaviors can be overruled at will by reasoning.

Squeaky and Sydney loosen their grip on their perch and let gravity pull them upside down, when an object is falling toward them from above. Once inverted, they are parallel to the falling object and avoid injury. Squeaky first demonstrated this behavior when she was 5 years old. There is no one she could have learned it from. Sydney first demonstrated this behavior when she was 3 years old, after she had moved in with Squeaky. None of the other parrots in the aviary do this.

This behavior demonstrates rapid reasoning and problem solving skills and the ability to act on them quickly. First, Squeaky and Sydney detect that an object is falling toward them and the rate at which it is falling. Second, Squeaky and Sydney determine that they can avoid the falling object by inverting themselves parallel to the path of the falling object. Third, Squeaky and Sydney understand that this maneuver is faster than trying to fly away; thus it has a greater chance of success. This maneuver also consumes less energy.

Early in life Squeaky figured out that she was seeing her reflection in a mirror. This occurred the first time Squeaky saw herself in a mirror. Squeaky demonstrated this ability regardless of which mirror she was held in front of. Squeaky discerned that she was looking at a reflection of herself because she noticed that the reflection moved the same direction at the same time she did. The ability to recognize a reflection of herself also indicates that Squeaky has an awareness of self.[189, 190]

Since she was a juvenile Squeaky has preferred to dismantle toys, especially puzzle toys, rather than play with them. Squeaky instantly removes the clappers from bells. The cylindrical blue metal bell is the one exception. Squeaky stares at a toy for a while, then dismantles it piece by piece. Squeaky does not break the toy apart—she reverse engineers it. To do so Squeaky must develop an understanding of how the various parts are connected and fit together. Some toy parts are common; some parts are unique to each toy. Mostly Squeaky does this with her beak and tongue; hence, a fair amount of dexterity is involved as well. The ability to reverse engineer toys demonstrates problem solving skills and an understanding of spatial relationships and rudimentary mechanical engineering principles. It is unlikely that wild Meyer's parrots spend much time dismantling toys.

When she was 5 years old Squeaky began imitating the sounds made by other birds in the aviary. By itself that would not be noteworthy. What made this behavior significant was the fact that Squeaky made a point of imitating the sound a particular bird made when I stood in front of its cage. In other words, Squeaky not only could imitate the sound a bird made, she associated that sound with the location of its aviary. Squeaky understood which birds lived in which aviaries and what sound each made. None of them are birds a Meyer's parrot would encounter in the wild.

Squeaky was never taught the English words for cardinal numbers. However, Squeaky acknowledges the concept of quantities in her beeps and whistles. When Squeaky hears an electro-mechanical device beep, she responds with the exact number of beeps. Squeaky doesn't always answer when an electro-mechanical device beeps; she may be busy doing something else. But, if Squeaky responds, she matches the number of beeps. Likewise, if I make a beeping sound or whistle, Squeaky matches the quantity of sounds. This practice demonstrates an understanding of quantities and the ability to count. Squeaky has matched the number of beeps up to seven. (She has not been tested at higher numbers.) Squeaky was not taught to do this; she learned it on her own. Repeating the same call (or beep) over and over is not part of the vocalization pattern of wild Meyer's parrots.

Squeaky's understanding of spatial relationships is another example of mathematical skills. Squeaky comprehends the size and shape of an item, the amount of space an item occupies, and the fact that two items cannot occupy the same space. When given an item that is too wide to pull straight through the aviary bars, Squeaky immediately adjusts her algorithm and pulls the item through at an angle. If the item still doesn't fit, Squeaky bites the item in half, thereby making it small enough to fit through the bars. These behaviors exhibit applied mathematical reasoning on Squeaky's part.

Most days Squeaky makes a special "beep-beep, beep-beep" sound to announce sunset. Squeaky does not make this sound mid-morning or mid-afternoon when it suddenly becomes cloudy. Squeaky does not make this sound when the sun lamps are turned on before sunset either. Squeaky has announced sunset since 2002, regardless of where she was living or which direction her cage was facing. In doing so Squeaky has demonstrated that she is in tune with the natural day/night cycle and understands the importance of sunset to her indoor flock, of which she is the senior member.

Table 9.3 summarizes the behaviors that were observed that indicate memory, reasoning, and problem solving skills. Because of their keen reasoning and problem solving skills Squeaky and Sydney need a steady stream of new experiences and challenging toys.

Table 9.3 Behaviors that Indicate Memory, Reasoning, and Problem Solving Skills.

Behavior	Squeaky	Sydney
• Pressure testing sunflower seeds before opening them	x	x
• Puncturing dry kernels of corn, then placing them in water to soften	x	
• Dodging falling objects by hanging upside down	x	x
• Figuring out mirrors	x	
• Dismantling toys, knotting chain on ring toss toy	x	x
• Associating the sound of a bird with the location of its aviary	x	
• Matching the number of beeps an item makes	x	
• Adjusting the angle to pull wide items through the cage bars	x	
• Announcing sunset	x	

Common Behaviors

Squeaky and Sydney exhibited several common behaviors, even while they were living in separate aviaries. These 12 common behaviors can be grouped into 5 categories:

- Play activities (3)
- Socialization (2)

- Growth and development (1)
- Memory, reasoning, and problem solving (5)
- Diet (1)

Both Sydney and Squeaky have a tendency to hang upside down when climbing or playing with a toy. Squeaky did this long before Sydney arrived in the aviary. Sydney started hanging upside down before she moved in with Squeaky or even paid any attention to her. Since Meyer's parrots and Red-bellied parrots are considered hanging parrots, this is a natural behavior. The short stocky body and short square tail of Poicephalus parrots is conducive to hanging upside down. (See Photograph 9.15.)

Photograph 9.15 Upside Down. Both Squeaky and Sydney started to hang upside down at an early age. (Right) Squeaky hangs upside down from a trellis in 2001 while she is a juvenile. (Left) Sydney hangs upside down from her front door in an identical posture six years later in 2007 when she is a juvenile. This behavior is common to Meyer's parrots and red-bellied parrots. © D. Herrmann

Squeaky and Sydney were given a wooden perch that had a spinning toy on one end. They were each given the perch at the same time when they were living in separate aviaries. The spinning end of the perch is filled with flexible paper prongs the size of a toothpick. The toy is designed with the intent that parrots will chew and bend the paper prongs, while spinning the end around slowly to pick out a favorite color. The other parrots in the aviary who received one of these perches played with it as intended. Neither Squeaky nor Sydney did so. Instead, they chewed up the block that held the spinning part on the perch. Then they threw the spinning part on the floor. Both Squeaky and Sydney did this within a matter of days after the new perch was introduced. It was almost as if the spinning part annoyed them or was in their way. This behavior is a variation of Squeaky dismantling puzzle toys.

When Squeaky and Sydney are out flying they play together and go exploring together. This involves challenging themselves to land on small narrow objects. They routinely ignore large items that are easy to land on. Squeaky and Sydney have landed on (or attempted to land on) air conditioning vents, smoke alarms, thermostats, calendars, and such. Often one bird will land on the small item; then the second bird will do likewise. Since the item is small it barely holds one bird; usually some shuffling around is required to accommodate the second bird. This common behavior seems to be some sort of game or challenge. (See Photograph 9.16.)

Photograph 9.16 A Perch is a Perch #2. From top left Sydney lands on an air duct, a thermostat, a smoke alarm, and the Oasis Sanctuary 2010 calendar. Squeaky and Sydney challenge themselves by trying to land on small narrow objects when they are out flying. One bird will land on an item, then the second. This play behavior seems to be a sort of game or challenge. Both birds did this as juveniles and continue to do so as adults. It is unknown if there is a parallel behavior in the wild. © D. Herrmann

Both Squeaky and Sydney did a circular dance and bowed upon arriving in the aviary. Another common socialization behavior relates to a neighbor of theirs. A grey Indian ringneck parakeet, Dusty (Section 10.6) lives in the aviary next to Squeaky and Sydney. Her grey feathers are the same color as those of Squeaky and Sydney. Of course her beak is a different color and her tail is a different shape. In spite of this, Squeaky and Sydney have adopted the Indian ringneck parakeet as one of their own. She can walk around on Squeaky and Sydney's cage and sit on their landing perch. Sydney frequently watches the grey Indian ringneck parakeet and imitates her behavior, like walking under her paper or playing with a particular toy. Squeaky doesn't do this, but she is amicable toward the grey Indian ringneck parakeet. The other Indian ringneck parakeets in the aviary are blue or green. They receive a chilly to hostile reception if they attempt to land on Squeaky and Sydney's aviary.

Both Squeaky and Sydney went through a terrible two's growth and development phase.

Common behaviors related to memory, reasoning, and problem solving skills were discussed earlier.

Another common behavior relates to diet. This behavior has been consistent whether Squeaky and Sydney were living together or in separate aviaries. Neither Squeaky nor Sydney chew cuttlebones or mineral blocks, regardless of the size, shape, or color. In 14 years Squeaky has never even sampled a cuttlebone or mineral block. Attempts were made to entice her by moving the cuttlebone and mineral block to a variety of different locations in her aviary, but to no avail. The closest Squeaky has come to sampling a cuttlebone is to remove it from its holder and drop it on the floor. She totally ignores mineral blocks. Sydney is equally as unenthusiastic about cuttlebones and mineral blocks; she has touched neither in 7.5 years.

Table 9.4 summarizes the behaviors observed that are common to both the Meyer's parrot and the Red-bellied parrot.

Table 9.4 Behaviors That Were Common to Both the Meyer's Parrot and the Red-bellied Parrot.

Type of Behavior	Behaviors Observed
• Play Activities	• Hanging upside down to climb and play • Chewing up, then throwing away spinning perch • Attempting to land on small narrow objects when out flying
• Socialization	• Circular dance and bowing upon arrival in the aviary • Accepting a grey Indian ringneck parakeet as one of their own
• Growth and Development	• Terrible two's developmental phase
• Memory, Reasoning, and Problem Solving	• Fear of large moving objects • Color sensitivity • Moving around while eating fruit and nuts • Dodging a falling object by inverting themselves • Pressure testing sunflower seeds before opening them
• Diet	• Not chewing on a cuttlebone or mineral block

Unique Behaviors

Squeaky has exhibited several unique behaviors, both before and after Sydney moved in. These 16 unique behaviors can be grouped into 4 categories:

- Play activities (1)
- Socialization (2)
- Memory, reasoning, and problem solving skills (7)
- Vocalization (6)

When she was young and living by herself Squeaky enjoyed challenging herself by flying full speed right toward a plant stand filled with cacti, only to veer off at the last second at a right angle away from the cacti. Squeaky did this 5 or 6 times a night, purposely testing and perfecting her precision flying skills. In doing so, Squeaky challenged herself mentally and physically. Sydney has not engaged in a similar behavior to date.

Squeaky formed an interspecies social play bond with two budgerigars, first Yellow Birdie, then Daffodil. Each of these social play bonds lasted several months. Sydney has not formed any interspecies social play bonds.

Squeaky's seven unique behaviors related to memory, reasoning, and problem solving skills were discussed earlier.

Squeaky has also exhibited unique behaviors in regard to vocalization. Wild Meyer's parrots have been described as making a high pitched "che-che-che" sound.[60] Squeaky does not make that sound. Instead, Squeaky makes 6 other types of sounds. The first is a "d-hee" sound that is repeated 4 to 6 times, often after receipt of a favorite food or treat. The second is a higher pitched "ps-tee" sound that is repeated 3 to 5 times. The call announcing sunset is similar to this sound. The third is a medium pitched "do-do, do-do" sound. Squeaky first made this sound during her circular dance upon arrival. Now Squeaky makes this sound infrequently. It appears to be a variation of a contact call. Squeaky makes this sound when she can see me and is interested in interacting. Fourth, Squeaky makes a "yyyo" sound. This brief sound is heard when Squeaky is very excited about something. Squeaky also makes a series of beeps and whistles that she has taught herself. If these vocalizations are played back to Squeaky, she will repeat them and bob her head up and down. It is unclear how many of these beeps and whistles occur in the wild. However, other reports have been made of indoor Meyer's parrots making beeps and whistles.[8] Meyer's parrot vocalizations can be heard on Disk 2 Track 2.

Table 9.5 summarizes the behaviors observed that were unique to the Meyer's parrot.

Sydney has exhibited several unique behaviors, before and after she moved in with Squeaky. These 7 unique behaviors can be grouped into 3 categories:

- Play activities (4)
- Socialization (2)
- Vocalization (1)

In contrast to Squeaky perfecting her aerobatic prowess, Sydney engages in dare devil flight patterns. Sydney flies full speed without really knowing where she is going or where she intends

Table 9.5 Behaviors that Were Unique to the Meyer's Parrot: Squeaky.

Type of Behavior	**Behaviors Observed**
• Play Activities	• Challenging herself and perfecting her flying skills by flying directly toward cacti
• Socialization	• Forming a social play bond with a budgerigar: Yellow Birdie • Forming a social play bond with a budgerigar: Daffodil
• Memory, Reasoning, and Problem Solving	• Puncturing dry kernels of corn, then placing them into water to soften • Figuring out mirrors • Dismantling toys • Associating the sound of a bird with the location of its aviary • Matching the number of beeps an item makes • Adjusting the angle to pull wide items through the aviary bars • Announcing sunset
• Vocalization	• "d-hee" • "ps-tee" • "do-do, do-do" • "yyyo" • multiple beeps • multiple whistles

Note: Meyer's parrot vocalizations can be heard on Disk 2 Track 2.

to land. Everything is a last minute decision. Needless to say, Sydney has had several close calls. Sometimes Sydney flies directly below the ceiling. Other times she cruises low between a table and a chair. Sydney is a strong fast flyer, but reckless.

Sydney sleeps under a cotton rope toy overnight. Sydney climbs inside a hollowed out kabob toy. Sydney walks on papers, floors, and carpeting without hesitation. Sydney even shreds her paper and walks underneath it.

Sydney accepts the grey Indian ringneck parakeet as one of her own and frequently imitates her play activities. However, Sydney has not formed any interspecies social play bonds. Sydney often seems unaware or oblivious to other birds and what they are trying to communicate when she is out flying. For example, Maku lunges at Sydney, with her beak open in a threatening manner, and nips at Sydney's feet whenever she lands on her aviary. Maku is larger than Sydney. All the other birds in the aviary understand her message and leave pronto. Not Sydney, she appears to be oblivious to it all.

Then again Sydney may be pursuing a different thought process. When Sydney was a juvenile the aviary next to her was occupied by a juvenile green Indian ringneck parakeet. They were both about the same age. The two young birds often mimicked each other's actions, played, and went exploring together. Later their two aviaries were moved to opposite sides of the room. Snapdragon (Section 10.4), the young green Indian ringneck parakeet, has since grown a black and rose neck ring since he is a male. Maku (Section 10.3) is a female green Indian ringneck parakeet; hence she does not have a neck ring. Perhaps Sydney thinks Maku is Snapdragon and expects her to reciprocate the way the two of them played together when they were both young. That might explain why Sydney doesn't retreat from Maku's aggressive behavior. Yes, parrots do have long memories.[99]

It is likely that Red-bellied parrots learn their vocalizations rather than inherit them. Wild Red-bellied parrots have been described as making a high pitched "screeeak" sound.[94] Sydney does not make that sound. In fact Sydney does not make any parrot-like sounds. Instead Sydney is very quiet. When Sydney vocalizes it sounds like she is mumbling English words and syllables quietly to herself. Sydney has picked up the phrase "it's ok", the standard phrase used to calm the birds after something scary. Only recently Sydney began trying to imitate Squeaky's "yyyo" sound and some of her beeps and whistles.

Table 9.6 summarizes the behaviors observed that were unique to the Red-bellied parrot.

Table 9.6 Behaviors that Were Unique to the Red-bellied Parrot: Sydney.

Type of Behavior	Behaviors Observed
• Play Activities	• Dare devil flying patterns • Sleeping under cotton rope toy overnight • Walking on papers, floors, and carpeting; Shredding paper, walking under paper • Climbs inside hollowed out kabob toy
• Socialization	• Being unaware or oblivious to what other non-Poicephalus birds are trying to communicate • Remembering a green Indian ringneck parakeet she lived next to as a baby; mistaking Maku for that bird years later
• Vocalization	• Mumbling English words and syllables

Temperament

Squeaky has been observed for 14 years. Sydney has been observed for 7.5 years. Both birds are relatively young considering their average life span is 20 to 25 years. Neither bird has had to learn to deal with issues associated with aging, the loss of a mate, an illness, or being rehomed. However, each bird has exhibited a variety of emotional states during this time.

Sydney was not in good shape emotionally when she arrived in the aviary. The traumatic experiences of her first 9 months were visible. As a result Sydney has exhibited some emotional states that Squeaky has not. For example, for the first few days Sydney displayed symptoms of being timid, unsure, anxious, fearful, and stressed out. The fact that Sydney did not eat at all the first day and had to be spoon fed like a nestling the second day (even though she was 9 months old), are evidence of her emotional state. In time and by being in the right environment Sydney was able to overcome this.

During their terrible two's Squeaky and Sydney's temperaments did a flip flop. Both became aggressive and impatient as they began to assert their independence and individuality. Sydney took longer to exit her terrible two's than Squeaky. However, both eventually returned to their earlier calm pleasant natures, a more mature version of themselves.

Both Squeaky and Sydney spent several years living by themselves. While Squeaky and Sydney enjoy flying and playing together outside their aviaries, an adjustment period was needed when they finally moved in together. Both birds had to learn the concepts of sharing and taking turns. Initially both birds displayed varying degrees of possessiveness. Eventually they worked everything out. Once in a while Squeaky and Sydney still have a tiff, but that is to be expected.

Other than these three scenarios, Squeaky and Sydney have exhibited a normal range of emotional states, as shown in Table 9.7.

Table 9.7 Spectrum and Prevalence of Emotional States Observed in an Indoor Meyer's Parrot and a Red-bellied Parrot.

	Squeaky	**Sydney**	**Neither**
I. Positive Emotions			
excited	x	x	
happy	x	x	
sad			x
confident/content	x	x	
timid/unsure[1]		x	
lonely			x
curious	x	x	
puzzled		x	
playful/eager	x	x	
anxious/fearful[1]		x	
aggressive/impatient[2]	x	x	
gentle/patient	x	x	
II. Negative Emotions			
anger			x
hatred			x
jealousy			x
grumpy			x
overwhelmed			x
stressed out[1]		x	
possessiveness[3]	x	x	
self-pity			x
depression			x
boredom			x

Notes:

1—Only immediately after arrival from pet store.
2—Only during "terrible two's".
3—Only briefly after Squeaky and Sydney moved in together.

Dietary Preferences

Since Squeaky and Sydney moved in together their dietary preferences have begun to converge. At the same time the variety of foods each bird eats has expanded. Some foods Squeaky never touched during the time she lived alone, she now eats because she saw Sydney eating them, and vice versa. The fact that one bird eats something is enough to pique the curiosity of the other into trying it at least once. For example Squeaky never touched millet sprays until 2010. Likewise Sydney used to drop almonds on the floor.

Some food preferences have remained static. Others have changed over the years. Squeaky readily ate raisins when she was young. Squeaky hasn't touched raisins since she entered her terrible two's. Also, Squeaky readily ate broccoli when she was young. She went several years without eating broccoli at all. Now Squeaky eats broccoli once in a while. Sydney ate natural pellets exclusively for several months when she first arrived. She did not eat pellets again until she was 5 years old.

Tables 9.8 and 9.9 list Squeaky's and Sydney's fruit, vegetable, grain, nut, seed, and legume preferences. Neither Squeaky nor Sydney consumes many of the standard ingredients of commercial

Table 9.8 Individual Meyer's Parrot and Red-bellied Parrot Fruit and Vegetable Preferences.

	Squeaky	Sydney	Neither
I. Fruits			
apple	x	x	
banana	x	x	
blueberries			x
cantaloupe	x		
cherries	x		
grapes		x	
kiwi berry			x
mango—dried mango—fresh	x	x	 x
orange	x	x	
papaya—dried papaya—fresh	x x	x x	
peaches	x	x	
raisins			x
strawberries			x
II. Vegetables			
baby spinach			x
broccoli	x		
carrots	x	x	
cherry tomato			x
corn—yellow sweet corn—white sweet	x	x	 x
dandelion leaves			x
green peas		x	
kale			x
red peppers—dried red peppers—fresh	x	x x	
squash	x	x	

Table 9.9 Individual Meyer's Parrot and Red-bellied Parrot Grain, Nut, Seed, and Legume Preferences.

	Squeaky	**Sydney**	**Neither**
I. Whole Grains and Legumes (cooked)			
barley	x	x	
buckwheat			x
lentils			x
long grain brown rice	x	x	
millet			x
quinoa	x		
rye			x
wheat, spelt, kamut			x
II. Tree Nuts (raw, shelled)			
almond	x	x	
Brazil nut			x
cashew	x		
filbert			x
pecan			x
pine nut			x
walnut	x		
III. Seeds and Legumes (raw)			
dried corn	x	x	
dried peas and beans			x
millet seeds (loose)			x
millet spray	x	x	
milo			x
oats			x
peanuts		x	
pumpkin seeds			x
safflower	x	x	
striped sunflower	x	x	

parrot seed mixes (like milo, millet, pumpkin seeds, and dried beans and peas). A check does not imply a food is a favorite. Rather, a check simply means Squeaky and/or Sydney eat these foods today. Most likely their food preferences will continue to evolve throughout their lives.

In summary, two poicephalus parrots were observed over a 14 year period. Both birds were female and acquired from a pet store before they were 1 year old. One was a Meyer's parrot (Squeaky); the other (Sydney) was a Red-bellied parrot. The two birds experienced an identical living environment here. Even so, Squeaky and Sydney are distinct individuals with different temperaments, vocalizations, daily activities, socialization patterns, toy preferences, and dietary preferences. Their 6 year age difference might explain some, but certainly not all of these differences. Simply put Squeaky and Sydney are unique individuals. Individual choices, preferences, and behaviors indicate intelligence and self-awareness.

10
Indian Ringneck Parakeet
Psittacula krameri manillensis

10.1 Background

Ringneck parakeets are also known as rose-ringed parakeets, green parakeets, and long-tailed parakeets.[41, 94] Ringneck parakeets were named by Scopoli in 1769 in honor of Austrian naturalist Wilhelm Heinrich Kramer.[203] Even so, some publications from the late 1800's and early 1900's refer to ringneck parakeets as *Palaeronis torquata* or "Little Alexandrine" instead of *Psittacula krameri*.[34, 81] Ringneck parakeets look like a diminutive version of an Alexandrine parakeet (*Psittacula eupatoria*). They are 17 inches (43 cm) in length versus 23 inches (58 cm) and lack the purple shoulder patches.[41, 202] However, DNA analysis indicates that ringneck parakeets are more closely related to the echo parakeets (*Psittacula eches*) of Mauritius Island.[203]

Little is published about ringneck parakeets even though humans have been aware of them for centuries. Ringneck parakeets have been kept as pets for a long time, even though they are somewhat rare today. The ancient Greeks purportedly kept Indian ringneck parakeets (*Psittacula krameri manillensis*) as pets, while the ancient Romans purportedly kept African ringneck parakeets (*Psittacula krameri krameri*) as pets.[84, 203] Archimedes is believed to be the first to document in 200 B.C. Indian ringneck parakeets being kept as pets.[84]

Reports from the late 1800's and early 1900's indicate that captive breeding was taking place in Europe at that time, but the dates cited don't always agree with contemporary publications. According to publications from 1891 and 1908:[34, 81]

- 1862 the first record of captive breeding of ringneck parakeets appeared on the London Zoological Society list
- 1889 the first yellow "aberration" was reported in England
- 1901 two ringneck parakeets hatched in Germany

In contrast, contemporary publications indicate that the first successful captive breeding of ringneck parakeets occurred in 1855 in Norfolk, then later in Essex and North Hampton in the 1930's.[58]

Wild ringneck parakeets are green. Several color mutations, like white, yellow, and blue, are available in the pet industry. The four subspecies of ringnecks are frequently mixed in captive breeding situations. That explains, in part, the number of hybrid mutations seen today.[194]

Ringneck parakeets have the most widespread distribution, spanning two continents, of any family of parrots.[41, 84] In 1908 ringneck parakeets were described as the most common and familiar birds in India.[34] The same publication made the observation that:[34]

> "At times it is imported in such numbers that the market becomes glutted, and specimens can be purchased at absurd prices ... as low as 5 shillings per bird."

Today the native south Asian populations of wild Indian ringneck parakeets are declining drastically, especially on the Indian subcontinent, due to trapping for the pet industry.[203] Crop destruction is frequently misused as a convenient excuse to justify trapping for the pet industry. However, recent studies of introduced populations of Indian ringneck parakeets have failed to uncover evidence of crop destruction.[58]

This situation needs to be monitored closely. The Indian ringneck parakeet's first cousin, the echo parakeet of Mauritius island, was described in 1908 as being "shy" and "gradually becoming extinct".[34] By the late 1980's only 8 to 12 birds could be accounted for.[93] The echo parakeet's range and habitat are severely limited, since it is native to a small island. Thanks to 30 years of intensive proactive conservation efforts by the Mauritian Wildlife Foundation, the Mauritius Government Conservation Unit, the World Parrot Trust, and others the echo parakeet did not become extinct.[93, 144] Instead the worldwide echo parakeet population reached 100 in 2001, 300 in 2006, and 500 in 2010.[93, 144]

10.2 Characteristics

Indian ringneck parakeets share several physical and behavioral characteristics. Let's examine these characteristics as they apply to wild and captive Indian ringneck parakeets, such as their:

- Range and habitat
- Diet
- Physical characteristics
- Vocalizations
- Socialization
- Family life

Range and Habitat

Wild ringneck parakeets are found in southern Asia (Pakistan, India, Nepal, Bangladesh, Afghanistan, Sri Lanka, and Myanmar) and a band across the middle of Africa from Senegal in the west to Somalia in the east.[41, 60, 84, 94, 202, 203] The ringneck parakeets' range in Africa overlaps with that of the Meyer's parrot (*Poicephalus meyeri*) discussed in Chapter 9.

In addition to the natural range of wild ringneck parakeets, several populations have been established through accidental and intentional releases. For example, populations have been established (i.e., population stable or increasing for 10 years) in the following countries:[58, 60, 94, 125, 147, 203]

- Australia
- Belgium (1987)
- Germany
- Hong Kong (1903)
- Japan
- Macao
- Netherlands (1968)
- Singapore
- Venezuela (1984)

- U.K. (1969)
- U.S.
  - California (1984)
  - Florida (1974)
  - Hawaii (1981), Kauai (1981), Oahu (1933), Maui (1990)

Unlike many parrots who are threatened by habitat destruction and human encroachment, ringneck parakeets have been able to successfully adjust to disturbed habitats, like farmland and urbanization.[202] Introduced ringneck parakeets have been found at altitudes up to 2,000 feet (609 m).[147] Most introduced ringneck parakeets that have established populations are believed to be from the *Psittacula krameri manillensis* subspecies.[203]

Ringneck parakeets live in a variety of different habitats in Asia and Africa. They can be found in lightly wooded lowland plains, lush vegetation, deciduous forests, savannah and scrub, up to 6,600 feet (2,011 m) in the Himalayas, even parks and gardens in urban areas.[41, 58, 60, 94, 202] Indian ringneck parakeets do not migrate. Their local movements are tied to the seasonal availability of food.[60, 94]

Diet

Wild ringneck parakeets eat a varied vegetarian diet composed of foods that ripen at different times, such as nuts, seeds, grains, and fruit.[41, 60, 94] Popular entrees include: maize, millet, milo, wheat, dates, figs, mango, guava, citrus, grapes, berries, sunflower seeds, mustard seeds, safflower seeds, almonds, flowers, buds, and nectar.[41, 58, 60, 94, 202] The fruit of baobab trees—the preferred place for Meyer's parrots to nest—is also a favorite.[58] Wild ringneck parakeets will eat leaves and sprouts if fruit is scarce.[75] Undoubtedly this varied seasonal diet has contributed to the success of the ringneck parakeets' survival.

Physical Characteristics

The physical characteristics of normal green wild Indian ringneck parakeets are discussed below. Differences between wild and captive Indian ringneck parakeets are noted here and in Sections 10.3 through 10.6.

Indian ringneck parakeets are medium sized parrots ranging from 16 to 17 inches (40.6 to 43.1 cm) in length, including a long narrow tail that is about 9 inches (22.8 cm) long.[41, 167, 194, 202, 203] Indian ringneck parakeets have a long slender body. They weigh 4.1 to 5.0 ounces (114.8 to 140 gr) and have a wingspan of 19 to 20 inches (48 to 50 cm).[41, 194, 203] Males and females are the same size.[203] (See Photograph 10.1.)

Indian ringneck parakeets are a medium green, with slight yellow highlights underneath.[34, 41, 94] Several sources indicated that Indian ringneck parakeets have pale yellow eyes. However, none of the birds discussed in this chapter have pale yellow eyes. Instead, the pupil is black and the iris consists of a narrow light blue band surrounded by white. The skin surrounding the eye is coral, with a narrow outer white band. Immature Indian ringneck parakeets have grey eyes.[192]

Indian ringneck parakeets have a rose to medium red colored beak. Some sources indicated that male ringneck parakeets have darker beaks than females. That is not the case with the three females discussed in this chapter. Other sources indicated that the bottom part of the beak or mandible is black. Three of the Indian ringneck parakeets discussed in this chapter have a black lower mandible. The fourth bird's upper and lower mandibles are the same rose color.

Indian ringneck parakeets have zygodactyl feet, like most parrots. The outside toes are about 0.75 inch (1.9 cm) long each. The inside toes are about 0.5 inch (1.27 cm) long each. Several sources indicated that the legs and feet are grey. However, all of the Indian ringneck parakeets discussed in this chapter have flesh colored legs and feet.

Photograph 10.1 Feather Comparison. Pictured left to right are: a zebra finch tail feather, a budgerigar tail feather, an Indian ringneck parakeet tail feather, and a mitred conure flight feather. Note the long narrow Indian ringneck parakeet tail feather. © D. Herrmann

Male ringneck parakeets have a distinctive neck ring or collar; hence the species common name. The ring is 0.125 inch (0.31 cm) wide and primarily black with slight hues of rose and light metallic blue, especially on the back of the neck. The rose and blue are most noticeable in direct sunlight. In contrast, female ringneck parakeets have a light green on green shadow effect neck ring.

The scientific sources indicated that male Indian ringneck parakeets do not acquire their neck rings until they are 3 years old and sexually mature. The pet bird hobbyist sources indicated that male Indian ringneck parakeets acquire their neck rings at 18 months and are sexually mature at that time, but do not breed until they are 3 years old.[192] The observations discussed in this chapter confirm the latter. The male Indian ringneck parakeet discussed in this chapter acquired his neck ring when he was 18 months old. He did not show any interest in the female ringneck parakeets until he was 3 years old.

Immature male Indian ringneck parakeets resemble females.[60, 203]

Male Indian ringneck parakeets and male echo parakeets resemble each other closely. Female Indian ringneck parakeets and female echo parakeets do not look alike. Female echo parakeets have pale yellow eyes and dusky black beaks.

There was a discrepancy among the sources consulted in regard to the average and maximum life spans for wild and captive Indian ringneck parakeets. The values given ranged from 15 to 30 years, with 25 years being cited most often. The oldest Indian ringneck parakeet discussed in this chapter is currently 20-plus years old. She shows no change in energy level, activity level, or food consumption from the time she was 10 years old. There is no evidence that she is in her senior years. So 25 years is probably the more accurate average life span.

Vocalizations

Indian ringneck parakeets are known for being loud and noisy.[60, 94] Unlike songbirds, Indian ringneck vocalizations are not pleasant to hear. In fact they have been described as being grating.[194] A quote from 1908 sums up this situation rather well:[34]

> "On the account of the facility with which it can be taught to speak, it is a great favorite with the ladies, although it gives them many a headache through its propensity to scream."

Wild Indian ringneck parakeets make a variety of shrieks, trills, whistles, and screams.[60, 94] These sounds are often repeated several times in a row.[94] Indian ringneck parakeets make a loud low pitched unpleasant distress call. The most common vocalizations associated with Indian ringneck parakeets are:[60, 94]

- che-ah che-ah che-ah (while in flight)
- kreh kreh kreh kreh
- chee chee chee chee chee
- ak ak ak ak ak (alarm call)

Indian ringneck parakeet vocalizations can be heard on Disk 2 Track 1.

Socialization

Wild Indian ringneck parakeets are rarely seen alone or as solitary pairs.[41] Instead Indian ringneck parakeets travel around in flocks of 100 to 1,000 birds.[41, 58, 94] Indian ringneck parakeets are gregarious, but feisty.[41, 94] They share communal roosts, often with other parrots and mynas.[69, 94] Wild Indian ringneck parakeets participate in joint flock defensive measures, such as mobbing would be predators.[41] They are not territorial.[94]

Indian ringneck parakeets are considered to be highly intelligent.[192, 194, 203] When kept as pets they need a lot of interaction and mental challenges. However, Indian ringneck parakeets are definitely not the cuddly type.

The flight of Indian ringneck parakeets is swift, direct, and strong, with rapid wing beats.[34, 41, 60, 194] Their flight is also very quiet. As a result the minimum recommended aviary size is 12 to 14 feet (3.65 to 4.26 m) in length.[194, 202]

The daily schedule of Indian ringneck parakeets is similar to other wild parrots. The morning hours (6 a.m. to 10 a.m.) are spent finding and consuming food and water.[125] The midday hours are spent resting and preening.[125] The evening hours are spent finding and consuming food and water before it is time to settle in for the night.[125]

Family Life

Indian ringneck parakeets mate for life, but renew their pair bonds annually.[41] During the courtship ritual the male makes a special vocalization and bows. Bonded pairs engage in allo-feeding, but not allo-preening.[194] This may explain in part why Indian ringneck parakeets usually do not like pats.

Indian ringneck parakeets nest in tree holes 10 to 30 feet (3 to 9 m) above ground.[34, 41, 58, 60, 84, 94] They often reuse and expand vacated woodpecker tree holes.[41, 60] Indian ringneck parakeets nest in winter, from late December through March, unlike other south Asian parrots.[34, 41, 84, 94, 202] This gives them an advantage over other birds in that they have the first choice of a nest site. The nest is lined with wood dust.[60]

Three to four eggs are laid 48 hours apart.[34, 41, 58, 60, 94] The eggs are almost round, white, and 1.125 x 0.9 inches (2.85 x 2.28 cm).[60] The eggs are incubated 22 to 24 days by the female.[41, 58, 60, 84] The male feeds the female while she is incubating the eggs.[58] The eggs hatch in the order they were laid.

The altricial hatchlings only weigh 0.2 ounce (5.6 gr).[202] Both parents feed the hatchlings.[41, 58, 84] The hatchlings eyes open when they are 10 days old.[41] The hatchlings fledge at 7 weeks of age.[41, 58, 60, 192, 202]

Now let's look at the characteristics of indoor Indian ringneck parakeets in comparison to their wild cousins, for example their behavior, daily activities, and life stages. Sections 10.3 through 10.6 describe the behavior of individual Indian ringneck parakeets, in particular play activities, socialization, vocalization, and reaction to changes and challenges throughout life. Behaviors are presented as they occurred. The analysis, comparison and interpretation of these behaviors are reserved for Section 10.7 Discussion. These observations took place at the locations and during the timeframes described in Annex C.

10.3 Maku

Maku arrived 12/20/2004. The local parrot adoption and rescue agency issued an emergency request a week earlier. A nameless nature exhibit was being shut down over the holidays. Three Indian ringneck parakeets that were part of the exhibit had to be adopted immediately or they would be euthanized. There are no regulations that govern organizations that operate nature exhibits. Obviously, there are no matters of conscience or ethics either. One pair of Indian ringneck parakeets and a single female Indian ringneck parakeet needed a home immediately. We were told that the single female's mate had died.

We agreed to take the single female who was promptly named Maku, after the mythical Indian princess Amaku. Maku was 10 years old when she arrived. She did not have a leg band. Her hatch date is unknown. She had spent her entire life in the now defunct nature exhibit. For ten years she had endured a non-stop parade of noisy children and adults waltzing in front of her cage all day long, tapping on the glass, sticking their fingers through the bars, taking flash photographs, and doing other annoying and scary things. But Maku had no place to retreat for privacy.

Maku is a normal green Indian ringneck parakeet. She has a light green shadow around her neck. Her upper mandible is a medium rose color. The lower mandible is dark. Her pupils are black. The iris is light blue, surrounded by white. The skin around her eyes is coral, with a white outer liner. Maku is very active, alert, and vocal.

Maku was nervous, high strung, change phobic, and terrified of people when she arrived. It took more than a year for Maku to accept and settle into her new environment. If a person looked at her for more than 10 seconds, walked in front of her cage (even 3 feet (1 m) away), or reached in to change her food and water dishes, she screeched a non-stop alarm call and had a panic attack, flying back and forth against the sides of her aviary. The panic attack and screeching, which could be heard 2 blocks away, lasted at least 15 minutes. In fact, in most of Maku's early photographs her beak is open because she is screeching her head off. After four months it was possible to walk in front of Maku's aviary without her screeching. Ironically, Maku did not have a panic attack when her aviary was moved or someone vacuumed around it. (See Photograph 10.2.)

One of Maku's favorite pastimes is to chew. She transformed a 12 inch (30.48 cm) long manu mineral peanut log into dust in less than a week. Maku scrapes sisal ropes until they split fiber by fiber. She is the only Indian ringneck parakeet discussed in this chapter to touch sisal. Maku enjoys

Photograph 10.2 Emoting. Maku has never been shy about expressing her displeasure about something—like having her photograph taken! Notice her green on green shadow neck ring, light blue eyes, coral eye ring, and dark lower mandible. © D. Herrmann

shredding palm leaf piñatas and balsa wood toys. Maku loves to chew cholla perches. She can demolish a 2 inch (5 cm) by 12 inch (30.48 cm) cholla perch in less than two weeks, shredding everything but the center nail. The cholla perch is replaced when nothing is left but the nail. Every time the cholla perch was replaced during the first three years that triggered a screeching panic attack that lasted 45 minutes. Today when a worn out toy or perch is replaced, Maku still screeches—just to let everyone know she is annoyed—but only for a couple minutes. This behavior infers a combination of change phobia and anger or frustration that she doesn't have any say or control over her immediate environment, and when and how things are done.

At first if Maku heard a whistle, but could not see the person whistling, she would repeat the whistle. If she saw the person whistling she did not whistle back. After 2 years she stopped whistling altogether. By then she probably decided there weren't any birds on the other end.

For the first couple years Maku stuck her head between the bars on the top of her aviary once or twice a week. Her head and neck fit through the bars, but not her body. Maku stuck her head through the bars, took a brief look around, and continued on about her business. She has not done that recently.

Maku has shown no hesitancy in exploring new foods, in spite of her change phobia. Apparently she was only fed a basic seed mixture at the nature center. Fresh fruits and vegetables, pellets, and other items were an entirely new experience for her. Initially Maku just looked at the new items or poked and prodded them. The second or third time Maku was exposed to a new item, she sampled it. Maku tried all of the new food she was exposed to: carrots, squash, sweet potato, broccoli, apples, oranges, banana, grapes, etc. Over the years Maku has decided there are certain things she likes, doesn't like, or will tolerate on occasion. For example, fresh sweet corn, green peas, and red peppers are favorites and can be served every day; while, broccoli is tolerated once in a while. Maku took 2 years to acquire a taste for popped popcorn, compared to the other parrots where it was an instant favorite. Maku also nibbles on African violets, jade trees, and spider plants on occasion when she is out flying.

Maku regularly chews on a peach colored mineral block. She has done that from her early days. Maku may have been originally attracted to the mineral block because it is similar in color to her beak. Maku is one of two Indian ringneck parakeets discussed in this chapter who chew peach colored mineral blocks. In contrast, Maku has consistently ignored cuttlebones.

A special approach was used to ease Maku's fears about people reaching into her cage. Instead of just yanking the old dishes out and shoving the new dishes in like she is an inanimate object, a 2-step process is used. First, I point to the old dishes and explain that I'm going to change them. Then I ask Maku if she wants to go "up high", that is up on a higher perch for security. After she flies to a higher perch the old dishes are removed. Second, when the new dishes are ready I show Maku the contents of each dish before replacing it. She particularly enjoys eyeing the fruit/vegetable dish. Through this process Maku knows what is going on and what to expect. Over time Maku has accepted this process so well that some days she doesn't even fly to a higher perch when her dishes are being changed.

A special routine was developed around peanuts, since they are one of Maku's favorite snacks. First Maku is shown the peanut. It is evident that she would like to come and take the peanut. But 10 years of mistreatment prevent that from happening. Instead Maku indicates that she is interested in the peanut. Then the peanut is placed on the ledge, a lower perch, or inside a toy. As soon as the person who deposited the peanut is 3 feet (1 m) away, Maku climbs or flies over to retrieve the peanut. This technique was used to increase Maku's confidence and entice her to explore new parts of her aviary. Even though Maku loves peanuts, she won't touch almonds, pecans, cashews, pistachios, or walnuts.

Maku has some unique habits in relation to water. For the longest time Maku insisted on bathing in her water dish even though she doesn't really fit. Only recently Maku started bathing in a swimming pool. She will only bathe in a swimming pool if it is attached to the side of her aviary. Maku won't touch a dish of water on the bottom of the aviary. Maku loves to be misted, almost to the point of being soaked. Only two of the Indian ringneck parakeets discussed in this chapter enjoy being misted. Most pet birds take a drink of water in the evening before retiring for the night. Not Maku, she takes a big gulp of water and gargles until she has splashed the water all over the floor. Maku has gargled every evening since the day she arrived.

Maku expressed no interest in swings over a period of several years. Eventually the swings were removed and replaced by a U-shaped cotton rope perch which was hung from the top of the aviary. Maku took to the cotton rope perch instantly. She preens it daily and takes naps there. Maku does not sleep on the cotton rope perch overnight, even though it is the highest perch in the aviary.

Maku absolutely does not like any bird, large or small, to land on her aviary or landing perch when they are out flying. Maku snaps and lunges at them, even if they are in a place she can't reach. The one exception was in 2010 when the oldest zebra finch nestling (*Poephila guttata castanotis*) fledged and accidentally ended up on Maku's landing perch. Maku seemed to understand that: (1) this was a new fledgling, and (2) the fledgling zebra finch was as surprised to be there as Maku was surprised to see him. (See Photograph 8.12.)

A cross town move took place 3.5 years after Maku arrived. That meant Maku had to spend 10 hours in a small travel cage. Surprisingly she did not resist being picked up in a towel and placed in the travel cage. However, once in the travel cage Maku had a severe panic attack. She did not make a sound or attempt to fly back and forth. Instead, Maku clung to the side of the travel cage terrified. She did not eat, drink any water, or move the entire time. She was misted several times in an attempt to calm her, but that did not help.

Maku was not terrified of the travel cage. When she is out flying Maku on occasion wanders into a room she is not supposed to be in and gets lost. Since Maku does not like to be held, a different process is used to get her back to her aviary. The bottom of a travel cage is removed and the top is placed over her. Maku hops up onto a perch in the travel cage. Then she is carried back to her aviary in the bottomless travel cage.

Instead, Maku was terrified of what the travel cage represented. The previous time Maku had been confined to a travel cage for a period of time was when she was removed from the nature center. Maku probably panicked because she thought she was being taken back to the nature center. Maku was fine after she was back in her aviary at the new location.

The opportunity to fly free increased Maku's confidence. At first she walked hesitantly toward the open door. Now she doesn't waste a second. Maku flies several laps around the room. She always visits the budgerigars (*Melopsittacus undulatus*). Maku seems to enjoy watching them fly around and play with their toys. Often she stays on the budgerigar aviaries for a half hour or more. This behavior, of course, is the opposite of when the budgerigars try to visit her aviary. (See Photograph 10.3.)

Photograph 10.3 Socializing. Maku visits Yellow Birdie. Maku likes to visit the budgerigars when she is out flying, even though they are much smaller than her. Perhaps it is because budgerigars have a similar shape and color. Or, perhaps the budgerigars' energetic, playful, gregarious nature intrigues Maku. She watches the budgerigars intently and attempts to play with their toys. © D. Herrmann

Maku is a strong accurate flier. Maku enjoys the freedom of flying and the ability to decide where she goes, who/what she visits, and how long she stays in each place. Maku returns to her aviary for a quick snack or a drink of water, then takes off flying again. Maku signals that she is through flying for the day by climbing up to the cotton rope perch.

Maku is very aware of her surroundings. Photographs of a male green Indian ringneck parakeet and a mitred conure (*Aratinga mitrata mitrata*) are on the wall next to her aviary. Photographs of other members of the parrot family are on the wall next to the budgerigar aviaries. One of the first things Maku does every time she is out is to try to fly into these photographs, particularly the two near her aviary. Maku flies directly toward the photograph of the Indian ringneck parakeet, hovers briefly, then veers off. Maku does this repeatedly for 10–15 minutes. Maku is not attacking the birds in the photographs. She is trying to interact with them.

Maku has a "crush" on Snapdragon, the male green Indian ringneck parakeet discussed in Section 10.4. It is important to note that Maku did not express any interest in Snapdragon until after he grew his neck ring. The intent was for Maku and Snapdragon to be buddies. Unfortunately, their first encounter did not go well. Both birds were let out to fly at the same time. Maku immediately hopped out of her aviary and into Snapdragon's. Maku proceeded to chase Snapdragon around his aviary and even pulled his tail in an attempt to win his attention. Alas, Snapdragon was not (and still is not) interested in Maku. He simply flew away. The 12-year difference in age may have something to do with it. Subsequent shared flying times have been equally unsuccessful.

In the fall of 2009, five years after her arrival, Maku suddenly began shredding her paper. This was the first time she had done this, even though she was 15 years old by this time. Maku had never even been on the bottom of the aviary beforehand. She is not one to retrieve pieces of dropped food. Maku did not tear a few pieces off the edge of the paper like most birds do. Instead Maku shredded her entire paper methodically all day everyday for 3 months, until only tiny pieces were left. For the first phase Maku stands on the paper with her left foot. Next she folds the paper with her beak to make a small hole. Then she scratches the paper with her right foot. The sound is somewhat rhythmical as she moves around the paper: step scratch scratch scratch, step scratch scratch scratch, step scratch scratch scratch. For the second phase Maku flies with a large piece of newspaper up to her perch and proceeds to shred it into strips. Lastly Maku pushes the shredded newspaper into the corners of her aviary, leaving the tray bottom exposed. (See Photograph 10.4.)

Maku is the only Indian ringneck parakeet discussed in this chapter to engage in this rather elaborate shredding behavior. The other two females, Bluebell (Section 10.5) and Dusty (Section 10.6) have different techniques. Snapdragon, the male, does not shred at all.

Since 2009 Maku has begun shredding every fall after the first cold snap and continues doing so for 3 to 3.5 months. This behavior is consistent with wild Indian ringneck parakeets nesting from late December through March.

Two unusual instances have occurred in relation to Maku's shredding activities. For a few days plain newsprint was placed in Maku's aviary. Maku stared at the plain newsprint from every angle for half an hour. Then she flew down to the paper. But as soon as she landed she flew back up to her perch, as if she wasn't sure that it was a solid surface. Maku flew down and back a couple more times before she began to shred. Maku went through this routine each time she was exposed to plain newsprint. This behavior is interesting considering the "normal" newspaper she sees everyday is different. The type, fonts, and colors are different. Maku is used to having a different printed newspaper every day. But, totally blank newsprint was a new experience that she recognized instantly and analyzed carefully.

Surprisingly, Maku began shredding her paper for a few days during August 2010. This shredding exercise was not in response to a temperature drop. Rather, once Maku heard three zebra finch hatchlings making a begging call, she started shredding. Maku could not see the zebra finch hatchlings; they were too far away. When Maku heard the hatchlings—even though they were zebra finches—she associated the sound with a nest and began shredding. Maku stopped shredding when the zebra finches fledged. (See Section 8.4 and Photograph 8.12.)

The power was out for 3.5 days during late January/early February in 2010 and again in 2011. The temperature in the aviary did not budge above 50°. That triggered more than paper shredding. For the first time Maku laid an egg on 3/7/2010 and a second on 3/9/2010. Again in 2011 Maku laid eggs on 2/9, 2/11, 2/13, and 2/15. The white eggs were almost round and about an inch (2.54 cm) long.

In 2010 Maku did not pay any attention to the eggs. In 2011 Maku sat on the eggs for 4 weeks. Each day the paper in her aviary was changed. The eggs were removed and placed on the clean newspaper. Afterward Maku carefully pushed the eggs together and sat on them. As long as Maku could see the eggs, she did not complain while the paper was being changed. After 4 weeks Maku decided the eggs were not going to hatch and abandoned them. This infers that Maku knows: (1) how long her eggs need to be incubated before they hatch, and (2) if the eggs haven't hatched within a certain timeframe they are unlikely to do so. During January/February of 2012 the power did not go out. Maku did not lay any eggs until 3/22 and 3/25. Again she sat on the eggs all day every day for four weeks. Maku even adopted an egg that was abandoned by another Indian ringneck parakeet. (See Photograph 10.5.)

Photograph 10.4 Shredding! Maku goes into a seasonal shredding frenzy each winter as soon as the temperature drops. This behavior and the timing of it coincide with wild Indian ringneck parakeets nesting in excavated tree holes from late December through March. Maku shreds as much paper as she is given on a daily basis. Maku tears the paper into smaller and smaller pieces throughout the day. She often carries a paper up to her top perch to shred, then pushes the pieces into the corners of the tray, as if the frame of the aviary were the boundaries of a tree hole. © D. Herrmann

2013 was the only year Maku laid two sets of eggs. The first set was laid at the usual time during late winter. Maku tended the (infertile) eggs for 4 weeks as usual before deciding they were not going to hatch. Then during summer a strange thing happened. When Maku heard the begging call of a common grackle nestling (*Quiscala quiscala*) that was with us from 6/17/2013–7/7/2013 as mentioned in the dedication, she began shredding her paper and laid 3 more eggs. This was the second

Photograph 10.5 Nesting #1. After a short break to eat Maku reassembles the three eggs she laid in 2013. Maku pushes the eggs gently into position one at a time with her beak. Once the eggs are in contact with her brood patch they are no longer visible. It is interesting to note that Maku's posture in the 4th photograph is the same posture Indian ringneck parakeet females assume when they are receptive to a male's attention. (See right inset.) Surprisingly, Maku did not complain when the eggs were removed and returned each day while the paper was being changed. 2013 was the first year Maku agreed to be photographed while tending her eggs. © D. Herrmann

time in 3 years that Maku engaged in extra nesting behaviors in response to hearing the begging call of another bird species. Maku, an Indian ringneck parakeet, would not hear the begging call of a zebra finch or a common grackle in her native habitat. These three birds live on different continents. This infers that Maku can identify the begging calls of species other than her own and associates a begging call with nesting behaviors.

In 2015 Maku laid eggs on 4/14, 4/16, and 4/18. This year, however, Maku nested under her paper for the first six days, probably after watching Bluebell walk under her paper. After the third egg was laid, Maku moved the eggs on top of the paper and resumed nesting there as usual.

When Maku arrived the only vocalizations she made were an alarm call, a distress call, and answering whistles. Other than that she was quiet. After Bluebell (Section 10.5) and Dusty (Section

10.6) arrived, Maku began making traditional Indian ringneck vocalizations. Maku had to learn how to make these new vocalizations. For the first few months she could be heard quietly practicing the new sounds. Maku was 14 years old at the time. This behavior infers that:

- Maku was able to recognize that Bluebell and Dusty were Indian ringneck parakeets even though they are different colors than she is, i.e., hybrids
- Other than alarm and distress calls, Indian ringneck parakeet vocalizations are learned from the family/social group

Maku is no longer terrified of people, but there are limits to her level of trust. Maku still refuses to be held; although she will take a peanut from my hand. Maku appreciates a little attention—especially someone bowing in front of her aviary or telling her she is pretty—but not too much. If someone pays too much attention to her, Maku gets nervous. All in all Maku is a success story. She has made significant progress since her arrival. It is a testament to her strength of character that she was willing to trust a human at all after being mistreated by them for 10 years. On 3/31/2014, for the first time since her arrival on 12/20/04, Maku solicited a pat. Since then she has agreed to head scratches and pats on an occasional basis. Maku is 20-plus years old at this point.

10.4 Snapdragon

Snapdragon is a male normal green Indian ringneck parakeet. His upper mandible is a medium rose color. The lower mandible is black. He arrived 11/20/06. Snapdragon was acquired from a pet store when he was 9 months old. His hatch date is 2/15/06.

Snapdragon is housed in a separate aviary near Maku. Maku totally ignored Snapdragon when he arrived. Surprisingly, Rudy, a male mitred conure (Section 11.3) was very excited about Snapdragon's arrival. Repeatedly for several weeks Rudy tried to feed and play with Snapdragon whenever he was out. Young mitred conures have a solid green head. It is likely that Rudy thought Snapdragon was a juvenile mitred conure.

During the months Snapdragon spent sitting in the pet store his cage was next to a pionus parrot. When Snapdragon arrived he did not make any traditional Indian ringneck parakeet vocalizations. Instead, Snapdragon made vocalizations normally associated with pionus parrots. Like Maku, this behavior also indicates that Indian ringneck parakeets learn vocalizations rather than acquire them through genetic intelligence.

Snapdragon and Maku are opposites in terms of temperament. Snapdragon is calm, content, playful, and curious. He has never been nervous, high strung, or fearful, not even the day he came home from the pet store. Almost all birds are disoriented when they arrive, since their world has been turned upside down with no warning or explanation. Even the hardiest birds take a day or two to settle in. Not Snapdragon. As soon as his travel cage was opened, Snapdragon hopped out, looked around, and began exploring. He didn't mind being held or carried from aviary to aviary to be introduced to the other birds.

Snapdragon's temperament has been consistent since the day he arrived. He did not transition from a cuddly juvenile to an aloof adult or go through the terrible two's like Squeaky (Section 9.3) or Sydney (Section 9.4). Snapdragon does not like to be patted and prefers to sit on a perch or a dish rather than a hand or finger. Snapdragon is always eager to interact whenever he is given the opportunity. If Snapdragon gets lost when he is out flying, becomes tangled in a plant, or ends up landing somewhere strange, he doesn't panic or screech. Snapdragon simply sits there quietly and waits for someone to come by with a dish to pick him up and carry him home.

In the years since Snapdragon arrived there was only one occasion when he wasn't calm and content. One day Snapdragon was out flying as usual. He landed on several aviaries as he visited the other birds. A couple hours later Snapdragon was sitting on top of his aviary. The feathers around his eyes were wet from his eyes tearing up, slowly one tear after another. Yet he didn't make a sound. Somehow while climbing up to the top of his aviary Snapdragon had shut the door and locked himself out. He couldn't get inside to eat or drink, or return to his favorite toys and perch. Snapdragon had never been in this situation before and didn't know what to do, so he cried quietly while waiting for help to come. As soon as Snapdragon was back inside his aviary he was fine.

Snapdragon becomes excited about the littlest things: a special treat, a new toy, or extra attention. When Snapdragon is excited he rocks back and forth on his feet or turns his head from side to side rapidly and makes a pleasant trilling sound. Snapdragon's other vocalizations are pleasant too and sound like a person laughing. He makes ha-ha-ha and hee-hee-hee sounds, with the third syllable being long and drawn out. Snapdragon also makes a giggling sound when he starts to eat fruit, as if he is delighted to have found some apples or oranges.

The power was out for 3.5 days in January/February 2010 and again 2011. The indoor thermometer never went above 50 degrees. Every morning at 5 a.m., long before it was light, Snapdragon made a long solitary mournful call. That sound summed up what everyone, bird and human, within in 10-mile radius was thinking: "Damn it's cold. I'm tired of being cold. Is it every going to get warm again?" These two occasions were the only time Snapdragon made this sound.

Early on Snapdragon was taught that if he wanted to be handed a treat he had to go to an upper perch. In other words, he would not be given the treat just anywhere. When told "on your perch" Snapdragon climbed to an upper perch. Then he received the treat.

Several months afterward the upper perches in Snapdragon's aviary were replaced. The first time Snapdragon was told "on your perch" he looked all over and climbed around and around from perch to perch, as if he were looking for something. Apparently Snapdragon had associated "on your perch" with a specific pink perch that had been replaced. In other words, Snapdragon learned more than we thought he had been taught. It took a couple days for Snapdragon to accept the fact that it was alright to sit on the new perch that was located where the pink perch had been, when he was told "on your perch".

Snapdragon is somewhat of an adventurous eater. He will try everything new at least once. Snapdragon readily eats most fruits; apples and oranges are favorites. Sometimes Snapdragon carries the fruit over to his dry food dish to eat it. Other times he doesn't. There doesn't seem to be a fixed pattern of behavior in this regard. When it comes to vegetables Snapdragon is a little more selective. He eagerly consumes fresh carrots, red peppers, yellow sweet corn, and green peas. Other items are consumed once in a while or totally ignored. Snapdragon ate fresh greens one leaf at a time when he was young—if he was hand fed. Now he rarely touches fresh greens. (See Photograph 10.6.)

Snapdragon is not overly enthusiastic about nuts either. He never touches almonds, walnuts, pecans, or pistachios. Snapdragon will eat shelled peanuts one at a time if they are handed to him. Unshelled peanuts are simply ignored or thrown on the floor. At first Snapdragon would only eat fruit flavored pellets; now he also eats natural pellets. Of all the budgerigars, poicephalus parrots, conures, and Indian ringneck parakeets discussed in this book, Snapdragon is the only one to eat fruit flavored pellets. The other birds prefer natural pellets. Millet sprays are accepted as a treat, as long as they are not offered more than once a week.

Popped popcorn is a favorite treat of his, like the other parrots. However, Snapdragon hasn't quite learned the correct technique for eating popcorn. Instead of picking up one piece at a time Snapdragon bites into a dish full of popcorn. Invariably his beak will pierce a piece, which becomes stuck on the upper mandible. Snapdragon interprets the stuck piece of popcorn as a would-be attacker and starts making a low pitched distress call. He shakes his head and rubs his beak against the rim of the dish—while making a distress call—until the popcorn falls off.

Snapdragon has never touched a mineral block. He sits on cuttlebones, but does not chew them.

Snapdragon plays with a variety of toys and has done so since he was very young. Instead of rushing to tear a palm leaf piñata apart in a day or two, Snapdragon methodically chews a piece here

Photograph 10.6 Growing Up. (L) Snapdragon was 9 months old and a juvenile when he arrived. (R) His neck ring did not appear until he was 18 months old. Snapdragon ate fresh greens regularly when he was young. But, he insisted on dunking them in water first even though they had just been rinsed. © D. Herrmann

and piece there over a period of weeks until nothing is left except the center rope. This approach is slower, but much more efficient. Cotton toys are Snapdragon's favorite, especially ones that have cotton ropes or strings 0.25 inch (0.63 cm) in diameter or less. Snapdragon chews the ropes to separate them into individual fibers. Snapdragon chews one cotton toy routinely every night before going to sleep. He has had this toy since he arrived. A second cotton toy is near his favorite perch. Snapdragon has chewed this one so often it has been replaced with an identical toy 3 times. Snapdragon rarely chews the wood components of these toys, but he enjoys pushing the toys until they swing back and forth. In contrast, Snapdragon ignores sisal, plastic, and metal toys. (See Photograph 10.7.)

Like Maku, Snapdragon has never demonstrated any interest in swings. He prefers to sit on toys or perches near the side of the aviary. Snapdragon often hangs on the side of the aviary like a woodpecker. Snapdragon hung on the side of his aviary frequently when he was very young. In contrast, Snapdragon rarely sits in the middle of a perch and hardly ever visits the lower perches.

Snapdragon is an agile climber and moves with ease over, under, or around obstacles that are in the way. He even hangs upside down from the top of his aviary every now and then. When Snapdragon was very young he often tried to imitate Squeaky (Section 9.3) and Sydney (Section 9.4) by hanging upside down. It was not uncommon to see his long green tail sticking straight up through the top of his aviary.

Photograph 10.7 Indian Ringneck Parakeet Toys Before and After. Snapdragon enjoys a variety of chewing toys. Snapdragon interacts with his toys in a steady systematic manner. Instead of racing to shred a palm leaf piñata in a few days like the other parrots, Snapdragon methodically chews a piece here and a piece there over a period of several weeks until nothing is left but the center rope. © D. Herrmann

Flying is not Snapdragon's best skill. In fact, Snapdragon is a rather clumsy flyer. That is why he regularly ends up in strange places and needs a lift home. Snapdragon takes off before he has decided where he is going, misses a few turns, miscalculates the landing approach, and plops down suddenly just in time to avoid a crash. Snapdragon's wings are strong enough. His close-up vision is fine. It is possible that his distance vision is not up to par.

Black spots started appearing around Snapdragon's neck when he was 18 months old. The first black spot was under his beak. The second and third black spots were under his ears. Over a period of 2.5 weeks the spots in front coalesced into a black ring 0.125 inch (0.317 cm) thick. The last pieces to be filled in were on the back of his neck. This part of the ring became a medium rose color with light blue metallic highlights.

Snapdragon suddenly began bowing the same time the neck ring appeared. He bows until he is parallel with the floor. At first Snapdragon bowed regardless of whether any other birds (or humans) were present. Today he only bows in Bluebell's direction. When Snapdragon bows he makes a long drawn out "ki-akh" sound once. Snapdragon is the only one of the four Indian ringneck parakeets to bow or make this sound. Both of these behaviors must be male-only behaviors. (See Photograph 10.8.)

Prior to the appearance of his neck ring Snapdragon was not overly concerned about his appearance. He went swimming infrequently—usually in his water dish—and went to great lengths to avoid being misted. Today Snapdragon readily bathes in a swimming pool and takes good care

Photograph 10.8 Bowing. (Top) Snapdragon's neck ring grew in when he became a young adult. The ring is black in front and rose in back, with a light blue tinge in direct sunlight. (Bottom) Snapdragon began bowing and making a special male-only vocalization shortly after his neck ring appeared. © D. Herrmann

of his feathers. But, he still runs away from being misted. Perhaps the need to take good care of his feathers was something Snapdragon learned as part of growing up and watching the other Indian ringneck parakeets.

Maku was the first female Indian ringneck parakeet Snapdragon met. During their first encounter Maku was too forward and pulled Snapdragon's tail in an attempt to get his attention. To this day Snapdragon remembers Maku pulling his tail and runs (or flies) in the opposite direction whenever she comes to visit.

Bluebell is Snapdragon's first and to date only "crush". (Her story is in the next section.) Bluebell and Snapdragon met in 10/2008, when Snapdragon was a little over 2.5 years old. Their aviaries were next to each other. At first Snapdragon did not pay any more attention to Bluebell than he did to any other bird. Over time that has changed.

Today Snapdragon and Bluebell routinely sit directly across from each other at the nearest point between their two aviaries. They often imitate each other's movements: preening at the same time, stretching at the same time, etc. When Snapdragon is out flying he spends most of his time visiting Bluebell. Snapdragon bows and calls to Bluebell, tries to feed her through the bars, and has even spent the night sleeping on top of her aviary.

Alas, like most relationships the relationship between Snapdragon and Bluebell is not exactly reciprocal. On occasion Bluebell snaps at Snapdragon's feet while is he visiting her aviary. When Bluebell is out flying she visits Snapdragon, but spends most of her time visiting other birds or flying laps. So far attempts to let Snapdragon and Bluebell out to fly at the same time have been unsuccessful. Like Maku, Bluebell is much too forward, which sends Snapdragon scurrying away. Eventually Snapdragon and Bluebell will probably work things out and move in together. They are both still very young. But for now theirs is a long distance relationship. (See Photograph 10.9.)

First thing on the morning of 5/27/2015, shortly after sunup Bluebell abruptly let herself out of her aviary and moved herself into Snapdragon's aviary. This was 4 years and 8 months after they first met. So far Bluebell and Snapdragon are getting along fine.

In summary Snapdragon is a sweet delightful bird. He has never exhibited any of the negative stereotypes associated with Indian ringneck parakeets: high strung, skittish, screaming. Snapdragon

Photograph 10.9 Long Distance Romance. Snapdragon's and Bluebell's aviaries are next to each other and they frequently imitate each other's behavior. Snapdragon visits Bluebell when he is out flying and vice versa. They appear to be interested in each other, but when let out at the same time their interactions have not been altogether friendly. For now Snapdragon and Bluebell remain in separate aviaries and have a long distance romance. © D. Herrmann

has never screamed the entire time he has been here. Snapdragon did not receive special treatment, training, or anything different from the other Indian ringneck parakeets. This is just who he is. Snapdragon celebrated his 9th hatch day 2/15/2015.

10.5 Bluebell

Bluebell is an elegant female hybrid white and light blue Indian ringneck parakeet. In direct sunlight her feathers resemble a pearl or an opal. Unlike other Indian ringneck parakeets she has light rose colored upper and lower mandibles. Bluebell's thighs are 0.5 inch (1.27 cm) longer than Maku or Snapdragon, which makes her taller than them. Bluebell is about 5% larger than the other Indian ringneck parakeets. Bluebell arrived 10/3/2008; at the time she was 2.4 years old. Her hatch date is 5/17/2006. (See Photograph 10.10.)

Photograph 10.10 Bluebell. Bluebell is an elegant hybrid white and light blue Indian ringneck parakeet. Note the light blue eyes and coral eye ring. Unlike other Indian ringneck parakeets, her upper and lower mandibles are the same rose color. © D. Herrmann

Bluebell is adopted. She had been purchased and returned to a pet store twice before being put up for adoption. Many pet stores have a bad habit of disposing of birds that reach 2 years of age and remain unsold.

Overall Bluebell has an even temperament. Surprisingly Bluebell is not fearful or timid nor does she harbor any phobias, in spite of being purchased and returned twice. Bluebell is not aggressive, although she protects her food and water dishes from people who dare to remove them. At the same time Bluebell accepts fruits, vegetables, and nuts directly from a person's hand and has done so from day one. Bluebell is not the cuddly type. Rather, she is independent and self-sufficient. Having been abandoned by humans multiple times, Bluebell probably isn't interested in them.

Bluebell is the only one who made vocalizations normally associated with Indian ringneck parakeets from the day she arrived. The other Indian ringneck parakeets learned these vocalizations from her. It is unclear from whom Bluebell learned her vocalizations. Perhaps one of the families that purchased her had other Indian ringneck parakeets. Bluebell's vocalizations are consistently pleasant. Like Snapdragon, she has yet to make an alarm call or a distress call.

Bluebell has eaten healthy from the beginning. She readily consumes most fruits and vegetables, along with natural pellets. Bluebell eats some seeds, mostly sunflower and safflower. Bluebell partakes of a millet spray on occasion. Peanuts are readily accepted, but not almonds, pistachios, walnuts, pecans, or cashews. Bluebell is the only one besides Maku to chew mineral blocks. She totally ignores cuttlebones.

Bluebell has a funny habit when it comes to eating fruits and vegetables. No matter what or how much is served, Bluebell always throws the first two pieces on the floor. Then she proceeds to eat what remains in the dish like normal. Wild birds do not eat the peel of fruits and vegetables. They seem to understand that the peel might be dirty or contaminated or contain unpleasant chemicals designed to protect a fruit from being eaten. The fruits and vegetables Bluebell are served have already been peeled and sliced. Yet, she may be throwing away the first two pieces as a substitute behavior for not eating the peel.

Bluebell will drink water from a water bottle or a water dish. If a water dish is used, Bluebell insists on placing the colorful plastic links of a nearby toy in the dish. This is a daily ritual. As soon as Bluebell takes her first drink from a fresh dish she plops the plastic links into the water. If the plastic links are removed from the water Bluebell puts them back in before the day is over. It is unclear whether this behavior is a personal preference or an imitation of a behavior that occurs in the wild. Bluebell is the only one of the four Indian ringneck parakeets to engage in this behavior.

Bluebell loves to be misted, unlike Snapdragon and Dusty who run away as if misting were a form of punishment or torture. Bluebell stands still or stretches her feathers while being misted and stays put until she is thoroughly soaked.

Bluebell eagerly bathes in a swimming pool attached to the side of her aviary. So far she has avoided swimming pools placed on the bottom of her aviary. Bluebell bathes in her water dish if she is in the mood and a swimming pool isn't available. When bathing in her water dish Bluebell holds on to the colorful plastic links with her beak. The reason for this is unclear. This behavior may be a personal preference or a safety measure intended to ensure that the water isn't too deep—an imitation of a wild behavior.

Bluebell is very neat, a characteristic rarely attributed to parrots. Budgerigars are the only other member of the parrot family that is ever described as being neat. Bluebell keeps her feathers immaculate. Her perches rarely ever need cleaning. Even her paper is usually neat, considering the circumstances. Every now and then Bluebell throws a dish on the floor; that is about the only thing she does that is messy. Oddly, Bluebell started throwing her dish on the floor during the three day power outage in 2010. Most likely it was a coincidence, but it seemed as if she were commenting on the cold.

Bluebell plays with almost all types of cotton, wood, plastic, cholla, palm leaf, cloth, and cardboard toys. Unlike the other birds Bluebell doesn't seem to have a favorite toy. She shakes, pokes, and chews them all equally. In contrast Bluebell has shown no interest in sisal toys, metal toys, or swings since she has been here. Bluebell is the only one of the four Indian ringneck parakeets to regularly sit on the perches in the lower half of her aviary and play with the toys there. (See Photograph 10.11.)

Bluebell is an inveterate chewer. It doesn't matter if the substance is wood, plastic, cotton, palm leaf piñatas, cholla pieces, or cardboard. Bluebell ignores cuttlebones, but she periodically chews on the metal bars of her aviary. This activity appears to be related to strengthening the beak muscles. In the process Bluebell has loosened one bar and caused another bar to fall out. This level of beak strength is not normally attributed to Indian ringneck parakeets.

Bluebell is a strong graceful precise flyer. When Bluebell is out she makes several laps around the room, usually making the che-ah vocalization all the while. Bluebell sits on the top and sides of the zebra finch and budgerigar aviaries. She watches them and attempts to play with their toys from the outside. Bluebell has even attempted to eat their millet sprays from the outside.

Photograph 10.11 Cotton Wreath. Bluebell enjoys a variety of cotton, wood, plastic, cholla, and palm leaf toys, like the cotton wreath shown here. Unlike the other parrots Bluebell does not: (1) have a favorite toy—she plays with them all equally, or (2) demolish her toys—she pokes, prods, and chews them gently. © D. Herrmann

As mentioned earlier Maku attempts to interact with the parrot photographs near her aviary by flying toward them and hovering. Bluebell isn't as polite or subtle. Bluebell leans over and punches the parrot photographs near her aviary until they are crooked. Once Bluebell has established her superiority, she leaves the photographs alone. Ironically, the photographs are of Snapdragon when he was very young.

Bluebell visits Snapdragon's aviary while she is out. She doesn't spend any more time there than she does visiting the other aviaries; sometimes she spends less time there. Bluebell is interested in Snapdragon. However, she is too aggressive when they are out together. Also, she is only interested in Snapdragon every now and then. Bluebell does not seem ready to commit to a permanent pair bond just yet, in spite of Snapdragon's attention.

Maku and Dusty have engaged in distinct nesting behaviors. Bluebell has not engaged in any prolonged nesting behaviors; nor did she lay any eggs until 2013. Each winter since 2010 Bluebell briefly chewed around the edge of her paper. A couple times a week Bluebell went under her paper, like it was a tent, and sat there for 5 minutes or so. That is the extent of her nesting behaviors to date.

In summary, Bluebell is a beautiful independent content confident pleasant bird and is likely to remain so. Of the four, Bluebell's vocalizations and behavior most closely match that of wild Indian ringneck parakeets. Bluebell celebrated her 9th hatch day 5/17/2015.

10.6 Dusty

Dusty is a female hybrid grey and black Indian ringneck parakeet. She has a cherry red upper mandible and a black lower mandible. Dusty arrived 10/3/2008; at the time she was 1.6 years old. Her hatch date is 2/26/2007. (See Photograph 10.12.)

Photograph 10.12 Dusty. Dusty is a female hybrid charcoal grey and black Indian ringneck parakeet. She has a cherry red upper mandible and a black lower mandible. Her eyes are light blue, with a coral eye ring, like the other Indian ringneck parakeets in this chapter. However, Dusty is about 10% smaller than they are. © D. Herrmann

Dusty is adopted. She could easily be a poster bird for the parrot rescue movement, illustrating why people who know nothing about parrots or how to take care of their physical, social, emotional, and nutritional needs should not be allowed to buy parrots, especially on impulse.

The first time I saw Dusty she was cowering in a back corner of her cage trembling. It was obvious that something was wrong with her wings. She avoided eye contact with humans and ran away from them if they came near her cage. Dusty made a loud non-stop distress call, as if she were in a death struggle, when she was picked up and put in a travel cage to be taken to her new home. I was told she would probably not ever be able to fly.

Dusty started life on her own in a pet store. She was purchased and returned twice. Other than the brief respite at the pet store, Dusty had been severely mistreated emotionally and physically. The only vocalizations Dusty made were a loud high pitched alarm call and a loud low pitched distress call. She alternated between these two calls nonstop if a human was in the same room with her.

For the first 3 months Dusty spent all her energy finding new ways to escape from her cage, several of which were ingenious and successful. Once out Dusty would climb or run fast toward the windows or plants. She could not fly. Her wings had been clipped, but not in the normal meaning of the term. All of the feathers on both wings had been cut off. All that remained were the skin and bones, like the wing stubs of a new hatchling. The tips of Dusty's wings were open wounds. Everywhere Dusty landed or tried to climb her wings left red marks. It was hard to comprehend the physical and emotional trauma Dusty had experienced in her short life.

Dusty was taken to the veterinarian immediately. She made a nonstop distress call throughout the exam. The veterinarian gave Dusty antibiotics for the wounds and a vitamin A, D, and E shot because she was so scrawny. Dusty is about 10% smaller than the other three Indian ringneck parakeets. Due to chronic malnutrition Dusty only weighed in at 95 grams, 17% below the normal weight for Indian ringneck parakeets.

While Maku was initially terrified of people, Dusty lived in sheer terror of them. She made an alarm call or distress call if anyone walked near her cage, changed the paper, or refreshed the food and water dishes. When she wasn't doing that, Dusty climbed around and around in her cage looking for a new escape route. If Dusty tired of that she hung vertically in a corner of her cage and shook her head back and forth nonstop or chewed on one of her toes. Dusty's emotional and physical scars run deep.

It was over a year before Dusty made any progress. Gradually, Dusty began soliciting attention. Instead of running away she began running forward and soliciting pats whenever her food, water, or paper was changed. Yet Dusty was (and still is) conflicted. Dusty leans forward to be patted, sometimes she even presses her body against my hand. At the same time Dusty often trembles while being patted, or lunges toward my hand, stopping short without biting. It is as if Dusty wants normal social interaction, but the painful memories of abuse are still very real.

On other occasions Dusty acts up to obtain attention, like a child acting up in school to get the teacher's attention. Dusty does something she knows is wrong, like throwing her water dish on the floor. Then she cowers in the corner.

Over time Dusty molted the chopped feather shafts and regrew new wing feathers. Her flight feathers are black, a nice contrast with her dark grey feathers. Like Maku, being let out to fly is a liberating experience for Dusty and helps increase her confidence. Dusty is not a strong or precise flyer, but she compensates by flapping twice as often as the other Indian ringneck parakeets. Dusty usually lands close to where she is going, but not always. Dusty has gotten herself tangled up in a sansevieria, African violet, and ponytail palm tree. (See Photograph 10.13.)

Being out gives Dusty a tremendous confidence boost, even if she is only out for a short while. Dusty generally half climbs, half flies wherever she is going. Dusty enjoys visiting other birds; she likes to sit or run on top of their aviaries. Dusty particularly likes to sit on the top of the zebra finch aviary and watch them buzzing around. One time Dusty got a little carried away. She leaned over from the zebra finch aviary and began chewing the woodwork. Fortunately the crunching sound was noticed and the zebra finch aviary was moved before Dusty did too much damage.

Dusty has even invited herself into Buster's aviary (Section 11.4) and the budgerigars' aviaries. Buster didn't mind Dusty visiting or checking out his toys and perches. But once Dusty tried to sample his food, Buster drew the line and shooed her out. Dusty has also explored the inside of the budgerigars' aviaries. Dusty played with their toys, sat on several perches, and sampled their food. Dusty even attempted to go swimming in their tiny 5-ounce (150 ml) stainless water dish. The budgerigars, being rather gregarious, didn't mind at all.

Dusty loves to swim in fresh water. This is one activity Dusty took to immediately, without hesitation. She runs down to her swimming pool before the door is closed. If her swimming pool isn't available, Dusty bathes in her water dish. Dusty doesn't just get her feet wet, she splashes until all the water is gone and she is soaked. Dusty is the only one of the four Indian ringneck parakeets to bathe in a dish on the bottom of the aviary. (See Photograph 10.14.)

Dusty is still skinny, partly because she is a finicky eater. Dusty will eat one almond, pistachio, or peanut at a time. If she is given two, the second one is usually thrown on the floor. Dusty nibbles

Photograph 10.13 Short Flights. Dusty learned how to make short flights, which boosted her confidence and morale. Dusty particularly enjoys visiting the zebra finches, on whose aviary she is sitting, and tries to play with their tiny toys. © D. Herrmann

on millet sprays. Sometimes she finishes a millet spray in an afternoon. Other times two or three days go by before she touches the millet spray. Dusty lunges energetically toward most fruits and vegetables. But in her enthusiasm Dusty drops most of the fruits and vegetables on the floor. Dusty prefers a seed mixture with natural pellets on top.

When Dusty is finished with her fruit/vegetable dish she picks up the dish and bangs it against the side of the aviary for a minute or two before dropping it on the floor. All the while Dusty makes a distress call, like she is locked in mortal combat with the dish. This behavior may be due to the fact that she sees her reflection in the bottom of the dish. However, she has also done this with opaque plastic dishes.

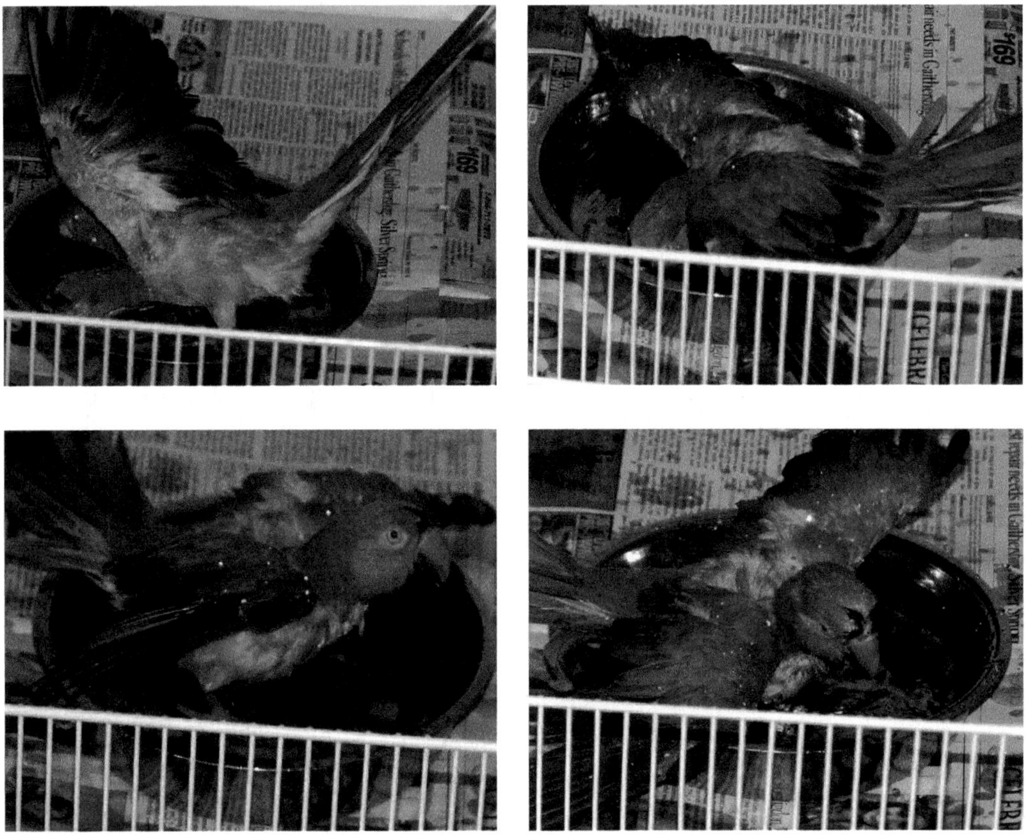

Photograph 10.14 Swimming. Dusty is an energetic swimmer. She jumped in enthusiastically the first time she had the opportunity, without any hesitation. Dusty is the only one who will swim in a dish on the bottom of the aviary. The other Indian ringneck parakeets will only swim in a dish attached to the side of the aviary. This sequence of four photographs was taken within an interval of 0.18 seconds. © D. Herrmann

At least once or twice a week Dusty positions herself under her water dish and taps the bottom with her beak, like she is interacting with another bird. Before long she can be seen pushing the dish upward with her head. For the grand finale Dusty picks up the dish with her beak, rocks it back and forth until all the water spills, then throws the dish on the floor. Dusty has even carried her water dish over on top of her fruit/vegetable dish before dropping it, as another demonstration of her mastery over the bird in the dish. (See Photograph 10.15.)

Dusty's door is left open when her food and water dishes are changed; she is calmer this way. One day as Dusty was being served breakfast there was a knock on the door. Dusty's fruit/vegetable dish was placed temporarily on the floor. That day Dusty climbed down the front of her aviary and walked over to the dish. Dusty sat there—on the floor—and ate the dried papaya, even though she avoids the floor when she is out flying. Then Dusty picked up the almond and climbed back up into her aviary. She waited patiently on her perch for the rest of her meal.

Through this behavior Dusty demonstrated that she knew it was her turn to be served and that the dish on the floor was hers. Dusty ate the papaya and returned home with the almond on her own. She made no attempt to fly or visit the other birds. Dusty understood that it was breakfast time, not flying time. This was the first time Dusty was presented with this situation and she interpreted it correctly. Dusty also learned how to use a water bottle in less than a minute. In spite of her horrific past Dusty is alert and intelligent, probably the most intelligent of the four Indian ringneck parakeets.

Photograph 10.15 Dish Battle. Dusty has a daily battle with her fruit and vegetable dish. As soon as she has finished eating Dusty pokes and shakes the dish while making a distress call. Then Dusty bangs the dish against the side of her aviary a few times while hanging upside down, before dropping the dish on the floor victorious! Dusty can see her reflection in the stainless dish. But that is not the only reason for this behavior. Dusty does the same thing with opaque colorful plastic dishes. © D. Herrmann

Dusty has shown no interest in cuttlebones in the years she has been here. She chews on mineral blocks once or twice every six months or so.

Dusty has made considerable progress since she arrived. Yet certain behaviors indicate that the emotional scars persist. Perhaps because Dusty experienced physical and emotional abuse while she was quite young—before she was 2 years old—these memories are too vivid and recent to be forgotten.

Dusty does not really play with toys; she attacks them. Dusty aggressively throws push pull toys against the side of her aviary. Dusty even broke a mirror toy. At first she talked to the mirror. When the bird in the mirror didn't respond, Dusty began to attack the mirror. One day Dusty pushed the mirror out of the aviary through the bars. The next day Dusty broke the mirror off the chain that was holding it. Then she proceeded to throw the mirror out of the aviary through the bars. To date Dusty has shown no interest in conventional swings. For a while she would sit on a triangle swing with her head bent low, under the ropes.

Dusty frequently sits with her tail tucked in, like a dog with its tail between its legs. This behavior is leftover from Dusty's early days when she used to cower trembling in a back corner of her cage.

Dusty hangs vertically in a corner of her aviary without moving for 20 to 30 minutes at a time. Dusty often bites or pings the same wire in front of her while there. She routinely sits on a perch facing the back wall, instead of facing forward and the other birds. Dusty also sits on a metal hoop and bites the same toe on the same foot for 20 to 30 minutes. Dusty can be distracted by a new palm leaf piñata toy. She will run over and shred the piñata for an hour or so. But, then it is back to biting her toe.

Cockatoos and macaws are known for plucking out their breast feathers when they are under stress. This destructive behavior has been attributed to boredom, and the lack of bird companions and the opportunity to engage in allo-preening.[84] Dusty does not engage in this self-destructive behavior. But, she regularly over preens the top of her shoulders, leaving them bare. Every now and then Dusty also chews the bottom of her flight feathers. Dusty's feathers generally look rumpled, even though she bathes every chance she gets.

Once in a while Dusty reverts back to making a non-stop alarm call. Sometimes this is in response to a sudden loud noise, like thunder. Other times there isn't an apparent reason. Dusty just starts making the exact same shrill sound over and over with the precision of a metronome, like a Chinese water torture on one's ear drums, for 30 to 45 minutes at a time. Neither talking softly nor offering a treat calms Dusty in this situation. Only turning off the lights quiets Dusty; that way it is dark and Dusty goes to sleep.

Like Maku, Dusty engaged in nesting behaviors during the winters of 2010 and 2011. However, Dusty does not shred her entire paper. Instead, she tears off pieces around the edge of her paper, using her beak and both feet. Then Dusty sits underneath the paper, like it is a tent. (See Photograph 10.16.)

Photograph 10.16 Nesting #2. Dusty's nesting behavior is more similar to Bluebell's than Maku's. Dusty nibbles around the edge of the paper and sits underneath it, as if the paper tent represents a tree hole. Unlike Bluebell, Dusty laid eggs in 2010 and 2011. She stayed under the paper with the eggs all day, except for a few quick trips to get some food and water. Dusty attempted to feed one of the eggs in an anticipatory behavior. © D. Herrmann

In 2010 Dusty laid two eggs, one on 4/5 and one on 4/7. She was ravenous beforehand. In 2011 Dusty laid three eggs: 2/17, 2/19, and 2/21. She went into a shredding frenzy 24 hours before the first egg was laid. Dusty pushed the eggs together and stayed underneath the paper with them; except for a brief interval when she ran up top for a quick meal. Dusty sang a short song each time she came up for a meal. Other than that she was quiet. Dusty did not over preen her shoulders or chew on her flight feathers while she was nesting. Dusty tended her eggs for 4 weeks both years. After that she abandoned them, deciding that they were not going to hatch.

In 2011 Dusty played with one of the eggs on top of the paper for a few days in a row. Dusty also tried to feed the egg, in an anticipatory behavior. One day an egg became temporarily stuck on Dusty's brood patch when she was trying to leave for a quick meal. Dusty made a distress call until the egg detached.

In 2012 Dusty laid one egg late on 4/1. She ignored the egg. The egg was given to Maku who was nesting at the time. Maku accepted the egg as one of her own. In 2013 Dusty laid 3 eggs; she cared for them non-stop for a month.

In summary Dusty has made significant progress. Because of her young age there is hope that eventually Dusty will be able to completely overcome her negative behaviors and emotional scars. Only recently Dusty began to take almonds directly from my hand, imitate normal Indian ringneck parakeet vocalizations, and solicit Snapdragon's attention. Earlier Dusty associated with Squeaky (Section 9.3) and Sydney (Section 9.4), probably since they are grey like she is. Today Dusty knows that she is an Indian ringneck parakeet. Dusty celebrated her 8th hatch day 2/26/2015.

10.7 Discussion

Now let's see what these observations tell us about the intelligence, behavior, and individuality of wild and captive Indian ringneck parakeets. In particular let's examine their:

- Temperament
- Vocalizations
- Unique behaviors
- Toy and dietary preferences
- Socialization

Temperament

Maku, Snapdragon, Bluebell, and Dusty have different temperaments, as depicted in Table 10.1. Maku overcame her initial skittishness, nervousness, and fear of people. Her change phobia is still present, but it is not as pronounced as it once was. Maku trusts certain people a little, but she will never be a cuddly bird. Maku explores new things, but cautiously after she has had some time to think about it. Snapdragon is calm, content, playful, and curious. He welcomes interaction with people and new experiences. Bluebell is aloof and self-sufficient. She is neither fearful of nor interested in people. Dusty has made progress in interacting with people and other birds, but slowly.

Vocalizations

Maku, Snapdragon, Bluebell and Dusty made different vocalizations when they arrived. Maku's primary vocalizations when she arrived were alarm calls. At the age of 14 she learned how to make normal Indian ringneck vocalizations. On rare occasions after receiving a special treat or extra attention Maku bows her head and makes a quiet "ahh" sound. She is the only one to make this sound.

Snapdragon's vocalizations are different than the three female Indian ringneck parakeets. Some of his vocalizations may be gender-linked. For example, Snapdragon bows and makes a long drawn out "ki-akh" sound. Snapdragon learned other non-Indian ringneck parakeet vocalizations while he was very young, sitting next to a pionus parrot in the pet store.

Bluebell is the only one who made traditional Indian ringneck vocalizations upon arrival, including trills and a "che-ah" sound while in flight. Maku and Dusty learned these vocalizations from her.

Table 10.1 Spectrum and Prevalence of Emotional States Observed in Indoor Indian Ringneck Parakeets.

	Maku	Snapdragon	Bluebell	Dusty
I. Positive Emotions				
excited	x	x		x
happy		x	x	
sad				
confident/content	x	x	x	
timid/unsure	x			x
lonely				
curious	x	x	x	x
puzzled				x
playful/eager	x	x	x	
anxious/fearful	x			x
aggressive/impatient			x	x
gentle/patient	x	x		x
II. Negative Emotions				
anger				
hatred				
jealousy				
grumpy				
overwhelmed[1]	x			x
stressed out[1]	x			x
possessiveness[2]			x	
self-pity				
depression				
boredom				

Notes:

1—Maku is change phobic. She does not like items in or near her aviary to be moved or changed. At first it took her an afternoon to adjust to changes. Now it is a matter of minutes.

2—Bluebell exhibits possessiveness when people remove and replenish her food and water dishes.

Dusty's primary vocalizations when she arrived were non-stop alarm calls and distress calls. She still goes into that stanza every now and then for no apparent reason. Recently Dusty has begun to imitate Bluebell's vocalizations.

Indian ringneck parakeet vocalizations can be heard on Disk 2 Track 1.

Unique Behaviors

Each of the four Indian ringneck parakeets goes about their daily activities on their own time table—or none at all. Perhaps it would be more accurate to say that each of the birds does what it wants to do when it wants to without a fixed schedule.

Several unique behaviors were observed, behaviors that only one or perhaps two of the birds engaged in. As shown in Table 10.2, 34 unique behaviors were observed. Of these only one bird engaged in 31 (or 91%) of these behaviors. Two of the four birds engaged in 3 (or 9%) of the unique behaviors. Seven (or 20%) of the behaviors were only attributed to the male.

Table 10.2 Unique Indian Ringneck Parakeet Behaviors.

Unique Behaviors	Others who engaged in this behavior	Mimics wild Indian ringneck behavior
Maku		
• Negative reaction to changes in her immediate environment		
• Chewing sisal ropes and toys		X
• Chewing cholla perches		X
• Answering a whistle		
• Nibbling African violets, jade trees, spider plants		X
• Chewing peach-colored mineral block	Bluebell	
• Gargling water in the evening before going to sleep		
• Flying into photographs of parrots near her aviary		X
• Shredding paper, flying around with the shredded paper		X
• Laying eggs 2010 and 2011 after a sudden temperature drop	Dusty	X
• Sitting on eggs on top of shredded paper		X
• Learning new vocalizations at 14 years of age		
• Enjoys misting	Bluebell	
Snapdragon		
• Giggling while eating fruit		
• Bowing		X
• Learned "on your perch" command		
• Waits to be picked up when he gets lost		
• Hangs on the side of his aviary like a woodpecker		
• Doesn't make traditional Indian ringneck vocalizations		
• Makes male-specific vocalizations		X
Bluebell		
• Putting plastic links in water dish, holding them while bathing		
• Throwing first two pieces of fruits or vegetables on the floor		X
• Made normal Indian ringneck vocalizations from the start		X
• Plays under her paper, but didn't lay any eggs until 2013		X
• Chewing on peach-colored mineral block	Maku	
• Rearranging parrot photographs near her aviary		
• Enjoys misting	Maku	
• Sits and plays on low perches		
• Removed two aviary bars		
Dusty		
• Soliciting pats		
• Going inside budgerigar and mitred conure aviaries		
• Bathing in dish on bottom of aviary		
• Battling and banging her dishes, before throwing them on the floor		
• Hanging vertically in a corner		
• Laying eggs 2010 and 2011 after a sudden temperature drop	Maku	X
• Sitting on the eggs underneath her paper		X
• Attempting to feed an egg (anticipatory behavior)		

Note: Indian ringneck parakeet vocalizations can be heard on Disk 2 Track 1.

Some of these behaviors mimic wild Indian ringneck behaviors; others are unique to life in captivity. For example, Maku, Bluebell, and Dusty all engaged in pseudo nesting behaviors in 2010, 2011, and 2013. Yet, their nesting behaviors were different.

Maku shreds her entire paper into tiny bits. Maku bites the paper with her beak to start a hole. Then she holds the paper with her left foot and scratches with her right foot. After she is finished, Maku pushes the shredded paper into the four corners of her aviary. Maku shreds her entire paper daily from late fall through early winter. In contrast, Bluebell and Dusty nibble a few bites around the edge of their papers, and then only for a few days. Bluebell and Dusty sat underneath their paper, like it was a tent. Maku doesn't do that.

Wild Indian ringneck parakeets nest in tree holes 10 to 30 feet (3 to 9 m) above ground. Nesting in captivity required a paradigm shift. Maku acknowledged the aviary frame as the "tree hole" and arranged the shredded paper accordingly. In contrast, Dusty and Bluebell used a paper tent to create an improvised tree hole.

Maku and Dusty laid eggs in 2010 and 2011; Bluebell did not. All three laid eggs in 2013. Maku placed the eggs on top of the shredded paper and sat on top of them. In contrast, Dusty pushed the eggs underneath the paper and sat on them there. In 2011 Dusty played with one of the eggs and attempted to feed it, as an anticipatory behavior.

Interestingly, Maku and Dusty would let their eggs be removed from the aviary when their paper was being changed. Everything was alright as long as the eggs were left in view until they were returned. Conversely, if an attempt was made to simply reach in and grab an egg Dusty went into attack mode and made a distress call.

Bluebell throws the first two pieces of fruit or vegetables on the floor, then eats the remaining pieces. This behavior mimics the habit of wild birds not eating the peels of fruit and vegetables. Bluebell is the only one of the four birds to do this.

Maku's penchant for chewing sisal rope and cholla pieces corresponds to the behavior of wild Indian ringneck parakeets, as does her tendency to nibble on African violets, jade trees, and spider plants.

Several of the more interesting behaviors are linked to life in captivity, as well as expressions of individuality, such as:

- Maku gargling water in the evening before going to sleep
- Bluebell putting her plastic link toy in her water dish every day and holding onto it while swimming in the dish
- Dusty's daily battles with her stainless dishes, before dropping them on the floor
- Snapdragon waiting for someone to come pick him up when he gets lost

Unique behaviors illustrate the birds' different temperaments. For example, Maku flies toward and hovers in front of the parrot photographs near her aviary in attempt to interact with them. In contrast, Bluebell knocks the photographs around until they are crooked. Dusty is the only one to solicit pats.

Toy and Dietary Preferences

The four Indian ringneck parakeets demonstrated distinct preferences in regard to toys and food. As shown in Table 10.3, all the birds played with natural fiber toys and shredding toys. But when it came to bells, shiny metal toys, and sisal ropes and toys, only one of the four birds played with them. Interestingly, toy color had no effect on toy preference. There were also differences in the time intervals each bird played with a toy. Maku and Snapdragon are more likely to play with a given toy for a long period of time. Bluebell and Dusty tend to play with toys for a shorter period of time. Bluebell and Dusty played with the largest variety of toys.

The four Indian ringneck parakeets demonstrated distinct diet preferences, as shown in Tables 10.4 and 10.5. Maku and Snapdragon have similar tastes. Bluebell eats the smallest variety of fruits (5). Snapdragon eats the largest variety of fruits (10). Only Snapdragon and Dusty eat banana. Vegetable preferences are somewhat similar. Only Maku and Snapdragon eat broccoli. Only Maku eats squash. Snapdragon is the only one that eats fruit flavored pellets. Maku and Bluebell prefer natural pellets. It took several years before Dusty would eat any pellets.

A greater spectrum of preferences is seen in the four birds' grain, nut, seed, and legume preferences. Other than long grain brown rice, cooked whole grains and legumes are not very popular.

Table 10.3 Individual Indian Ringneck Parakeet Toy Preferences.

Toy	Maku	Snapdragon	Bluebell	Dusty
bells				x
cardboard chewing toys			x	
cholla blocks, wood pieces	x		x	
cotton ropes and toys	x	x	x	x
foot toys				
ladders				
mirrors, shiny metal				x
natural fiber/palm leaf toys	x	x	x	x
push pull toys		x	x	x
puzzle toys				
rattles				
shredding toys, piñatas, yucca pieces	x	x	x	x
sisal ropes and toys	x			
swings				
wiffle balls				

Note: Toy color did not affect toy preferences.

Snapdragon and Dusty will also eat cooked barley, but only once in a while. Contrary to popular lore and advertising, raw tree nuts are not very popular either. Bluebell, Maku, and Snapdragon avoid tree nuts altogether. Surprisingly Dusty is the most adventurous when it comes to tree nuts. She will eat a small quantity of almonds, cashews, and pistachios. Seed and legume preferences are fairly similar with two exceptions. Maku and Dusty are the only ones to sample dried corn, although all four relish popped popcorn. And, when it comes to peanuts, Snapdragon and Dusty will only eat them if they are shelled. Snapdragon chews on peanuts in the shell as if they were a shredding toy, then drops them on the floor. Dusty simply throws peanuts in the shell on the floor. Oddly, Maku will not eat roasted unsalted human grade peanuts; she only eats raw parrot grade peanuts. Bluebell will eat any type of shelled or unshelled peanuts.

Socialization Patterns

Likewise, the four Indian ringneck parakeets displayed different socialization patterns, with humans and other birds. Maku likes a little attention from humans, but not too much. Maku has consistently displayed an interest in socializing with Snapdragon, after he grew his neck ring. Maku has also displayed a consistent interest in socializing with the budgerigars, even though they are less than half her size. At the same time, Maku avoids the other two female Indian ringneck parakeets, the mitred conures, the Meyer's parrot, and the red-bellied parrot (*Poicephalus rufiventris*).

Snapdragon is always ready to interact with humans. He prefers to socialize with Bluebell and usually doesn't venture too far from her.

Bluebell has no interest in humans. She visits Snapdragon, but spends more time socializing with the budgerigars and zebra finches. Like Maku, she does not socialize with the other two female Indian ringneck parakeets.

At this point in time Dusty craves human attention and pats. Dusty socializes briefly with the budgerigars and mitred conure that are near her, but she avoids the other Indian ringneck parakeets. Early on Dusty socialized with the Meyer's parrot and red-bellied parrot; possibly because they are

Table 10.4 Individual Indian Ringneck Parakeet Fruit, Vegetable, and Pellet Preferences.

	Maku	Snapdragon	Bluebell	Dusty
I. Fruits				
apple	x	x	x	x
banana		x		x
blueberries				
cantaloupe	x	x	x	
cherries	x	x	x	
grapes	x	x	x	x
kiwi berry				
mango—dried mango—fresh	x	x		x
orange	x	x	x	x
papaya—dried papaya—fresh	x x	x x		x
peaches	x	x		
strawberries				
II. Vegetables (fresh)				
broccoli	x	x		
carrots	x	x	x	x
cherry tomato				
corn—yellow sweet	x	x	x	x
dandelion leaves				
green peas	x	x	x	x
kale				
red peppers	x	x	x	x
spinach				
squash	x			
III. Pellets				
fruit flavored		x		
natural	x		x	x

grey like she is. While Dusty enjoys visiting the top and inside of the aviaries of other bird species, attempts for her to socialize and move in with other Indian ringneck parakeets has been disastrous. Dusty viciously attacks the Indian ringneck parakeets without provocation.

Wild Indian ringneck parakeets are known to travel around in flocks of 100 to 1,000 birds. Hence, Maku, Bluebell, and Dusty's tendency to avoid interacting with each other is unusual.

In summary, four Indian ringneck parakeets, three female and one male, were observed over a 10 year period. Three were adopted at various ages; one was acquired from a pet store as a juvenile. While their previous history is different, they experienced an identical living environment here. Even so, Maku, Snapdragon, Bluebell, and Dusty are distinct individuals with different temperaments, vocalizations, daily activities, toy and dietary preferences, and socialization patterns. Their 12 year age difference might explain some, but certainly not all of these differences. Simply put Maku, Snapdragon, Bluebell and Dusty are unique individuals. Individual choices, preferences, and behaviors indicate intelligence and self-awareness.

Table 10.5 Individual Indian Ringneck Parakeet Grain, Nut, Seed, and Legume Preferences.

	Maku	**Snapdragon**	**Bluebell**	**Dusty**
I. Whole Grains and Legumes (cooked)				
barley		x		x
buckwheat				
lentils				
long grain brown rice	x	x	x	x
millet				
quinoa				
rye				
wheat, spelt, kamut				
II. Tree Nuts (raw, shelled)				
almond				x
Brazil nut				
cashew				x
filbert				
pecan				
pistachio				x
walnut				
III. Seeds and Legumes (raw)				
dried corn	x			x
dried peas and beans				
millet seed (loose)				
millet spray		x	x	x
milo				
oats				
peanuts	x	x	x	x
pumpkin seeds		x	x	
safflower	x	x	x	x
striped sunflower	x	x	x	x

11
Mitred Conure
Aratinga mitrata mitrata

11.1 Background

Forshaw, a leading authority on parrots worldwide, was the first to point out that very little has been published about mitred conures.[60] Articles in the scientific literature are practically non-existent. The few articles that are published merely repeat what is in Forshaw's books. Two of the four pet bird hobbyist books dedicated to conures do not even mention mitred conures at all. The other two books have a scant one page description of a mitred conure's appearance and note that they are a bit noisy; as if anyone who has ever dealt with conures wouldn't know that? One veterinarian asked what kind of parrot Buster (Section 11.4) was. Another veterinarian asked if he could take a photograph of Buster since he had never seen a mitred conure. Both veterinarians are Board Certified Avian Practice. Even the omnipotent Microsoft has never heard of mitred conures. Microsoft Word™ flags both "mitred" and "conure" as spelling errors.

The lack of publications correlates to a lack of firsthand field observations. The few publications that exist claim that mitred conures have bluish tinged feathers on the back of their head. Neither of the two mitred conures discussed in this chapter have any feathers anywhere on their body that could remotely be considered to be blue. The disconnect between the published literature and firsthand field observations of mitred conures was noted in the Winter 2006 newsletter of the International Conure Association.[196]

Mitred conures, *Aratinga mitrata mitrata*, were named by Tschudi in 1844.[203] Because so little is known about mitred conures the debate continues about the number and alignment of *Aratinga mitrata* subspecies.[60, 94, 196, 203] Mitred conures are closely related to Wagler's conure, *Aratinga wagleri*, and their ranges overlap.[175] Outside the United States, mitred conures are occasionally referred to as mitred parakeets.[203]

Mitred conures are generally not available in the pet industry, in part because they do not breed well in captivity.[175, 196] To illustrate, the first captive breeding report in Germany was made in 1982 and there was only one hatchling.[175] The second captive breeding report in Germany did not occur until 1984.[175]

The majority of mitred conures living in captivity are found in the United States, primarily in the southwest.[202] Most of these are wild caught birds that were brought across the border legally or illegally. In the 1980's before parrot conservation laws went into effect, large numbers of mitred conures were captured in Bolivia and traded on the international market.[94] Neither of the two birds discussed in this chapter arrived with leg bands. It is possible that one or both birds were wild caught.

Although mitred conures are rare in the pet industry, they do pop up in the parrot adoption arena. A recent scan of petfinders.com revealed a dozen such birds awaiting adoption. Many of them were severely plucked. It would appear that knowledge of how to provide for a mitred conure's physical, nutritional, social, and intellectual needs in captivity is as rare as knowledge about mitred conures in general.

The current IUCN conservation status of mitred conures is rated "least concern". That rating is conveniently given to all birds that are classified as "agricultural pests". This label relieves a bird's native country of any responsibility for protecting the species from smuggling or extinction. But, one must question how mitred conures could be classified as agricultural pests when so little is known about them and even less is published. As we will see later, the diet of wild mitred conures is unknown.[196] So, where is the peer-reviewed scientific evidence that proves: (1) mitred conures are agricultural pests, and (2) mitred conures even exist in large enough numbers to pose a threat to agriculture?

11.2 Characteristics

Mitred conures share several physical and behavioral characteristics. Let's examine these characteristics as they apply to wild and captive mitred conures, such as:

- Range and habitat
- Diet
- Physical characteristics
- Vocalizations
- Socialization
- Family life

Range and Habitat

Mitred conures home territory is a narrow stretch of land that starts in central and southern Peru and runs through Bolivia and into northwest Argentina.[60, 94, 196, 202, 203]

Mitred conure populations have been introduced in California (1991), Florida (1989), and Hawaii (1990).[147, 202, 203] Undoubtedly all of these birds were wild caught. The populations in California and Florida were counted regularly from 1991 through 2006. Both populations started as a small group of individuals. Over the years there have been significant fluctuations in the population totals:[147]

- California
  1995: 175 birds
  2000: 50 birds
  2006: 325 birds

- Florida:
  1994: 50 birds
  1999: 0 birds
  2000: 225 birds
  2004: 175 birds
  2006: 25 birds

These fluctuations tend to point to the release of new birds, rather than an indigenous population sustaining itself. In comparison the population of mitred conures in Hawaii (Maui) was reported to be 200 in 2003.[203]

Mitred conures live in a variety of habitats within their native range, such as dry subtropical vegetation, deciduous forests, valleys, and semi-cultivated open grasslands.[94, 196, 203] Mitred conures tend to stay in areas that are 3,000 to 13,300 feet (914 to 4,053 m) in altitude.[94, 196] Mitred conures do not migrate, but there is some seasonal movement based on the availability of food.[94, 196]

Diet

The exact diet of wild mitred conures is unknown, given the limited number of field observations.[196] According to current speculation mitred conures eat grain, nuts, and berries.[94, 196, 203] This speculation is based on observing captive mitred conures and other types of wild conures.

Physical Characteristics

The unique physical characteristic of all conures are their wedge-shaped tails. In fact publications in the 1800's referred to conures as "wedge-tailed parrots".[81] Mitred conures are 15 inches (38.1 cm) in length.[60, 94, 175, 202, 203] A mitred conure's long wedge-shaped tail accounts for 6.75 inches (17.1 cm) of its length.[60] The wingspan varies from 18 to 18.5 inches (45.7 to 47 mm). The normal weight for a mitred conure is 8.0 to 8.6 ounces (224 to 240.8 grams).[175, 203]

The front, back, wings, and tail of a mitred conure are bright green and reflective in direct sunlight. The underside of the tail displays slight yellow tones. Mitred conures have a bright red forehead. An irregular red pattern continues around the eyes and over the ears. When a mitred conure molts, these small red feathers do not always grow back in the same pattern. Mitred conures also have a narrow band of red feathers around the top of each leg. Every now and then mitred conures grow a couple small red feathers on the bend of their wings.

Their large eyes are a dark orangish brown, surrounded by a white ring. The thick hooked bill is a yellowish bone color and about 1.5 inches (3.81 cm) in length. The legs and feet are flesh colored. The outer toes are about 1 inch (2.54 cm) in length, while the inner toes are about 0.75 inch (1.9 cm) in length. Mitred conures have long black nails and a stout build. Because of their large head and beak mitred conures have thick neck muscles.

Male and female mitred conures have the same appearance. Immature mitred conures have less red feathers on their forehead and their eyes are a lighter color.[94] The average life span of mitred conures in captivity is believed to be 25 years. While the maximum life span of mitred conures in captivity is believed to be 30 years.[203]

Vocalizations

The few publications that exist agree that mitred conures can be noisy. That assertion is consistent with the observations discussed in this chapter! Other adjectives used to describe mitred conure vocalizations include: "harsh", "strident", "loud", and "raucous".[94, 196, 203] Three vocalizations have been attributed to wild mitred conures:[94, 203]

- "wee-wee-quiee-quiee-wee-wee"
- "cherah-chcrah"
- "ker-EET ker-EET"

However, neither of the mitred conures discussed in this chapter made any of these vocalizations. Mitred conure vocalizations can be heard on Disk 2 Track 3.

Socialization

Wild mitred conures are known to travel around in two's, three's, and small groups.[60, 94, 196] They are noticeable primarily because of the noise they make rather than visual sighting.[196] Mitred conures associate with other conures, in particular Corriker's conure, *Aratinga wagleri minor*, and the sharp-tailed conure, *Aratinga acuticaudata acuticauda*.[196] Mitred conures have been observed flying long distances and at considerable heights when they are looking for food.[196]

Reports from the pet bird hobbyist community are inconsistent in regard to mitred conures interaction with humans and socialization with other birds. Some sources described mitred conures as being "playful, active, and intelligent".[203] Other sources described mitred conures as being shy around new people and new things.[175] Mitred conures were also described as being nippy and having a strong personality.[196] As we will see later the first description matches Rudy (Section 11.3), while the third description matches Buster (Section 11.4). The second description does not match either bird discussed in this chapter.

Family Life

Since they live in the southern hemisphere, wild mitred conures breed between November and February.[196, 203] For the most part mitred conures nest in cliff hollows in Peru and Bolivia and in holes in dead trees in Argentina.[94, 175, 196, 203] In either case, the nests are up high.

Two or three eggs are laid, 48 hours apart.[60, 175, 203] The white oval eggs measure 1.2 x 1.0 inches (3.0 x 2.54 cm) and weigh 1 ounce (28 gr).[60, 175, 203] The mother incubates the eggs for about 23 days and is fed by the father.[196, 203]

The eggs hatch in the order they were laid. The hatchlings weigh in at 0.4 ounces (11.2 gr).[203] Both parents feed the hatchlings.[196] The hatchlings eyes open when they are two weeks old.[175] They are ready to fledge around 9 weeks of age.[196, 203]

From the time they fledge until they are 2 years old, mitred conures are considered to be juveniles. Captive juvenile mitred conures are considered to be playful, open to exploring new things, and interested in interacting with new people.[196]

From the time they are 2 years old until they are 4 years old, mitred conures are considered to be adolescents. Captive adolescent mitred conures are reported to be less open to new people and new experiences than they were as juveniles; they are also known to become a bit stubborn and nippy at this time as they develop a sense of self.[196] This behavior is similar to the terrible two's that Meyer's parrots and red-bellied parrots go through (Chapter 9).

Mitred conures over 4 years of age are considered to be adults and they begin to mellow over time.[196] Since wild mitred conures are known to travel in two's, three's, and small groups, it is likely that young mitred conures stay with their families until they are adults.

Now let's look at the characteristics of indoor mitred conures in comparison to their wild cousins, for example their behavior, daily activities, and life stages. Sections 11.3 and 11.4 describe the behavior of individual mitred conures, in particular play activities, socialization, vocalization, and reaction to changes and challenges throughout life. Behaviors are presented as they occurred. The analysis, comparison, and interpretation of these behaviors are reserved for Section 11.5 Discussion. These observations took place at the locations and during the timeframes described in Annex C.

11.3 Rudy

Rudy is a male mitred conure. He was adopted 6/1/2006. This is Rudy's second home. Rudy had the good fortune to spend 20-plus years in a stable loving home environment. The only reason Rudy was put up for adoption was that his caretaker was no longer able to take care of Rudy (or his other birds) due to a debilitating illness and advanced age. This is a situation that everyone who has pet birds worries about, dreads, and hopefully plans ahead for. Rudy does not have a leg band.

Rudy's first afternoon here was memorable, but not for the right reasons. The local parrot adoption organization dropped Rudy off—let Rudy loose in the room—and ran out the door. No attempt was made to orient Rudy toward his new home or calm him; hence the transition was very abrupt. Rudy ended up on top of a bookcase. He was scared and disoriented and began wheezing. Eventually Rudy found his way to his new aviary and sat on top of it.

Rudy's immense sense of sadness, rejection, and abandonment were palpable. He had no idea why he had been taken away from his steady home of 20-plus years, his loving caretaker, or the other birds he had lived with. Rudy did not understand why he had been deposited in this strange room and so abruptly. Rudy just sat still, sad, and dejected with his head hanging low all afternoon, all evening, and the next morning. Rudy didn't eat or drink anything; nor did he make a sound. Rudy's behavior was identical to that of a bird going through a mourning period for a lost mate.

Rudy was given a lot of extra attention in an attempt to make him feel at home. He looked up briefly but that was about all, since we were still strangers. Something dramatic was needed to make Rudy feel welcome in his new home.

Saturday is the normal swimming time for all the birds in the aviary. I made sure Rudy saw me putting bird baths in each aviary. Then I put a bird bath on the bottom of Rudy's aviary. Rudy watched the other birds swimming for a few minutes, then he ran down to the bottom and began splashing away too! From then on Rudy decided his new home would be alright. He began eating and interacting with his new bird neighbors.

Swimming remains one of Rudy's favorite activities. He swims in a 10.0 x 2.0 inch (25.4 x 5.08 cm) olive green plant saucer. Rudy is in the water before the door to his aviary is closed. Rudy steps in the middle of the water, dunks his front a couple times, then goes completely under and starts flapping his wings to spread the water on his back. In the process Rudy becomes completely drenched and water goes flying every which way. Rudy does not take quick baths. He spends 10 to 15 minutes getting soaked, climbs up to a top perch for a minute or two, then climbs back down for a repeat performance. Because Rudy is a parrot and not a water bird, his feathers are not waterproof. When Rudy swims his bright green feathers take on a brownish hue. (See Photograph 11.1.)

Rudy quickly learned that Saturday was swimming day. If it were Saturday and Rudy felt it was past time for the swimming pool to be placed in his aviary, he would hold his head underwater in his water dish and make a mournful sound to remind us! Sometimes Rudy even wanted to go swimming before breakfast. Somehow Rudy knows when it is Saturday and he equates Saturday with swimming. Only about once a year Rudy confuses Friday with Saturday and indicates that he wants to go swimming then. Unlike all the other parrots discussed in this book Rudy never drinks water out of the swimming pool. He understands that water is only for swimming. The other birds take a drink first.

One Saturday afternoon there was a knock on the door while Rudy's swimming pool was being prepared. Rudy's swimming pool was sat on the floor in the middle of the room while the door was answered. Rudy was out roaming around at the time. Rudy recognized his swimming pool and decided that he was meant to swim on the floor in the middle of the room that day, instead of in his aviary. Rudy walked over to the pool and began to splash away as usual. In short, Rudy recognized and interpreted the new situation correctly—more or less!

Rudy's olive green swimming pool is made of plastic. As a result it is flexible. To prevent the possibility of the green dish from twisting, spilling, or flipping over on Rudy, a stainless steel dish the same size was acquired. Even though Rudy's food and water dishes are stainless steel, he refuses to swim in the stainless swimming pool. For some reason the shiny steel swimming pool scares Rudy. The steel dish was quickly retired and the green plastic dish was promptly reinstated.

Because he came from a good home Rudy was used to a healthy diet. Rudy loves his fruit and vegetable dish. He jumps on it every morning before it is placed in the hoop. Broccoli, carrots, fresh yellow sweet corn, dried papaya, apples, oranges, and bananas are his favorites. Sometimes Rudy picks up a piece of broccoli, which is moist since it was just rinsed, and uses it like a powder puff to wash his beak. Other days Rudy plays a game with his broccoli. He picks up the broccoli first and

Photograph 11.1 Swimming. Rudy's favorite pastime is swimming and he never misses an opportunity to get completely soaked. His bright green feathers take on a brownish hue when wet, unlike Canada geese and mallard ducks whose feathers are waterproof and don't change color. Rudy starts to swim with a couple quick dunks to get the water churning. Then Rudy does a belly flop and flaps his wings to spread the water on his back. For several months Rudy would not let anyone photograph him while he was swimming. © D. Herrmann

drops it on the floor. Later, after he has eaten everything in his fruit and vegetable dish, Rudy climbs down to the tray and is delighted to "find" the broccoli. Rudy eats the broccoli florets one tiny bite at a time, then shreds the stalk with gusto. On occasion Rudy also hides pieces of dried papaya to "find" later. (See Photograph 11.2.)

Photograph 11.2 Vegetable Break. Rudy takes a break from flying to enjoy some fresh yellow sweet corn. Neither Rudy nor the other parrots will eat white or bicolor sweet corn. Perhaps they associate white kernels with unripe corn. Note that Rudy stands on his right foot and holds the sweet corn in his left foot. The small band of red feathers on the top of his legs is just visible. © D. Herrmann

Rudy relishes natural pellets. Natural pellets are Rudy's favorite treat when he is out flying. He likes to be hand fed pellets then, one at a time. Rudy will eat up to eight pellets this way at a time.

Rudy is not interested in nuts or peanuts at all. Rudy treats peanuts like shredding toys. If Rudy is hand fed a peanut, he will shred the shell quickly, letting the peanuts fall to the floor. Shelled almonds, walnuts, cashews, and such are simply ignored.

Rudy is the only large bird discussed in this book to consume cuttlebones. About a year after Rudy arrived he learned about cuttlebones by accident. One day he was climbing and took a bite of cuttlebone by mistake. Rudy decided that cuttlebone was alright. After that Rudy began taking a bite of cuttlebone after dinner most days.

Conversely, Rudy has shown no interest in mineral blocks, not even manu mineral logs. The latter was surprising since manu mineral logs are made from material found in South American clay licks that are popular with many parrots indigenous to that continent. Some theories purport that these clay licks contain minerals and micro-nutrients that are essential to the diets of South American parrots like Rudy. However, as Lee points out clay binds (or absorbs and neutralizes) toxins in the body so they can be expelled.[102] Consequently, South American parrots may be consuming this clay for therapeutic rather than nutritional reasons.[102] That might explain why Rudy ignores manu mineral logs.

Rudy's vision in his left eye is not as accurate as his right eye, particularly when he is trying to view things up close. Before taking a treat from my hand Rudy turns his head from left to right to view the item separately from each eye, as if he is triangulating the coordinates. If Rudy gets excited and forgets to do this, he often misses the treat by an inch (2.54 cm) and/or bites a finger by mistake. Whenever this happens Rudy realizes his error instantly and pulls his head and beak back. Rudy then lowers his head and backs away as if to apologize and say he is sorry. Rudy has to be encouraged to come back a second time for the treat. When Rudy eats a treat he stands on his right foot and holds the treat with his left foot. (See Photograph 11.3.)

Photograph 11.3 Snack Time. Rudy munches a piece of egg bread. Rudy enjoys a healthy diet and is open to trying a variety of new foods. Note that Rudy stands on his right foot and holds the bread with his left foot. © D. Herrmann

Rudy's favorite treat is warm popped plain popcorn. Rudy devours his popcorn with energy and enthusiasm. Rudy gets very excited as soon as he hears and smells the popcorn popping. Rudy starts rocking back and forth doing a little dance, but he doesn't make a sound. Rudy watches and waits patiently as all the birds are served popcorn. He is confident knowing that his turn will come and doesn't feel any need to make a fuss.

One day out of the blue Rudy flew over and landed on the back of a chair near me while I was having a salad. Rudy seemed really interested in something. Perhaps it was the salad. Or, perhaps it was the social aspect of sharing a meal. Normally I would have shared the salad with Rudy. But I had put a lot of salt on the salad, which was not good for him. Instead, I made Rudy a plate of carrots and broccoli and sat it in front of him. Rudy looked at the plate for a couple seconds, then leaned over and took a big bite out of the flowering snapdragon, *Antirrhinum majus,* that was on the table! Rudy broke off a stem at a time and munched away. Fortunately snapdragons are a bird safe plant. That is the only time Rudy landed on a chair back or ate the snapdragon (or any other plants). It is unclear what prompted this unique behavior on that particular day.

Rudy loves chilled filtered mineral water. He understands that the water is coolest when it is served and usually takes several large drinks right away. Unlike most parrots, Rudy keeps his water dish clean.

Rudy is not a big fan of toys. He engages in plenty of other activities to keep busy. His favorite toys are the three that came with him from his original home. All three toys are well worn, but he prefers them to new toys, probably because of the long and pleasant memories associated with them.

The first toy consists of a 3.0 x 5.0 inch (7.62 x 12.7 cm) semi-shiny metal plate that hangs from a chain. A 4.0 x 0.5 inch (10.16 x 1.27 cm) piece of manzanita also dangles from the chain. Rudy pokes, prods, and chews this toy gently every day. Sometimes Rudy partially hides behind this toy when he takes a nap. The metal part of the toy is store bought. The piece of manzanita was added by hand. When I first unpacked this toy and began to hang it in Rudy's new aviary, I got a glimpse of how much Rudy's previous caretaker had cared for him. The O-ring that attaches the piece of manzanita and the top link in the chain had been soldered by hand to ensure that Rudy didn't damage his beak or cut his tongue on any irregular or sharp edges. Wow! Of the thousands of people around the world who have pet parrots, how many would have thought of that? Commercial toys are not even that safe.

The second toy consists of a metal chain, some pieces of cholla, and a bell on the bottom. Rudy bites the pieces of cholla four or five times a day and pushes the toy around gently. Once in a while Rudy bites through a cholla piece and it has to be replaced.

The third toy goes with Rudy's favorite perch: a hand cut piece of hardwood. This perch also reflects the handiwork of Rudy's original caretaker. This perch is special because it arches up about 45° before leveling off. Hence it is much more interesting than the typical flat store bought perches. Overtime Rudy has chewed the bark off, so it is hard to tell what kind of wood it is. The third toy hangs above the middle of this perch. This toy has been around a long time too and is missing a few pieces. Faded yellow, purple, and red wooden pegs remain, all of which are well chewed.

Rudy loves his naps and as a senior bird he is entitled to. His favorite place to nap is on the top of his aviary on the back slope near the wall. Rudy curls his head around under his left wing for a serious nap. Rudy raises one feather about 0.5 inch (1.27 cm) long from the top of his wing to cover his eye. If something disturbs Rudy while he is sleeping, he lowers that feather and peers out. Considering that most birds have about 3,000 feathers on their body, Rudy demonstrates an incredible ability to isolate an individual muscle to raise and lower this one feather. When Rudy is in this position his left or bad eye faces the wall, while his right or good eye faces outward. This behavior indicates that Rudy understands his own limitations. (See Photograph 11.4.)

Rudy also takes naps on the bottom of his aviary. He lies on the paper with his feet flat and his head and beak snuggled up in either the right or left front corner of the tray. Rudy sleeps soundly in this position. His feet are relaxed and he doesn't have to hold up his head and beak—the heaviest part of a parrot. However, Rudy doesn't want anyone to know he is down there. If someone walks in the room when Rudy is napping on the paper, Rudy awakens quickly and scampers up to a perch. This behavior is similar to the naps Uzziel (Section 7.3) took on her paper during her senior years.

Rudy acts like he is the papa bird in the aviary, probably because he is the largest and the oldest bird. Rudy loves to visit all the other birds, flying and hopping from the top of one aviary to another. Rudy watches the other birds, talks to them, and checks out their toys. His favorite place to hang out is on top the zebra finch aviary. Rudy even naps there and rubs his beak on the wall. As noted in Chapter 8, the zebra finches are not fazed at all by having a big green bird five times their size snooze on top of their aviary. (See Photograph 8.11.)

Rudy is generally very prompt about bedtime. Usually before 9 p.m. Rudy checks in for the evening. Most of the time Rudy finds his way home by then and climbs up to his sleeping perch—the handmade hardwood perch. Once in a while I have to remind him by ringing his bell and saying "bedtime Rudy". Rudy also understands it is time to head home if I flip the lights on and off briefly. If it is Friday or Saturday night, sometimes I relent and let him stay out all night. This tickles Rudy to no end. However, the next day he is very sleepy and hungry. On the other hand, if Rudy is already settled in on his sleeping perch for the night and he thinks the other birds or I are being too noisy, Rudy rings the bell as if to say "quiet down, it is bedtime".

Rudy regularly participates in flying attacks two or three times a day, even if he has just been out flying. To do so Rudy climbs down to the center bottom back wall of his aviary. Next Rudy carefully stretches out his wings to make sure they fit in the available space without bumping anything. Once he is satisfied with that, Rudy starts flapping his wings as hard as he can for 30 to 40 seconds at a time. While doing so Rudy holds onto the aviary with his beak and feet. Sometimes Rudy has one

Photograph 11.4 Nap Time. Rudy curls up for a nap in a comfortable yet safe place. Rudy acknowledges his visual limitations even while napping. His left (weak) eye faces the wall, while his right (good) eye faces forward. Note the small feather in front of his eye that is partially lowered so he can check out the commotion the camera is making. Rudy is sleeping on his left foot. If need be he can lean against the wall for balance. This is something senior parrots tend to think about more than the younger ones. © D. Herrmann

flying attack, pauses for a few seconds, then starts flapping his wings again. Rudy bobs his head up and down in sheer joy after he is finished. Flying attacks are a conditioning exercise to keep the wing muscles in shape, since a bird can beat its wings harder and faster than it can when flying in a confined space. Rudy kicks up a fair amount of turbulence when he is having a flying attack. Even the vertical blinds 6 feet (1.82 m) away from his aviary move! (See Photograph 11.5.)

Rudy is a strong, fast, precise flyer. Rudy quickly learned the schedule for flying time, just like he learned that Saturday's meant swimming time. As flying time approaches, Rudy gets excited and starts doing a little dance on the highest perch. Rudy races out his aviary door as soon as it is opened. He climbs from his low landing perch to his high landing perch to the top of his aviary in seconds. Next Rudy fluffs all of his feathers once or sometimes twice. Then Rudy stretches out his right wing and off he flies. Rudy is airborne in less than 10 seconds after his door is opened. For some unknown reason Rudy never stretches his left wing before takeoff.

Rudy flies happily from place to place, partly to go visiting and partly for fun and a chance to burn up some excess energy. He understands that flight and feathers are what makes a bird a bird. Rudy flies straight lines, at right angles, and even makes U-turns depending on where he wants to

Photograph 11.5 Flying Attack. Rudy prepares for a flying attack by slowly stretching out both wings to make sure there is enough room before he starts flapping. Rudy holds onto the horizontal cross bar with his beak and the vertical bars with both feet. Through this anticipatory behavior Rudy demonstrates that he understands spatial relationships. © D. Herrmann

land. On his first flight after a cross town move Rudy flew the entire 24 foot (7.31 m) length of the new aviary. Rudy even headed through the doorway and into the next room as if he knew his way around. The new floor plan may have reminded Rudy of his original home. At the same time Rudy began doing new things he had never done before, like letting himself out through the small access door, when he didn't want to wait for flying time.

Nothing fazes Rudy. One time when Rudy was flying home he forgot to triangulate the coordinates between his good eye and his bad eye. He crashed and slid down the wall about a foot (30.48 cm) from his aviary. Rudy didn't make a sound. He just stood up, shook himself off, and removed the three broken flight feathers. Then Rudy climbed up into his aviary and went to sleep for the night as if nothing had happened.

Rudy likes to visit the other birds when he is out flying. However, if an Indian ringneck parakeet *Psittacula krameri manillensis*, Meyer's parrot *Poicephalus meyeri*, or red-bellied parrot *Poicephalus rufiventris* is out flying at the same time Rudy retreats to the back corner on top of his aviary. In this situation Rudy makes a quiet but steady "caw-caw-caw" sound. This sound is similar but lower pitched than the hawk alarm call the cherry-headed conures, *Psittacara erythrogenys* make in Chapter 8 of the Wild Parrots of Telegraph Hill DVD.[21, 187] Rudy also makes this sound if he sees a peregrine falcon *Falco peregrinus* flying outside the window. Rudy probably feels vulnerable because of his age and poor eye sight, even though there has never been any aggression when he was out flying. Most likely Rudy cannot tell who is flying rapidly toward him. Rudy does not retreat or make this sound when the budgerigars *Melopsittacus undulatus* are out flying.

Rudy gets excited at the littlest things. He even finds having his paper changed exciting. Rudy runs back and forth on the grill and watches his tray being cleaned, as if this is special. After Rudy had been here a year he began exploring on foot, in addition to his flying and climbing. Rudy slides down the outside of his aviary, hops on the floor, and roams around on foot for a half hour or so. Rudy climbs around and under everything in the room. He doesn't seem to have any plan or purpose in mind other than exploring. When Rudy is finished he climbs up the ladder and back into his aviary.

Rudy does not appreciate pats or head scratches; he is too independent for that. Instead, Rudy likes company, to be in the same room with people, and for people to talk to him. In other words, Rudy prefers peer-to-peer social interaction, rather than parent-to-child social interaction. Rudy tolerates an occasional short ride on someone's arm or shoulder; but he does not want to spend an extended period of time there.

When Rudy is extremely content he stands on his left foot, turns his head upside down, and scratches the top of his head with his right foot, while his eyes are closed. If someone talks softly while he is engaging in this activity, Rudy is in his utmost state of contentment. Rudy also loves to sit in direct sunlight for an hour or so first thing each morning. (See Photograph 11.6.)

Photograph 11.6 Head Scratch. Rudy turns his head upside down and scratches the top and around his eyes, slowly and gently with his foot. Rudy does this when he is most content. Note that his eyes are partially closed, indicating a relaxed state. This behavior mimics the social allo-preening Rudy would engage in if he were in the wild and/or part of a bonded pair. While Rudy enjoys this, he does not permit humans to scratch his head. Rudy stands on his left foot and scratches with his right foot. This is the reverse footing of that which Rudy uses to eat. The small band of red feathers that circles the top of his legs is just visible. © D. Herrmann

Rudy bows whenever I enter the room, and has done so since the second day after he arrived. When I bow in return Rudy reciprocates. Then to change the pace Rudy nods his head quickly. I nod back. We repeat this exchange a few times. Then all of a sudden Rudy does a quick double nod—probably just to see if I can keep up! This half greeting ceremony, half game delights Rudy each time we do it, just like the first time. If for some reason I have to enter the aviary after it is dark and

Rudy has been asleep for a while, I speak softly to let the birds know it is me so they won't panic. Rudy nods a couple times after he is awake to acknowledge that he recognizes me. Rudy frequently combines winks with his bows and head nods.

In November 2006 a baby normal green Indian ringneck parakeet named Snapdragon (Section 10.4) joined the aviary. The other Indian ringneck parakeet, Maku (Section 10.3), ignored Snapdragon. In contrast Rudy was ecstatic! For the next several weeks Rudy went nuts whenever Snapdragon was carried over to meet him. Rudy couldn't get close enough to Snapdragon and kept trying to feed him. Rudy probably thought Snapdragon was a mitred conure hatchling, since they have solid green heads.

Rudy is meticulous about his appearance and has large brown expressive eyes. Rudy is active, intelligent, and inquisitive. Rudy figures out new toys, new situations, or where things are after they have been rearranged immediately without missing a beat. Rudy also has superb communication skills. It is uncanny how well he senses what a person is saying or feeling. For example, when Rudy is out flying I point to him, then point to his aviary. Rudy understands immediately that it is time for him to go home. Rudy is the only bird discussed in this book to do that. And, I only have to do this once. (See Photograph 11.7.)

Photograph 11.7 At Home. Rudy rests on the top landing perch outside of his aviary. Rudy uses the landing perches to climb up to the top of his aviary. Note that Rudy's large brown eyes almost fill the white elliptical eye ring. The inner front toes are shorter than the outer front toes. The underside of the tail feathers is yellow. © D. Herrmann

Rudy is equally as effective at communicating his wants. Most of the time Rudy is quiet and communicates via body language: changes in posture, activity level, and of course his eyes. Other times Rudy vocalizes, a prolonged "wha—at". Rudy varies the pitch, inflection, and timing to indicate that he is happy, excited, wants to come out, wants a snack, and so forth. In spite of his eagerness, Rudy is never loud or demanding. He may be persistent on occasion, but he is always gentle, patient, and appreciative in return.

In summary, Rudy is a sweet, loving, affectionate, alert, active, intelligent bird. Every interaction with him is special and makes him happy. Rudy arrived here a fully polished gem, as a tribute to his original long-term caretaker. All of the birds discussed in this book have been special, each in their own way. But the extent of communication, interaction, and trust with Rudy is on a totally different level.

11.4 Buster

Buster is a male mitred conure. He was adopted on 9/27/2008. This is Buster's third home. Before arriving here Buster had been relinquished to the local parrot adoption agency, adopted out, and returned three years later. Buster's age and hatch date are unknown. The best guess is that Buster was between 5 and 7 years old when he arrived. Buster does not have a leg band.

Some alarm bells were triggered upon seeing Buster for the first time. Buster had been billed as a mitred conure, yet he was only two-thirds the size of Rudy. Buster's size was that of a cherry-headed conure, not a mitred conure. Buster had the coloring of a mitred conure, including the irregular red feather pattern around the eyes and ears. Yet, his eyes were half the diameter of Rudy's. The white eye ring was the same size; just Buster's eyes were smaller. In addition Buster was very inactive which is unusual for conures, especially young ones. Furthermore Buster could not fly—he simply coasted to the ground.

Initially Buster was overwhelmed by the size of his new aviary. For the first month he did not venture beyond two perches near the right side of the aviary. Obviously Buster had been kept in a small tabletop parakeet-size cage. This explained why he couldn't fly and was so inactive. Buster had never had enough room to move around or fly before. The toys that came with him were filthy and soon found their way to the junk pile. Gradually Buster began exploring other parts of his aviary and his new toys. But he was very clumsy. Buster never went to the perches or toys in the bottom half of his aviary, unless he fell there by accident. Finally in December 2012 Buster began to explore this part of his home.

Besides the inappropriate housing, Buster had not been fed properly either. The first time Buster was given a parrot seed/pellet mix he devoured the entire dish rapidly without spilling any and screamed for more. Buster ate two or three dishes of the parrot seed/pellet mix in the morning and another dish in the evening! In other words, Buster ate four times as much as Rudy. It was as if Buster were starved. Obviously whatever Buster had been fed before he arrived was not the right type or quantity of food. Buster continued his ravenous campaign for 3.5 months. After that Buster slowed down and began to eat an ordinary amount of food. It was a year before Buster filled out and became a normal-sized mitred conure.

While Buster was in his ravenous phase he would readily sample new foods. It was clear that Buster had never seen fruits, vegetables, or pellets before. At first Buster would nibble on anything he was presented. Sometimes Buster would gorge on one particular item, like pumpkin seeds or squash, for several days in a row as if he were compensating for a deficiency in a particular nutrient. Later, Buster wouldn't touch that particular item for several weeks.

Today Buster has established preferences. There are fruits and vegetables Buster: (1) readily eats on his own, (2) can be persuaded to eat if he is hand fed them, and (3) won't touch no matter what. Today Buster readily eats apples, oranges, dried papaya, and fresh red peppers. Buster can be persuaded to eat a small amount of shredded carrots and sweet corn. On the other hand, Buster ignores broccoli. His adventurous eating days are over.

Like the other parrots, Buster does not eat oranges per se; he sucks the juice out, then discards the pulp. Buster goes one step further than the other parrots. He frequently presses orange slices against a perch to obtain the juice. This activity qualifies as an example of a parrot using a tool. Buster was not taught to do this; he figured this out on his own. (See Photograph 11.8.)

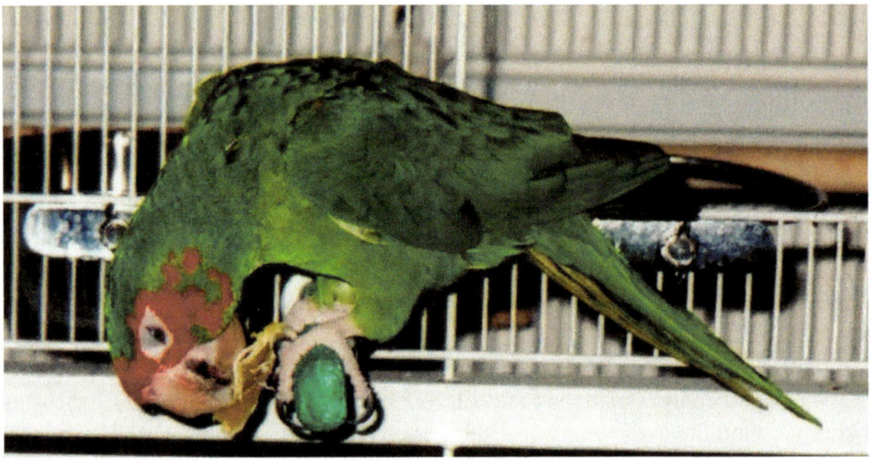

Photograph 11.8 Using a Perch as a Tool. Buster presses an orange slice against a perch to facilitate extracting the juice. In this case the perch functions as a tool. Parrots do not consume citrus fruit. Instead they suck the juice and then discard the pulp. Buster had not been exposed to fresh fruits and vegetables before he arrived here. Yet, Buster learned to do this on his own less than a year after his arrival. © D. Herrmann

Buster's favorite treat is pistachios in the shell. Buster expects four roasted unsalted pistachios every morning. He routinely raises a loud fuss until he receives his pistachios. Breakfast means pistachios to Buster; everything else is secondary. Buster knows in which cabinet the pistachios are stored. If that cabinet is opened, Buster starts doing a dance. Buster also knows what type of container the pistachios are stored in. If a different container is removed from that cabinet, Buster ceases to be excited. Buster has also learned that pistachios rattle around in the container and make a specific noise. Once Buster hears that sound he goes into hyper drive! The cabinet is about 12-feet (3.65 m) away from Buster's aviary and the container the pistachios are in is only 6-inches (15.24 cm) tall. Hence, Buster's eye sight and hearing are rather good.

Popped popcorn was an instant favorite treat of Buster's. Another favorite treat is miniature shredded wheat squares—the kind sold as breakfast food for humans! This dietary preference was discovered by accident. Buster goes nuts whenever he sees or smells the small shredded wheat biscuits and has to have one or two immediately. Perhaps Buster was fed shredded wheat instead of a parrot seed/pellet mix in one of his previous homes. Shredded wheat is not harmful to parrots—unless it is the sugary kind. But, shredded wheat hardly offers the full range of vitamins, minerals, protein, or other nutrients parrots need.

Like the majority of parrots discussed in this book Buster has consistently avoided cuttlebones and mineral blocks the entire time he has been here. Buster also ignores pumice and coral blocks that are supposed to help keep his beak trimmed.

It is counterproductive to fill Buster's food and water dishes too full. If the dry food dish is filled more than a third full Buster stirs it with his beak in a random pattern until he has spilled half of the food on the floor. Buster does not appear to be looking for anything in particular. He is just playing with his food and seems to enjoy the sound the seeds and pellets make when they hit the newspaper!

If the 20-ounce (600 ml) stainless water dish is filled to the brim, Buster won't go near it. He appears to be afraid of the depth of the water—it is deeper than his beak is long. If the water dish is filled half full, Buster immediately goes swimming in it—regardless of the time of day—and splashes away until all the water is gone and there is nothing left to drink. Buster expends a considerable amount of energy in this activity, although he doesn't even come close to fitting in the dish. If the water dish is filled a third full, Buster will drink the water and use it to wash or moisten his food. (See Photograph 11.9.)

Photograph 11.9 Preening. Buster is messy when eating or swimming. In contrast, he takes preening seriously. Even with a large beak Buster is able to preen small feathers carefully. © D. Herrmann

In spite of his fondness for swimming Buster refuses to swim in a larger more appropriately sized swimming pool that is placed on the bottom of his aviary. He doesn't even look at the swimming pool. Likewise, Buster has shown no inclination to learn how to use a water bottle.

The years of chronic malnutrition did more than stunt Buster's growth. Buster's nails and beak grew rapidly and he expelled excess clear water with his droppings. In senior birds the rapidly growing beak would be a sign of diminished liver function, while expelling excess water would be a sign of diminished kidney function. But Buster was a young adult.

Buster's initial (and subsequent) physical and lab tests did not indicate that anything was wrong. Over a period of two years Buster made multiple visits to different veterinarians. All sorts of exotic tests were ordered; all of which yielded normal results. All sorts of exotic remedies were prescribed: red palm oil, milk thistle, antibiotics, omega-3 fatty acids, etc.; none of which had any effect on the rate at which Buster's beak or nails grew.

After two years and over $4,000 in veterinarian bills no progress had been made. Unfortunately some veterinarians have fallen into the same trap as some human physicians, in that they are always looking for an exotic disease with an equally exotic remedy. In doing so, they often fail to consider basic life style issues.

Buster's beak was growing 0.33 inch (0.83 cm) a month. That is normal for mitred conures. His beak had to be trimmed every 3 to 4 months and his nails had to be trimmed every 4 to 5 months. Buster had special sand and pumice perches which were supposed to help keep his nails trimmed, but didn't.

A decision was made to change Buster's environment to make it more like what mitred conures experience in the wild. All of his perches were replaced with wood perches: a mixture of hard wood perches, soft wood perches, and cholla perches. New toys with wooden and cholla pieces were added to his aviary. At the same time it was learned from reading the pet bird hobbyist literature that the more frequently a parrot's beak is trimmed, the faster it grows.[117]

Overnight Buster began chewing on his wooden toys and rubbing his beak back and forth on the cholla perches. Buster had never rubbed his beak on any perch before. Buster didn't need hundreds of dollars of veterinarian bills. He just needed a more natural living environment.[51, 70] Now Buster only needs a beak trim every 6 months and a nail trim once a year. In time, with a proper diet and the proper environment Buster probably won't need any trims at all.

As far as expelling excess clear water, anyone who consumes large quantities of coffee, tea, or soda knows that the more fluid one drinks, the more fluid one expels. Buster drinks a lot of water and he eats a lot of fruit. He enjoys the fresh chilled filtered mineral water. Sometimes Buster takes eight drinks at once. Hence, he expels a lot of clear water. This is not an organic problem either.

Multiple veterinarian visits meant multiple journeys in the travel cage. The first few times Buster resisted going into the travel cage, possibly because he associated a travel cage with being rehomed. Before long Buster learned that the travel cage meant he was going outside where he could see and hear the trees, flowers, birds, and squirrels. The travel cage also meant Buster was going to receive lots of extra attention for the next three hours. As a result, the travel cage quickly became a welcome sight.

While walking to the subway, waiting on the platform, or riding the train Buster peers out the grill and chatters away. This causes all sorts of people to come over to see him and shower attention on him. At the veterinarian's office Buster puts on quite a show. Usually the receptionist, veterinarian technicians, and veterinarians have been treating sick cats and dogs all day. They welcome the sight of Buster strutting and clowning around and he gets even more attention.

One of the few times Buster attempts to fly is when he is getting into or out of the travel cage. At those times Buster jumps off the top of the travel cage (or the chair the cage is sitting on) and flaps his wings a few times as he coasts to the floor. Buster walks around on the floor for a while, going around and under various items. Eventually Buster comes running over and climbs up on an arm or shoulder to indicate that he is ready to go back in his aviary.

If Buster's front door is left open, he does not come out to fly. Instead Buster sits on his landing perch or climbs around the top of his aviary. Buster only stays out for 20 or 30 minutes, then he goes back inside.

In the last six months Buster has begun to have flying attacks once or twice a day. He hangs on the side of his aviary and flaps his wings rapidly for 30 to 40 seconds at a time. His flapping is becoming stronger and creates more turbulence. With enough practice, hopefully Buster will be able to fly soon.

Like most parrots Buster has toys that are his favorites and toys that he plays with every now and then. When Buster first arrived his favorite toy was a 6-inch (15.24 cm) cotton rope wreath. Buster would chew on the cotton ropes for a while. Then he tossed the cotton wreath on his back, like it was a blanket. After a few months Buster abandoned that toy for a new one: a single braided cotton rope with 4.0 x 0.5 inch (10.16 x 1.27 cm) manzanita pegs every two inches (5 cm). Buster chewed the dark brown bark off the manzanita peg that is eye level. Buster nibbles on the cotton rope and other pegs a bit. Mostly he chews on the peg that is eye level, while the entire toy is draped over his back. Buster even takes naps under the cotton rope toy. (See Photograph 11.10.)

A large bright green whiffle ball is another of Buster's favorite toys. The whiffle ball is dangling at eye level near the end of a perch. Buster pokes and prods the whiffle ball with his beak, knocking

Photograph 11.10 Rope Buddy. A favorite toy often becomes a pet parrot's "best friend", especially when the people in the house are gone all day and the parrot is not part of a bonded pair. Buster took to this new toy immediately after he arrived. He chewed the bark off the top manzanita peg within a few days after his arrival. Seven years later this is still Buster's favorite toy. He often naps with the rope draped over his back. A Meyer's parrot exhibits a similar behavior in Photograph 9.5. © D. Herrmann

it around. Sometimes Buster grabs the whiffle ball with his foot and shakes it. There is an orange rubber ball with cutouts on the other side of the aviary. It dangles from a chain too. Buster bites and punches the orange rubber ball around just like the green whiffle ball. Near the top perch there is a chain filled with cholla pieces; it has a small green bell on the bottom. Buster chews on the cholla pieces or grabs hold of the bell and shakes the entire toy. When Buster isn't napping or eating or swimming in his water dish, he keeps himself busy playing with these four toys. (See Photographs 11.11 and 11.12.)

Pets can be grouped into categories based on whether or not they want their human caretaker's attention and/or need their approval. This of course has ramifications on what type of pets people select and, conversely, how much interaction pets expect or will tolerate from their human caretakers.

	Want Attention	Need Approval
Dogs	yes	yes
Parrots	yes	no
Cats	no	no
Tropical fish	no	no
Turtles	?	probably not

Photograph 11.11 Toys #1. Buster prefers colorful dangling toys that he can poke, shake, and knock around with his beak and feet, like this bell and rubber ball. It is unclear what, if any, wild behavior this activity mimics. © D. Herrmann

Most captive parrots want some degree of human attention, especially if they are the only bird in the family. The current assumption in regard to parrots is that the larger the bird, the more attention it needs; i.e., a scarlet macaw needs four hours of human interaction a day, while a budgerigar only needs 15 minutes of human interaction a day. That thinking is inconsistent with the observations discussed in Chapters 7 through 11.

Photograph 11.12 Toys #2. (Top) A recently molted flight or tail feather is often a favorite toy of many a pet parrot, at least for a brief period of time. Parrots seem intrigued by being able to examine a complete unattached feather. Smaller parrots will even explore a feather from a larger parrot. It is unclear whether wild parrots engage in this behavior. (Bottom) Buster checks out the bell's clapper. © D. Herrmann

First, the amount of attention a bird wanted or was comfortable with had nothing to do with its size. Rather the attention quotient reflected the preference of an individual bird. Some birds like or want a lot of attention. Other birds do not need or want a lot of attention. In addition, some birds are comfortable with a lot of attention one day, then just a little attention for several days in a row.

In addition, the amount of attention a bird wanted or was comfortable with depended on whether or not they were part of a bonded pair.

Second, the amount of attention a bird wanted or was comfortable with corresponded to events that took place early in its life and group (flock) socialization experiences. For example, young budgerigars are taken from their family and clutchmates and put into a group cage in a pet store. Budgerigars learn the norms of behavior and how to interact in a group situation from an early age. In contrast parrots are taken from their family and clutchmates at an early age and stuck in a cage by themselves in a pet store. And there they sit for months until they are purchased or they reach their second hatch day and the pet store disposes of them. Parrots are not likely to acquire very good socialization or vocalization skills in this environment. Many parrots crave attention and social interaction but don't know how to act when they get it.

Buster falls into this category. Buster wants lots of human interaction and attention constantly. Buster will do most anything to get attention, from turning his head upside down like a clown to screaming. Buster has no concept of what is and isn't acceptable behavior. Buster does not understand the idea of living in a group environment or taking turns. Whenever it is time to come out to play, be served breakfast, or receive a special treat, Buster expects to be first. If he is not the first one, Buster throws a tantrum. On other occasions Buster goes into a nervous nelly whining mode when he wants attention. While the sound is not as loud as his screaming it is equally as unpleasant. In contrast, Buster is quite a showoff when it is time to get in the travel cage and journey to the veterinarian's office. Buster will sit on an arm or shoulder once in a while, but he actively avoids pats and head scratches. (See Photograph 11.13.)

Buster socializes with some of the other birds. Buster visits the budgerigars and doesn't mind them popping in and out of his aviary. Early on Buster saw the budgerigars being given millet sprays and he insisted on having one, even though his beak is too large to eat millet. Today Buster will nibble on a millet spray, but only if a person holds it for him. Buster tolerated Dusty (Section 10.6) visiting his aviary and playing with his toys. But he drew the line and shooed Dusty out when she tried to eat some of his food. When any of the other birds are out flying and try to land on top of his aviary, Buster gets annoyed and chases them away. Unlike Rudy, Buster has never visited the zebra finches. (See Photograph 7.15.)

In summary, Buster has made progress in terms of his health and nutrition. Through environmental enrichment, Buster takes better care of his grooming and explores new toys. Buster still has more to learn in the areas of group social norms and acceptable vocalizations. If Buster keeps practicing his flying attacks, he should be strong enough to fly soon. The increased mobility flying brings and the ability to make decisions about where he goes and who he visits, will improve Buster's social and communication skills.

11.5 Discussion

Now let's see what these observations tell us about the intelligence, behavior, and individuality of wild and captive mitred conures.

Temperament

Rudy and Buster have quite different temperaments, as depicted in Table 11.1. Rudy is gentle, patient, calm, content, and even tempered. Rudy is never frazzled or aggressive. Rudy gets excited about the littlest things and is an all-around happy playful parrot. In contrast, Buster is a high strung impatient attention seeking nervous nelly.

Photograph 11.13 Upside Down Clown. Buster clowns around by hanging upside down in hopes of attracting attention. Buster craves attention, even from strangers. In comparison, Rudy relishes and is content with whatever amount of attention he receives. Note the chevron-shaped tail characteristic of conures. © D. Herrmann

Vocalization

Neither Rudy nor Buster makes any of the vocalizations attributed to wild mitred conures. There are two plausible explanations for this. First, it is possible that an accurate and complete set of wild mired conure vocalizations has not yet been identified given the dearth of field observations. Second, it is possible that mitred conure vocalizations are learned versus inherited. Most likely Rudy and Buster were taken from their families at an early age before they had a chance to learn a traditional mitred conure vocabulary.

Furthermore, there is no commonality between Rudy's and Buster's vocalizations. Rudy makes two main vocalizations. His primary vocalization is a prolonged "wha—at", which he varies to convey different messages. Rudy also makes a "caw-caw-caw" alarm call when he spots a peregrine

Table 11.1 Spectrum and Prevalence of Emotional States Observed in Indoor Mitred Conures.

	Rudy	Buster
I. Positive Emotions		
excited	x	x
happy	x	x
sad[1]	x	
confident/content	x	
timid/unsure		x
lonely		
curious	x	x
puzzled		x
playful/eager	x	x
anxious/fearful		
aggressive/impatient		x
gentle/patient	x	
II. Negative Emotions		
anger		
hatred		
jealousy[2]		x
grumpy		
overwhelmed		
stressed out		
possessiveness		
self-pity		
depression		
boredom		

Notes:

1—Only during the first day after being adopted and rehomed.
2—Buster has to be the center of attention, let out to fly first, served breakfast or treats first, etc. or he throws a tantrum.

falcon flying outside the window. Rudy's alarm call is similar to that made by cherry-headed conures. Buster's vocalizations are random. Buster screams when he is annoyed about something or he wants attention. At times Buster goes into a nervous nelly whining routine. Frequently when the other birds are vocalizing Buster joins in and adds a squawk here and there as if he is commenting on the dialog.

Mitred conure vocalizations can be heard on Disk 2 Track 3.

Unique Behaviors

Several individual and unique behaviors were observed. These behaviors also illustrate differences between Rudy and Buster, as captured in Table 11.2.

Rudy and Buster engaged in some common activities, but they went about them differently. Rudy only swims in an olive green dish on the bottom of his aviary. Buster only swims in a stainless water dish attached to the side of his aviary. Rudy knows better than to drink from the dish in which he swims. Buster swims and drinks from the same dish.

Table 11.2 Unique Mitred Conure Behaviors.

Behavior Context	**Rudy**	**Buster**
swimming	• Only swims in olive green dish on bottom of aviary • Does not drink from swimming pool • Swims on Saturdays	• Only swims in stainless dish attached to side of aviary • Drinks from swimming pool • Swims any time of day any day of the week
flying and mobility	• Has regular flying attacks • Strong precise flyer • Goes in bottom half of aviary • Explores around and under items in the room by foot • Climbs up ladders	• Has regular flying attacks • Does not fly well • Did not go in bottom half of aviary until 4 years after his arrival
food	• Hides broccoli and dried papaya to "find" later • Uses broccoli to wash his beak • Consumes cuttlebone • Ate flowering snapdragon plant one day • Keeps water dish clean	• Does not consume cuttlebone • Does not keep water dish clean
intelligence	• Aware of visual limitations, triangulates coordinates, naps with weak eye facing wall • Covers eye with single feather while napping • Knows Saturday is swimming day • Knows schedule for flying time • Acknowledges his mistake when he misses a treat and accidentally bites • Figures out new things quickly	• Uses perch as a tool to squeeze juice out of an orange • Knows what cabinet and container pistachios are stored in
communication	• Understands when told it is bedtime (spoken, bell, lights, or pointing) • Rings bell when he wants other birds to quiet down for the night • Holds his head underwater in water dish and makes a mournful sound when he wants to go swimming • Quiet, communicates via body language • Makes "wha—at" sound to indicate specific wants • Makes "caw-caw-caw" alarm call	• Loud random vocalizations • Screaming • Nervous nelly whining
socialization	• Likes company, peer-to-peer social interaction • Bowing and nodding greeting ceremony • Understands the concept of taking turns • Socializes with zebra finches and other birds	• Craves attention • Shows off in front of strangers • Has to be first one let out, served a meal, given a treat, etc. • Socializes with budgerigars

***Note:** Mitred conure vocalizations can be heard on Disk 2 Track 3.

Both Rudy and Buster have flying attacks. Rudy methodically climbs down to the back wall in the bottom half of his aviary. Rudy stretches out his wings slowly to make sure there is enough room before he starts flapping. In contrast, Buster climbs up the side wall in the top half of his aviary. Buster starts flapping without checking to see if he or his wings will fit. Rudy is a strong precise flyer. Buster cannot fly. Rudy explores around and under things in the room on foot and climbs up the ladder when he is finished. Buster does not explore on foot or climb up ladders. Rudy goes into the bottom half of his aviary. Buster did not start exploring the bottom half of his aviary until he had been here for four years.

Rudy hides broccoli and dried papaya so he can "find" it later. Buster does not. Rudy consumes cuttlebone. Buster does not. One day Rudy ate a flowering snapdragon plant. This behavior mimicked that of wild mitred conures. Buster has shown no interest in any plants in the aviary. Rudy keeps his water dish clean. Buster's water dish looks like a hazmat scene before noon each day.

Rudy is an intelligent bird and has demonstrated this in several ways. Rudy is aware of his visual limitations. To compensate he triangulates the coordinates between his left (bad) and right (good) eyes before taking a treat or flying to a destination. In addition, Rudy naps with his left (bad) eye facing the wall. Rudy learned the schedule for flying time and that Saturday is swimming day. Rudy figures out things quickly. Rudy also acknowledges his mistake (and regret?) when he misses while reaching for a treat and accidentally bites.

Buster independently learned how to use a perch as a tool to help squeeze juice out of an orange slice. Buster also learned what cabinet and container pistachios (his favorite treat) are stored in.

Rudy has superb communication skills. For the most part Rudy is quiet and communicates via body language. Rudy understands that it is time to head home for the evening when he hears "bedtime" or his bell ringing, or he sees the lights flip on and off or someone pointing to him, then to his aviary. Likewise, Rudy rings his bell to tell the other birds to quiet down for the night if he thinks they are being too noisy. Rudy holds his head underwater in his water dish and makes a mournful sound when he wants to go swimming.

Buster's loud random indiscriminate vocalizations make it difficult to discern what he is attempting to communicate, other than he wants something.

Another interesting observation reflects different thinking patterns. If I point to the right, Rudy moves to the right (his left). In contrast, if I point to the right, Buster goes to the left (his right).

Socialization Patterns

Rudy has mature socialization skills and understands how to live in a group situation. Rudy likes company and enjoys peer-to-peer social interaction with humans. Rudy engages in a combined bowing and nodding greeting ceremony when I enter the room. Rudy understands the concept of taking turns and waits patiently until he is served. Rudy socializes with the other birds, particularly the zebra finches. Rudy is reserved around strangers.

Buster craves attention and acts up in the hopes of getting it. Buster shows off in front of strangers. This behavior is also designed to garner attention. Buster has to be first, whether it is being let out, served a meal, or given a treat. If not, Buster throws a tantrum. Buster socializes with the budgerigars.

Toy and Dietary Preferences

Rudy and Buster demonstrated distinct preferences in regard to toys, as shown in Table 11.3. Both Rudy and Buster like to chew on wooden toys, particularly those containing cholla pieces. This behavior mimics that of wild mitred conures. That was the only type of toy in which they both shared an interest. Since Rudy explores on foot he is open to foot toys and ladders. Shredding toys are the next best thing to chewing wooden toys, at least Rudy thinks so. Rudy's favorite toy is the semi shiny metal plate with the manzanita peg that he has had for years. In contrast, Buster prefers dangling toys that he can bite, shake, and knock around, like whiffle balls, bells, and rubber balls.

Table 11.3 Individual Mitred Conure Toy Preferences.

Toy	Rudy	Buster
acrylic and plastic toys		
bells		x
cardboard chewing toys		
cholla blocks, wood pieces	x	x
cotton ropes and toys		x
foot toys	x	
ladders	x	
mirrors, shiny metal	x	
natural fiber toys (palm, sisal)		
push pull toys		x
puzzle toys		
rattles		
rubber ball on a chain		x
shredding toys, piñatas, yucca pieces	x	
sisal ropes and toys		
swings		
wiffle balls		x

*Toy color did not affect toy preferences.

Table 11.4 Individual Mitred Conure Fruit and Vegetable Preferences.

	Rudy	Buster
I. Fruits		
apple	x	x
banana	x	x
blueberries		
cantaloupe	x	
cherries	x	
grapes		x
kiwi berry		
mango—dried mango—fresh	x	x
orange	x	x
papaya—dried papaya—fresh	x x	x x
peaches	x	x
strawberries		
II. Vegetables (fresh)		
broccoli	x	
carrots	x	
cherry tomato		
corn—yellow sweet	x	x
dandelion leaves		
green peas		
kale		
red peppers	x	x
spinach		
squash	x	
III. Pellets		
fruit flavored		
natural	x	x

Rudy and Buster have different dietary preferences as well, as illustrated in Tables 11.4 and 11.5. In regard to fruits and vegetables, Rudy and Buster have similar tastes. However Buster eats a more restricted diet. Rudy enjoys bananas, cantaloupe, cherries, broccoli, and squash, which Buster mostly ignores. Natural pellets are Rudy's favorite treat when he is out flying. Buster did not start eating pellets regularly until he had been here for four years.

A similar pattern holds true in regard to Rudy's and Buster's grain, nut, seed, and legume preferences. As expected they both like safflower, striped sunflower, and pumpkin seeds. But at that point their tastes diverge. Rudy is open to a few cooked grains, like barley, long grain brown rice, and quinoa. Buster ignores cooked grains. Rudy ignores tree nuts. In comparison Buster does a dance every morning in anticipation of receiving four pistachios. Rudy eats a small amount of dried corn. Buster does not. Rudy treats peanuts in the shell like a shredding toy. Buster eats peanuts, but only if they are shelled.

Table 11.5 Individual Mitred Conure Grain, Nut, Seed, and Legume Preferences.

	Rudy	**Buster**
I. Whole Grains and Legumes (cooked)		
barley	x	
buckwheat		
lentils		
long grain brown rice	x	
millet		
quinoa	x	
rye		
wheat, spelt, kamut		
II. Tree Nuts (raw, shelled)		
almond		
Brazil nut		
cashew		
filbert		
pecan		
pistachio		x
walnut		
III. Seeds and Legumes (raw)		
dried corn	x	
dried peas and beans		
millet seeds (loose)		
millet spray		only if hand fed
milo		
oats		
peanuts		only if shelled
pumpkin seeds	x	x
safflower	x	x
striped sunflower	x	x

In summary, two male mitred conures were observed over a 9-year period. Both were adopted. One mitred conure, Rudy, was a senior parrot (20-plus years old) when he arrived. The other mitred conure, Buster, was a young adult (5 to 7 years old) when he arrived. While their previous history is different, they experienced an identical living environment here. Even so, Rudy and Buster are distinct individuals with different temperaments, vocalizations, daily activities, socialization patterns, and toy and dietary preferences. Their 15-year age difference might explain some, but certainly not all of these differences. Simply put Rudy and Buster are unique individuals. Individual choices, preferences, and behaviors indicate intelligence and self-awareness.

12
Avian Communication Patterns

This chapter provides background information for the two audio disks that accompany the book.

12.1 Background

Birds have a more compact and efficient style of communication than humans.[14] This is due in part to their unique syrinx organ, through which two sounds can be produced at the same time.[18, 74, 137] Birds also have more complex hearing than humans; they can hear a range of frequencies that is both higher and lower than humans can hear.[14, 18, 74, 154] Birds hear sounds from a greater distance than humans. Birds have more intricate abilities for the time resolution of sound.[14] To illustrate, birds have the ability to isolate or distinguish individual sounds arriving in rapid succession ten times faster than humans.[14] What we hear as a single sound, a bird may hear as ten discrete tonal variations.[14] Feathers cover the ear canal to protect a bird's ears and hearing from turbulence on windy days and during flight.[74] This is similar to putting a shield over a microphone to prevent it from picking up wind or a person breathing.

Birds vary their pitch, volume, and frequency (high to low frequency, wide to narrow band) depending on whom they want to communicate with and the environment they are in.[74, 137] By varying these parameters a bird determines whether or not the source and location of the message can be determined and how far the message will travel.[74] For example, long distance vocalizations can travel 55–219 yards (50–200 m).[74] By adjusting these three parameters a bird compensates for water, vegetation, and other environmental factors that might cause interference, absorption, distortion, scattering, and attenuation of the message being vocalized.[74, 137] This process is not unlike how modern telecommunication systems adjust signal strength and the signal to noise ratio (SNR) to ensure reliable communications.

By varying pitch, duration, volume, timbre, and intensity birds communicate complete thoughts in a few notes. What appears to be a single note or two to our ears embeds a complete idea, like today's computer function keys and mouse clicks or the hieroglyphics of ancient cultures. Instead of reflecting a vowel or consonant, a note reflects a complete sentence or thought. Bird vocalizations are a more rapid, efficient, and accurate form of communication; one that is rarely misunderstood or mistranslated due to cultural or stylistic nuances, or over use of idiomatic expressions, acronyms, jargon, and slang.

Today the common practice is to record a species' signature call. Generally a single bird is recorded in isolation vocalizing one or two key phrases. These recordings are about 15 to 20 seconds in duration or less. This approach provides essential information for identifying a particular species in the field. The two audio disks that accompany this book were made for a different purpose; hence, a different approach was followed.

A key goal of this book is to explore the variety and complexity of avian communication patterns within and among species. Avian communication models are rarely 1-to-1. Rather they are mostly 1-to-many or many-to-many group conversations. Furthermore, other birds (not the message's intended recipients) are always listening in. Since avian communication takes place outdoors, there are multiple simultaneous sources of audio information that a bird must isolate, distinguish, and act upon in addition to direct vocalizations. As a result, it is important to understand the context of avian communication patterns—what else is going on—and how these communication patterns fit within the interactions of multiple birds and species.

To this end this book captures original recordings of the full range of vocalizations and complete conversations among multiple birds. In doing so many new vocalizations have been identified. The recordings feature dialogs among multiple birds of the same species and their interaction with other avian species in their midst.

For example on Disk 1 Track 1 a few wood duck (*Aix sponsa*) solos can be heard in the middle of mallard duck (*Anas platyrhynchos*) bantering, as well as the mallard ducks and wood ducks response to a blue jay (*Cyanocitta cristata*) alarm call. On Disk 1 Track 3 the peeping of Canada goslings can be heard in sequence with occasional commentary by their parents (*Branta canadensis*).

A second goal of this book is to explore which vocalizations are inherited versus learned among the five species of parrots examined. There is a concern that not only are parrots being lost from the wild, but those species that have learned-only or combined inherited and learned vocabularies are losing their natural vocabularies in captivity. As explained in Chapters 7 through 11:

- Young pet parrots often end up speaking a form of parrot gibberish since they have no one from which to learn their native language
- The native vocabulary of pet parrots increased dramatically in proportion to the number of parrots present, especially in mixed age groups
- The tendency of parrots to mimic human speech is not a cute trait, but rather an attempt by a solitary parrot to learn its language and communicate

The four Indian ringneck parakeets (*Psittacula krameri manillensis*) in Chapter 10 arrived with different vocabularies. Over time the three females' vocalizations converged in spite of differences in age. The male makes gender-specific vocalizations in addition to gender-neutral vocalizations. Hence, it appears that Indian ringneck parakeet vocalizations reflect a combination of inherited and learned phrases, some of which are gender-specific. (Refer to Disk 2 Track 1.)

The Meyer's parrot (*Poicephalus meyeri*) in Chapter 9 named Squeaky made a variety of vocalizations since she was a juvenile. Squeaky also taught herself new vocalizations by imitating sounds she heard around her. Most likely Meyer's parrot vocalizations reflect a combination of inherited and learned phrases. In contrast the Red-bellied parrot (*Poicephalus rufiventris*) in Chapter 9 named Sydney did not make any vocalizations when she arrived even though she was 9 months old. For the most part Sydney's vocalizations are limited to quiet babbling—a mishmash of English words and phrases she picked up on her own—or imitations of Squeaky's vocalizations. Hence, it appears that Red-bellied parrots learn most all of their vocalizations. (Refer to Disk 2 Track 2.)

The two male mitred conures (*Aratinga mitrata mitrata*) in Chapter 11 made very different vocalizations. Rudy's vocalizations are somewhat similar to those of wild mitred conures. In contrast, Buster's vocalizations appear to be improvised. He had limited opportunities to hear and interact with other mitred conures. This would indicate that mitred conure vocalizations are primarily learned rather than inherited. (Refer to Disk 2 Track 3.)

As noted in Chapter 7 the variety and complexity of budgerigar (*Melopsittacus undulatus*) vocalizations increased dramatically once there were 10 or more budgerigars in the aviary. Ten is usually the smallest flock size of budgerigars seen in the wild. Hence, budgerigar vocalizations appear to reflect a combination of inherited and learned phrases. (Refer to Disk 2 Track 4.)

Lastly, it is interesting to note the interaction among these five species of parrots, particularly during the long budgerigar track. Even though these five species of parrots live on four different continents and would not encounter each other in the wild, they interact and participate in each other's conversations.

The sections that follow describe the individual tracks on each disk. A common format is used to describe the tracks. Each section opens with a header that summarizes the date, place, and number of birds of that species recorded. This information is followed by a discussion of the unique vocalizations, dialogs, and conversations heard during the track and which birds participated in them. Background sounds are also identified in order to put the communication in context. Sample audio spectral graphs are provided for each track.

12.2 Mallard Duck Ice Medley (*Anas platyrhynchos*), Disk 1 Track 1

Date recorded:	January 2011
Location recorded:	Pond B
Number of mallard ducks present:	36 to 48
Track duration:	12:33:33 minutes

This recording captures the vocalizations of the Pond B flock of mallard ducks during a harsh winter day when they are in survival mode, crowded together in a small area of the pond that didn't freeze. This recording was made on a day similar to that depicted in Photographs 4.24 and 4.25.

These vocalizations are different than those normally heard on warm sunny summer and fall days. (Refer to Table 4.7.) The most common vocalization is a clicking sound. Usually one duck starts the clicking; then the others slowly join in. The clicking sound is repeated for a few minutes. Next, everyone goes silent. Then the chorus starts over again.

Various sounds can be heard in the background, like snow crunching as the ducks walk, water splashing when a duck swims or bathes, and a few songbirds; in particular a blue jay. An occasional wood duck call can also be heard. The local wood ducks stay put at Pond B throughout the winter too.

A sample audio spectral graph is presented in Figure 12.1.

12.3 Canada Geese Winter Song (*Branta canadensis*), Disk 1 Track 2

Date recorded:	January 2011
Location recorded:	Pond C
Number of Canada geese present:	36 to 80
Track duration:	24:20:01 minutes

This recording captures the vocalizations of a large flock of Canada geese that returned to Pond C during January 2011 in response to unusual snowstorms in the southeast U.S. The flock was spread out across the ice on top of Pond C, much like Photograph 3.38. It was a cold day with the wind whipping around. The stronger the wind gusts, the more vocal the geese were. These vocalizations are different than those normally heard on warm sunny summer and fall days. (Refer to Table 3.2.) For one thing, many of the vocalizations are higher pitched. In addition multiple geese are vocalizing at the same time asynchronously.

The sound of ice cracking can be heard about halfway through the track, as individual geese walk cautiously across the frozen pond. At this time solitary geese can be heard making quizzical vocalizations which appear to express uncertainty. Once in a while the ice gives way and the geese can be heard splashing and fluffing their feathers. In the middle of the track and again at the end a young quartet can be heard making some low raspy vocalizations. In general the geese held their

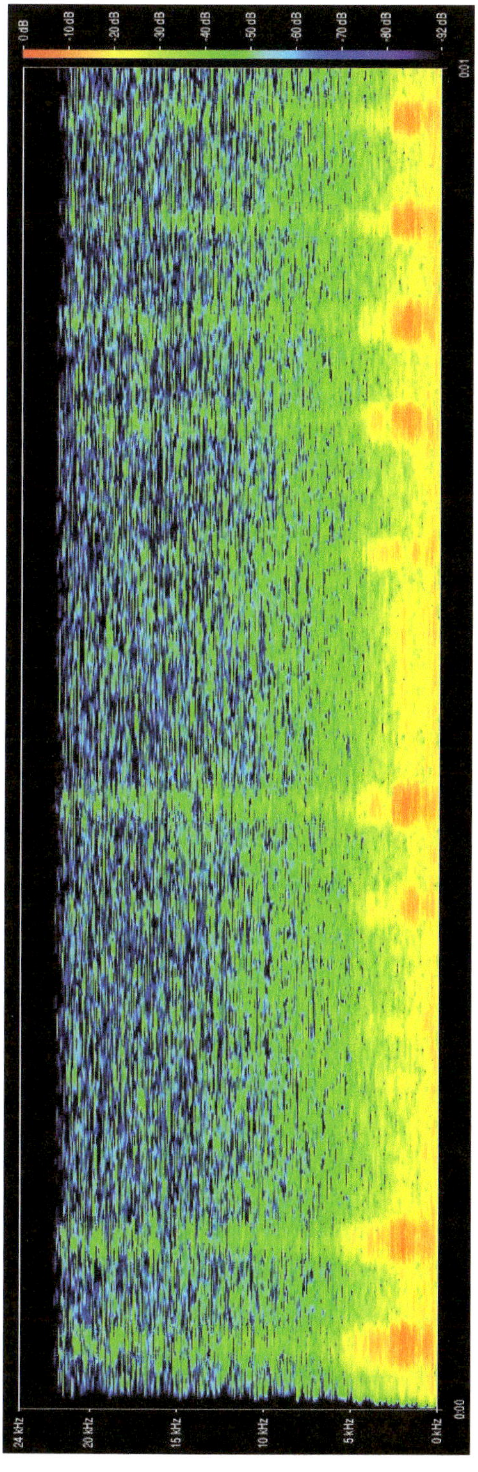

a. Clicking (0:15–0:16)

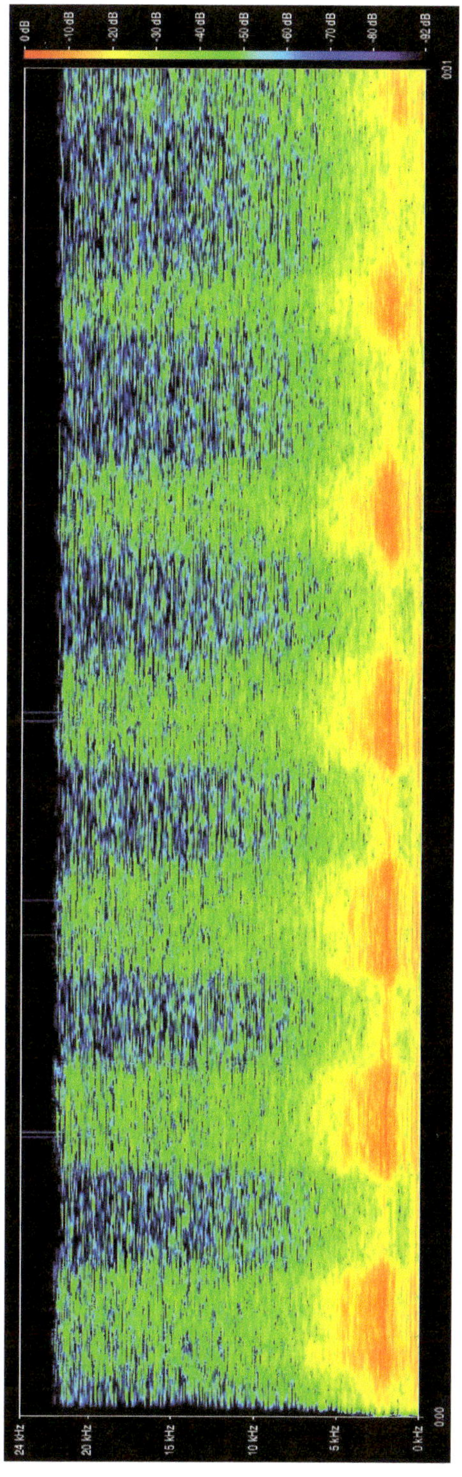

b. Quacking (3:40–3:41)

Figure 12.1 contd....

Figure 12.1 contd.

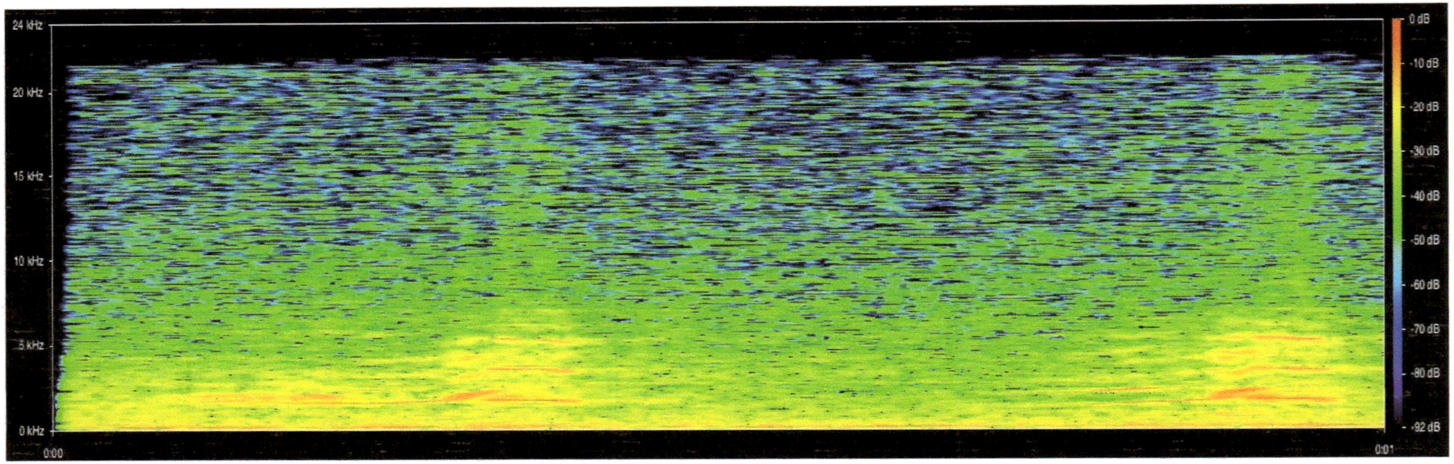

c. Wood Duck (4:12–4:13)

Figure 12.1 Mallard Duck Audio Spectral Graph. The times are shown for three segments from Disk 1 Track 1. The scale on the left is frequency (kHz). The scale on the right is volume (db). The scale on the bottom is in seconds. Spikes indicate the full spectrum of sound within the vocalization. Black reflects background noise. If there is no horizontal scale, the segment is one second or less. © D. Herrmann head and neck upright when making a high-pitched sound and bent over, with their head and neck closer to the ground, when making low-pitched sounds.

A sample audio spectral graph is presented in Figure 12.2.

12.4 Canada Goslings (*Branta canadensis*), Disk 1 Track 3

Date recorded:	June 26, 2012
Location recorded:	Pond C
Number of Canada goslings present:	8
Track duration:	6:07:23 minutes

This recording captures the vocalizations of eight Canada goslings at Pond C. One family of four goslings was 4-weeks old. A second family of four goslings was 5-weeks old. (See Photographs 3.18 through 3.21.) The two families, including all four parents, joined forces when the youngest goslings were 2-weeks old. 2012 was the first year during this study that Canada goslings hatched at Pond C.

The goslings' vocalizations consist of a variety of high-pitched whistling sounds, often from multiple goslings at the same time. This vocalization is used: (1) as a signal for the goslings to come together; and (2) to summon the parents when an individual gosling becomes lost. Canada goslings begin making this vocalization within their first week and continue to make this sound until they are 10 to 12 weeks old. Mallard ducklings make a very similar sound.

Various sounds can be heard in the background, such as a crackling sound when the goslings are eating grain, an occasional low-pitched "hunh" sound made by a parent, a red-winged blackbird (*Agelaius phoeniceus*) call, and a pinging sound when a gosling tapped the microphone. About two minutes into the track one of the parents makes a brief hissing sound when I purposely stepped too close to the goslings.

A sample audio spectral graph is presented in Figure 12.3.

12.5 Zebra Finch Phonics (*Poephila guttata castanotis*), Disk 1 Track 4

Date recorded:	April 2011
Location recorded:	Indoor free flight aviary
Number of zebra finches present:	6
Track duration:	9:48:61 minutes

This recording captures the vocalizations of three adult male zebra finches (Tiny Guy, Little Bald Guy, and Little Boy) and three adult female zebra finches (Lady Finch's Sister, Naomi, and Little Girl) in their indoor aviary. (Refer to Table 8.1.) A large mirror was placed in front of the aviary to encourage curiosity and vocalizations, particularly those of the males.

All of the gender-specific and gender-neutral zebra finch vocalizations identified in Table 8.2 and variations of them can be heard during this recording. These vocalizations are from two generations of the same zebra finch family. The vocalizations are slightly louder at first as the zebra finches attempt to communicate with the birds they see in the mirror. Generally multiple zebra finches vocalize at the same time. They tend to have many-to-many group conversations.

Various sounds can be heard in the background, such as the zebra finches flying, landing on a perch, and jumping off a swing.

A sample audio spectral graph is presented in Figure 12.4.

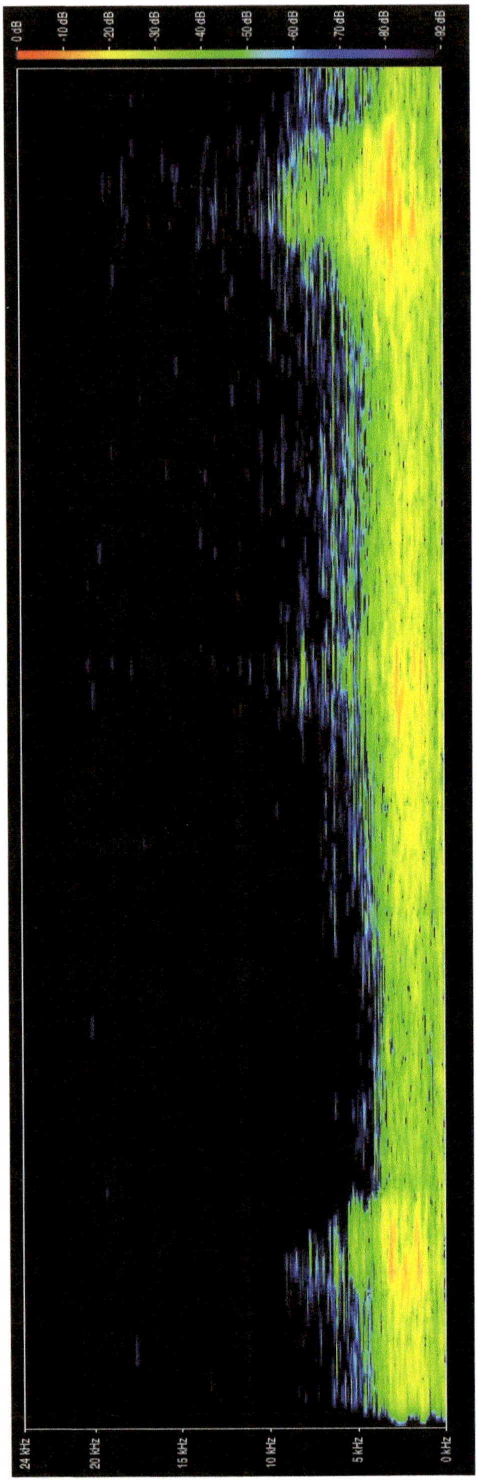

a. High-pitched (0:18–0:19)

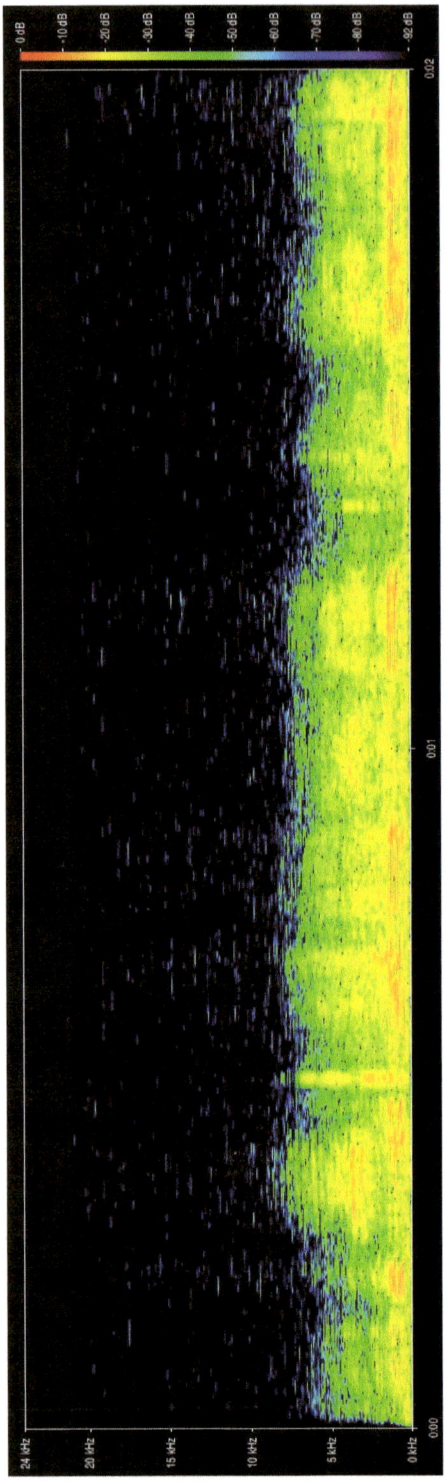

b. Low-pitched (17:05–17:08)

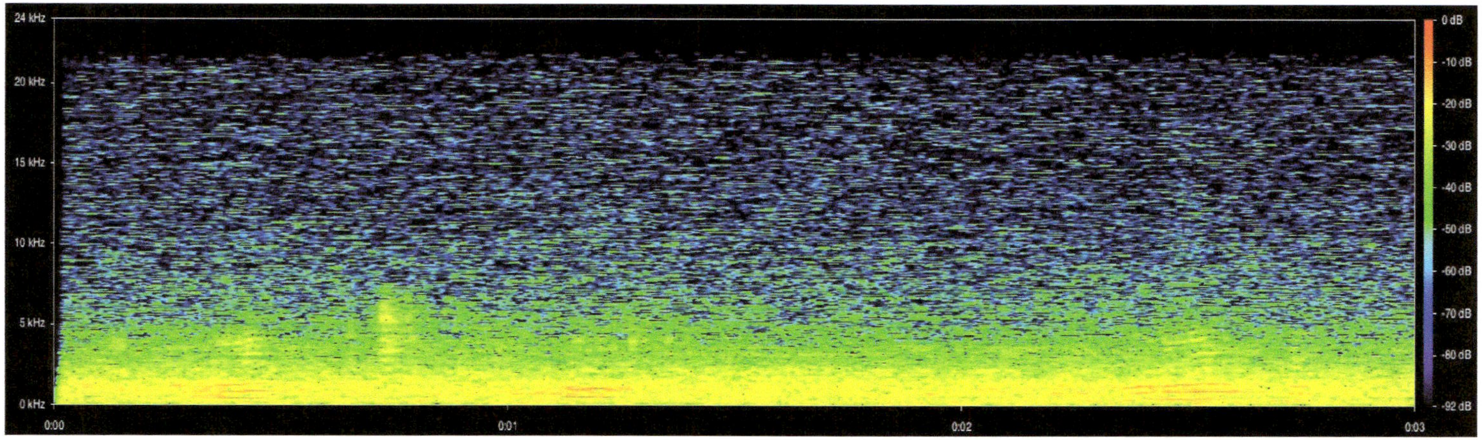

c. Solitary quizzical honk on ice (10:49–10:52)

Figure 12.2 Canada Geese Audio Spectral Graph. The times are shown for three segments from Disk 1 Track 2. The scale on the left is frequency (kHz). The scale on the right is volume (db). The scale on the bottom is in seconds. Spikes indicate the full spectrum of sound within the vocalization. Black reflects background noise. If there is no horizontal scale, the segment is one second or less. © D. Herrmann

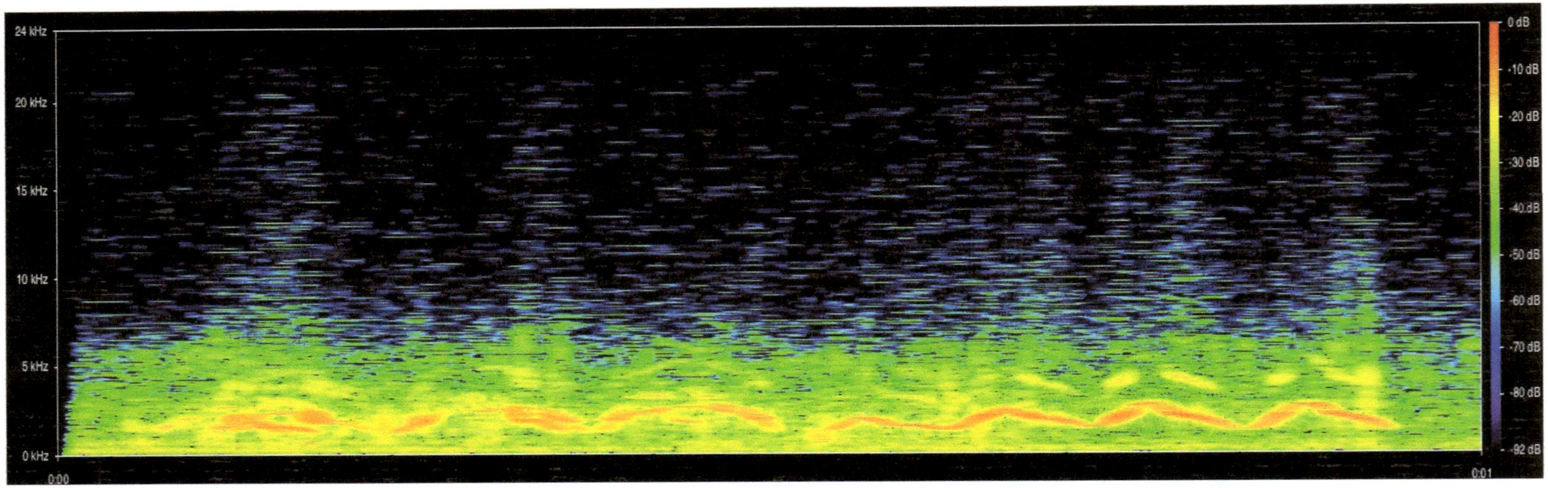

a. Whistling (2:07–2:08)

Figure 12.3 Canada Gosling Audio Spectral Graph. The times are shown for one segment from Disk 1 Track 3. The scale on the left is frequency (kHz). The scale on the right is volume (db). The scale on the bottom is in seconds. Spikes indicate the full spectrum of sound within the vocalization. Black reflects background noise. If there is no horizontal scale, the segment is one second or less. © D. Herrmann

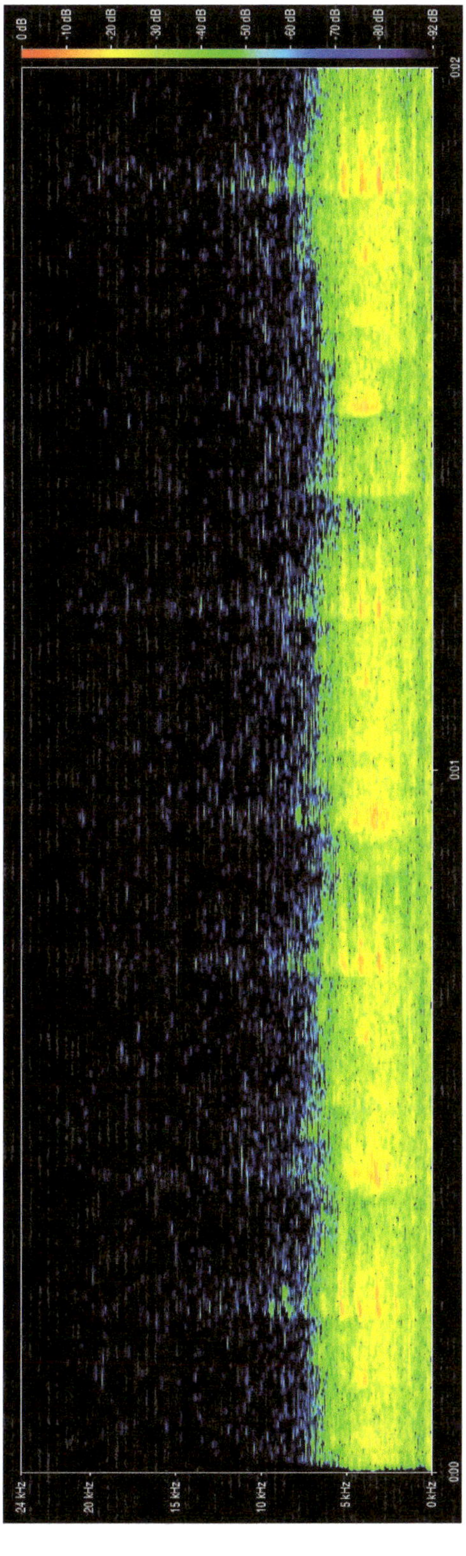

a. T-du-du (male, 3:05–3:08)

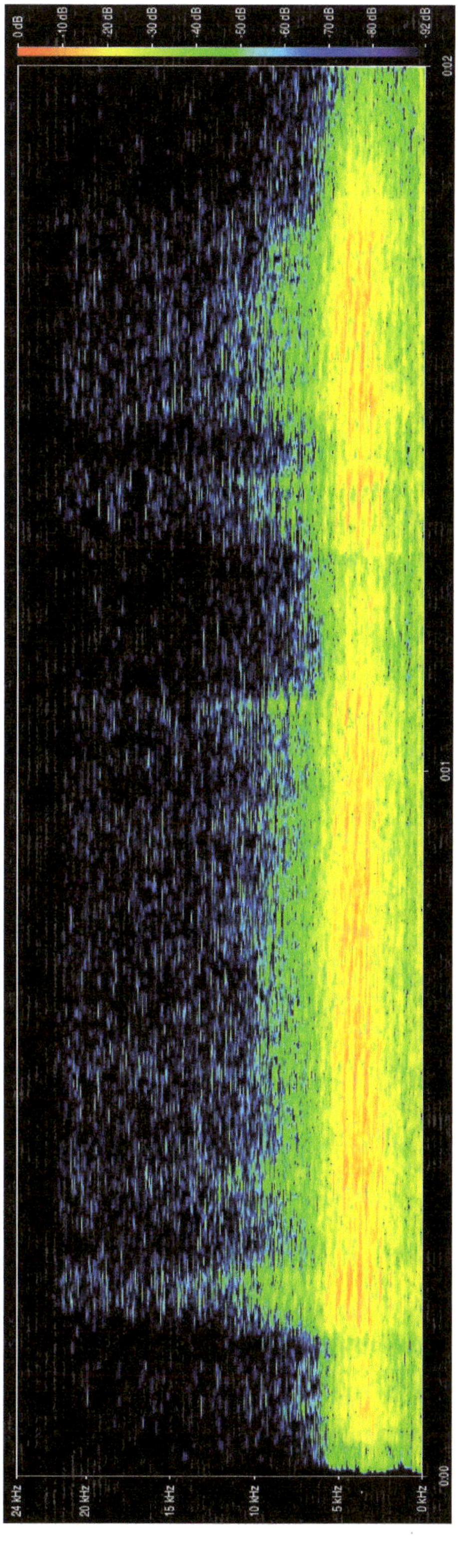

b. Duh-duh-duh (male, 1:59–2:01)

Figure 12.4 cond....

Figure 12.4 contd.

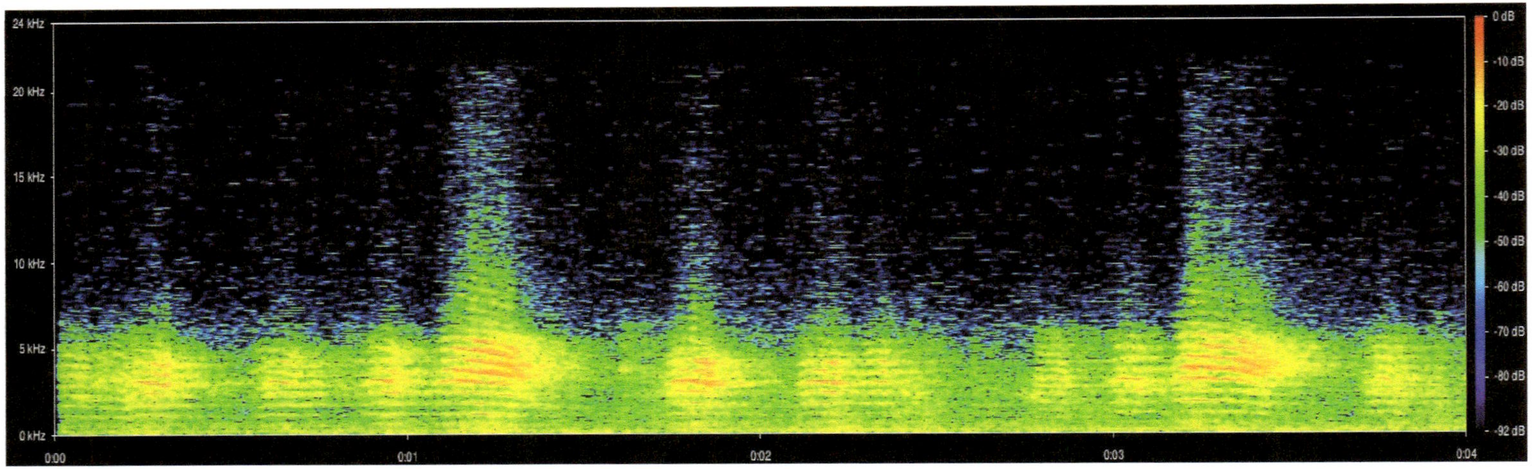

c. Female (0:18–0:22)

Figure 12.4 Zebra Finch Audio Spectral Graph. The times are shown for three segments from Disk 1 Track 4. The scale on the left is frequency (kHz). The scale on the right is volume (db). The scale on the bottom is in seconds. Spikes indicate the full spectrum of sound within the vocalization. Black reflects background noise. If there is no horizontal scale, the segment is one second or less. © D. Herrmann

12.6 Indian Ringneck Parakeet Roundtable (*Psittacula krameri manillensis*), Disk 2 Track 1

Date recorded:	April 2011
Location recorded:	Indoor free flight aviary
Number of Indian ringneck parakeets present:	4
Track duration:	5:17:61 minutes

This recording captures the vocalizations of four adult Indian ringneck parakeets in an indoor free flight aviary. They are from 4 to 18 years of age. Three of the Indian ringneck parakeets are female (Maku, Bluebell, Dusty). One of the Indian ringneck parakeets is male (Snapdragon).

Indian ringneck parakeets make a variety of interesting sounds. Generally one Indian ringneck parakeet vocalizes at a time; the others listen and/or answer. Indian ringneck parakeets vocalize in short sequences, which are often repeated two or three times with pauses in between.

At the beginning of this recording Snapdragon can be heard making male-only vocalizations, such as the "ch-ee ch-ee ch-ee", "cluck cluck cluck", and "hu hu hu" sounds that are made when a male bows to a female. The "hu hu hu" vocalization almost sounds like a person laughing.

In the middle of this recording Bluebell can be heard making a "che-ah che-ah che-ah" sound. This is the signature Indian ringneck vocalization. Bluebell makes this call when she is flying or when she spots something new—like the microphone. Dusty answers with the same "che-ah" sound. Dusty's vocalizations can be distinguished from Bluebell's because they are more intense and shrill. Maku is the quietest of the three.

The "che-ah" vocalization is gender-neutral, as are the variety of trilling whistles and "ruh-o-o" vocalizations heard throughout the recording.

Zebra finches, budgerigars, a mitred conure, and toys bouncing off the aviary bars can be heard in the background.

A sample audio spectral graph is presented in Figure 12.5.

12.7 Meyer's Parrot Beeps and Whistles (*Poicephalus meyeri*), Disk 2 Track 2

Date recorded:	April 2011
Location recorded:	Indoor free flight aviary
Number of Meyer's parrots present:	1
Track duration:	6:41:56 minutes

This recording captures the vocalizations of an adult female Meyer's parrot (Squeaky) in an indoor free flight aviary. Meyer's parrots are known for being quiet in captivity and in the wild. Meyer's parrots make short vocalizations, rather than long continuous sequences.

Squeaky seemed to be intrigued by the microphone and proceeded to vocalize almost her entire repertoire, with discrete pauses between each entry as if she were waiting for a response. (Refer to Table 9.5.) A variety of quiet clicking sounds, beeps, and whistles can be heard. At one point Squeaky performs a counting sequence. Toward the end Squeaky makes her signature "do-do" vocalization three times.

Zebra finches and budgerigars can be heard chattering and flying around in the background.

A sample audio spectral graph is presented in Figure 12.6.

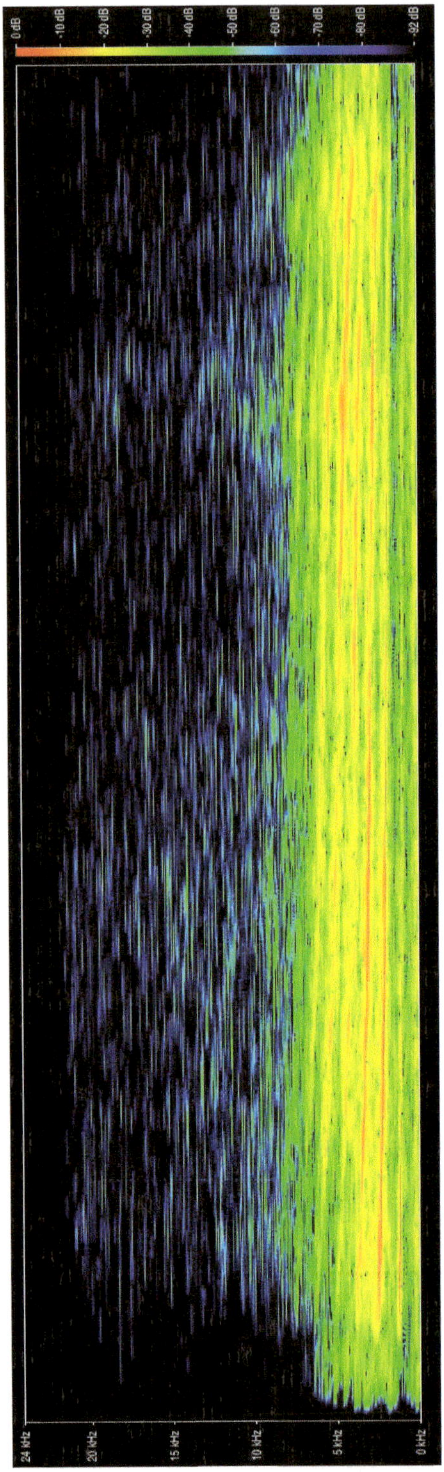

a. Male bowing (0:16–0:17)

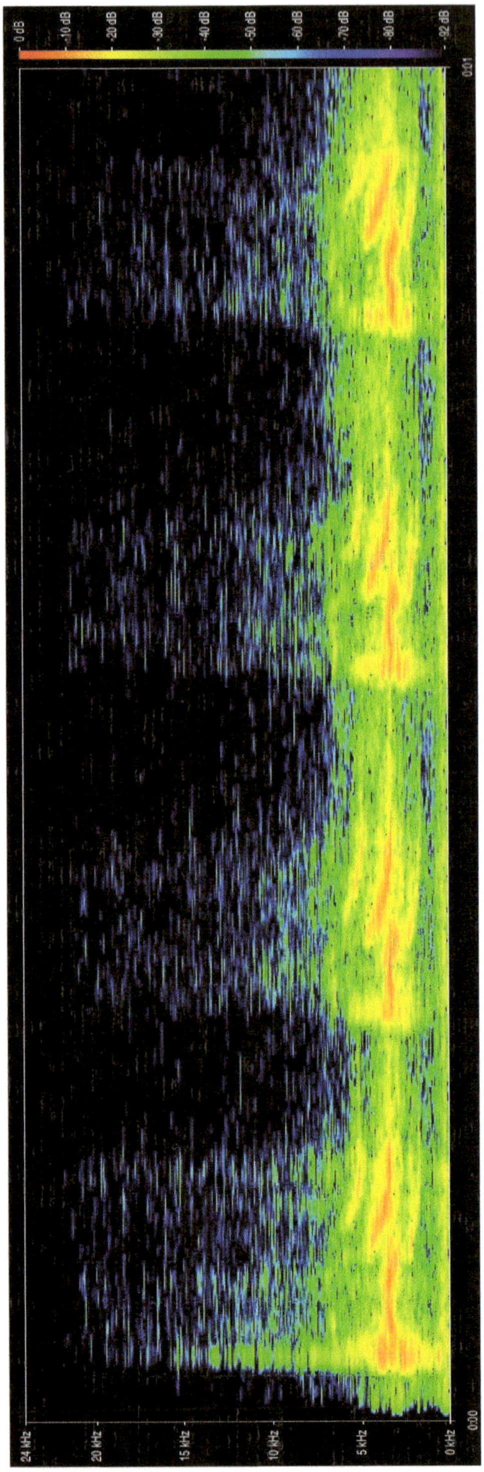

b. Che-ah (2:01–2:02)

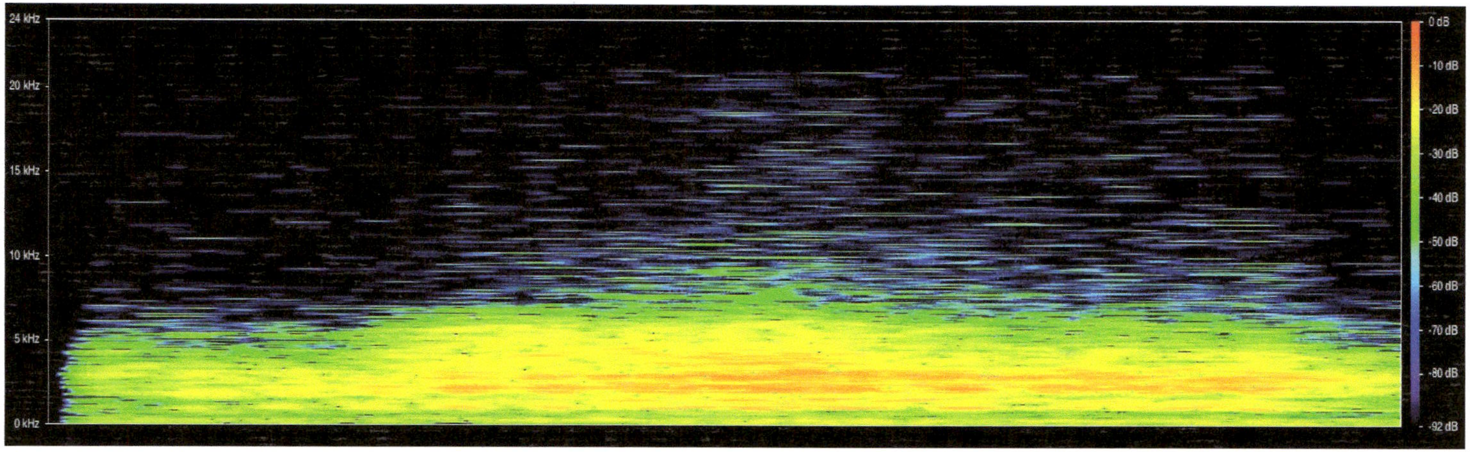

c. Trilling Whistle (3:36–3:27)

Figure 12.5 Indian Ringneck Parakeet Audio Spectral Graph. The times are shown for three segments from Disk 2 Track 1. The scale on the left is frequency (kHz). The scale on the right is volume (db). The scale on the bottom is in seconds. Spikes indicate the full spectrum of sound within the vocalization. Black reflects background noise. If there is no horizontal scale, the segment is one second or less. © D. Herrmann

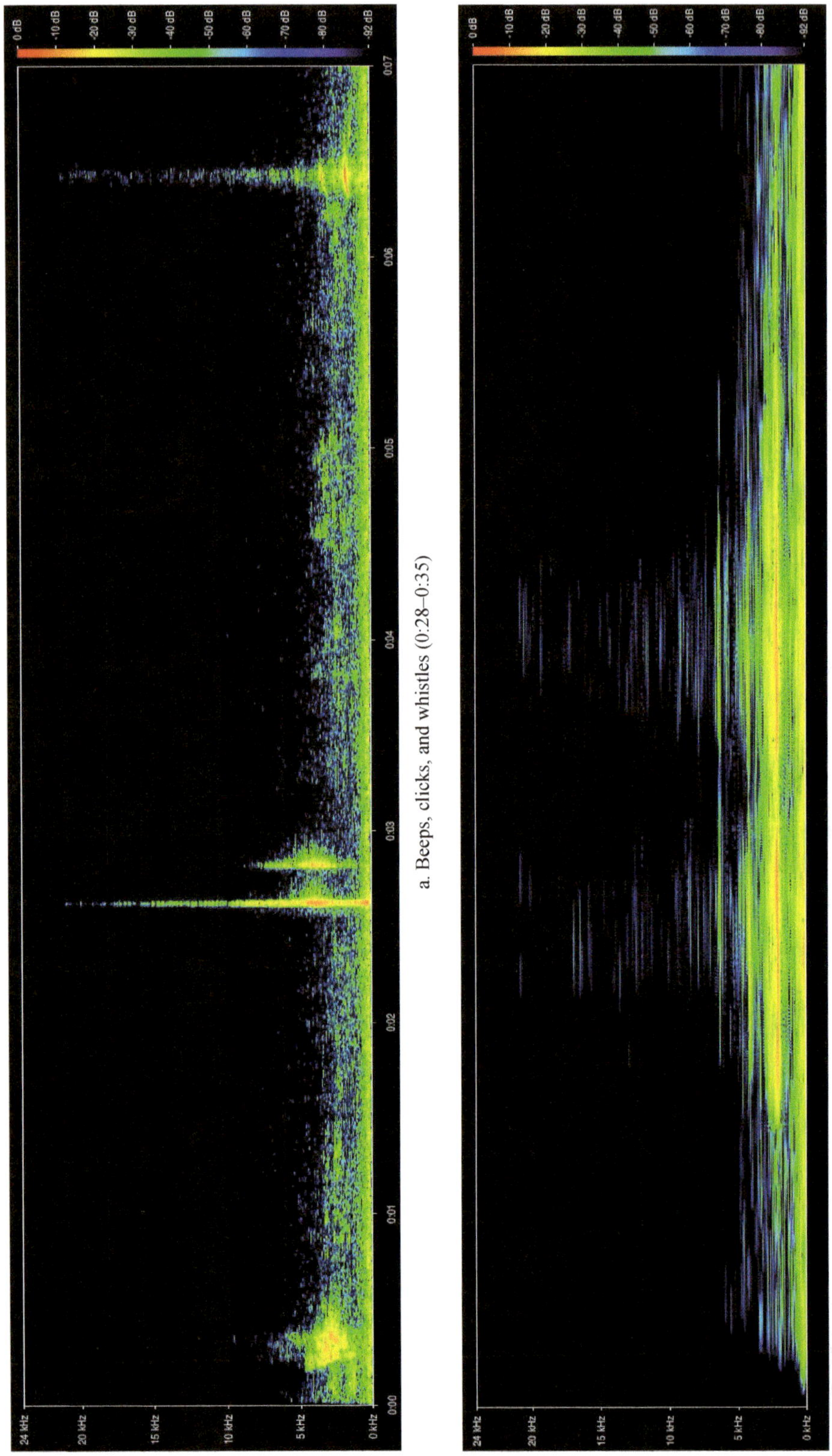

a. Beeps, clicks, and whistles (0:28–0:35)

b. Do-do (6:21–6:22)

Figure 12.6 Meyer's Parrot Audio Spectral Graph. The times are shown for two segments from Disk 2 Track 2. The scale on the left is frequency (kHz). The scale on the right is volume (db). The scale on the bottom is in seconds. Spikes indicate the full spectrum of sound within the vocalization. Black reflects background noise. If there is no horizontal scale, the segment is one second or less. © D. Herrmann

12.8 Mitred Conure Chortling (*Aratinga mitrata mitrata*), Disk 2 Track 3

Date recorded:	April 2011
Location recorded:	Indoor free flight aviary
Number of mitred conures present:	1
Track duration:	8:26:23 minutes

This recording captures the vocalizations of an adult male mitred conure (Buster) in an indoor free flight aviary. Conures in general are known for being rather noisy and Buster is no exception. Buster's vocalizations do not match those normally attributed to wild mitred conures; perhaps because mitred conure vocalizations are learned rather than inherited.

Buster makes a series of one to three syllable vocalizations while varying the pitch and volume. Some of the vocalizations sound like an "annh-annh"; others sound like an "ought oh". Buster's vocalizations appear to be in response to the vocalizations and activity of the nearby zebra finches, budgerigars, and Meyer's parrot, as if he is joining or commenting on their dialog. Buster's vocalizations are louder and faster as he becomes more excited—not unlike humans. About 6 minutes into the track and again at about 7:30 minutes Buster can be heard throwing his food on the floor. This is one of his favorite activities when he is excited.

A sample audio spectral graph is presented in Figure 12.7.

12.9 Budgerigar Morning Chorus (*Melopsittacus undulatus*), Disk 2 Track 4

Date recorded:	April 2012
Location recorded:	Indoor free flight aviary
Number of budgerigars present:	12
Track duration:	54:10:64 minutes

This recording captures the vocalizations of four female (Uzziel II, Sunflower II, Slate II, and Daisy) and eight male (Lilac II, Yellowhead, Dandelion, Daffodil, Perky II, Greenie III, Blue II, and Sky II) budgerigars as they engage in their morning chorus at sunrise. The budgerigars range in age from 1 year to 9 years. The budgerigars are in an indoor free flight aviary. As a result the sounds of the birds flying and jumping on and off toys and perches can be heard.

The budgerigars' vocalizations consist of a considerable variety of rapid fire high-pitched multi-syllable sounds. There is also quite a variety in pitch and volume, from a quiet quizzical whistle, at the beginning of the track as the budgerigars get used to the microphone, to a noisy clicking sound. The budgerigars become more animated as time goes on. At times only one budgerigar is vocalizing. Other times multiple birds are engaged in a rhythmical conversation. It is interesting to note that all of the birds become quiet at the same time. Several vocalizations were captured by this recording that a human ear does not normally isolate and that have not yet been attributed to wild budgerigars.

The budgerigars' vocalizations are in the foreground. A few other birds can be heard in the background. This interspecies interaction is interesting as well, especially since these five species of parrots would not meet in the wild—they live on different continents.

A sample audio spectral graph is presented in Figure 12.8.

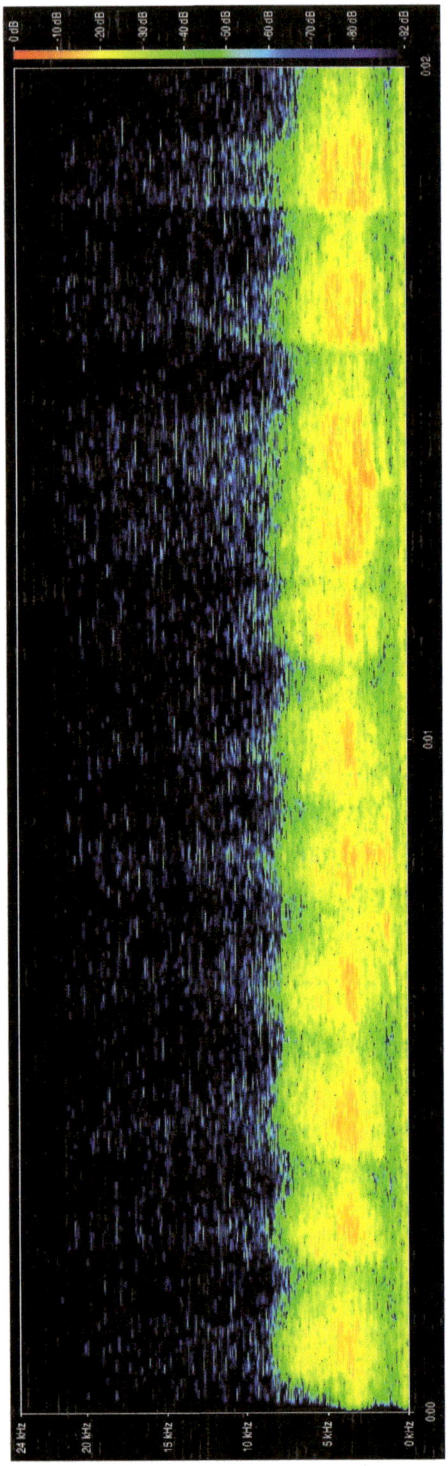

a. Annh-annh (5:57–6:00)

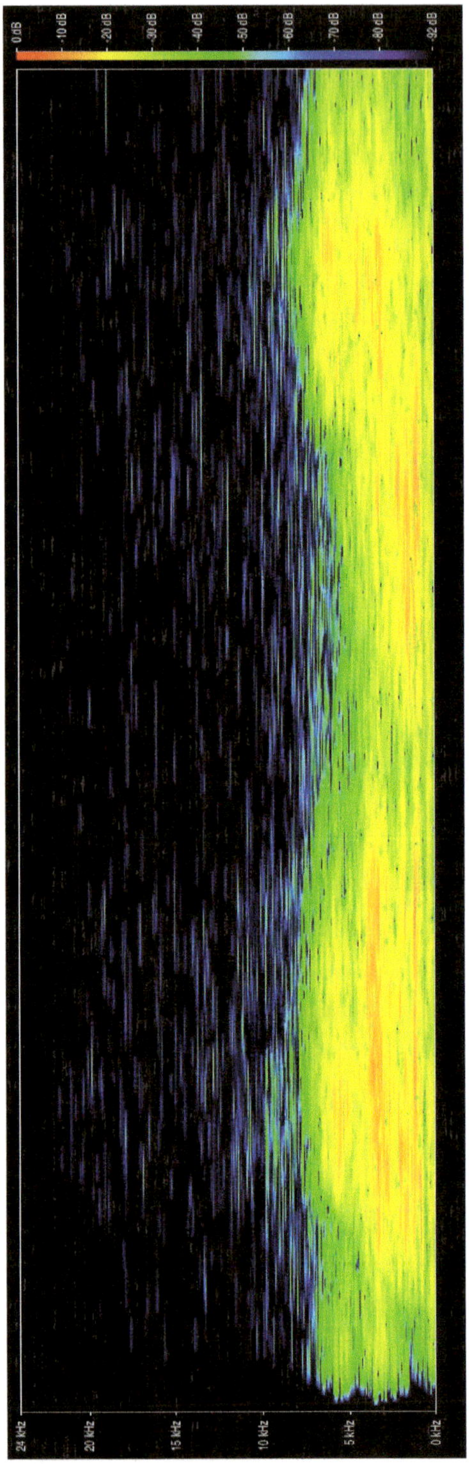

b. Ought-oh (4:05–4:06)

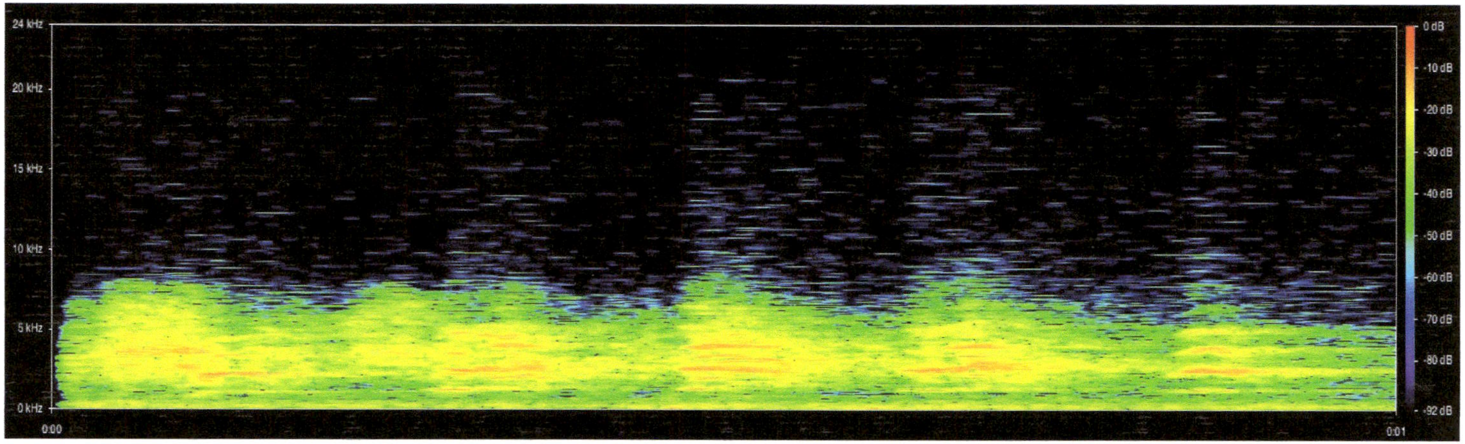

c. Laughing (0:01–0:02)

Figure 12.7 Mitred Conure Audio Spectral Graph. The times are shown for three segments from Disk 2 Track 3. The scale on the left is frequency (kHz). The scale on the right is volume (db). The scale on the bottom is in seconds. Spikes indicate the full spectrum of sound within the vocalization. Black reflects background noise. If there is no horizontal scale, the segment is one second or less. © D. Herrmann

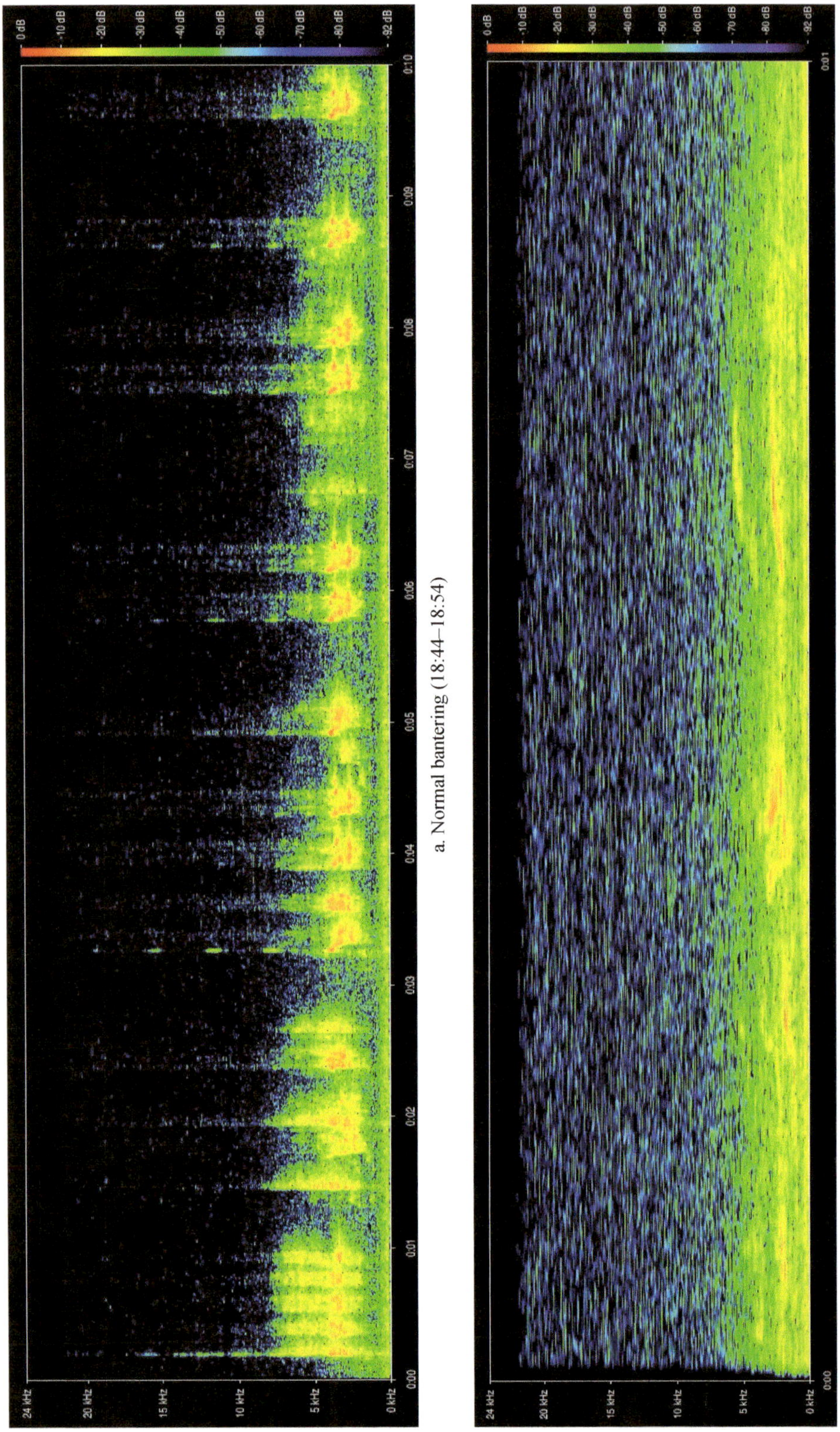

a. Normal bantering (18:44–18:54)

b. Quiet quizzical whistle (0:46–0:47)

Figure 12.8 Budgerigar Audio Spectral Graph. The times are shown for two segments from Disk 2 Track 4. The scale on the left is frequency (kHz). The scale on the right is volume (db). The scale along the bottom is seconds. Spikes indicate the full spectrum of sound within the vocalization. Black reflects background noise. If there is no horizontal scale, the segment is one second or less. © D. Herrmann

Conclusion

The previous 12 chapters discussed firsthand observations of 15 species of birds and more than 300 individual birds that are native to 6 continents which were made over a span of 30 years. Many new vocalizations and behaviors were identified—including behaviors that are not supposed to happen. Numerous examples of biological intelligence, genetic intelligence, and applied intelligence were presented throughout these chapters.

Whether the birds lived life as individuals (like herons and hummingbirds) or as part of a flock (like Canada geese and Mallard ducks), they demonstrated a high degree of individual reasoning, individual choices, and individual preferences while going about their daily lives. The birds showed resourcefulness when responding to novel situations, adversity, and physical disabilities, as well as the ability to apply newly acquired knowledge to future situations. Evidence of rapid decision-making, curiosity, and a tendency to engage in play activities was abundant.

The birds exhibited a range of temperaments within the same species and flock. The birds displayed sophisticated audio and visual communication systems. The birds living indoors acknowledged as one of their own species hybrid colors that do not occur in the wild. It is unknown whether or not wild birds would do that. The birds devised age-related compensating behaviors in their senior years. The birds experienced a ~ 2-week mourning period following the loss of a nest, hatchling, or mate. This is a significant amount of time given the average life spans of these birds.

All of this points to a strong sense of self-awareness, as does the existence of alarm calls and distress calls. After all, what is the point of making a distress call if a bird does not: (1) have an awareness of self and the likelihood of impending injury or death; or (2) understand that others nearby might be able to help? Likewise, what is the point of making an alarm call, except to warn others? Usually only one bird makes an alarm call. An alarm call does not benefit the bird making it. On the contrary, an alarm call may give away that bird's location.

Self-awareness has long been considered a sign of consciousness and intelligence. Beyond a sense of self, the birds demonstrated a keen awareness of others, friend and foe. The birds exhibited a strong identification with their species, flock, and family. For example, the Canada geese, mallard ducks, and wood ducks each acknowledged which birds were:

- Members of their family, even within multi-generational families
- Members of their flock
- Members of their species that were passing through the area during the biannual migrations
- Members of other bird species (water birds and songbirds) that lived at the same pond and were not a threat
- Predators

Even the 52 birds that lived in the indoor aviary understood who belonged in their "neighborhood". They acknowledged:

- When a new bird joined the group
- When a bird was missing from the group due to an illness, death, or a routine trip to the veterinarian

The birds demonstrated this awareness even when the new or missing bird did not reside in their aviary. Days-old zebra finch hatchlings noticed when one of their siblings was removed from the nest. Furthermore, the birds living in the indoor aviary learned to share toys and engage in group play activities, even in aviaries that contained multiple species, species that would not encounter each other in the wild.

The birds awareness of self and distinction between self and others often extended to behavior considered acts of altruism. Sometimes these acts of altruism were directed toward one or a few birds, such as family members. Other times these acts of altruism were directed toward an entire flock or all the birds in the immediate vicinity. Wild birds and birds living in the indoor aviary engaged in altruistic behaviors. Some of the more notable examples include:

- Ami feeding, grooming, and staying near Tzvi during his final days (Chapter 8). This behavior contradicts current thinking that birds chase off or attack sick or weak members of their flock in order to not attract predators.
- Precious feeding, grooming, and staying near Mr. Green Birdie during his final days (Chapter 7). This behavior contradicts current thinking that birds chase off or attack sick or weak members of their flock in order to not attract predators.
- A male cardinal charging two blue jays to protect his fledgling (Chapter 6).
- The young male mallard duck warning me to flee from an approaching peregrine falcon (Chapter 4)
- Canada geese flocks on the ground calling out to other flocks overhead to land, join them, and share the limited resources they had found during mid-winter partial migrations north, which were undertaken in response to freak weather in the south (Chapter 3).
- Mr. and Mrs. Speckleneck accepting four new goslings that had hatched elsewhere into their family even though they were not orphans (Chapter 3).
- Helpers (goslings from the previous year) and other adults providing a security perimeter around a pair of Canada geese with young goslings when they were chased by a dog. Neither the parents nor the goslings could fly at this time. The helpers could have flown away, but they did not (Chapter 3).
- Male mallard ducks creating a security barrier around the red-eared slider turtles to keep them away from a family of ducklings (Chapter 4).
- A Canada goose family accepting an orphaned wood duckling into its group of goslings (Chapter 4).

Common grackles provided some of the more poignant examples of altruistic behavior. One winter a sudden severe snowstorm left 14 inches (35.5 cm) of snow and temperatures in the single digits (Fahrenheit). Overnight all of the wild birds food sources were buried. As soon as a ground-feeding station was filled more than 30 birds descended on it. The birds were ravenous. They packed together closely and ate the food so energetically that the feeder needed to be refilled every 2 hours. None of the birds were aggressive. None of the birds tried to hoard food or prevent other birds from eating. Instead, they only consumed what they needed. The food included mixed seeds and grain, peanuts, and bread cubes soaked in peanut oil. One elderly common grackle stood a few feet from the crowd. It tried, but could not keep up with the other birds energetic movements. The elderly common grackle stood back and waited, obviously hungry. After about 15 minutes a young common grackle noticed what was happening. The young common grackle used its beak like a hockey stick and hit some food down the snow drift to the elderly common grackle.

Another incident occurred during spring. A red-tailed hawk treed a common grackle at the edge of the woods near Pond D. The common grackle could not leave the tree without being snatched by the red-tailed hawk. The common grackle stayed put and made a loud distress call while raising and lowering its head. Several pairs of common grackles live in the area. Within a few minutes a second common grackle landed on a branch the same height as the first, but on the opposite side of the red-tailed hawk. Then other common grackles arrived and perched nearby. Before long the red-tailed hawk was surrounded and could not attempt an attack without being mobbed; eventually it left. The second common grackle in particular put itself at risk; it was probably the first common grackle's mate.

Acquiring new knowledge quickly is a sign of intelligence. The rapidity with which birds learn new life skills, vocalizations, and other information is impressive. Humans on average spend 20 years, or 25% of their average life span, learning to become responsible for themselves. In contrast, birds spend on average 3–5% of their maximum life span learning to become responsible for themselves, and they face more dangerous and persistent threats to their survival than humans. The ability to plan ahead is another sign of intelligence. Several examples of acquiring knowledge quickly and planning ahead were given throughout Chapters 2–11.

Perhaps the most interesting was watching the indoor birds learning from each other. Different species experimented with toys, food, or movements—like climbing under their paper—that they had never done before, after seeing another bird in another aviary do something. Even the baby grackle began climbing up the side of its temporary home using its beak, after seeing the budgerigars do this. This is not something a baby grackle would learn or do in the wild.

Birds, particularly the parrot family, are known for having superb long term memories and they don't have to write things down in order to remember them. Unlike humans, birds are constantly learning new things, including new vocalizations, throughout their lives. This ability is facilitated in part by neuronal replacement which allows birds to delete old or obsolete information that is no longer useful.[92]

Birds have more complex hearing than humans; they can hear a range of frequencies that is both higher and lower than humans can hear.[14, 18, 74, 154] Birds hear sounds from a greater distance than humans. Birds have more intricate abilities for the time resolution of sound.[14] To illustrate, birds have the ability to isolate or distinguish individual sounds arriving in rapid succession ten times faster than humans.[14] What humans hear as a single sound, a bird may hear as ten discrete tonal variations.[14] Birds can see ultraviolet light; humans cannot. Avian visual perception is superior to humans and most vertebrates. Birds can see a larger volume of visual data, more detailed visual data, and from greater distances. To do so birds have more sophisticated cognitive systems and algorithms to process, interpret, store, and act upon this auditory and visual information. A mallard duck can identify an individual mallard duck in a flock. A Canada gosling can identify its parents in a flock. How many humans can identify an individual mallard duck or Canada goose in a flock? A budgerigar named Sparky learned to repeat 531 words and 383 sentences in English.[70] Alex, an African grey parrot, learned how to answer questions correctly in English; he understood the question and formulated correct answers.[134, 135] How many humans can learn to speak budgerigar or converse in an African grey dialect? Most humans have difficulty learning another human language, especially one that is written in a different alphabet or script.

Evidence of intelligence can also be gleaned from what birds do not do. Unlike many mammals, birds that are herbivores, like ducks and geese, do not overgraze an area and kill the vegetation. Birds that are predators do not kill off an entire population through over hunting. The green herons, great blue herons, and black-crowned night herons did not eat all the fish and tadpoles in one pond. Instead, they alternated between ponds. When a red-tailed hawk attacked a gosling, it only took one gosling from a family. The hawk did not return day after day to consume all the goslings or ducklings in the same family. This behavior implies that birds understand the need to balance what they consume today with the need to conserve food sources for the future. Birds that are predators only kill for food that they need to survive. They don't kill for the sake of killing or so that they can hang a stuffed head on the wall.

Annex A

Glossary

abandonment: the feeling of being left behind and out of the flock. If severe or prolonged, abandonment can trigger emotional problems in pet birds, even death.[9, 10]

adaptation: gradual and ongoing modification of an organism, its physiology and/or behavioral traits that optimize it for surviving, prospering, and progressing in an ever changing environment.

alarm call: a warning of a new danger that has just entered the immediate area that is broadcast to and understood by all types of birds within hearing distance. The danger could be a perceived predator or sudden environmental change like an earthquake, avalanche, or tornado. An alarm call differs from a distress call in sound and purpose.

almond eyes: this term, generally applied to budgerigars, refers to the practice of pet birds exhibiting a relaxed state by letting their eyelids droop a bit, which makes their round eyes look like they are almond-shaped.

allo-feeding: when two birds feed each other. This act establishes a social bond between two birds and is often a precursor to becoming a bonded pair.

allo-preening: when two birds preen each other, usually around the head and neck—areas that are difficult for a bird to preen itself. In addition to essential grooming, this act often establishes a social bond between two birds.

altricial: hatchlings that are helpless and dependent on their parents for food, warmth, and protection; for example, songbird hatchlings who cannot walk, see, fly, or feed themselves and must stay in the nest for several days or weeks. Also referred to as nidicolous.

amygdala: part of the brain that supervises the processing of stimuli that trigger emotions, the emotional response to these triggers, and the storage of memories related to emotional events.

anisodactyl: a toe arrangement common to Passerines and other birds whereby three toes face forward and one toe faces backward.

barbs: (1) any of the side branches of the shaft of a feather;[179] (2) parallel branches extending from each side of a feather shaft.[18, 137]

barbules:	(1) minute barbs, especially ones that fringe the barbs of a feather;[179] (2) branchlets off both sides of the barbs of a feather at right angles to the barbs. Adjacent barbules hook together, except in down feathers.[18, 137]
beak:	same as bill; projecting jaws of a bird, including the upper and lower mandibles that a bird uses to gather food, preen, climb, feed its young, etc.[18, 41]
beak boxing:	when two birds deliberately strike each other's beaks in a semi-aggressive playful manner that is not intended to inflict bodily harm. Pet birds also engage in beak boxing with a favorite toy for fun.
beaking:	a parrot testing the feel and firmness of various substances with its beak.
bill:	see beak.
biological tool:	a biological system or structure that has become specialized over time to facilitate a species ability to survive, prosper, and progress within its particular eco-system niche.
blood feather:	a newly emerging immature feather containing blood vessels.[51] If this feather is broken or pulled out, the bird will bleed.
BSCE:	Bird Strike Committee Europe.
call notes:	short simple bird sounds that convey a specific message, such as hunger, danger, and the caller's location.[137]
cere:	a waxy featherless protuberance at the base of the bill of a bird which encompasses the nares. Adult budgerigars have different color ceres: blue for males and beige for females. In most other birds the cere is the same color for both sexes.
cholla:	any of several arborescent very spiny cacti of the southwestern U.S. and Mexico.[126, 179] After they die, cholla branches are dried and cut into perches for pet birds. The spines are removed, creating cavities that encourage birds to chew; however the cavities also collect debris.
cloaca:	see vent.
clumping:	the habit of several small birds, such as sparrows and wrens, to gather and roost close together in order to jointly retain body heat.[3] This process may allow individual birds to retain up to 17% of their body heat.[158] The expression of this behavior is similar to contact sitting, but it is undertaken for a different purpose. (See contact sitting.)
clutch:	a complete set of eggs laid by a single female bird during a single nesting cycle.
cognitive ethology:	a new area of study that focuses on how the minds of animals work; Dr. Donald R. Griffin was a pioneer in this field.
consciousness:	the state of being that is characterized by sensation, awareness, emotion, volition, and thought.[179]
contact sitting:	several small birds, like zebra finches, siting side by side with their bodies touching. The birds may or may not all be facing the same direction. Contact

sitting is seen year round and in captive birds; hence this behavior is engaged in primarily for social and security reasons. (See clumping.)

coverts: small feathers covering a bird's main flight feathers on its wings and tail.[18]

crest: set of feathers on the top of the head that is used to express emotions; common to cardinals, cockatiels, cockatoos, and green herons among others.

crop: dilation of the lower esophagus that stores food and performs preliminary digestion.

cuttlebone: the dried shell of cuttlefishes that is used to supply pet birds with calcium and other minerals. Pet birds chew on a cuttlebone to keep their beaks trim, and in the process ingest the minerals.

CVT: certified veterinarian technician.

dimorphic: physical characteristics that distinguish male and female birds from each other that usually appear at maturity.[137, 182] For example, male Indian ringneck parakeets are easily discernible from female Indian ringneck parakeets because of the black band around their necks.

distress call: a loud unpleasant sound made by a bird when it is in a severe state of danger, physical agony, and/or suffering, such as when a bird is under direct attack by a predator. A distress call communicates sheer terror combined with a plea for help. Some pet birds make a distress call if they are handled roughly. A distress call differs from an alarm call in sound and purpose.

down: under layer of small feathers whose primary purpose is to regulate body temperature and conserve heat.

drake: a male duck.

droppings: the combination of urates, water, and fecal material excreted from the vent.[11]

duck: any of various swimming birds with a short neck and legs; a member of the family Anatidae; a female duck.[179]

emotion: complex psycho-physiological experiences of an individual's state of mind interacting with biochemical and environmental influences.[46, 203]

fledge: (1) the time when a young bird has acquired a complete set of flight feathers and is ready to leave the nest;[137] (2) the process of learning to fly for the first time.[9, 10]

fledgling: a young bird that has a complete set of flight feathers and is ready to leave the nest.[137]

flock: a bird's extended family or community; members of a flock may or may not all be related to each other.

flying attack: about once a day pet birds hold onto a perch or the side of their cage and flap their wings vigorously, as if flying full speed, for exercise. This activity also boosts a bird's morale and is referred to as flying attacks.

gander: an adult male goose.

geophagy:	the practice of many species of parrots eating dirt (soil and clay) in various parts of the world; the most famous examples are the clay licks in South America that are frequented by parrots.
goose:	any of numerous large swimming birds with long necks; intermediate waterfowl between ducks and swans; a member of the family Anatidae; a female goose.[179]
hanging parrot:	members of the parrot family that enjoy spending part of their day hanging upside down, such as Meyer's and red-bellied parrots. In general hanging parrots are short and stocky and have square tails.
hallux:	the back toe of birds that have three toes facing forward and one toe facing backward.
hatch day:	birds are not born, they hatch. The hatch day is the day a hatchling pecks its way out of its shell.
HEPA:	high efficiency particulate air, refers to a specific type of high-efficiency air filter.
IBSC:	International Bird Strike Committee.
ICSBM:	International Center for the Study of Bird Migration.
imprinting:	a rapid learning process that takes place early in the life of some birds.
incubation patch:	area on the lower breast of a hen where the feathers are thinner so the eggs can come in contact with the skin.
information center:	a midday communal roost where birds gather to pass on information about good feeding sites and other topics of common interest to one another.[18]
infrasound:	very low frequency sounds that birds can hear, but not humans; perhaps used during homing and migration.[18]
instinct:	knowledge that has been acquired through the process of adaptation, stored genetically, and is available for decision-making at the subconscious level.
intelligence:	the ability to acquire, discern, assimilate, generate, retain, and apply facts and knowledge, some of which are constants and some of which are variables, that are needed for problem solving skills, especially when dealing with novel situations, for an individual and species to survive, prosper, and progress in its ever changing environment; includes the ability to distinguish irrelevant data and trivia from useful information.
intuition:	instant insight, perception, or cognition that is arrived at without any rational thought or inference.[179]
keratin:	any of various sulfur containing fibrous proteins that form the chemical basis of horny epidermal tissues, like a bird's beak, cere, and nails.[179]
lamellae (pl.):	thin comb-like membrane that lines the inner rim of a duck or goose's bill and functions like a sieve.
landing perch:	a perch that is installed on the outside of a pet bird's cage, near the main entrance. This perch functions as a front porch and offers birds a place to rest

before and after flying. Landing perches also provide a neutral place for other birds to stop by for a visit.

leg bands or rings: numbered metal bands or rings placed on a newly hatched bird's leg for identification and tracking purposes.

mandible: (1) either the upper or lower segment of the bill of a bird; (2) lower bill or beak.

maxilla: the upper beak; the notched protuberance that gives hookbills their name.

migrant: a bird that undertakes regular seasonal movements, breeding in one locality and overwintering in another.[3]

mineral block: minerals such as calcium and iodine that are pressed into a firm shape so that pet birds can chew them to supplement their diet while trimming their beak.

mobbing: the cooperative security practice of a flock of small birds crowding around, attacking, and otherwise annoying a larger bird to keep it away from their nests and prevent the larger bird from harming eggs and hatchlings.

molting: the cyclical process of shedding all or part of the feathers and replacing them with new growth.[137]

mourning period: the approximately 2-week mourning period pet birds and some outdoor birds go through after losing their mate, a hatchling, or a clutch. The signs are consistent: a profound sense of sadness and loss, reduced appetite, reduced activity level, searching for and calling out to the lost bird. After the mourning period, a pet bird slowly returns to their normal routine. Some eventually bond with a new bird, others choose to remain single. Often the process of being re-homed causes a pet bird to go through a mourning period, as it grieves for the loss of its former buddies and caretaker.

nares: openings of the nasal cavity located in the upper beak, near its base.[137]

nesting dispersal: the practice of some flock birds of distributing their nests a safe distance apart from each other. The intent is to ensure that, should one nest experience a calamity (predation, flooding, etc.), the entire flock's hatchlings are not lost. This behavior is the opposite of birds who nest close together to provide community safety.

nidicolous: see altricial.

nidifugous: see precocial.

night fright: unexplained thrashing around the cage in the night that is sometimes seen in cockatiels and other companion parrots.[9, 10]

nippy: a bird that is prone to accidental, unintentional, or nonaggressive pinching that is not intended to cause damage.[9, 10]

normal or nominate green budgerigar: wild budgerigars are only green with yellow heads. They are referred to as normal or nominate green budgerigars. The other colors of budgerigars seen in pet stores and bird shows are the result of genetic manipulation by domestic breeders.

Let's look at the key problems facing humans around the world today:

- High unemployment following the 2008 recession
- High taxes
- High crime rates
- Environmental degradation
- Shortage of sustainable energy sources
- Ideological conflict and warfare
- Economic disparity
- High stress sedentary lifestyles

All of these problems are manmade—they are not part of nature. These problems are the outcome of economic, political, and social structures humans on all continents have created over the centuries. For example, there are over 8 million mallard ducks living in North America.[210] None of them are unemployed. In fact, mallard ducks would laugh at the idea that one mallard duck could "fire" another duck or prevent it from working.

Birds on the other hand do not face any equivalent bird-made problems. The complex societies in which birds live have not created any problems for themselves or other species. On the contrary, the primary problems birds face today are:

- Environmental degradation and habitat destruction caused by *humans*
- Poaching by *humans* for the pet bird trade
- Predation by house cats that *humans* let run loose

All other species living on Earth give as much back to nature as they take from it. Humans are the only species that continually withdraws from its environmental account without making any deposits. Humans are the only species that has failed to learn to live with nature as part of it. This is evident from humans lack of biological intelligence, lack of genetic intelligence, and failure at adaptive radiation.

In summary, humans have many more ideas and inventions to learn from birds.

parakeet:	any of a number of small slender parrots that have a long thin tail, a subset of the parrot family.
parrot:	any of a number of birds of the order Psittaformes; their common distinguishing features include zygodactyl feet, a hooked beak, and brightly colored plumage.
pectinate:	the middle toe of heron which is longer than the other two forward toes, the nail of the middle or pectinate toe has a small comb-like feature that is used to distribute a heron's powder down. (See powder down.)
pinch:	a behavior designed to get a human's attention when a bird puts a person's skin between their beak and squeezes hard enough to be noticed, without breaking the skin.[9, 10]
powder down:	instead of a uropygial or preening gland herons and a few other birds have powder down. These powder down feathers periodically dissolve into a powder that is used to clean a heron's feathers. (See pectinate.)
precocial:	hatchlings that are capable of a high degree of independence and responsibility, such as ducklings and goslings that are able to walk, swim, and feed themselves within 24 hours of hatching. Precocial hatchlings need the protection of their parents until they can fly. Also referred to as nidifugous.
preen:	to groom feathers with the beak.
preening invitation:	a bird bowing its head and/or ruffling its feathers to induce another bird or human to preen it.
quarantine:	to remove or isolate birds in a separate room for a specific period of time in order to lessen the risk of spreading disease to other birds on the premises. A 30-day quarantine period is commonly recommended for new pet birds.
quick links:	small oval shaped metal clasps that are used to attach pet bird toys and swings to the top of a cage.
senescence:	the process of aging or becoming old. Birds have the longest life span of any vertebrates with the equivalent body mass, body temperature, and metabolic rate.[137]
sleeping perch:	preferred location where a pet bird likes to sleep; similar to a roost for outdoor birds.
speculum:	an iridescent patch of color on the secondary wing feathers of mallard ducks and some other birds that resemble military insignia.
torpidity, torpor:	a state of dormancy reflected by motor inactivity and a lowered body temperature and metabolic rate, usually undertaken to conserve energy and combat possible starvation in the face of adverse environmental conditions.[3, 18, 179]
toys:	pet birds need a variety of toys to challenge and stimulate them physically and intellectually. There are a variety of pet bird toys, such as puzzle toys, chew toys, foot toys, preening toys, and shredding toys, which vary in size, color, texture, and composition.
undulate:	to rise and fall in volume, pitch, and cadence.[179]

uropygial:	preening gland on the lower back of most birds, near the tail feathers, that secretes an oily fluid which is used to preen the feathers.
vent (or cloaca):	a common external opening in birds into which the intestinal, excretory, and reproductive systems discharge. Some sources differentiate between the cloaca and the vent; whereby the cloaca is the internal duct and the vent is the external opening.
wingspan:	the distance from the tip of one wing to the tip of the opposite wing.
zygodactyl:	a perching grip common to parrots and other birds whereby two toes face forward and two toes face backward.

Annex B

Bibliography

B.1 Publications

1. Able, K. (ed.). Gathering of Angels: Migrating Birds and Their Ecology, Cornell University Press, 1999.
2. Ainley, D. The Adelie Penguin: Bellwether of Climate Change, Columbia University Press, 2002.
3. Alderton, D. The World Encyclopedia of Birds and Birdwatching, JG Press, World Publications Group, 2008.
4. Animal Lover, Montgomery County Humane Society, Summer 2010, p. 3.
5. Armstrong, S. Bird Display and Behavior: An Introduction to the Study of Bird Psychology, Dover Publications, Inc., 1965.
6. Arnal, L. and Keymer, I. Bird Diseases: An Introduction to the Study of Birds in Health and Disease, T.F.H. Publications, Inc., 1975.
7. Athan, M. Guide to Behavior, Budgies: Celebrating the World's Favorite Parakeet. Bow Tie Magazine, 2008, pp. 76–87.
8. Athan, M. and Deter, D. Guide to the Senegal Parrot and Its Family, Barron's Educational Series, Inc., 1998.
9. Athan, M. Guide to Companion Parrot Behavior, Barron's Educational Series, 1999.
10. Athan, M. Guide to a Well-Behaved Parrot, 2nd edition, Barron's Educational Series, Inc., 1999.
11. Athan, M. and Deter, D. The Second-hand Parrot, Barron's Educational Series, Inc., 2002.
12. Attenborough, D. The Life of Birds, BBC Books, 1998.
13. Ayala-Guerrero, F. Sleep Patterns in the parakeet Melopsittacus undulatus, Physiology & Behavior, Volume 46, Issue 5, November 1989, pp. 787–791.
14. Barber, T. The Human Nature of Birds: A Scientific Discovery with Startling Implications, St. Martin's Press, 1993.
15. Barcott, B. The Last Flight of the Scarlet Macaw: One Woman's Fight to Save the World's Most Beautiful Bird, Random House, 2008.
16. Bates, H. and Busenbark, R. Finches and Soft-billed Birds, T.F.H. Publications, Inc., 1970.
17. Beissinger, S. and Snyder, N. (eds.). New World Parrots in Crisis: Solutions from Conservation Biology, Smithsonian Institution Press, 1992.
18. Bird, D. The Bird Almanac: A Guide to Essential Facts and Figures of the World's Birds, Firefly, 2004.
19. Bird Safe House Plants: A Shoppers Guide, RSS, 2000.
20. Birkhead, T. The Wisdom of Birds: An Illustrated History of Ornithology, Bloomsbury, 2008.
21. Bittner, M. The Wild Parrots of Telegraph Hill, Three Rivers Press, 2004.
22. Blanchan, N. Bird Neighbors: An Introductory Acquaintance with One Hundred and Fifty Birds Commonly Found in the Gardens, Meadows, and Woods About Our Homes, Doubleday, Page, and Company, 1918.
23. Boehrer, B. Parrot Culture: Our 2500-Year Old Fascination with the World's Most Talkative Birds, University of Pennsylvania Press, 2004.
24. Boruchowitz, D. The Guide to Owning a Conure, T.F.H. Publications, Inc., 2003.
25. Boyes, S. The Meyer's Parrot—an African parrot, PsittaScene, November 2005, pp. 14–15.
26. Brinker, B. and Friedman, S.G. The Nature of Greys, Bird Talk, November 1999.

27. Brittan-Powell, E., Dooling, R. and Farabaugh, S. Vocal Development in budgerigars (Melopsittacus undulatus): Contact calls, Journal of Comparative Psychology, Volume 111, Issue 3, September 1997, pp. 226–241.
28. Bull, E., Nielsen-Pincus, N., Wales, B. and Hayes, J. The influence of disturbance events on pileated woodpeckers in Northeastern Oregon, Forest Ecology and Management, Volume 243, Issues 2-3, May 2007, pp. 320–329.
29. Bull, J. and Farrand, Jr., J. The Audubon Society Field Guide to North American Birds: Eastern Region, Alfred A. Knopf, 1977.
30. Burger, J. Birds: A Visual Guide, Firefly Books, 2006.
31. Burgmann, P. Feeding Your Pet Bird, Barron's Educational Series, Inc., 1993.
32. Butler, C. Population Biology of the Introduced Rose-ringed Parakeet (Psittacula krameri) in the UK, thesis, University of Oxford, Department of Zoology, Edward Grey Institute of Field Ornithology, November 2003.
33. Butler, A. Foreign Birds for Cage and Aviary, Volume 1, The Feathered World, London, 1908.
34. Butler, A. Foreign Birds for Cage and Aviary, Volume 2, The Feathered World, London, 1909.
35. Caro, T. Anti-Predator Defenses in Birds and Mammals, University of Chicago Press, 2005.
36. Cayley, N. What Bird is That? A Guide to the Birds of Australia, 5th edition, Angus and Robertson Publishers Pty. Ltd., 1971.
37. Chan, G. Old Birds Sing the Hits, Epoch Times, p. B7, 25 August 2011.
38. Chan, G. Brainy Birds Use Tools in Different Ways, Epoch Times, p. B7, 23 June 2011.
39. Chan, G. Pigeons, Like Monkeys Can Order Numbers, Epoch Times, p. B6, 5 June 2012.
40. Chu, M. Songbird Journeys: Four seasons in the lives of migratory birds, Walker, 2007.
41. Clements, K. The Encyclopedia of Birds, Facts on File, Volumes 1 and 2, 2007.
42. Coborn, J. The Professional Book of Conures, T.F.H. Publications, Inc., 1991.
43. Cokinos, C. Hope is the Thing with Feathers, Penguin Books, 2009.
44. Couzens, D. Extreme Birds: The World's Most Extraordinary and Bizarre Birds, Firefly, 2008.
45. Cruickshank, A. Birds of America, Dover Publications, Inc., 1977.
46. Crystal, D. (ed.). The Cambridge Encyclopedia, 4th edition, Cambridge University Press, 2000.
47. Dearborn, N. Bird Houses and How to Build Them, Farmers' Bulletin 609, U.S. Department of Agriculture, April 1918.
48. Deutsch, R. The Healthy Bird Cookbook: A Lifesaving Nutritional Guide and Recipe Collection, T.F.H. Publications, Inc., 2004.
49. Doane, B. Parrot Training: A Guide to Taming and Gentling Your Avian Companion, Wiley Publishing, Inc., 2001.
50. Dooling, R. Perception of vocal signals by budgerigars (Melopsittacus undulatus), Experimental Biology, Volume 45, Issue 3, 1986, pp. 195–218.
51. Dorenkamp, B. Natural Health Care for Your Bird, Barron's Educational Series 1998 (English translation).
52. Dunne, L. Nutrition Almanac, 3rd edition, McGraw-Hill Publishing Company, 1990.
53. Earle, K. and Clarke, N. The Nutrition of the Budgerigar (Melopsittacus undulatus), The Journal of Nutrition, Volume 121, 1991, pp. 5186–5192.
54. Erickson, L. Where Have All the Birds Gone? BirdScope, Cornell Lab of Ornithology, Vol. 22(4), Autumn 2008, p. 3.
55. Ettlinger, R. On Feathered Wings, Abrams, 2008.
56. FAA, Wildlife Strikes to Civil Aircraft in the United States, 1990–2008, National Wildlife Strike Database, Serial Report Number 15, Washington, D.C., September 2009.
57. Farabaugh, S., Linzenbold, A. and Dooling, R. Vocal Plasticity in Budgerigars (Melopsittacus undulatus): Evidence for social factors in the learning of contact calls, Journal of Comparative Psychology, Volume 108, Issue 1, March 1994, pp. 81–92.
58. Feare, C. Rose-ringed parakeet Psittacula krameri: a love-hate relationship in the making? The introduction and naturalization of birds, J.S. Holmes and J.R. Simons (ed.), London, HMSO, 1996, pp. 107–112.
59. Forshaw, J. "The Wild Ones", Budgies: Celebrating the World's Favorite Parakeet, Bow Tie Magazine, 2008, pp. 52–61.
60. Forshaw, J. and Cooper, W. Parrots of the World, T.F.H. Publications, Inc., 1977.
61. Friedman, S. G. and Brinker, B. Early Socialization: A Biological Need and the Key to Companionability, Original Flying Machine, Issue 2, September/October 2000.

62. Friedman, S. G. and Brinker, B. The Facts About Punishment, Original Flying Machine, Issue 4; January/February 2001.
63. Friedman, S. G. and Brinker, B. The Struggle for Dominance: Fact or Fiction? Original Flying Machine, Issue 6, May/June 2001.
64. Friedman, S. G. The ABCs of Behavior, Original Flying Machine, Issue 9, November/December 2001.
65. Friedman, S. G. Alternatives to Breaking Parrots, presentation at the Stop PDD Conference 2002.
66. Friedman, S. G. Parrots in Temporary Shelters: The Groundwork for Empowerment and Trust, Avian Welfare Coalition Shelter Manual, February 2004.
67. Friedman, S. G. Straight Talk about Parrot Behavior, presentation at the Stop PDD Conference, November 2004.
68. Frischman, C. Conures, T.F.H. Publications, Inc., 2007.
69. Gadgil, M. and Ali, S. Communal roosting habits of Indian birds, Journal of Bombay Natural History Society, Volume 72, Issue 3, 1974, pp. 716–727.
70. Gallerstein, G. The Complete Bird Owner's Handbook, Howell Book House, 1994.
71. Gallup, G. and Capper, S. Preference for mirror-image stimulation in finches (Passer domesticus) and parakeets (Melopsittacus undulatus), Animal Behavior, Volume 18, Part 4, November 1970, pp. 621–624.
72. Gardner, H. Frames of Mind: Theory of Multiple Intelligences, Basic Books, 2004.
73. Gedney, C. Foreign Cage Birds, Volume 2: Waxbills, Finches, Weavers, Orioles, and Other Aviary Birds, A. Bradley London and County Printing Works, 1879.
74. Gill, F. Ornithology, W. H. Freeman and Company, 3rd edition, 2006.
75. Gokula, V. and Venkatraman, C. Foraging and breeding biology of the Malabar Parakeet (Psittacula columboides) in the Siruvani foothills, Tamil Nadu, India, Forktail, Volume 19, 2003, pp. 138–139.
76. Goodfellow, P. Avian Architecture: How Birds Design, Engineer, and Build, Princeton University Press, 2011.
77. Gore, A. An Inconvenient Truth: the Planetary Emergence of Global Warning and What We Can Do About It, Rodale, 2006.
78. Gould, J. and Gould, C. Animal Architects: Building and the Evolution of Intelligence, Basic Books, 2007.
79. Gramza, A. Vocal Mimicry in Captive Budgerigars (Melopsittacus undulatus), 1970 (paper posted on web without complete citation details).
80. Greene, E. A Lifetime with Birds: An Ornithological Logbook, Edwards Brothers, Inc., 1966.
81. Greene, W. Favourite Foreign Birds for Cages and Aviaries, A. Bradley London and County Printing Works, 1891.
82. Griffin, D. Animal Minds, University of Chicago Press, 1992.
83. Grzimek, B. (ed.). Animal Life Encyclopedia, Volume 7: Birds I, Van Nostrand Reinhold Co., 1975.
84. Grzimek, B. (ed.). Animal Life Encyclopedia, Volume 8: Birds II, Van Nostrand Reinhold Co., 1975.
85. Grzimek, B. (ed.). Animal Life Encyclopedia, Volume 9: Birds III, Van Nostrand Reinhold Co., 1973.
86. Hansel, M. Built by Animals: The natural history of animal architecture, Oxford University Press, 2007.
87. Harrison's Bird Foods, Handbook for a Healthier Bird, HBD International, 2004.
88. Heaton, J. and Brauth, S. Effects of Deafening on the Development of Nestling and Juvenile Vocalizations in Budgerigars (Melopsittacus undulatus), Journal of Comparative Psychology, Volume 113, Issue 3, September 1999, pp. 312–320.
89. Heidenreich, B. The Parrot Problem Solver: Finding Solutions to Aggressive Behavior, T.F.H. Publications, Inc., 2005.
90. Hile, A., Burley, N., Coppersmith, C., Foster, V. and Striedter, G. Effects of Male Vocal Learning on Female Behavior in the Budgerigar, Ethology, Volume 111, Issue 10, October 2005, pp. 901–923.
91. Hile, A., Plummer, T. and Striedter, G. Male vocal imitation produces call convergence during pair bonding in budgerigars (Melopsittacus undulatus), Animal Behavior, Volume 59, Issue 6, June 2000, pp. 1209–1218.
92. Janka, L. Birds Change Their Minds—Literally: Our changing understanding of bird brains, BirdScope, Cornell Lab of Ornithology, Vol. 22(4), Autumn 2008, p. 1.
93. Jones, C. Back from the Brink: the Echo Parakeet Story, PsittaScene, August 2010, pp. 3–5.
94. Juniper, T. and Parr, M. Parrots: A Guide to Parrots of the World, Yale University Press, 1998.
95. Kaufman, K. Lives of North American Birds, Houghton-Mifflin, 1996.
96. Knight, G. and Johnson, A. Recent Research About the Birds At Our Feeders, Winter Bird Highlights, Cornell Lab of Ornithology and Bird Studies Canada, pp. 4–5, 2008.

97. Koepff, C. and Romagnano, A. The Finch Handbook, Barron's Educational Series, 2001.
98. Kroodsma, D. The Singing Life of Birds: The Art and Science of Listening to Birdsong, Houghton-Mifflin, 2005.
99. Lafeber, T. Tender Loving Care for Pet Birds, Dorothy Products, 1977.
100. Lam, Stephanie. Adoption Not Limited to Social Animals. The Epoch Times, June 24–30, 2010, p. B6.
101. LeChene, E. Silent Heroes: the Bravery and Devotion of Animals in War, Souvenir Press, 1994.
102. Lee, A. Discovering Clay Licks, PsittaScene, November 2009, pp. 14–17.
103. Leshem, Y. Proposal to Develop a Global Network to Predict Bird Movements on a Real-Time and Daily Scale Using Radar, paper presented to BSCE-23/WP50, London, 1996.
104. LeShem, Y. Establishment of an International Center for the Study of Bird Migration and Flight Safety at Latrun, Israel, as a Model for Regional Activity, paper presented to BSCE-23/WP53, London, May 1996.
105. LeShem, Y. Evaluating the Cost of Bird-Related Damage to Civilian and Military Flights as a Vital Tool to Increase Flight Safety, paper presented to BSCE-23/WP10, May 1996.
106. LeShem, Y., Shamoun-Baranes, J., Yanai, M., Tamir, R. and Yom-Tov, Y. The Development of a Global Database on Bird Movements and Bird Strikes in Military and Civilian Flight, paper presented to IBSC 24, Slovakia, September 1998.
107. Leshem, Y. and Bahat, O. Flying with the Birds, Yedioth Ahronoth Chemed Books, 1999.
108. Lint, K. and Lint, A. Diets for Birds in Captivity, Blanford Press, 1981.
109. Lohr, B. and Dooling, R. Detection of Changes in Timbre and Harmonicity in Complex Sounds by Zebra Finches (Poephila guttata) and Budgerigars (Melopsittacus undulatus), Journal of Comparative Psychology, Volume 112, Issue 1, March 1998, pp. 36–47.
110. Long, K. Squirrels: A Wildlife Handbook, Johnson Books, 1995.
111. Lovette, I. Are Birds Just Feathered Dinosaurs? BirdScope, Cornell Lab of Ornithology, Vol. 22(4), Autumn 2008, p. 3.
112. Lowell, M. Your Pet Bird: A Buyer's Guide, Henry Holt and Company, Inc., 1994.
113. Luescher, A. (ed.). Manual of Parrot Behavior, Blackwell Publishing, 2006.
114. Ly, M. Songbirds Know to Use Grammar When Singing, Epoch Times, p. B7, 7 July 2011.
115. Mancini, J. Why Does My Bird Do That: A Guide to Parrot Behavior, Wiley Publishing, Inc., 2007.
116. Marshall, A. (ed.). Biology and Comparative Physiology of Birds, Volume II, Academic Press, 1961.
117. Martin, R. Encyclopedia of Aviculture: Keeping and Breeding Birds, Arco Publishing, Inc., 1983.
118. McCluggage, D. and Higdon, P. Holistic Care for Birds: A Manual of Wellness and Healing, Simon & Schuster, 1999.
119. McKinley, M. How to Attract Birds, Ortho Books, 1983.
120. Morrison, R. Australia: Land Beyond Time, New Holland Publishers, Pty. Ltd., 2002.
121. Mottley, K. and Heyes, C. Budgerigars (Melopsittacus undulatus) Copy Virtual Demonstrators in a Two-Action Test, Journal of Comparative Psychology, Volume 117, No. 4, 2003, pp. 363–370.
122. Moustaki, N. Zebra Finch Care, T.F.H. Publications, Inc., 2004.
123. Moustaki, N. Parakeet Care, T.F.H. Publications, Inc., 2004.
124. Moustaki, N. Parakeets, T.F.H. Publications, Inc., 2006.
125. Nebot, J. First Report on the Rose-ringed Parakeet (Psittacula krameri) in Venezuela and Primary Observations of its Behavior, The Neotropical Ornithological Society, Volume 10, 1999, pp. 115–117.
126. Nobel, P. (ed.). Cacti: Biology and Uses, University of California Press, 2002.
127. Nozedar, A. The Secret Language of Birds: A Treasury of Myth, Folklore, and Inspirational True Stories, Harper Element, Great Britain, 2006.
128. O'Brien, S. Wesley the Owl: the Remarkable Love Story of an Owl and His Girl, Free Press, 2008.
129. Oil Spill Recovery, AP photo and caption, The Gazette, p. 9A, 27 January 2010.
130. Okanoya, K. and Dooling, R. Perception of Distance Calls by Budgerigars (Melopsittacus undulatus) and Zebra Finches (Poephila guttata), Journal of Comparative Psychology, Volume 105, Issue 1, March 1991, pp. 60–72.
131. Pakenham, T. The Remarkable Baobab, W. W. Norton & Company, 2004.
132. Peake, E. Budgies from the Beginning. Budgies: Celebrating the World's Favorite Parakeet, Bow Tie Magazine, 2008, pp. 102–107.

133. Pearn, S., Bennett, A. and Cuthill, I. Ultraviolet vision, fluorescence and mate choice in a parrot, the budgerigar Melopsittacus undulatus (paper posted on web without complete citation details).
134. Pepperberg, I. The Alex Studies: Cognitive and Communicative Abilities of Grey Parrots, Harvard University Press, 1999.
135. Pepperberg, I. Alex and Me: How a Scientist and A Parrot Uncovered a Hidden World of Animal Intelligence—and Formed a Deep Bond in the Process, Harper Collins, 2008.
136. Perrins, C. and Elphick, J. Complete Encyclopedia of Birds and Bird Migration, Chartwell Books, Inc., 2003.
137. Podulka, S., Rohrbuagh, Jr. R. and Bonney, R. (eds.). Handbook of Bird Biology, Cornell Lab of Ornithology, in association with Princeton University Press, 2nd edition, 2004.
138. Rach, J. and Gallerstein, G. First Aid for Birds, Howell Book House, 1999.
139. Rach, J. The Budgie, Simon & Schuster, 1997.
140. Reddy, V. Damage to ber (zizyphus mauritiana) fruits by rose-ringed parakeet (Psittacula krameri) in Hyderabad, Journal of Ecotoxicology & Environmental Monitoring, Volume 16, Issue 6, November 2006, pp. 587–589.
141. Reed, C. The Bird Book: Illustrating in Natural Colors More Than Seven Hundred North American Birds; Also Several Hundred Photographs of Their Nests and Eggs, Doubleday, Page, and Company, Garden City, New York, 1915.
142. Reyman, J. Millions of Gifts for the Gods, PsittaScene, November 2009, pp. 6–9.
143. Reynolds, M. The Parrot Phenomenon: The past, present, and future of the world's most charismatic birds, PsittaScene, August 1995, pp. 6–7.
144. Richards, H. The 500 Mark: a Landmark Season, PsittaScene, August 2010, pp. 6–10.
145. Ritter, S. Using Barn Owls and Kestrels as Biological Pest Controllers Becomes a National Project, A Bird's Eye View of the Holy Land, No. 8, Spring 2009, p. 5.
146. Robbins, C., Bruun, B. and Zim, H. A Guide to Field Identification of North America Birds, Golden Press, 1983.
147. Runde, D., Pitt, W. and Foster, J. Population Ecology and Some Potential Impacts of Emerging Populations of Exotic Parrots, Proceedings Managing Vertebrate Invasive Species, Witmer, G., Pitt, W. and Fagerstone, K. (eds.). USDA/APHIS, National Wildlife Research Center, Fort Collins, CO, 2007, pp. 338–360.
148. Sarao, A. Bamboo Wrens Coordinate Their Duets, Epoch Times, p. B7, 9 November 2011.
149. Schappert, P. The Last Monarch Butterfly: Conserving the Monarch Butterfly in a Brave New World, Firefly, 2004.
150. Science in Pictures: Broadclub Cuttlefish, Epoch Times, 27 October 2011, p. B6.
151. Shifflet, T. Dolphin Talk: Learning from Marine Mammals, Watermarks, Summer 2010, pp. 6–7.
152. Sibley, C. and Monroe, B. Distribution and Taxonomy of Birds of the World, Yale University Press, 1990.
153. Sibley, D. The Sibley Guide to Bird Life and Behavior, Alfred Knopf, 2009.
154. Skutch, A. The Minds of Birds, Texas A&M University Press, 1996.
155. Soderberg, P. Foreign Birds for Cage and Aviary, Volume II: Waxbills, Weavers, and Whydahs, Cassell, London, 1956.
156. Snyder, N., Wiley, J. and Kepler, C. The Parrots of Luquillo: Natural History and Conservation of the Puerto Rican Parrot, The Western Foundation of Vertebrate Zoology, 1987.
157. Soothill, E. and Whitehead, P. Wildfowl of the World, Blanford Press, 1978.
158. Sparks, J. Bird Behavior, Grosset & Dunlap, Inc., 1970.
159. Stein, B. On Her Own Terms: Annie Montague Alexander and the Rise of Science in the American West, University of California Press, 2001.
160. Stokes, D. A Guide to Bird Behavior, Volume 1, Little, Brown, and Company, 1979.
161. Stokes, D. and Stokes, L. A Guide to Bird Behavior, Volume 3, Little, Brown, and Company, 1983.
162. Stutchbury, B. and Morton, E. Behavioral Ecology of Tropical Birds, Academic Press, San Diego, 2001.
163. Su, C. Hummingbirds Use Snap-buckling Beaks to Catch Flies, Epoch Times, p. B7, 28 July 2011.
164. Sullivan, B. and Liguori, J. Active Flight Feather Molt in Migrating North American Raptors, Birding, Vol. 41(4), July 2009, pp. 34–45.
165. Temple, S. (ed.). Endangered Birds: Management Techniques for Preserving Threatened Species, The University of Wisconsin Press, 1977.
166. Threatened Birds of the Americas, The ICBP/IUCN Red Data Book, International Council for Bird Preservation/The World Conservation Union, Smithsonian Institution Press, 3rd edition, Part 2, 1992.

167. Tilford, T. The Cage & Aviary Bird Handbook, New Holland, 2001.

168. Timbs, J. Manual of Cage Birds, British and Foreign: With Directions for Breeding, Rearing, and Keeping Them, David Bogue, London, 1847.

169. Tudge, C. The Bird: A Natural History of Who Birds Are, Where They Came From, and How They Live, Crown Publishers, 2008.

170. Venuto, V., Ferraiuolo, V., Bottoni, L. and Massa, R. Distress call in six species of African Poicephalus parrots, Ethology, Ecology, and Evolution, Volume 13, 2001, pp. 49–68.

171. Vezo, T. and Hagner, C. Wings of Spring: Courtship, Nesting, and Fledging, Stackpole Books, 2006.

172. Vornberger, C. Birds of Central Park, Harry N. Abrams, Inc., 2004.

173. Vriends, M. Gouldian Finches: Everything About Purchase, Housing, Care, Nutrition, Breeding, and Diseases, Barron's Educational Series, Inc., 1991.

174. Vriends, M. The Zebra Finch, Simon & Schuster, 1997.

175. Vriends, M. Conures: Everything About Purchase, Care, Nutrition, and Behavior, Barron's Educational Series, Inc., 2000.

176. Walker, S. Animal Thought, Routledge & Kegan Paul, 1983.

177. Walters, M. A Shadow and A Song: The Struggle to Save An Endangered Species, Chelsea Green, 1992.

178. Watson, B. and Hurley, M. The Ultimate Parrot, Howell Book House, 2000.

179. Webster's Ninth New Collegiate Dictionary, Merriam-Webster, Inc., 1984.

180. Whitaker, J. National Audubon Society Field Guide to Mammals, revised edition, Alfred A. Knopf, 1996.

181. Wilkinson, R. The Conservation Status of African Parrots: A Review for the World Parrot Trust—Part 1, May 1998, pp. 6–9.

182. Williamson, S. Hummingbirds of North America, Houghton Mifflin, 2001.

183. Wyndham, E. Environment and food of the budgerigar, Australian Journal of Ecology, Volume 5, Issue 1, March 1980, pp. 47–61.

184. Zann, R. The Zebra Finch: A Synthesis of Field and Laboratory Studies, Oxford University Press, 1996.

185. Zerden, S. A Cure for Heart Disease: Researchers find an association with hypothyroidism, Epoch Times, February 2, 2011, p. B3.

B.2 Videos

186. Attenborough, D. The Life of Birds, BBC Video, 2002.

187. Bittner, M. (author) and Irving, J. (director) The Wild Parrots of Telegraph Hill, Pelican Media, 2005.

188. March of the Penguins, narrated by Morgan Freeman, Warner Brothers, 2005.

189. Nature: Animal Minds—Are Animals Intelligent? Questar, 2003.

190. Nature: Animal Minds—Do Animals Have Emotions? and Animal Consciousness, Questar, 2003.

B.3 Websites

Individual web pages are listed in the order referenced, by chapter. Given the fleeting nature of web pages, the date of access is provided.

191. Audubon Society, www.audubon.org/field-guide accessed 11/10/2009

192. Central Pets, www.centralpets.com

http://www.centralpets.com/pages/critterpages/birds/parrots/PRT4103.shtml accessed 12/14/2004

193. Cornell Lab of Ornithology, www.birds.cornell.edu
http://www.allaboutbirds.org/guide/green_heron/lifehistory accessed 10/15/2009
http://www.allaboutbirds.org/guide/great-blue_heron/lifehistory accessed 10/15/2009
http://www.allaboutbirds.org/guide/Black-crowned_Night-heron/lifehistory accessed 10/15/2009
http://www.allaboutbirds.org/guide/Yellow-crowned_Night-Heron/lifehistory accessed 10/15/2009
http://www.allaboutbirds.org/guide/Canada_goose/lifehistory accessed 10/15/2009
http://www.allaboutbirds.org/guide/Mallard/lifehistory accessed 10/15/2009

http://www.allaboutbirds.org/guide/Northern_Cardinal/lifehistory accessed 10/15/2009
http://www.allaboutbirds.org/guide/ruby-throated_hummingbird/lifehistory accessed 10/15/2009

194. David McKain, Indian Ringneck, www.netpets.com
http://www.netpets.com/birds/reference/species/ringneck.html accessed 12/14/2004
http://www.netpets.com/birds/reference/species/nwparrotsid.html accessed 3/4/2011

195. Honolulu Zoo, www.honoluluzoo.org

196. International Conure Association, www.conure.org, Winter 2006 newsletter

197. New Jersey Fish, Game, and Wildlife Service, www.nj.gov/dep/fgw

198. New York Natural History Program, www.acris.nynhp.org/guide.php?id=6752
accessed 11/15/2011

199. Steve Boyes, parrot blog: Of Elephants and Parrots: The uncanny relationship between African elephants and Meyer's parrots, posted June 1, 2008

http:/www.parrots.org/index/php/blog_boyes accessed 2/22/2011

200. U.S. Geological Survey, www.usgs.gov
http://www.mbr-pwrc.usgs.gov/id/framlst/i1940id.html accessed 7/26/2010
http:///pwrc.usgs.gov/bioeco/gbheron.htm accessed 7/26/2010
http://www.nwrc.usgs.gov/wdb/pub/hsi/hsi-099.pdf accessed 7/26/2010
http://www.mbr-pwrc.usgs.gov/bbs.hthtmp/a1940.html accessed 7/26/2010
http://www.mbr-pwrc.usgs.gov/infocenter/i2020id.html accessed 7/26/2010
http://www.pwrc.pwrc.usgs.gov/bioeco/bcheron.htm accessed 7/26/2010
http://www.umcsc,usgs.gov/documents/publications/2007/custerT_c_2007.htm
accessed 7/26/2010

201. U.S. National Zoo, www.nationalzoo.si.edu/Animals/Birds/Facts/fact-blknightheron.cfm accessed 11/15/2011

202. World Parrot Trust, www.parrots.org
http://www.parrots.org/index.php/encyclopedia/profile/budgerigar accessed 2/20/2011
http://www.parrots/org/index.php/encyclopedia.profile/senegal_parrot accessed 2/22/2011
http://www.parrots.org/index.php.encyclopedia.profile/ringneck_parakeet accessed 2/22/2011
http://www.parrots.org/index.php/encyclopedia/profile/mited_conure accessed 2/22/2011

203. Wikipedia, www.en.wikipedia.org
http://en.wikipedia.org/wiki/Intelligence accessed 9/26/2010
http://en.wikipedia.org/wiki/Emotional_intelligence accessed 9/26/2010
http://en.wikipedia.org/wiki/Consciousness accessed 9/26/2010
http://en.wikipedia.org/wiki/Emotion accessed 9/26/2010
http://en.wikipedia.org/wiki/Amygdala accessed 9/23/2010
http://en.wikipedia.org/wiki/Heron accessed 8/30/2010
http://en.wikipedia.org/wiki/Green_Heron accessed 9/12/2011
http://en.wikipedia.org/wiki/Great_Blue_Heron accessed 9/12/2011
http://en.wikipedia.org/wiki/Black-crowned_Night_heron accessed 9/12/2011
http://en.wikipedia.org/wiki/Yellow-crowned_Night_heron accessed 9/12/2011
http://en.wikipedia.org/wiki/Canada_Goose accessed 7/23/2010
http://en.wikipedia.org/wiki/Mallard accessed 8/13/2010
http://en.wikipedia.org/wiki/Wood_duck accessed 5/13/2011
http://en.wikipedia.org/wiki/Hooded_Merganser accessed 5/13/2011
http://en.wikipedia.org/wiki/Cardinal_(bird) accessed 8/13/2010
http://en.wikipedia.org/wiki/Ruby-throated_Hummingbird accessed 8/13/2010
http://en.wikipedia.org/wiki/Budgerigar accessed 8/13/2010
http://en.wikipedia.org/wiki/Spinifex/(genus) accessed 3/2/2011
http://en.wikipedia.org/wiki/HEPA accessed 8/15/2010
http://en.wikipedia.org/wiki/Zebra_Finch accessed 8/13/2010
http://en.wikipedia.org/wiki/Meyer's_Parrot accessed 8/13/2010
http://en.wikipedia.org/wiki/Red-bellied_Parrot accessed 8/13/2010
http://en.wikipedia.org/wiki/Rose-ringed_Parakeet accessed 8/13/2010
http://en.wikipedia.org/wiki/Mitred_Conure accessed 8/13/2010

B.4 Correspondence

204. Jennifer Tinnell, Nonnative Species Special Project Leader, Florida Fish and Wildlife Conservation Commission, 3 February 2011.
205. Geo-Technology Associates, Inc., Poplar Run Environmental Background and Disclosure, 12 January 2010.
206. Paul Schmidt, Acting Assistant Director, Migratory Bird Program, U.S. Fish & Wildlife Service, 11 May 2011. The following reports were included in that correspondence.
207. Migratory Birds at Risk from Gulf Coast Oil Spill, U.S. Fish & Wildlife Service, 16 August 2010.
208. North American Waterfowl Management Plan 2004—Implementation Framework: Strengthening the Biological Foundation, Canadian Wildlife Service, U.S. Fish & Wildlife Service, Secretaria de Medio Ambiente y Recursos Naturales.
209. North American Waterfowl Management Plan 2004—Strategic Guidance, Canadian Wildlife Service, U.S. Fish & Wildlife Service, Secretaria de Medio Ambiente y Recursos Naturales.
210. Trends in Duck Breeding Populations, 1955–2010, U.S. Fish & Wildlife Service, 2 July 2010.

Annex C

Observation Locations and Populations

The birds discussed in this book were observed over a 30-year period in three distinct locations. The bird populations and the locations in which they were observed are described below.

C.1 Heron, Canada Geese, and Mallard Ducks (Chapters 2, 3, and 4)

Herons, Canada geese, and mallard ducks were observed at four waterways in the mid-Atlantic states that are near a metropolitan area. For simplicity of reference the waterways are referred to as Ponds A, B, C, and D. The four waterways are located within an area that is 3 miles (4 km) long east to west and 1 mile (1.33 km) wide north to south. Two of the waterways, Ponds A and D, are manmade ponds designed to catch surplus runoff during heavy rains. Pond B was a natural combined pond and creek. It has been modified over time to support enhanced watershed management. Pond C is a natural pond. Ponds A and D are on the eastern side of the observation area. Ponds B and C are on the western side of the observation area. And, in order to make this an ideal area for observing Canada geese, a golf course with a water trap is between Ponds C and D.

All the ponds are at the same altitude and have essentially the same amount and type of vegetation and trees around them. Pond A was constructed about 1990. Pond A is a triangle shaped pond. The point of the triangle faces west, while the base of the triangle faces east. The normal size of the pond is about 35 feet (10.66 m) wide at the widest point by about 65 feet (19.81 m) long. However, following heavy rains the length and width of the pond may increase by 50% or more and the water level may rise 4 to 5 feet (1.21 to 1.52 m). During very dry conditions Pond A may shrink to two thirds its normal size. In contrast, the other three ponds stay essentially the same size year round. The north side of Pond A is more sloped than the other ponds. Pond B is oblong, running north to south. Pond B is about 300 feet (91.44 m) long and about 36 feet (10.97 m) wide. Pond C is circular shaped and about 200 feet (60.91 m) in diameter. Pond D was constructed in the 1970's and redesigned beginning in the fall of 2010. The original Pond D was circular shaped and about 60 feet (18.28 m) in diameter. The new Pond D is oblong and about 20 feet (6.09 m) wide and 80 feet (24.38 m) long.

Over the years a variety of birds, plants, animals, insects, and reptiles have made these ponds and the surrounding area their home, such as: herons, Canada geese, mallard ducks, wood ducks, black ducks, hooded mergansers, belted kingfishers, egret, bittern, swallows, red-winged blackbirds, robins, mourning doves, cardinals, house sparrows, Downey woodpeckers, blue jays, gold finches, turtles, snakes, frogs, snails, squirrels, rabbits, deer, chipmunks, minnow, bass, butterflies, and cattails.

An historic creek with abundant wildlife, which has been documented since the 1700's, is less than 0.5 mile (0.66 km) away from Ponds A and D. The land both ponds occupy was originally part of the creek's ecosystem. The wildlife that resides in and around Ponds A and D roamed over from the creek at some point in time.

Herons *Butorides virescens, Ardea herodias, Nycticorax nycticorax, Nyctanassa violacea*

Pairs and individual green herons, great blue herons, black-crowned night herons, and yellow-crowned night herons were observed on a daily basis from Spring 2009 through Fall 2012. Two to four individuals of each heron species were observed each year. The herons reside in the observation area most of the year. Most of the herons migrate. A few great blue herons do not migrate, depending on the yearly fall/winter weather patterns. Even though there are four ponds in the observation area, the herons reside and nest at Pond B. Green herons, great blue herons, and black-crowned night herons visit Pond C during the summer and fall. Green herons were also spotted at the original Pond D a few times prior to it being redesigned in 2010. None of the herons visited Pond A or the new Pond D.

Canada Geese *Branta canadensis*

A flock of Canada geese was observed on a daily basis from Spring 2005 through Fall 2012. This flock resides in the observation area most of the year. In addition several flocks of Canada geese pass through the area during the fall and spring migrations. The local flock size ranged from 36 to 48 adult geese.

Mallard Ducks *Anas platyrhynchos*

A flock of mallard ducks was observed on a daily basis from Spring 2005 through Fall 2012. This flock resides in the observation area most of the year. In addition several flocks of mallard ducks pass through the area during the fall and spring migrations. The local flock size averaged 36 adult ducks.

C.2 Cardinals and Ruby-Throated Hummingbirds (Chapters 5 and 6)

The ruby-throated hummingbirds and cardinals lived in a natural outdoor setting in a small wooded nature preserve in the mid-Atlantic states. The nature preserve followed a winding creek for several miles (kilometers). The creek varied from 6 to 20 feet (1.82 to 6.09 m) wide at any given point. Dense plant growth and a mixture of deciduous trees about 48 feet (14.63 m) or more deep bordered each side of the creek. The largest and oldest trees were about 3 feet (0.91 m) in diameter. They were surrounded by a random pattern of younger trees at various stages of development. Amidst the non-stop urban sprawl this narrow preserve provided an ideal habitat and home for a vast array of birds, squirrels, chipmunks, ground hogs, rabbits, raccoon, fox, deer, turtles, frogs, and snakes. In addition to the ruby-throated hummingbirds and cardinals, the bird inhabitants included: sparrows, wrens, goldfinches, chickadees, robins, mourning doves, nuthatches, tufted titmice, blue jays, blackbirds, pileated woodpeckers, and an occasional hawk. Every now and then small fish could be seen swimming in the creek.

Ruby-throated Hummingbirds *Archilochus colubris*

Four to six ruby-throated hummingbirds were observed each year from Spring 1985 through Fall 1994.

Cardinals *Cardinalis cardinalis*

Four to six cardinals were observed each year from Spring 1985 through Fall 1994 at this location, in particular one male cardinal and his multi-generational family. Six to eight cardinals were observed each year at Pond B, described in Section C.1, from Spring 2005 through Fall 2012.

C.3 Budgerigar, Zebra Finch, Meyer's Parrot, Red-Bellied Parrot, Indian Ringneck Parakeet, and Mitred Conure (Chapters 7 through 11)

The budgerigars, zebra finches, Meyer's parrots, red-bellied parrots, Indian ringneck parakeets, and mitred conures live in an indoor multi-bird, multi-aviary setting. The aviary features many bird safe plants, full spectrum lighting, and several windows so that the birds can see:

- the birds, plants, animals, and trees outdoors
- sunshine and the normal day/night pattern
- daily weather conditions
- the changing seasons

Like other birds the metabolism of these birds is synchronized with the weather.[99] As a result, during winter the temperature in the aviary is kept between 62°F and 65°F. During summer the temperature in the aviary is kept between 74°F and 76°F. Except for the zebra finches, all the birds have regular flying time outside their aviaries during which they can socialize with birds from other aviaries. The birds are not kept in a breeding situation. However, as discussed in Chapters 7–11, they do form lasting social and pair bonds. All the birds can see and hear each other. The birds have daily contact with humans.

The housing, diet, furnishings (toys, perches, swings, dishes, etc.), and daily routine of contemporary birds is the same. Whenever a new item is introduced, it is introduced in all the aviaries of a species at the same time. The birds are observed throughout the day, from just before sunrise as they awaken, until the lights are turned out in the evening and they go to sleep. None of the birds were taught human language words or phrases so that their native vocalizations could be studied. Likewise, none of the birds were taught tricks so that their natural behaviors would emerge.

None of the birds in this study had their wings clipped. A few young birds acquired from pet stores had clipped wings when they arrived, but they were allowed to grow back. Flight and feathers are what makes a bird a bird, even though some species have become flightless over the centuries. Flight is an essential part of the emotional, intellectual, and physical wellbeing of any bird.

Budgerigars *Melopsittacus undulatus*

Twenty-eight budgerigars were observed over a 30-year period from Fall 1982 through Fall 2012. To maximize the opportunity to see as many common and individual behaviors as possible, a spectrum of budgerigars were observed: young, middle-aged, and elderly, adopted and not adopted, male and female. Likewise a variety of colors of budgerigars were observed because of their different genetic make-up. Specifically, twenty-one budgerigars were observed their entire lives; seven were adopted at various ages. Nine budgerigars were normal green budgerigars, like those living in the wild. Nineteen budgerigars were various hybrid colors. None of the budgerigars were show budgerigars or the larger English budgerigars, since the smaller budgerigars are genetically closer to wild budgerigars. One budgerigar was acquired directly from a breeder. The twenty birds that were not adopted were acquired from various pet stores. Seventeen of the budgerigars were males. Eleven of the budgerigars were females.

Zebra Finches *Peophila guttata castanotis*

Sixteen zebra finches were observed over a 12-year period from Fall 2001 through Spring 2014. To maximize the opportunity to see as many common and individual behaviors as possible, a spectrum of zebra finches were observed: young, middle-aged, and elderly, adopted and not adopted, male and female. Specifically, nine zebra finches were observed their entire lives; seven were adopted at various ages. Eight of the finches were normal gray zebra finches, like those living in the wild. Two were white and four were fawn. In addition one fawn pied zebra finch and one gray pied zebra finch were observed. Of the nine zebra finches that were not adopted, six were acquired from various pet stores and three hatched during the study. Six of the zebra finches were male. Ten of the zebra finches were female.

Meyer's Parrot and Red-Bellied Parrot *Poicephalus meyeri and Poicephalus rufiventris*

Two members of the Poicephalus family, a Meyer's parrot and a Red-bellied parrot, were observed over a 14-year period from Fall 2001 through Spring 2015. Initially the parrots resided in separate aviaries next to each other. Now they reside in the same aviary. Both the Meyer's parrot and the Red-bellied parrot are females. Both birds were acquired from a pet store before they were 1-year old. Neither bird has ever had their wings clipped. Members of the Poicephalus family are known for becoming lethargic if they are cage bound and not given enough time to fly and play outside their cage.[112]

Indian Ringneck Parakeets *Psittacula krameri manillensis*

Four Indian ringneck parakeets were observed over a 10-year period from Fall 2004 through Fall 2014. To maximize the opportunity to see as many common and individual behaviors as possible, a spectrum of Indian ringneck parakeets were observed: young and middle-aged, adopted and not adopted, male and female, normal coloration and hybrids. Specifically, one Indian ringneck parakeet was acquired from a pet store and has been observed its entire life; three were adopted at various ages. Two of the Indian ringneck parakeets are normal green Indian ringneck parakeets, like those living in the wild. One is a light blue and white hybrid. One is a dark grey and black hybrid. One of the Indian ringneck parakeets is male. Three of the Indian ringneck parakeets are female.

Mitred Conures *Aratinga mitrata mitrata*

Two mitred conures were observed over a 9-year period from Summer 2006 through Fall 2014. Both mitred conures are males. One was adopted at 20 years of age; the other was adopted at 5 years of age.

Annex D

Budgerigars and Children: A Tribute to Perky

I obtained my first budgerigar, Perky, when I was in fourth grade. In many ways Perky was the genesis of the idea for this book. Perky was a male normal green budgerigar. He was not a present or a spur of the moment purchase, far from it. Every time my family went to the local Woolworth's to shop I skipped the toy department and visited the pet department instead. Early on I realized that goldfish, turtles, canaries, and budgerigars were alive, while toys were not, and that fascinated me. The pets looked at me when I looked at them and responded when I moved closer or farther away or held them (goldfish excluded). Their curiosity seemed to match mine and I always wondered what they were thinking.

One day I asked if I could have a budgerigar. My parents did not say yes or no right away. Instead, they reminded me that a budgerigar is a living being and that if we bought one, I would be responsible for taking care of it. I agreed immediately, of course. What child would not? My parents were a little more practical. Instead of buying me a budgerigar, they bought me a small book that explained how to take care of a budgerigar. After I read the book they asked me some questions. Next, it was my job to get Perky's new home ready. Since my allowance would not cover the cost of a new bird cage, it was agreed that if I pasted the green stamps each week, I could use them to "buy" a cage. From that point on I eagerly looked forward to trips to the grocery store, especially double stamp days. Finally I had enough green stamps. The cage was ordered and arrived!

The cage came with perches and dishes, but no toys, food, or gravel. For the next two months I saved up my allowance. Every other week we went back to Woolworth's to look at the birds. I systematically bought a swing, toys, an orange cardboard box of food, an orange cardboard box of gravel, a cuttlebone, a treat jar, growing greens, egg biscuits, and a bag of millet sprays until the cage was fully equipped. I convinced my parents that I knew how to change the food, water, and paper. We sat the cage in my room on my desk by the window. Then it was time to bring Perky home.

I loved that little bird. Perky sat on my finger, my shoulder, and my head. His favorite place to sit was on my pencil when I was doing my homework, because it was always moving. He would slide off, fly around the room and land on my pencil again and again. Perky was the first "person" I said hello to every morning and the last "person" I said good night to every evening. I took Perky (in his cage) everywhere: upstairs, downstairs, outside to sit in the grass, down the street to visit friends and neighbors, even to school one day. Perky was never too busy to talk to me or too tired to play with me. He was a happy reliable consistent presence in the ever changing world of a growing child. Perky stayed with me through my first year in college.

Annex E

External Factors to Consider When Giving a Bird a Medical Exam

As part of the mind/body connection, here are some additional factors to consider when examining a bird. Often what initially appears as a symptom of disease, may be caused by external factors like these.

1. Like humans, not all birds have perfect hearing or perfect eyesight. A bird that appears to be clumsy may have a vision problem. Likewise, a bird that is overly loud, overly quiet, or not responsive to verbal commands may have a hearing deficiency.
   - Sky (Chapter 7) occasionally flew into the top of his aviary.
   - Snapdragon's (Chapter 10) close-up vision was fine, but his long distance vision was off a bit which caused him on occasion to overshoot his landing when flying.
   - Rudy's (Chapter 11) right eye was good, but his left eye was not. When Rudy triangulated between his left and right eyes, everything was fine. If Rudy forgot to triangulate he would crash when flying or accidentally bite someone when reaching for a treat.
   - Two goslings (Chapter 3) appeared to have hearing limitations that their fellow goslings did not.
2. Consider that environmental issues may be the root cause of the symptoms rather than disease:
   - Lack of opportunity to exercise
   - Cage is too small
   - Not enough sleep
     - Budgerigars and other parrots often vocalize to drown out noise when they are trying to sleep. (Chapter 7)
   - Not enough wood to chew, etc. (Buster Chapter 11)
   - Inadequate diet and nutrition
   - Not enough sunshine (UVA, UVB).
     - Two cockatiels *Nymphicus hollandicus* and two zebra finches *Poephila guttata castanotis* were chronic egg layers. When their aviaries were moved to a sunny window the chronic egg-laying ceased.

3. Consider that emotional and/or social issues may be the root cause of the symptoms rather than disease:
    - The bird has recently been re-homed, lost a mate, lost a clutch, lost its primary caregiver, and/or relocated to a new city and house. (Rudy Chapter 11)
    - The bird needs more daily human interaction, mental challenge, and stimulation.
    - The bird is kept alone in the home and doesn't have any other birds with which to interact.
        - A 6-month old rescued Canary-winged parakeet *Brotogeris versicolorus* was entirely plucked. After being put in an aviary with another Canary-winged parakeet her feathers grew back right away. She spent the first few days constantly preening the other bird. Three years later these two birds still do everything together, including sleeping on the same swing overnight.
4. Remember that clipping the wings of a bird that has been free-flighted its entire life is a very demoralizing experience. Only do this if absolutely necessary. (Slate Chapter 7)
5. While cats and dogs may enjoy a ride in the car to/from the veterinarian's office, most birds do not like being removed from their large flight cage (and buddies) and confined in a small travel cage for a trip to/from the veterinarian's office in noisy traffic. For many birds, especially the smaller ones, this journey is very stressful.
    - Sky (Chapter 7) had a panic attack upon being confined in a travel cage and broke a blood feather on the way to the veterinarian's office.

Also, check your records before automatically sending out postcard or email reminders that a bird's annual exam is due. Nothing is worse than receiving such a reminder after a bird has died.

Index

G

H

I